p. 33 Eq. 2–19 The constant is not arbitrary but is proportional to the magnitude of any surface dipole layer (Eq. 1–61).

. 39–40 Properly, an isolated molecule is in a state of definite parity, such as one of its rotational eigenstates, and therefore has no permanent electric dipole moment (the expectation value of an operator of odd parity, such as the electric dipole moment, in a state of definite parity is zero). However, a proper quantum mechanical calculation gives the same result at the classical calculation here. The details are in P. Debye's 1929 book *Polar Molecules*, but not in most modern textbooks.

p. 40 Eq. 2–45 fails for many polar liquids, like H_2O, because of short-range correlations resulting from strong angle-dependent (covalent) intermolecular forces.

p. 48 It is clearer to say that the triangles with sides rld and $l'r'd'$ are similar.

p. 85 Eq. 5–16 the last $=$ should be a $-$.

p. 92 Eq. 5–42 e_0 should be ϵ_0.

p. 93 Exc. 3 Pv should be P_n.

p. 99 The derivation of the symmetry of $\kappa_{\alpha\beta}$ (last paragraph, Eqs. 6–18 and 6–19) is incorrect. The symmetry of $\kappa_{\alpha\beta}$ follows from defining dielectric susceptibility in a thermodynamically consistent manner, as the second derivative tensor of the free energy with respect to the electric field; see Landau and Lifschitz *Electrodynamics of Continuous Media* §11.

p. 113 Eq. 6–76 the upper limit should be B.

p. 117 Exc. 7 (6–44) should be (6–45).

p. 128 Eq. 7–42 $\oint$ should be $\int$.

p. 168 Exc. 9–4 can be solved if the orbital radius is taken to be constant, which is inconsistent with the conditions given. It is possible to solve this problem if the central force field is ignored, the initial $\vec{B}$ is nonzero, and its variation is slow compared to the gyroperiod. The second part of Exc. 9–7 shows that for a special spatial distribution of $\vec{B}$ and no central force field the orbital radius remains constant.

p. 168 Exc. 9–7 (first part) requires the assumption that $\vec{B}$ is independent of φ for *any* axis z. Do only the second part.

p. 174 Eq. 10–14 $\frac{\partial}{\partial t}$ should be $\frac{d}{dt}$.

p. 321 Eq. 17–73 $-2c^2\mathbf{p}_1 \cdot \bar{\mathbf{p}}_1$ should be $+2c^2\mathbf{p}_1 \cdot \bar{\mathbf{p}}_1$.
      Eq. 17–74 should have a $-$ sign on one side.

p. 364 Eq. 20–35 $(u^2/c)^2$ should be (u^2/c^2).

p. 365 Eq. 20–38 should have a factor m_0c^2 on the right.

p. 370 The reason real steady currents flowing through wires don't radiate significantly is their low speed and the Pauli exclusion principle; the current distribution is not microscopically continuous, contrary to the assertion here.

p. 378 One line above 21–9, (17–32) should be (18–32).
      Eq. 21–9 The $-$ sign should be $+$.

p. 389 Eq. 21-56 The first $-$ sign on the left hand side should be $+$.

p. 401 Eq. 22–1 In the middle and right hand side (two places) x should be $\dddot{x}$.

p. 404 One line after 22–20, (19–20) should be (20–18).

p. 412 After the unnumbered equation (22–9) should be (22–19).

p. 419 Second line from bottom, $\epsilon|\mathbf{H}|^2$ should be $\epsilon|\mathbf{E}|^2$.

p. 420 Eq. 22–76 The first exponential is $e^{-ikr(1-\cos\theta)}$.

p. 422 Eq. 22–89 This equation appears to give the erroneous result that for $\kappa_2 = 0$ (as is true to high accuracy for optical glass) but $\kappa_1 \neq 1$ the scattering cross-section must be zero. This paradox arises from the use of the infinite-medium relation between $\mathbf{p}$ and $\mathbf{E}$. In the case of a small particle (or interface) the proper relation must include radiation damping, implying $\kappa_2 \neq 0$ (with its actual value determined by 22–89).

p. 465 1 farad $= 9 \times 10^{11}$ cm (cgs capacitance).

1 ohm $= \frac{1}{9} \times 10^{-11}$ sec/cm (cgs resistance; = esu).

CLASSICAL ELECTRICITY AND MAGNETISM

SECOND EDITION

Wolfgang K. H. Panofsky
Stanford University

Melba Phillips
Washington University

Dover Publications, Inc.
Mineola, New York

Bibliographical Note

This Dover edition, first published in 2005, is an unabridged republication of
the second (1962) edition of the work first published by the Addison-Wesley
Publishing Company, Inc., Reading, Massachusetts, in 1955. The authors have
provided a new Errata list specially for the Dover edition.

Library of Congress Cataloging-in-Publication Data

Panofsky, Wolfgang Kurt Hermann, 1919–
    Classical electricity and magnetism / Wolfgang K.H. Panofsky, Melba
Phillips.
       p. cm.
    Originally published: 2nd ed. Reading, Mass. : Addison-Wesley Pub. Co.,
1962, in series: Addison-Wesley series in physics. With new errata list.
    Includes bibliographical references and index.
    ISBN 0-486-43924-0 (pbk.)
    1. Electricity. 2. Electromagnetic theory. 3. Relativity (Physics) I. Phillips,
Melba, 1907– II. Title.

QC518.P337 2005
530.14'1—dc22

                                                                              2004058269

Manufactured in the United States of America
Dover Publications, Inc., 31 East 2nd Street, Mineola, N.Y. 11501

PREFACE TO THE FIRST EDITION

This book is designed to emphasize those aspects of classical electricity and magnetism most useful to the modern student as a background both for experimental physics and for the quantum theory of matter and radiation. We have made no attempts at novelty beyond those inherent in looking at subject matter that has become a part of the foundations of physics, and has thus gained in usefulness as it has lost in immediacy. While no rigid adherence to historical development is attempted, the emphasis is on physical theory as evolved from fundamental empirical laws rather than on mathematics and strict internal logic. Thus Maxwell's equations are derived from the experimental laws of Coulomb, Ampère, and Faraday, instead of being postulated initially. In the opinion of the authors the physical concepts emerge more clearly in this way, and the approach represents the manner in which physical theory evolves in practice. The field formulation is preferred to the action-at-a-distance viewpoint even in electrostatics, however, since for the conventional treatment it is more readily extended to the nonstatic case. This despite the fact that it is possible, both for static and for nonstatic phenomena, to formulate an entirely consistent electromagnetic theory based on the delayed-action-at-a-distance principle.

The climax of 19th century electrodynamics was the theory of electromagnetic waves and its confirmation, and it is inevitable that any treatment of the subject today includes the principles of recent applications involving metallic boundaries. The introduction of the electrodynamic potentials and the Hertz solution of the wave equation are treated in the conventional way, but we have chosen to introduce the special theory of relativity before undertaking the theory of the electron. Historically the evidence was building up simultaneously along two separate lines, and many of the early difficulties in the derivation of radiation theory as applied to elementary charges were clarified in a very simple way by relativistic considerations. This approach has the advantage that the other problems of classical electron theory, especially those which have taken on added significance with the advent of quantum theory, can be exhibited more clearly.

Rationalized mks units are used throughout, simply because the majority of modern reference books and papers are now written in this system. Especially in the consideration of the electron, all quantities are so written that they can be immediately translated into Gaussian units. In

Appendix I will be found a discussion of the units in current use, and tables contain the fundamental relations of electrodynamics expressed in various systems as well as numerical conversion factors.

The text is based on graduate course lectures given by one of us (Panofsky) at the University of California and Stanford University. Early mimeographed notes on much of the subject matter were prepared with the aid of Howard Chang, Roger Wallace, Richard Madey, and Lee Aamodt, whose help is gratefully acknowledged. The editorial help of Miss Laurose Becker is also acknowledged with thanks.

The reader is assumed to have had courses in advanced calculus, differential equations, vector analysis, and, at least for the latter portions, is assumed to be familiar with classical mechanics on the graduate level. Prior knowledge of tensor analysis would be helpful, but is not necessary. References to appropriate collateral and background material are included at the end of each chapter, with some indication of what relevant material is to be found in each reference, and a full bibliographical list is given at the end.

The presentation is designed to be somewhat flexible, depending on the organization of course material. For purely theoretical courses Chapters 4 and 5, together with portions of other chapters dealing with particular applications of potential theory, etc., may be omitted entirely. Some of the material in Chapter 12 is often covered in optics courses. And if a course in relativity theory is given separately Chapters 15–18 may be omitted, since we have endeavored to make Chapter 19 continuous with Chapter 14, insofar as the theory of radiation is concerned.

A final word about problems: for the most part they are designed to supplement the text. It had been our intention to give credit to original sources for those we did not invent ourselves, but in almost every case this turns out to be impossible: like discoveries, problems are rarely made singly, and in a subject as old as this ingenuity mainly recreates old ideas. And despite our adherence to the exhortation used by Becker, "be ye doers of the word and not hearers only, deceiving your own selves," we have not concentrated primarily on problem solving. The heart of the matter, we believe, lies in the ideas and their development.

W. K. H. P.
M. P.

PREFACE TO THE SECOND EDITION

The second edition of *Classical Electricity and Magnetism* is intended principally to remedy errors and inadequacies of the first edition. We have attempted to correct errors and make extensive revisions without changing the basic approach to the material; we hope that in so doing we have responded to the many helpful comments we have received from users of the book without introducing too many departures. The only radical change is in the treatment of radiation reaction, which has been completely rewritten and introduces new concepts. New material has been added in several instances: there is a new chapter on the basic principles of magnetohydrodynamics; the use of "superpotentials" for obtaining symmetric expansion of electric and magnetic wave-fields has been introduced; the material on the classical radiation of electrons moving in a circle has been expanded; the motion of particles with spin is treated; and the classical forms of such theorems as the dispersion relation and the "optical" or "shadow" theorem are now included.

We have not attempted to make the methods used in this book uniform; on the contrary, we believe that there is a great deal of educational value in the demonstration that many of the methods used are equivalent. As before, we stress physical ideas rather than mathematical techniques.

Without the generous help of many correspondents, who have pointed out errors or transmitted comments, this revision would not have been possible. This help has been so extensive that we cannot acknowledge each contribution; we are, however, particularly grateful to F. Rohrlich for a helpful exchange of correspondence. We are also much indebted to Mrs. Laurose Richter for assistance in preparing the manuscript and to Mrs. Adèle Panofsky for preparing the index.

W. K. H. P.
M. P.

Stanford and St. Louis

CONTENTS

Errata xv

Chapter 1. The Electrostatic Field in Vacuum 1

1-1 Vector fields 1
1-2 The electric field 7
1-3 Coulomb's law 8
1-4 The electrostatic potential 10
1-5 The potential in terms of charge distribution 11
1-6 Field singularities 13
1-7 Clusters of point charges 13
1-8 Dipole interactions 19
1-9 Surface singularities 20
1-10 Volume distributions of dipole moment 23

Chapter 2. Boundary Conditions and Relation of Microscopic
to Macroscopic Fields 28

2-1 The displacement vector 28
2-2 Boundary conditions 31
2-3 The electric field in a material medium 33
2-4 Polarizability 38

Chapter 3. General Methods for the Solution of Potential
Problems 42

3-1 Uniqueness theorem 42
3-2 Green's reciprocation theorem 43
3-3 Solution by Green's function 44
3-4 Solution by inversion 47
3-5 Solution by electrical images 49
3-6 Solution of Laplace's equation by the separation of variables . 53

Chapter 4. Two-dimensional Potential Problems 61

4-1 Conjugate complex functions 61
4-2 Capacity and field strength 63
4-3 The potential of a uniform field 64
4-4 The potential of a line charge 64
4-5 Complex transformations 66
4-6 General Schwarz transformation 67
4-7 Single-angle transformations 70
4-8 Multiple-angle transformations 71
4-9 Direct solution of Laplace's equation by the method of harmonics 73
4-10 Illustration: Line charge and dielectric cylinder 74
4-11 Line charge in an angle between two conductors 77

CHAPTER 5. THREE-DIMENSIONAL POTENTIAL PROBLEMS 81

5–1 The solution of Laplace's equation in spherical coordinates . . 81
5–2 The potential of a point charge 82
5–3 The potential of a dielectric sphere and a point charge . . . 83
5–4 The potential of a dielectric sphere in a uniform field . . . 84
5–5 The potential of an arbitrary axially-symmetric spherical
     potential distribution 86
5–6 The potential of a charged ring 87
5–7 Problems not having axial symmetry 88
5–8 The solution of Laplace's equation in cylindrical coordinates . 88
5–9 Application of cylindrical solutions to potential problems . . 91

CHAPTER 6. ENERGY RELATIONS AND FORCES IN THE ELECTRO-
             STATIC FIELD 95

6–1 Field energy in free space 95
6–2 Energy density within a dielectric 98
6–3 Thermodynamic interpretation of U 100
6–4 Thomson's theorem 101
6–5 Maxwell stress tensor 103
6–6 Volume forces in the electrostatic field in the presence
     of dielectrics 107
6–7 The behavior of dielectric liquids in an electrostatic field . . 111

CHAPTER 7. STEADY CURRENTS AND THEIR INTERACTION 118

7–1 Ohm's law 118
7–2 Electromotive force 119
7–3 The solution of stationary current problems 120
7–4 Time of relaxation in a homogeneous medium 122
7–5 The magnetic interaction of steady line currents 123
7–6 The magnetic induction field 125
7–7 The magnetic scalar potential 125
7–8 The magnetic vector potential 127
7–9 Types of currents 129
7–10 Polarization currents 129
7–11 Magnetic moments 130
7–12 Magnetization and magnetization currents 134
7–13 Vacuum displacement current 135

CHAPTER 8. MAGNETIC MATERIALS AND BOUNDARY VALUE
             PROBLEMS 139

8–1 Magnetic field intensity 139
8–2 Magnetic sources 140
8–3 Permeable media: magnetic susceptibility and boundary
     conditions 144
8–4 Magnetic circuits 145

8–5 Solution of boundary value problems by magnetic scalar potentials 146
8–6 Uniqueness theorem for the vector potential 147
8–7 The use of the vector potential in the solution of problems . . 148
8–8 The vector potential in two dimensions 151
8–9 The vector potential in cylindrical coordinates 153

CHAPTER 9. MAXWELL'S EQUATIONS 158

9–1 Faraday's law of induction 158
9–2 Maxwell's equations for stationary media 159
9–3 Faraday's law for moving media 160
9–4 Maxwell's equations for moving media 163
9–5 Motion of a conductor in a magnetic field 165

CHAPTER 10. ENERGY, FORCE, AND MOMENTUM RELATIONS IN THE ELECTROMAGNETIC FIELD 170

10–1 Energy relations in quasi-stationary current systems 170
10–2 Forces on current systems 172
10–3 Inductance 174
10–4 Magnetic volume force 177
10–5 General expressions for electromagnetic energy 178
10–6 Momentum balance 181

CHAPTER 11. THE WAVE EQUATION AND PLANE WAVES 185

11–1 The wave equation 185
11–2 Plane waves 187
11–3 Radiation pressure 191
11–4 Plane waves in a moving medium 193
11–5 Reflection and refraction at a plane boundary 195
11–6 Waves in conducting media and metallic reflection 200
11–7 Group velocity 202

CHAPTER 12. CONDUCTING FLUIDS IN A MAGNETIC FIELD (MAGNETOHYDRODYNAMICS) 205

12–1 "Frozen-in" lines of force 205
12–2 Magnetohydrodynamic waves 207

CHAPTER 13. WAVES IN THE PRESENCE OF METALLIC BOUNDARIES . 212

13–1 The nature of metallic boundary conditions 212
13–2 Eigenfunctions and eigenvalues of the wave equation . . . 214
13–3 Cavities with rectangular boundaries 218
13–4 Cylindrical cavities 219
13–5 Circular cylindrical cavities 222
13–6 Wave guides 223
13–7 Scattering by a circular cylinder 226

13–8 Spherical waves 229
13–9 Scattering by a sphere 233

CHAPTER 14. THE INHOMOGENEOUS WAVE EQUATION 240

14–1 The wave equation for the potentials 240
14–2 Solution by Fourier analysis 242
14–3 The radiation fields 245
14–4 Radiated energy 248
14–5 The Hertz potential 254
14–6 Computation of radiation fields by the Hertz method . . . 255
14–7 Electric dipole radiation 257
14–8 Multipole radiation 260
14–9 Derivation of multipole radiation from scalar superpotentials . 264
14–10 Energy and angular momentum radiated by multipoles . . . 267

CHAPTER 15. THE EXPERIMENTAL BASIS FOR THE THEORY OF
              SPECIAL RELATIVITY 272

15–1 Galilean relativity and electrodynamics 272
15–2 The search for an absolute ether frame 274
15–3 The Lorentz-Fitzgerald contraction hypothesis 278
15–4 "Ether drag" 279
15–5 Emission theories 280
15–6 Summary 283

CHAPTER 16. RELATIVISTIC KINEMATICS AND THE LORENTZ
              TRANSFORMATION 286

16–1 The velocity of light and simultaneity 286
16–2 Kinematic relations in special relativity 288
16–3 The Lorentz transformation 293
16–4 Geometric interpretations of the Lorentz transformation . . . 297
16–5 Transformation equations for velocity 301

CHAPTER 17. COVARIANCE AND RELATIVISTIC MECHANICS 305

17–1 The Lorentz transformation of a four-vector 305
17–2 Some tensor relations useful in special relativity 307
17–3 The conservation of momentum 311
17–4 Relation of energy to momentum and to mass 313
17–5 The Minkowski force 316
17–6 The collision of two similar particles 318
17–7 The use of four-vectors in calculating kinematic relations
       for collisions 320

CHAPTER 18. COVARIANT FORMULATION OF ELECTRODYNAMICS . . . 324

18–1 The four-vector potential 324
18–2 The electromagnetic field tensor 327
18–3 The Lorentz force in vacuum 331

18–4 Covariant description of sources in material media 332
18–5 The field equations in a material medium 334
18–6 Transformation properties of the partial fields 336

CHAPTER 19. THE LIÉNARD-WIECHERT POTENTIALS AND THE FIELD
            OF A UNIFORMLY MOVING ELECTRON 341

19–1 The Liénard-Wiechert potentials 341
19–2 The fields of a charge in uniform motion 344
19–3 Direct solution of the wave equation 347
19–4 The "convection potential" 348
19–5 The virtual photon concept 350

CHAPTER 20. RADIATION FROM AN ACCELERATED CHARGE 354

20–1 Fields of an accelerated charge 354
20–2 Radiation at low velocity 358
20–3 The case of u̇ parallel to u 359
20–4 Radiation when the acceleration is perpendicular to the
     velocity (radiation from circular orbits) 363
20–5 Radiation with no restrictions on the acceleration or velocity . 370
20–6 Classical cross section for bremsstrahlung in a Coulomb field . 371
20–7 Čerenkov radiation 373

CHAPTER 21. RADIATION REACTION AND COVARIANT FORMULATION
            OF THE CONSERVATION LAWS OF ELECTRODYNAMICS . . 377

21–1 Covariant formulation of the conservation laws of vacuum
     electrodynamics 377
21–2 Transformation properties of the "free" radiation field . . . 379
21–3 The electromagnetic energy momentum tensor in material media 380
21–4 Electromagnetic mass 381
21–5 Electromagnetic mass—qualitative considerations 383
21–6 The reaction necessary to conserve radiated energy 386
21–7 Direct computation of the radiation reaction from the
     retarded fields 387
21–8 Properties of the equation of motion 389
21–9 Covariant description of the mechanical properties of the
     electromagnetic field of a charge 390
21–10 The relativistic equations of motion 392
21–11 The integration of the relativistic equation of motion . . . 394
21–12 Modification of the theory of radiation to eliminate divergent
     mass integrals. Advanced potentials 394
21–13 Direct calculation of the relativistic radiation reaction . . . 398

CHAPTER 22. RADIATION, SCATTERING, AND DISPERSION 401

22–1 Radiative damping of a charged harmonic oscillator 401
22–2 Forced vibrations 403
22–3 Scattering by an individual free electron 404

22–4　Scattering by a bound electron 407
22–5　Absorption of radiation by an oscillator 407
22–6　Equilibrium between an oscillator and a radiation field . . . 409
22–7　Effect of a volume distribution of scatterers 411
22–8　Scattering from a volume distribution. Rayleigh scattering . 414
22–9　The dispersion relation 416
22–10 A general theorem on scattering and absorption 419

CHAPTER 23. THE MOTION OF CHARGED PARTICLES IN ELECTRO-
　　　　　　MAGNETIC FIELDS 425

23–1　World-line description 425
23–2　Hamiltonian formulation and the transition to three-
　　　　dimensional formalism 427
23–3　Equations for the trajectories 430
23–4　Applications 433
23–5　The motion of a particle with magnetic moment in an
　　　　electromagnetic field 437

CHAPTER 24. HAMILTONIAN FORMULATION OF MAXWELL'S EQUATIONS 446

24–1　Transition to a one-dimensional continuous system 446
24–2　Generalization to a three-dimensional continuum 448
24–3　The electromagnetic field 451
24–4　Periodic solutions in a box. Plane wave representation . . . 454

APPENDIX I. UNITS AND DIMENSIONS IN ELECTROMAGNETIC THEORY . 459

Tables: I–1. Conversion Factors 465
　　　　I–2. Fundamental Electromagnetic Relations Valid in vacuo as
　　　　　　They Appear in the Various Systems of Units 466
　　　　I–3. Definition of Fields from Sources (mks system) . . . 468
　　　　I–4. Useful Numerical Relations 469

APPENDIX II. USEFUL VECTOR RELATIONS 470

Table II–1. Vector Formulas 470

APPENDIX III. VECTOR RELATIONS IN CURVILINEAR COORDINATES. . 473

Table III–1. Coordinate Systems 475

BIBLIOGRAPHY 479

INDEX 485

ERRATA

p. 33 Eq. 2–19 The constant is not arbitrary but is proportional to the magnitude of any surface dipole layer (Eq. 1–61).

pp. 39–40 Properly, an isolated molecule is in a state of definite parity, such as one of its rotational eigenstates, and therefore has no permanent electric dipole moment (the expectation value of an operator of odd parity, such as the electric dipole moment, in a state of definite parity is zero). However, a proper quantum mechanical calculation gives the same result at the classical calculation here. The details are in P. Debye's 1929 book *Polar Molecules*, but not in most modern textbooks.

p. 40 Eq. 2–45 fails for many polar liquids, like H_2O, because of short-range correlations resulting from strong angle-dependent (covalent) intermolecular forces.

p. 48 It is clearer to say that the triangles with sides rld and $l'r'd'$ are similar.

p. 85 Eq. 5–16 the last $=$ should be a $-$.

p. 92 Eq. 5–42 e_0 should be ϵ_0.

p. 99 The derivation of the symmetry of $\kappa_{\alpha\beta}$ (last paragraph, Eqs. 6–18 and 6–19) is incorrect. The symmetry of $\kappa_{\alpha\beta}$ follows from defining dielectric susceptibility in a thermodynamically consistent manner, as the second derivative tensor of the free energy with respect to the electric field; see Landau and Lifschitz *Electrodynamics of Continuous Media* §11.

p. 113 Eq. 6–76 the upper limit should be B.

p. 117 Exc. 7 (6–44) should be (6–45).

p. 128 Eq. 7–42 $\oint$ should be $\int$.

p. 168 Exc. 9–4 can be solved if the orbital radius is taken to be constant, which is inconsistent with the conditions given. It is possible to solve this problem if the central force field is ignored, the initial $\vec{B}$ is nonzero, and its variation is slow compared to the gyroperiod. The second part of Exc. 9–7 shows that for a special spatial distribution of $\vec{B}$ and no central force field the orbital radius remains constant.

p. 168 Exc. 9–7 (first part) requires the assumption that $\vec{B}$ is independent of φ for *any* axis z. Do only the second part.

p. 174 Eq. 10–14 $\frac{\partial}{\partial t}$ should be $\frac{d}{dt}$.

p. 321 Eq. 17–73 $-2c^2\mathbf{p}_1 \cdot \bar{\mathbf{p}}_1$ should be $+2c^2\mathbf{p}_1 \cdot \bar{\mathbf{p}}_1$.
Eq. 17–74 should have a $-$ sign on one side.

p. 364 Eq. 20–35 $(u^2/c)^2$ should be (u^2/c^2).

p. 365 Eq. 20–38 should have a factor m_0c^2 on the right.

p. 370 The reason real steady currents flowing through wires don't radiate significantly is their low speed and the Pauli exclusion principle; the current distribution is not microscopically continuous, contrary to the assertion here.

p. 378 One line above 21–9, (17–32) should be (18–32).
Eq. 21–9 The $-$ sign should be $+$.

p. 401 Eq. 22–1 In the middle and right hand side (two places) x should be $\dddot{x}$.

p. 404 One line after 22–20, (19–20) should be (20–18).

p. 412 After the unnumbered equation (22–9) should be (22–19).

p. 419 Second line from bottom, $\epsilon|\mathbf{H}|^2$ should be $\epsilon|\mathbf{E}|^2$.

p. 420 Eq. 22–76 The first exponential is $e^{-ikr(1-\cos\theta)}$.

p. 422 Eq. 22–89 This equation appears to give the erroneous result that for $\kappa_2 = 0$ (as is true to high accuracy for optical glass) but $\kappa_1 \neq 1$ the scattering cross-section must be zero. This paradox arises from the use of the infinite-medium relation between $\mathbf{p}$ and $\mathbf{E}$. In the case of a small particle (or interface) the proper relation must include radiation damping, implying $\kappa_2 \neq 0$ (with its actual value determined by 22–89).

p. 465 1 farad = 9×10^{11} cm (cgs capacitance).

1 ohm = $\frac{1}{9} \times 10^{-11}$ sec/cm (cgs resistance; = esu).

CHAPTER 1

THE ELECTROSTATIC FIELD IN VACUUM

The interaction between material bodies can be described either by formulating the action at a distance between the interacting bodies or by separating the interaction process into the production of a *field* by one system and the action of the field on another system. These two alternative descriptions are physically indistinguishable in the static case. If the bodies are in motion, however, and the velocity of propagation of the interaction is finite, it is both physically and mathematically advantageous to ascribe physical reality to the field itself, even though it is possible to replace the field concept by that of "delayed" and "advanced" direct interaction in the description of electromagnetic phenomena. We shall formulate even the electrostatic interactions as a field theory, which can then be extended to the consideration of nonstatic cases.

1-1 Vector fields. Field theories applicable to various types of interaction differ by the number of parameters necessary to define the field and by the symmetry character of the field. In a general sense, a field is a physical entity such that each point in space is a degree of freedom. A field is therefore specified by giving the behavior in time at each coordinate point of a quantity suitable to describe the physical content.

The types of fields possible are restricted by various considerations. Fields are classified according to the number of parameters necessary to define the field and by the "transformation character" of the field quantities under various coordinate transformations. A "scalar" field is described by the time dependence of one quantity at each point in space, a "three-dimensional vector field" by three such quantities. In general, an "nth-rank tensor field" requires the specification of d^n components, where d is the dimensionality of the space in which the field is defined. A scalar field is a zero-rank tensor field, and a vector field is a first-rank tensor field.

The field description of a physical entity is independent of the particular choice of coordinate system used. This fact restricts the transformation properties of the field components under coordinate transformations. We consider two types of transformations of coordinates. "proper" and "improper" transformations. Proper transformations are those which leave the cyclic order of the coordinates invariant (i.e., do not transform a right-handed into a left-handed coordinate system in three dimensions);

1

translation and rotation are proper transformations. Improper transformations, such as inversion of the coordinate axes and reflection of the coordinate system in a plane, change the cyclic order of coordinates.

A basic vector is the distance $\mathbf{r}$ connecting two points; the components of $\mathbf{r}$ may be designated by r_α. The components V_α of a *vector* field $\mathbf{V}$ transform like the components r_α under both proper and improper transformations. A *scalar* is invariant under proper and improper transformations. The components P_α of a *pseudovector* field $\mathbf{P}$ transform like the components r_α under proper transformations, but change sign relative to r_α under improper transformations. A *pseudoscalar* is invariant under proper transformations but changes sign under improper transformations.

The electric field is a three-dimensional vector field, i.e., a field definable by the specification of three components. The theory of vector fields was developed in connection with the study of fluid motion, a fact which is betrayed repeatedly by the vocabulary of the theory. We shall consider some general mathematical properties of such fields before specifying the physical content of the vectors.

All vector fields in three dimensions are uniquely defined if their circulation densities (curl) and source densities (divergence) are given functions of the coordinates at all points in space, and if the totality of sources, as well as the source density, is zero at infinity. Let us prove this theorem formally. Consider a three-dimensional vector field $\mathbf{V}(x, y, z)$ such that

$$\nabla \cdot \mathbf{V} = s, \tag{1-1}$$

$$\nabla \times \mathbf{V} = \mathbf{c}. \tag{1-2}$$

Equation (1–2) is self-consistent only if the circulation density $\mathbf{c}$ is irrotational, i.e., if

$$\nabla \cdot \mathbf{c} = 0. \tag{1-2'}$$

We shall first show that if

$$\mathbf{V} = -\nabla\phi + \nabla \times \mathbf{A}, \tag{1-3}$$

where

$$\phi(x_\alpha) = \frac{1}{4\pi} \int \frac{s(x'_\alpha)}{r(x_\alpha, x'_\alpha)}\, dv' \tag{1-4}$$

and

$$\mathbf{A}(x_\alpha) = \frac{1}{4\pi} \int \frac{\mathbf{c}(x'_\alpha)}{r(x_\alpha, x'_\alpha)}\, dv', \tag{1-5}$$

then $\mathbf{V}$ satisfies Eqs. (1–1) and (1–2).

It is necessary to examine the notation of Eqs. (1–4) and (1–5) before proceeding with the proof. The symbol x_α stands for x, y, z at the *field point;* the symbol x'_α stands for x', y', z' at the *source point;* the function $r(x_\alpha, x'_\alpha)$ is the symmetric function

$$r(x_\alpha, x'_\alpha) = \left| \sqrt{\sum_{\alpha=1}^{\alpha=3} (x_\alpha - x'_\alpha)^2} \right|$$

representing the positive distance between field and source point. The reader should note carefully the functional relationships explicit in Eqs. (1–4) and (1–5). In integrals of this type these functional dependences will often not be fully stated; for example, we may write the volume integrals

$$\phi = \frac{1}{4\pi} \int \frac{s}{r}\, dv', \tag{1–4'}$$

$$\mathbf{A} = \frac{1}{4\pi} \int \frac{\mathbf{c}}{r}\, dv', \tag{1–5'}$$

as a short notation. We shall sometimes use $\mathbf{R}$ for the radius vector from an origin of coordinates to the field point x_α, and $\boldsymbol{\xi}$ for that of a source point x'_α; then $r = |\mathbf{R} - \boldsymbol{\xi}|$.

Let us demonstrate that $\mathbf{V}$ as expressed by Eq. (1–3) is a solution of Eqs. (1–1) and (1–2):

$$\boldsymbol{\nabla} \cdot \mathbf{V} = -\nabla^2 \phi + \boldsymbol{\nabla} \cdot (\boldsymbol{\nabla} \times \mathbf{A}) = -\nabla^2 \phi$$

$$= -\frac{1}{4\pi} \nabla^2 \left\{ \int \frac{s}{r}\, dv' \right\}.$$

The Laplacian operator ∇^2 operates on the field coordinates; hence

$$\boldsymbol{\nabla} \cdot \mathbf{V} = -\frac{1}{4\pi} \int s \nabla^2 \left(\frac{1}{r} \right) dv'. \tag{1–6}$$

Now we can show that

$$\nabla^2 \left\{ \frac{1}{r(x_\alpha, x'_\alpha)} \right\} = -4\pi\, \delta(\mathbf{r}), \tag{1–7}$$

where $\delta(\mathbf{r})$, the Dirac δ-function, is defined by the functional properties

$$\delta(\mathbf{r}) = 0, \qquad \mathbf{r} \neq 0, \quad \text{i.e.,} \quad x_\alpha \neq x'_\alpha, \tag{1–8}$$

$$\int \delta(\mathbf{r})\, dv' = 1, \tag{1–9}$$

if the point $\mathbf{r} = 0$ is included in the volume of integration, and by

$$\int f(x'_\alpha)\, \delta(\mathbf{r})\, dv' = f(x_\alpha), \tag{1-10}$$

for any arbitrary function f so long as the volume of integration includes the point $\mathbf{r} = 0$. The δ-function is not an analytic function but essentially a notation for the functional properties of the three defining equations. It will always be used in terms of these properties.

Since it is evident by direct differentiation that $\nabla^2(1/r) = 0$ for $r \neq 0$, we have only to prove that

$$\int \nabla^2(1/r)\, dv' = -4\pi \tag{1-11}$$

in order to verify Eq. (1-7). [In Eq. (1-11) the point $r = 0$, that is, $x_\alpha = x'_\alpha$, is included in the volume of integration.] By the application of Gauss's divergence theorem, applicable to any vector $\mathbf{V}$,*

$$\int \nabla \cdot \mathbf{V}\, dv = \int \mathbf{V} \cdot d\mathbf{S},$$

it is seen that

$$\int \nabla^2 \left(\frac{1}{r}\right) dv' = \int \nabla \left(\frac{1}{r}\right) \cdot d\mathbf{S}'$$

$$= -\int \frac{\mathbf{r} \cdot d\mathbf{S}'}{r^3} = -\int d\Omega,$$

where Ω is the solid angle subtended at x_α by the surface of integration S' over the variables x'_α. Since S' includes x_α, we have simply $\int d\Omega = 4\pi$, and Eq. (1-11) is verified. Hence from Eqs. (1-6) and (1-10),

$$\nabla \cdot \mathbf{V} = -\frac{1}{4\pi} \int s\nabla^2 \left(\frac{1}{r}\right) dv' = \int s(x'_\alpha)\, \delta(\mathbf{r})\, dv' = s(x_\alpha), \tag{1-12}$$

which was to be proved.

---

* Strictly speaking, Gauss's divergence theorem is not necessarily applicable, since the function $\mathbf{V} = \nabla(1/r)$ is singular at $r = 0$. If, however, we remove the singularity by substituting for $1/r$ the function $(1 - e^{-r/a})/r$, for example, where a is an arbitrarily small radius, then

$$\nabla \left[\frac{1}{r}(1 - e^{-r/a}) \right] = -\frac{\mathbf{r}}{r^3}(1 - e^{-r/a}) + \frac{\mathbf{r}}{r^3}\left(\frac{r}{a} e^{-r/a}\right).$$

Since the magnitude of the second term varies only as r^{-1}, its surface integral over a small sphere surrounding the point $r = 0$ will vanish as the radius of the sphere goes to zero.

Similarly,

$$\mathbf{\nabla} \times \mathbf{V} = -\mathbf{\nabla} \times \mathbf{\nabla}\phi + \mathbf{\nabla} \times (\mathbf{\nabla} \times \mathbf{A}) = \mathbf{\nabla}(\mathbf{\nabla} \cdot \mathbf{A}) - \nabla^2\mathbf{A}$$

$$= \frac{1}{4\pi}\left\{\int (\mathbf{c} \cdot \mathbf{\nabla})\mathbf{\nabla}\left(\frac{1}{r}\right) dv' - \int \mathbf{c}\nabla^2\left(\frac{1}{r}\right) dv'\right\}. \tag{1–13}$$

We shall be able to show that the first integral vanishes if $\mathbf{c}$ is bounded in space. If we anticipate this result, we see immediately, from Eq. (1–7), that

$$\mathbf{\nabla} \times \mathbf{V} = \int \mathbf{c}(x'_\alpha)\, \delta(\mathbf{r})\, dv' = \mathbf{c}(x_\alpha), \tag{1–14}$$

so that Eq. (1–2) is also satisfied.

To prove that the first term of Eq. (1–13) vanishes, let us examine the coordinate variables involved in the integrand. The operator $\mathbf{\nabla}$ has the components $\partial/\partial x_\alpha$. If we introduce the operator $\mathbf{\nabla}'_\alpha = \partial/\partial x'_\alpha$, operating on the source coordinates, then for any arbitrary function $g[r(x_\alpha, x'_\alpha)]$, we have

$$\mathbf{\nabla}g = -\mathbf{\nabla}'g. \tag{1–15}$$

Therefore the first integral of Eq. (1–13) may be written

$$\mathbf{I} = \int (\mathbf{c} \cdot \mathbf{\nabla})\mathbf{\nabla}\left(\frac{1}{r}\right) dv' = \int (\mathbf{c} \cdot \mathbf{\nabla}')\mathbf{\nabla}'\left(\frac{1}{r}\right) dv'.$$

The differential operators now operate on the variables of integration and we may integrate by parts. Each component of $\mathbf{I}$ becomes

$$I_\alpha = \int (\mathbf{c} \cdot \mathbf{\nabla}') \frac{\partial}{\partial x'_\alpha}\left(\frac{1}{r}\right) dv'$$

$$= \int \mathbf{\nabla}' \cdot \left\{\mathbf{c} \frac{\partial}{\partial x'_\alpha}\left(\frac{1}{r}\right)\right\} dv' - \int (\mathbf{\nabla}' \cdot \mathbf{c}) \frac{\partial}{\partial x'_\alpha}\left(\frac{1}{r}\right) dv'. \tag{1–16}$$

The second integral vanishes because the divergence of $\mathbf{c}$ is zero [Eq. (1–2′)]. The first term can be transformed to a surface integral by means of Gauss's theorem; if $\mathbf{c}$ is bounded in space the surface may be taken sufficiently large so that $\mathbf{c}$ is zero over the entire integration. Hence Eq. (1–16) is zero, and the proof is complete.

We have thus proved that if the source density s and the circulation density $\mathbf{c}$ of a vector field $\mathbf{V}$ are given everywhere, then a solution for $\mathbf{V}$ can be derived from a *scalar potential* ϕ and a *vector potential* $\mathbf{A}$. The potentials ϕ and $\mathbf{A}$ are expressed as integrals over the source and circulation densities.

It can be proved that this system of solutions is unique if the sources are bounded in space, i.e., there are no sources at infinity, and thus the fields themselves vanish at sufficiently large distance from the sources. Suppose that there are two functions, $\mathbf{V}_1$ and $\mathbf{V}_2$, which satisfy Eqs. (1–1) and (1–2). Their difference, the function $\mathbf{W} = \mathbf{V}_1 - \mathbf{V}_2$, obeys the conditions

$$\mathbf{\nabla} \cdot \mathbf{W} = 0, \tag{1–17}$$

$$\mathbf{\nabla} \times \mathbf{W} = 0, \tag{1–18}$$

at every point in space and is zero at infinity. If we now show that $\mathbf{W}$ vanishes everywhere, we shall have proved that for finite sources there is only one solution for Eqs. (1–1) and (1–2). To prove this we note that if Eq. (1–18) is satisfied we can always put

$$\mathbf{W} = -\mathbf{\nabla}\psi \tag{1–19}$$

and, from Eq. (1–17),

$$\nabla^2\psi = 0 \tag{1–20}$$

everywhere. If we apply Gauss's divergence theorem to the vector $\psi\mathbf{\nabla}\psi$, we obtain

$$\int \psi\mathbf{\nabla}\psi \cdot d\mathbf{S} = \int [\psi\nabla^2\psi + (\mathbf{\nabla}\psi)^2]\, dv. \tag{1–21}$$

The left side vanishes if the boundary is taken at sufficiently large distance from the sources, since ψ tends to zero at least as $1/r$, and the first term on the right is identically zero because of Eq. (1–20). Therefore Eq. (1–21) reduces to

$$\int (\mathbf{\nabla}\psi)^2\, dv = \int (\mathbf{W})^2\, dv = 0, \tag{1–22}$$

and hence $\mathbf{W} = \mathbf{V}_1 - \mathbf{V}_2 = 0$ everywhere. Thus $\mathbf{V}$ as given by Eq. (1–3) is unique.

We have gone into this formal proof in great detail not only because the theorems are of fundamental importance but also because the methods are of general usefulness throughout the study of electromagnetic fields. For convenience, let us summarize the results obtained:

(a) If the source density s and the circulation density $\mathbf{c}$ of a vector field $\mathbf{V}$ are given for a finite region of space and there are no sources at infinity, then $\mathbf{V}$ is uniquely defined.

(b) If $\mathbf{V}$ has sources s but no circulation density $\mathbf{c}$, $\mathbf{V}$ is derivable from a scalar potential ϕ.

(c) If $\mathbf{V}$ has circulation density $\mathbf{c}$ but no sources s, $\mathbf{V}$ is derivable from a vector potential $\mathbf{A}$.

(d) **V** is always derivable from a scalar and a vector potential.

(e) At points in space where s and **c** vanish, **V** is derivable from a scalar potential ϕ for which $\nabla^2\phi = 0$, or from a vector potential **A** for which $\nabla \times \nabla \times \mathbf{A} = 0$. We may add that at such points the field is said to be *harmonic*.

(f) If s and **c** are identically zero everywhere, **V** vanishes everywhere.

(g) The unique solution for **V** in terms of s and **c** is given by means of the potentials as expressed by the integrals (1–4) and (1–5).

(h) We have established a systematic notation for source and field coordinates. If we add the convention that the vector **r** points *from* source *to* field point we may extend our list of useful mathematical relations:

$$\nabla^2(1/r) = -4\pi\,\delta(\mathbf{r}),$$

$$\nabla[g(r)] = -\nabla'[g(r)],$$

$$\nabla r = -\nabla' r = \mathbf{r}/r,$$

$$\nabla \cdot \mathbf{r} = +3.$$

These properties of general vector fields will be indispensable in the physical considerations which follow. We shall have a consistent field theory representing the empirical laws of electricity and magnetism when we have written these laws as a set of equations giving the source and circulation densities, i.e., the divergence and curl, of the field vectors representing the electromagnetic fields. This is the fundamental program of classical electromagnetic theory.

1–2 The electric field. We shall first consider the electrostatic field in vacuum. The electric field is defined in terms of the force produced on a test charge q by the equation

$$\lim_{q \to 0} \frac{\mathbf{F}}{q} = \mathbf{E}, \tag{1–23}$$

where **F** is the force (newtons) on the test charge q (coulombs). The definition is entirely independent of the system of units, but in the mks system* the electric field **E** defined by this equation is in volts per meter. The limit $q \to 0$ is introduced in order that the test charge will not influence the behavior of the sources of the field, which will then be independent of the presence of the test body. The requirement that the test charge be vanishingly small compared with all sources of the field raises a funda-

---

* See Appendix I for a discussion of units and the relations between units of various systems.

mental difficulty, since the finite magnitude of the electronic charge does not permit the limit $q \to 0$ to be carried out experimentally. This restriction therefore limits the practical validity of the definition to cases where the sources producing the field are equivalent to a large number of electronic charges. Definition (1–23) is thus entirely suitable only for macroscopic phenomena, and we shall have to exercise great care in applying it to the treatment of the elementary charges of which matter is actually composed. For microscopic processes the field cannot be defined "operationally" in terms of its effect; it must be *described* in terms of its sources, *assuming* that the macroscopic laws of the field sources are still valid.

The field **E** is a three-dimensional vector (not pseudovector) field if we adopt the convention that the charge q is a scalar (not a pseudoscalar). In terms of this convention the transformation character of both the electric and (later) the magnetic fields is defined.

1–3 Coulomb's law. The experimentally established law for the force between two point charges *in vacuo* was originally formulated as an action at a distance:

$$\mathbf{F}_2 = \frac{q_1 q_2 \mathbf{r}}{4\pi\epsilon_0 r^3} = -\frac{q_1 q_2}{4\pi\epsilon_0} \boldsymbol{\nabla}\left(\frac{1}{r}\right). \tag{1–24}$$

Here $\mathbf{F}_2$ is the force on charge q_2 due to the presence of charge q_1, and **r** is the radius vector position of charge q_2 measured from an origin located at charge q_1; ϵ_0 is a constant, $10^7/4\pi c^2$ farads/meter in this system of units; c is the experimental value of the velocity of propagation of plane electromagnetic waves in free space (see Appendix I), and all distances are measured in meters. In the mathematical identity on the right the gradient operator acts on the coordinates of the charge q_2. The law applies equally to positive and negative charges, and indicates that like charges repel, unlike charges attract. If the test charge in Eq. (1–23) is assumed positive, comparison with Eq. (1–24) yields immediately

$$\mathbf{E} = \frac{q}{4\pi\epsilon_0}\frac{\mathbf{r}}{r^3} = -\frac{q}{4\pi\epsilon_0}\boldsymbol{\nabla}\left(\frac{1}{r}\right), \tag{1–25}$$

giving the electric field **E** at the position **r** due to a charge q at the origin of the radius vector. Here q corresponds to q_1 of Eq. (1–24).

We can prove Gauss's flux theorem

$$\int_S \mathbf{E} \cdot d\mathbf{S} = \frac{q}{\epsilon_0} \tag{1–26}$$

as a direct consequence of Coulomb's law: Consider an element of sur-

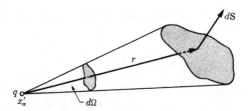

FIG. 1–1. Elements of surface and solid angle contributing to the total electric flux of Gauss's theorem.

face $d\mathbf{S}$, expressed as a vector directed along the outward normal of the element as shown in Fig. 1–1, at a distance r from a charge q at a point x'_α. By taking the scalar product of $d\mathbf{S}$ and both sides of Eq. (1–25) we secure

$$\mathbf{E} \cdot d\mathbf{S} = \frac{1}{4\pi\epsilon_0} \frac{q}{r^3} \mathbf{r} \cdot d\mathbf{S} = \frac{q}{4\pi\epsilon_0} d\Omega. \qquad (1\text{--}26')$$

Since the integral of $d\Omega$ over a closed surface which includes the point x'_α is just 4π, Gauss's theorem follows immediately. The principle of superposition enables us to sum the separate fields of any number of point charges, so that q of Eq. (1–26) is the total charge inside the boundary surface S.

If we apply the divergence theorem to $\mathbf{E}$,

$$\int_S \mathbf{E} \cdot d\mathbf{S} = \int \boldsymbol{\nabla} \cdot \mathbf{E} \, dv,$$

and make use of the fact that the total volume integral of the charge density ρ is simply the total charge q, the application of the flux theorem enables us to put Eq. (1–25) into the form

$$\boldsymbol{\nabla} \cdot \mathbf{E} = \frac{\rho}{\epsilon_0}. \qquad (1\text{--}27)$$

Here ρ is the charge per unit volume at the point where the electric field is $\mathbf{E}$. Since the curl of the gradient of a scalar is zero, it further follows from Eq. (1–25) and the principle of superposition that

$$\boldsymbol{\nabla} \times \mathbf{E} = 0. \qquad (1\text{--}28)$$

The electrostatic field is thus irrotational. That the electrostatic field is completely defined by a charge distribution then follows from the theorem that a vector field is uniquely determined by the curl and the divergence of the field.

It is instructive to note that Eqs. (1–27) and (1–28) follow directly from Coulomb's law in the form

$$\mathbf{E}(x_\alpha) = \frac{1}{4\pi\epsilon_0} \int \rho(x'_\alpha) \frac{\mathbf{r}}{r^3} \, dv' \tag{1–29}$$

$$= -\frac{1}{4\pi\epsilon_0} \int \rho \nabla \left(\frac{1}{r}\right) dv'. \tag{1–30}$$

It follows that

$$\nabla \cdot \mathbf{E} = -\frac{1}{4\pi\epsilon_0} \int \rho \nabla^2 \left(\frac{1}{r}\right) dv' = \frac{1}{\epsilon_0} \int \rho(x'_\alpha) \, \delta(\mathbf{r}) \, dv' = \frac{\rho(x_\alpha)}{\epsilon_0}. \tag{1–27'}$$

Also

$$\nabla \times \mathbf{E} = -\frac{1}{4\pi\epsilon_0} \int \rho \nabla \times \nabla \left(\frac{1}{r}\right) dv' = 0, \tag{1–28'}$$

since the curl of a vector expressible as a gradient vanishes identically.

For a point source q we have directly from Eq. (1–25):

$$\nabla \cdot \mathbf{E} = \frac{-q}{4\pi\epsilon_0} \nabla^2 \left(\frac{1}{r}\right) = q \frac{\delta(\mathbf{r})}{\epsilon_0}. \tag{1–27''}$$

1–4 The electrostatic potential. Since the static field is irrotational, it may be expressed as the gradient of a scalar function. We may define an electrostatic potential ϕ by the equation

$$\mathbf{E} = -\nabla\phi. \tag{1–31}$$

In Cartesian coordinates the components of the field parallel to the x_α axes respectively are given by

$$E_\alpha = -\frac{\partial \phi}{\partial x_\alpha}. \tag{1–32}$$

The application of the general vector relation known as Stokes' theorem,

$$\int (\nabla \times \mathbf{E}) \cdot d\mathbf{S} = \oint \mathbf{E} \cdot d\mathbf{l}, \tag{1–33}$$

where $d\mathbf{l}$ is the infinitesimal vector length tangent to a closed path of integration, leads to

$$\oint \mathbf{E} \cdot d\mathbf{l} = 0, \tag{1–34}$$

since the curl of $\mathbf{E}$ is everywhere zero. This shows that the electrostatic field is a conservative field: no work is done on a test charge if it is moved around a closed path in the field. Since the work done in moving a test charge from one point to another is independent of the path, we can uniquely define the work necessary to carry a unit charge from an infinite distance to a given point as the potential of that point. If one considers fields of less than three dimensions, i.e., sources extending to infinity in one or more directions, this definition will lead to difficulties and a point other than infinity must be taken as a reference point. So long as only finite sources are considered, however, this definition of potential is both adequate and convenient.

The substitution of Eq. (1–31) in Eq. (1–27) leads at once to Poisson's equation

$$\nabla^2 \phi = -\rho/\epsilon_0, \tag{1–35}$$

and in a region of zero charge density to Laplace's equation,

$$\nabla^2 \phi = 0. \tag{1–36}$$

The fundamental problem of electrostatics is to determine solutions of Poisson's equation appropriate to the conditions under particular consideration.

1–5 The potential in terms of charge distribution. The electrostatic potential at a given point was defined by Eq. (1–31) in terms of the electric field at that point. Since the source density of the electrostatic field is just ρ/ϵ_0, we know from Eq. (1–4) that the potential is

$$\phi(x_\alpha) = \frac{1}{4\pi\epsilon_0} \int \frac{\rho(x_\alpha') \, dv'}{r(x_\alpha, x_\alpha')}$$

in terms of the charge density at all points in space. Field theory, however, permits us to find a solution even if $\rho(x_\alpha')$ is known only within an arbitrary surface S; the effect of the other sources is then replaced by the knowledge of the *boundary values* of the potentials or their derivatives on the surface S.

To obtain an explicit expression for $\phi(x_\alpha)$ in terms of ρ within S and ϕ and $\nabla\phi$ on S, we make use of Green's theorem, which states

$$\int (\phi\nabla^2\psi - \psi\nabla^2\phi) \, dv = \int (\phi\nabla\psi - \psi\nabla\phi) \cdot d\mathbf{S}. \tag{1–37}$$

Here ϕ and ψ are scalar functions of position that are continuous with continuous first and second derivatives in the region of integration and on its boundary. The integration will be taken over the primed coordinates x_α', so that the field point x_α is simply a parameter. Let ϕ be the electrostatic potential defined in Eq. (1–31), and let $\psi = 1/r$, the point source solution of Laplace's equation. Since $\nabla'(1/r) = \mathbf{r}/r^3$, and we have from

Eq. (1–7) that $\nabla'^2(1/r) = \nabla^2(1/r) = -4\pi\delta\,(\mathbf{r})$, substitution of these expressions into Eq. (1–37) gives

$$-4\pi\int\left[\phi(x'_\alpha)\,\delta(\mathbf{r}) - \frac{1}{r}\,\frac{\rho(x'_\alpha)}{4\pi\epsilon_0}\right]dv' = \int\left[\phi\,\frac{\mathbf{r}}{r^3} - \frac{\nabla'\phi}{r}\right]\cdot d\mathbf{S}'.$$

Hence

$$\phi(x_\alpha) = \frac{1}{4\pi\epsilon_0}\int\frac{\rho(x'_\alpha)}{r}\,dv' + \frac{1}{4\pi}\int\left(-\phi\,\frac{\mathbf{r}}{r^3} + \frac{\nabla'\phi}{r}\right)\cdot d\mathbf{S}'. \quad (1\text{–}38)$$

Note that we have made use of the functional properties of the δ-function. Clearly, Eq. (1–38) could also have been obtained without the aid of the δ-function by a limiting process. Such a process was actually implicit in the derivation of Eq. (1–7), however, so that it is unnecessary here.

Let the integrals be carried out over a volume v bounded by a surface S. The first integral of Eq. (1–38) is simply the contribution of the volume charge distribution within v. Note that the distance r is a function of the coordinates both of the point of observation x_α and of the point of integration x'_α, in both the volume and the surface integrals.

The surface integrals of Eq. (1–38) summarize the effect of any charge distribution outside the region v and thus not contained in the first term. We therefore conclude that the potential at any point within S is uniquely determined by the charge distribution within S and by the values of ϕ and the normal component of $\nabla\phi$ at all points on the surface S. In particular, the potential within a charge-free volume is uniquely determined by the potential and its normal derivative over the surface enclosing the volume. What we have shown here is that knowledge of the potential and its normal derivative over the surface is *sufficient* to determine the potential inside, but we have not shown that both these pieces of information are *necessary*. We shall see later that it is, in fact, sufficient in a charge-free region to have either the potential *or* its normal derivative over a surface in order to determine the potential at every point within the surface except for an arbitrary additive constant. The reason is that ϕ and $\nabla\phi$ may not be independently specified over the surface, since ϕ must be a solution of Laplace's equation.

In later sections we shall be led to a physical interpretation of the surface integrals as equivalent to a charge and dipole distribution on the surface S. We shall therefore be able to conclude that the potential can be calculated by the direct superposition of the individual potentials of all the volume charge distribution, but that we can, if we wish, replace any part of the distribution by an equivalent surface charge layer and dipole layer distribution.

The volume term of Eq. (1–38) can be looked on as being a particular integral of Poisson's equation, while the surface terms are complementary integrals of the differential equation in the sense that they are general solutions of the homogeneous equation, i.e., Laplace's equation.

1–6 Field singularities. We have written the solution of the potential problem as a sum of boundary contributions and a volume integral extending over the source charges. These volume integrals will not lead to singular values of the potentials (or of the fields) if the charge density is finite. If, on the other hand, the charges are considered to be surface, line, or point charges, then singularities will result as shown in Table 1–1. Note that if either surface or line charges are infinite in spatial extent (i.e., the fields are considered one- or two-dimensional) the potential cannot be referred to infinity. Although these singularities do not actually exist in nature, the fields that do occur are often indistinguishable, over much of the region concerned, from those of simple geometrical configurations. The idealizations of real charges as points, lines, and surfaces not only permit

TABLE 1–1

FIELD SINGULARITIES

Type of charge distribution	Behavior of potential near distribution	Behavior of field near distribution
Surface	r	Constant
Line	$\log r$	r^{-1}
Point	r^{-1}	r^{-2}
Point 2^n pole	r^{-n-1}	r^{-n-2}

great mathematical simplicity, they also give rise to convenient physical concepts for the description and representation of actual fields. For these reasons we shall now discuss in more detail the nature of potentials corresponding to such sources.

1–7 Clusters of point charges. The potential ϕ at the point x_α, due to the charge q located at the point x'_α, is given by

$$\phi = \frac{1}{4\pi\epsilon_0} \frac{q}{r} = \frac{1}{4\pi\epsilon_0} \left\{ \frac{q}{|[(x - x')^2 + (y - y')^2 + (z - z')^2]^{1/2}|} \right\}$$

$$= \frac{1}{4\pi\epsilon_0} \left\{ \frac{q}{|[\Sigma(x_\alpha - x'_\alpha)^2]^{1/2}|} \right\}. \tag{1–39}$$

It has a first-order singularity at the point x'_α corresponding to $r = 0$.

Singularities of higher order can be generated by superposing on this potential a potential corresponding to an equal charge of opposite sign, displaced a distance $\Delta x'$ from the original charge. This process is equivalent to differentiating Eq. (1–39) with respect to x'_α. We have noted in Eq. (1–15), however, that differentiation of a function of the relative coordinates only with respect to x'_α gives the same result as differentiation with respect to x_α except for sign, and at the field point Laplace's equation holds. Since the derivative of a solution of Laplace's equation is also a solution, the process of differentiation with respect to the source point as physically described above will generate new solutions with successively higher order singularities near $r = 0$. Such potentials are called *multipole* potentials.

For a single differentiation, we obtain

$$\phi^{(2)} = \frac{\partial \phi}{\partial x'} \, \Delta x' = \frac{q \, \Delta x'(x - x')}{4\pi\epsilon_0 r^3} = (q \, \Delta x') \frac{\cos \theta}{4\pi\epsilon_0 r^2} \, . \qquad (1\text{–}40)$$

If we let

$$\lim_{\substack{q \to \infty \\ \Delta \mathbf{x}' \to 0}} q \, \Delta \mathbf{x}' = \mathbf{p}^{(1)} \qquad (1\text{–}41)$$

be the dipole moment of the distribution (positive from $-q$ to $+q$ as indicated in Fig. 1–2), we can write this potential as

$$\phi^{(2)} = \frac{1}{4\pi\epsilon_0} \mathbf{p}^{(1)} \cdot \mathbf{\nabla}' \left(\frac{1}{r} \right) = -\frac{1}{4\pi\epsilon_0} \mathbf{p}^{(1)} \cdot \mathbf{\nabla} \left(\frac{1}{r} \right) = \frac{1}{4\pi\epsilon_0} \frac{\mathbf{p}^{(1)} \cdot \mathbf{r}}{r^3} \, . \tag{1–42}$$

(The sign conventions for $\mathbf{r}$ and $\mathbf{\nabla}$ have been discussed in Section 1–1.) This solution is the *dipole* potential.

The potential distribution, and consequently the nelds of higher moments of the charge, or multipoles, can be generated by the same method of geometrical construction. For example, the potential field of a 2^{n+1} pole is generated by taking the potential of a 2^n pole and subtracting from

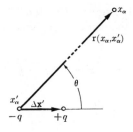

Fig. 1–2. The generation of an electric dipole.

it the potential of another 2^n pole that is displaced infinitesimally in an arbitrary direction (or by superposing the potential of the displaced 2^n pole with opposite sign).

The general form of the potential corresponding to a 2^n pole is thus

$$\phi^{(2n)} = \frac{p^{(n)}}{4\pi\epsilon_0 n!} \frac{\partial^n}{\partial x'_\alpha \partial x'_\beta \ldots} \left(\frac{1}{r}\right) = \frac{(-1)^n p^{(n)}}{4\pi\epsilon_0 n!} \frac{\partial^n}{\partial x_\alpha \partial x_\beta \ldots} \left(\frac{1}{r}\right), \quad (1\text{–}43)$$

where in general $p^{(n)}$ would be an nth-rank tensor in three dimensions. The displacement by which a multipole is generated from one of lower order need not be along coordinate axes, but the derivative corresponding to an oblique displacement can be written as a sum of derivatives with respect to x, y, z, having the direction cosines of the displacement as coefficients. A few examples of simple multipoles are shown in Fig. 1–3. In the special case where all the displacements are in one direction the problem has axial symmetry and only one angle is needed to specify the position of the field point. For a linear 2^n pole,

$$\phi^{(2n)}(x, y, z) = \frac{p^{(n)}}{4\pi\epsilon_0 n!} \frac{\partial^n}{\partial x'^n} \left(\frac{1}{r}\right) = \frac{p^{(n)}}{4\pi\epsilon_0} \frac{P_n(\cos\theta)}{r^{n+1}}, \quad (1\text{–}44)$$

where $P_n(\cos\theta)$ is the Legendre polynomial of order n which may be defined by the relation

$$\frac{P_n(\cos\theta)}{r^{n+1}} = \frac{(-1)^n}{n!} \frac{\partial^n}{\partial x^n} \left(\frac{1}{r}\right), \quad (1\text{–}45)$$

in which θ is the angle between x and r. In the linear case $p^{(n)}$ is defined by the recurrence relation

$$\lim_{\substack{\Delta x'_n \to 0 \\ p^{(n-1)} \to \infty}} n p^{(n-1)} \Delta x'_n = p^{(n)},$$

where $\Delta x'_n$ is the displacement leading to the 2^n pole.

Equation (1–43) can be recognized as the nth term of a general Taylor expansion of $1/r$ in terms of the source coordinates. The coefficients are the multipole moments as defined in terms of the specific charge distribution referred to above. We shall now show that the potential of an arbitrary charge distribution $\rho(x'_\alpha)$ of finite extent can at large distances always be expressed as the sum of multipole potentials where the coefficients are certain integrals (moments) of the charge distribution.

To facilitate the proof of this statement, we shall choose the origin of coordinates in or near the charge distribution (Fig. 1–4). Let R be the distance from the origin O to the field point, i.e., let the components of

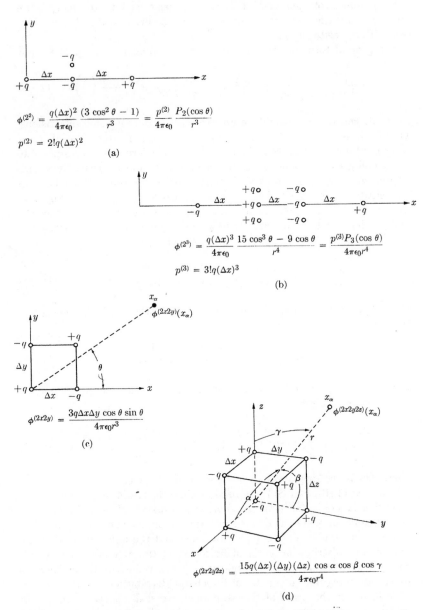

$$\phi^{(2^2)} = \frac{q(\Delta x)^2}{4\pi\epsilon_0} \frac{(3\cos^2\theta - 1)}{r^3} = \frac{p^{(2)}}{4\pi\epsilon_0} \frac{P_2(\cos\theta)}{r^3}$$

$$p^{(2)} = 2!q(\Delta x)^2$$

(a)

$$\phi^{(2^3)} = \frac{q(\Delta x)^3}{4\pi\epsilon_0} \frac{15\cos^3\theta - 9\cos\theta}{r^4} = \frac{p^{(3)}P_3(\cos\theta)}{4\pi\epsilon_0 r^4}$$

$$p^{(3)} = 3!q(\Delta x)^3$$

(b)

$$\phi^{(2x2y)} = \frac{3q\Delta x\Delta y \cos\theta\sin\theta}{4\pi\epsilon_0 r^3}$$

(c)

$$\phi^{(2x2y2z)} = \frac{15q(\Delta x)(\Delta y)(\Delta z)\cos\alpha\cos\beta\cos\gamma}{4\pi\epsilon_0 r^4}$$

(d)

FIG. 1–3. Examples of multipoles: (a) linear quadrupole; (b) linear octupole; (c) two-dimensional quadrupole; (d) three-dimensional octupole.

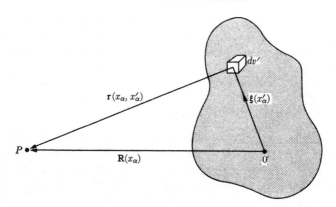

FIG. 1-4. Geometry for the multipole expansion of the potential at P in terms of moments of the charge distribution ρ about the origin O.

$\mathbf{R}$ be x_α. The coordinate vector whose components are x'_α is designated by $\boldsymbol{\xi}$. We may then expand $1/r$ in powers of x'_α about O, assuming that $a/R \ll 1$, where a is a limiting dimension of the bounded distribution of charge. By Taylor's theorem,

$$\frac{1}{r} = \frac{1}{R} + x'_\alpha \left[\frac{\partial}{\partial x'_\alpha}\left(\frac{1}{r}\right)\right]_{r=R} + \frac{1}{2!} x'_\alpha x'_\beta \left[\frac{\partial^2}{\partial x'_\alpha \, \partial x'_\beta}\left(\frac{1}{r}\right)\right]_{r=R} + \cdots. \quad (1\text{-}46)$$

In Eq. (1-46) we are employing the "summation convention" which we shall continue to use throughout: when indices are repeated in the same term summation over these indices is implied. Upon substituting Eq. (1-46) in the general potential, we obtain the "multipole expansion":

$$\phi(x_\alpha) = \frac{1}{4\pi\epsilon_0} \int \frac{\rho(x'_\alpha)}{r}\, dv'$$

$$= \frac{1}{4\pi\epsilon_0}\left\{\frac{1}{R}\int \rho\, dv' + \left[\frac{\partial}{\partial x'_\alpha}\left(\frac{1}{r}\right)\right]_{r=R}\int x'_\alpha \rho\, dv'\right.$$

$$\left. + \frac{1}{2!}\left[\frac{\partial^2}{\partial x'_\alpha \, \partial x'_\beta}\left(\frac{1}{r}\right)\right]_{r=R}\int x'_\alpha x'_\beta\, \rho\, dv' + \cdots\right\}. \quad (1\text{-}47)$$

The coefficients represent the moments of the charge distribution: $\int \rho\, dv'$ is the total charge q, $\int x'_\alpha \rho\, dv'$ is the α-component of the dipole moment, etc. The radial and angular dependence of each term in Eq. (1-47) is clearly identical with that given by Eq. (1-43).

Note that the words "dipole," "quadrupole," etc., are being used in two ways: first to describe a specific charge distribution, and secondly to designate moments of an arbitrary charge distribution. Both physical quantities give rise to the same potential distribution.

The symmetric quadrupole tensor

$$Q_{\alpha\beta} = \int x'_\alpha x'_\beta \, \rho \, dv' \tag{1–48}$$

has six components when referred to arbitrary axes, and has three diagonal components when referred to its principal axes; in the latter case three parameters are needed to specify the orientation of the principal axes relative to an arbitrary coordinate system. We shall now show that the quadrupole potential $\phi^{(4)}$, which is the third term in Eq. (1–47), depends on only two quantities when referred to principal axes.

We have

$$\phi^{(4)} = \frac{1}{2} \left\{ \left[\frac{\partial^2}{\partial x'_\alpha \, \partial x'_\beta} \left(\frac{1}{r} \right) \right]_{r=R} Q_{\alpha\beta} \right\},$$

which is equivalent to

$$\phi^{(4)} = \frac{1}{6} \left\{ \left[\frac{\partial^2}{\partial x_\alpha' \partial x_\beta'} \left(\frac{1}{r} \right) \right]_{r=R} (3Q_{\alpha\beta} - \delta_{\alpha\beta} Q_{\gamma\gamma}) \right\}, \tag{1–49}$$

where $\delta_{\alpha\beta} = 0$ for $\alpha \neq \beta$ and $\delta_{\alpha\beta} = 1$ for $\alpha = \beta$ is the unit second-rank tensor. (By the summation convention $Q_{\gamma\gamma}$ is the sum of the diagonal terms of the quadrupole tensor.) The added term in Eq. (1–49) does not affect the potential, since

$$\delta_{\alpha\beta} \frac{\partial^2}{\delta x'_\alpha \, \delta x'_\beta} \left(\frac{1}{r} \right) = \nabla^2 \left(\frac{1}{r} \right) = 0. \tag{1–50}$$

Hence $Q_{\alpha\beta}$ in Eq. (1–49) can be replaced by $D_{\alpha\beta}/3$, where

$$D_{\alpha\beta} = 3Q_{\alpha\beta} - \delta_{\alpha\beta} Q_{\gamma\gamma}. \tag{1–51}$$

When reduced to principal axes it is evident that $D_{\alpha\beta}$ has only two independent components, since $D_{\alpha\alpha} = 0$. If Eq. (1–51) is written out in Cartesian components, it is seen that $D_{\alpha\beta}$ (conventionally called the "quadrupole moment") depends on differences of the second moments Q_{xx}, Q_{yy}, Q_{zz} when referred to principal axes.

If the charge distribution has an axis of symmetry, say the x-axis, the potential depends on the single quantity

$$D = 3Q_{xx} - (Q_{xx} + Q_{yy} + Q_{zz}) = \int \rho (3x_\alpha'^2 - \xi^2) \, dv', \tag{1–52}$$

which is a direct measure of the eccentricity of the distribution.

1–8 Dipole interactions. Consider an electric dipole **p** in an electric field **E** of arbitrary spatial variation. From the definitions of electric field and the electric dipole, we have

$$F_\alpha = p_\beta \frac{\partial}{\partial x'_\beta} E_\alpha \quad \text{or} \quad \mathbf{F} = (\mathbf{p} \cdot \boldsymbol{\nabla}')\mathbf{E}, \quad (1\text{–}53)$$

since for a dipole of length $\Delta \mathbf{x}'$ the αth component of the force **F** on each end of the dipole changes from F_α to $F_\alpha + (\partial F_\alpha / \partial x'_\beta)\,\Delta x'_\beta$; the limiting process (1–41) generates Eq. (1–53).

Equation (1–53) can be written as

$$\mathbf{F} = (\mathbf{p} \cdot \boldsymbol{\nabla}')\mathbf{E} = \boldsymbol{\nabla}'(\mathbf{p} \cdot \mathbf{E}) - \mathbf{p} \times (\boldsymbol{\nabla}' \times \mathbf{E}). \quad (1\text{–}54)$$

Hence in an electrostatic field (for which $\boldsymbol{\nabla}' \times \mathbf{E} = 0$) the force **F** can be derived from an energy U such that

$$\mathbf{F} = -\boldsymbol{\nabla}'U, \quad (1\text{–}55)$$

where

$$U = -(\mathbf{p} \cdot \mathbf{E}). \quad (1\text{–}56)$$

This gives the energy of a dipole of fixed moment **p** as a function of its orientation and position in the field **E**. If the field is not irrotational (as in the presence of time-varying magnetic fields), the force expression (1–53) remains correct but cannot be derived from a potential energy.

The torque **L** acting on the dipole can be obtained by differentiation of Eq. (1–56) with respect to the angle between **p** and **E** or by direct consideration of the forces. One obtains

$$\mathbf{L} = \mathbf{p} \times \mathbf{E}. \quad (1\text{–}57)$$

The direction of **L** tends to align the positive directions of **p** and **E**.

Let us now consider the interaction force and energy between two dipoles, such as those shown in Fig. 1–5, whose moment vectors are oriented in space at an arbitrary angle with each other. On combining the force

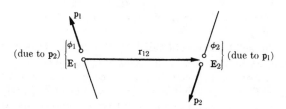

FIG. 1–5. Interaction of two dipoles.

equation above with the potential equation (1–42) and the field equation (1–31), we have for the force $\mathbf{F}$ on dipole 1 in the field $\mathbf{E}$ of dipole 2,

$$\phi = \frac{-1}{4\pi\epsilon_0}\,\mathbf{p}_2 \cdot \boldsymbol{\nabla}\left(\frac{1}{r}\right); \qquad \mathbf{E} = -\boldsymbol{\nabla}\phi = \frac{1}{4\pi\epsilon_0}\,\boldsymbol{\nabla}\left[\mathbf{p}_2 \cdot \boldsymbol{\nabla}\left(\frac{1}{r}\right)\right];$$

$$\mathbf{F} = (\mathbf{p}_1 \cdot \boldsymbol{\nabla})\mathbf{E} = \frac{1}{4\pi\epsilon_0}\,(\mathbf{p}_1 \cdot \boldsymbol{\nabla})\left\{\boldsymbol{\nabla}\left[\mathbf{p}_2 \cdot \boldsymbol{\nabla}\left(\frac{1}{r}\right)\right]\right\}. \tag{1–58}$$

The interaction energy between the two dipoles may be obtained by inserting the field $\mathbf{E}$ above into the expression for the energy U. We find for the energy U_{12} of dipole 1 in the field of dipole 2, and conversely for U_{21}:

$$U_{12} = -\mathbf{p}_1 \cdot \mathbf{E} = -\frac{(\mathbf{p}_1 \cdot \boldsymbol{\nabla})}{4\pi\epsilon_0}\left[\mathbf{p}_2 \cdot \boldsymbol{\nabla}\left(\frac{1}{r}\right)\right],$$

$$U_{12} = \frac{1}{4\pi\epsilon_0}\left[\frac{\mathbf{p}_1 \cdot \mathbf{p}_2}{r^3} - \frac{3}{r^5}\,(\mathbf{p}_1 \cdot \mathbf{r})(\mathbf{p}_2 \cdot \mathbf{r})\right]. \tag{1–59}$$

This is the general expression for the interaction energy of two dipoles.

1–9 Surface singularities. Surface singularities of the second order or dipole form are of particular interest in both electrostatics and magneto-statics. Let us consider a double layer charge arrangement with a dipole moment per unit area designated by τ. The potential arising from such a distribution is given by

$$\phi = \frac{1}{4\pi\epsilon_0}\int \frac{\boldsymbol{\tau} \cdot \mathbf{r}}{r^3}\,dS. \tag{1–60}$$

This expression reduces, in the case when τ is uniform and directed along the normal to the surface outward from the observation point, to

$$\phi = \frac{-|\tau|}{4\pi\epsilon_0}\int \frac{\mathbf{r} \cdot d\mathbf{S}}{r^3} = \frac{-|\tau|}{4\pi\epsilon_0}\,\Omega. \tag{1–61}$$

Here Ω is the solid angle subtended by the dipole sheet at the point of observation, as in Fig. 1–6. The solid angle subtended by a nonclosed surface jumps discontinuously by 4π as the point of observation crosses the surface. This means that in the ideal case of an infinitely thin dipole charge layer the potential function will have a discontinuity of magnitude $|\tau|/\epsilon_0$, but it will have a continuous derivative at the dipole sheet.

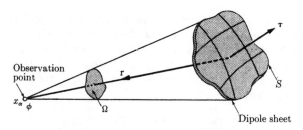

FIG. 1-6. Potential due to dipole layer.

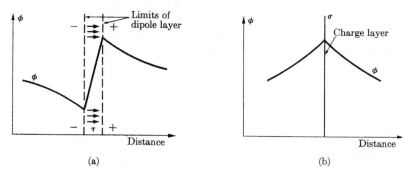

FIG. 1-7. Behavior of the potential at a dipole layer, (a), and at a layer of charge, (b).

On the other hand, a simple surface charge layer will not result in a discontinuity in potential, but will produce a discontinuity in the normal derivative of the potential, the magnitude of discontinuity being σ/ϵ_0, where σ is the surface charge density of the layer. A comparison between the two cases is shown in Fig. 1-7(a) and (b). Since surface charge layers and dipole layers enable us to introduce arbitrary discontinuities in the potential and its derivatives at a particular surface, we can make the potential vanish outside a given volume by surrounding the volume with a suitably chosen charge layer and dipole layer. This is a further explanation of the significance of the surface terms in Eq. (1-38), which was derived from Green's theorem. These terms, when ϕ and $\nabla\phi$ are properly evaluated on the surface in terms of τ and σ, are precisely those necessary to cancel the field of those charges inside the surface S in the region outside of S. This can be seen by writing Eq. (1-38) as

$$\phi = \frac{1}{4\pi\epsilon_0}\left(\int\frac{\rho\,dv}{r} + \int\tau\frac{\mathbf{r}\cdot d\mathbf{S}}{r^3} + \int\sigma\frac{dS}{r}\right), \qquad (1-62)$$

where $\tau = \epsilon_0\phi$ and σ is ϵ_0 times the normal derivative of ϕ.

As an example of a combined surface charge and dipole layer that will just cancel the field outside a given surface, yet leave the field inside the surface unchanged, consider a point charge q located at the point $R = 0$, and the surface $R = a$ surrounding this charge. If we place a surface charge density $\sigma = -q/4\pi a^2$ per unit area on the sphere $R = a$, it will give rise to a potential:

$$\phi_\sigma = -\frac{q}{4\pi\epsilon_0 a}, \quad \text{for} \quad R < a,$$

$$\phi_\sigma = -\frac{q}{4\pi\epsilon_0 R}, \quad \text{for} \quad R > a.$$

If, in addition, a surface dipole layer of moment $\tau = q\mathbf{R}/4\pi aR$ per unit area is placed on the sphere, it will make a contribution

$$\phi_\tau = \frac{q}{4\pi\epsilon_0 a}, \quad R < a,$$

$$\phi_\tau = 0, \quad R > a.$$

The potential of the original charge q is $\phi_0 = q/4\pi\epsilon_0 R$ for all R. The total of all these potentials is

$$\phi = \phi_0 + \phi_\sigma + \phi_\tau = \frac{q}{4\pi\epsilon_0 R}, \quad \text{for} \quad R < a,$$

$$\phi = 0, \quad \text{for} \quad R > a.$$

The electric field produced by a dipole layer of area S as shown in Fig. 1–8 can be derived as follows. Consider a change in the potential cor-

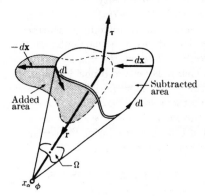

Fig. 1–8. Illustrating the derivation of the electric field produced by a dipole layer.

responding to a displacement of the point of observation x_α by a distance $d\mathbf{x}$,

$$d\phi = -\mathbf{E} \cdot d\mathbf{x}. \tag{1-63}$$

The change in solid angle, $d\Omega$, subtended by the dipole layer at the field point is the same whether the field point moves a distance $d\mathbf{x}$ or the layer moves through $-d\mathbf{x}$. The latter case is shown in the figure. Since in this displacement an element $d\mathbf{l}$ of the boundary sweeps over an area $d\mathbf{x} \times d\mathbf{l}$, the total change in solid angle is

$$d\Omega = \oint \frac{(d\mathbf{x} \times d\mathbf{l}) \cdot \mathbf{r}}{r^3} = \oint \frac{(d\mathbf{l} \times \mathbf{r}) \cdot d\mathbf{x}}{r^3}. \tag{1-64}$$

The change in potential corresponding to this change in solid angle is, from Eq. (1-61),

$$d\phi = -\frac{\tau}{4\pi\epsilon_0} d\Omega. \tag{1-65}$$

(The negative sign follows from the fact that the negative side of the dipole layer is toward the observer at x_α.) Equating the two expressions for $d\phi$, we obtain

$$-\mathbf{E} \cdot d\mathbf{x} = \frac{-\tau}{4\pi\epsilon_0} \oint \frac{(d\mathbf{l} \times \mathbf{r}) \cdot d\mathbf{x}}{r^3}. \tag{1-66}$$

Since $d\mathbf{x}$ is an arbitrary displacement and this last expression holds for all possible $d\mathbf{x}$, it is permissible to write

$$\mathbf{E} = \frac{\tau}{4\pi\epsilon_0} \oint \frac{d\mathbf{l} \times \mathbf{r}}{r^3} = -\frac{\tau}{4\pi\epsilon_0} \oint d\mathbf{l} \times \boldsymbol{\nabla}\left(\frac{1}{r}\right). \tag{1-67}$$

That the potential due to a dipole layer is double valued at the surface is strictly true only in the limit in which the dipole layer has zero thickness (see Fig. 1-7); for this reason the discontinuity does not actually have physical reality. Nevertheless, the method of generating nonconservative potentials by means of such discontinuities is a useful one, particularly in the theory of magnetic fields due to currents, where the corresponding potential does have a multivalued behavior completely analogous to the properties of the surface dipole moment.

1-10 Volume distributions of dipole moment. The potential due to the volume distribution of dipole moment is found by considering the dipole moment in Eq. (1-42) as a volume density and integrating over the volume. If $\mathbf{P}$ is the dipole moment per unit volume,

$$\phi = \frac{1}{4\pi\epsilon_0} \int \mathbf{P} \cdot \boldsymbol{\nabla}'\left(\frac{1}{r}\right) dv'. \tag{1-68}$$

This can be changed into a form that is physically more revealing by means of Gauss's divergence theorem and the relation

$$\mathbf{\nabla}' \cdot \left(\frac{\mathbf{P}}{r}\right) = \frac{1}{r}\,\mathbf{\nabla}' \cdot \mathbf{P} + \mathbf{P} \cdot \mathbf{\nabla}'\left(\frac{1}{r}\right). \tag{1-69}$$

We obtain

$$\phi = \frac{1}{4\pi\epsilon_0}\left[\int \mathbf{\nabla}' \cdot \left(\frac{\mathbf{P}}{r}\right) dv' - \int \frac{1}{r}\,\mathbf{\nabla}' \cdot \mathbf{P}\, dv'\right]$$

$$= \frac{1}{4\pi\epsilon_0}\left[\int \frac{\mathbf{P} \cdot d\mathbf{S}}{r} - \int \frac{\mathbf{\nabla}' \cdot \mathbf{P}}{r}\, dv'\right]. \tag{1-70}$$

This expression can be interpreted as follows. The first term, a surface integral, is a potential equivalent to that of a surface charge density, while the second term is a potential equivalent to that of a volume charge density. The charge densities which have potentials equivalent to those produced by the volume polarization of a region of space are

$$\sigma_P = P_n, \qquad \rho_P = -\mathbf{\nabla}' \cdot \mathbf{P}. \tag{1-71}*$$

Note that if $\mathbf{P}$ does not have discontinuities the volume charge $\rho_P = -\mathbf{\nabla}' \cdot \mathbf{P}$ is sufficient to describe the source; it can easily be proved that the expression for σ_P in Eq. (1-71) will result if $\mathbf{\nabla}' \cdot \mathbf{P}$ becomes infinite at a boundary.

The relations between these surface and volume charges and the polarization can be derived from purely geometrical considerations. If, for example, we have an inhomogeneous dipole moment per unit volume, ρ_P will represent the charge density that accumulates from incomplete cancellation of the ends of the individual dipoles distributed in the volume. The quantity σ_P, on the other hand, represents the charge density on the surface produced by the lack of neighbors for the dipoles which lie with their ends on the surface. It is evident that ρ_P will vanish in a homogeneous medium; in fact, a sufficient condition for its vanishing is that the dipole moment per unit volume have a zero divergence. In general, however, the potential due to the two forms of polarization charges is

$$\phi = \frac{1}{4\pi\epsilon_0}\left[\int \frac{\sigma_P\, dS'}{r} + \int \frac{\rho_P\, dv'}{r}\right]. \tag{1-72}$$

Note that the "equivalence relations" of Eq. (1-71), although derived above by means of the electric field which the respective terms produce, are

---

* Since $\rho_P = -\mathbf{\nabla} \cdot \mathbf{P}$ is a field equation, the prime on $\mathbf{\nabla}$ can be dropped without ambiguity. The prime on the $\mathbf{\nabla}$ is only necessary in integral expressions which relate a field quantity to an integral over a source quantity.

actually simple geometrical quantities. We can see this formally by considering the total dipole moment **p** of a distribution. According to the "equivalence relations," we should have

$$\mathbf{p} = \int \mathbf{P}\,dv' = \int \boldsymbol{\xi}\rho_P\,dv' + \int \boldsymbol{\xi}\sigma_P\,dS' = -\int \boldsymbol{\xi}(\boldsymbol{\nabla}'\cdot\mathbf{P})\,dv' + \int \boldsymbol{\xi}(\mathbf{P}\cdot d\mathbf{S}'),$$

$$(1\text{-}73)$$

where $\boldsymbol{\xi}$ is the vector whose components are x_α'. The αth component of **p** can be integrated by parts from the identity

$$\boldsymbol{\nabla}'\cdot(x_\alpha'\mathbf{P}) = x_\alpha'\boldsymbol{\nabla}'\cdot\mathbf{P} + P_\alpha.$$

Therefore,

$$p_\alpha = \int P_\alpha\,dv' = -\int x_\alpha'\boldsymbol{\nabla}'\cdot\mathbf{P}\,dv' + \int \boldsymbol{\nabla}'\cdot(x_\alpha'\mathbf{P})\,dv'$$

$$= -\int x_\alpha'\boldsymbol{\nabla}'\cdot\mathbf{P}\,dv' + \int x_\alpha'\mathbf{P}\cdot d\mathbf{S}'.$$

Equation (1-73) is thus proved as a geometrical relationship without reference to any interaction.

Suggested References*

M. Abraham and R. Becker, *The Classical Theory of Electricity and Magnetism* (Volume 1). The standard reference on a high intermediate level, beginning with an excellent summary of vector analysis and the properties of vector fields. Gaussian units are used throughout Abraham-Becker and in most other classical texts on electrodynamics. For a comparison of units, see Appendix I of the present book.

G. P. Harnwell, *Principles of Electricity and Electromagnetism*. Intermediate text with technical applications.

J. H. Jeans, *Electricity and Magnetism*. A comprehensive treatment of electrostatics, although one which avoids vector notation.

W. R. Smythe, *Static and Dynamic Electricity*. The most comprehensive modern treatment (in English) of electrostatics and the solution of potential problems.

J. A. Stratton, *Electromagnetic Theory*. This work is developed from Maxwell's equations as postulates, so that an account of the electrostatic field is considerably postponed. The theory of multipoles is explicit for nonorthogonal displacements.

The electrostatic field is also treated in well-known textbooks of theoretical physics:

P. G. Bergmann, *Basic Theories of Physics: Mechanics and Electrodynamics*.

G. Joos, *Theoretical Physics*.

---

* Full bibliographical description of the references listed at the end of each chapter will be found in the bibliography at the end of the book.

L. PAGE, *Introduction to Theoretical Physics.*

J. SLATER AND N. FRANK, *Introduction to Theoretical Physics* or *Electromagnetism.*

The properties of Legendre functions are summarized by Jeans and by Smythe, for example, and are also to be found in useful form in such mathematical references as:

H. MARGENAU AND G. M. MURPHY, *The Mathematisc of Physics and Chemistry.*

L. A. PIPES, *Applied Mathematics for Engineers and Physicists.*

EXERCISES

1. If $\mathbf{A}$, $\mathbf{B}$, $\mathbf{C}$, $\mathbf{D}$ are vectors, show that
   (a) $\mathbf{A} \cdot \mathbf{B}$ is a scalar,
   (b) $\mathbf{A} \times \mathbf{B}$ is a pseudovector,
   (c) $\mathbf{A} \cdot (\mathbf{B} \times \mathbf{C})$ is a pseudoscalar,
   (d) $(\mathbf{A} \times \mathbf{B}) \cdot (\mathbf{C} \times \mathbf{D})$ is a scalar.

2. Show that the mean value of the potential over a spherical surface is equal to the potential at the center of the sphere, provided that no charge is contained within the sphere.

3. From the results of Exercise 2, show that a charge cannot be held in equilibrium in an electrostatic field. (This is Earnshaw's theorem.)

4. Show that Green's theorem, Eq. (1–37), follows from Gauss's divergence theorem.

5. We shall need the "one-dimensional" δ-function defined by

$$\int \delta(x - a) \, dx = 1,$$

$$\int f(x) \, \delta(x - a) \, dx = f(a)$$

when $x = a$ is in the interval of integration, and both integrals vanish otherwise. Show by means of the Fourier integral theorem that

$$\lim_{L \to \infty} \int_{-L}^{L} \cos kx \, dk = 2\pi \, \delta(x)$$

in the sense that both behave the same way as factors in an integrand.

6. Prove by considering the axial point on a disk that the potential undergoes no sudden change from one side to the other of a charge layer, and that the same statement holds for the normal derivative of the potential in the case of a dipole layer.

7. Functions of the type $\phi = x$, or indeed $\phi = x^2 + 2y^2 - 3z^2$, satisfy Laplace's equation at all points of space. Does this mean that such potentials have no sources? Discuss in detail the significance of such solutions, and their bearing on the uniqueness proof for potentials.

8. If the electric field of a point charge q were proportional to $qr^{-2-\delta}\hat{\mathbf{r}}$, where $\hat{\mathbf{r}}$ is a radial unit vector and $\delta \ll 1$, (a) calculate $\boldsymbol{\nabla} \cdot \mathbf{E}$ and $\boldsymbol{\nabla} \times \mathbf{E}$, for $r \neq 0$;

(b) and if two concentric spherical conducting shells of radii a and b were connected by a very thin wire, with q_a on the outer shell, prove that a charge

$$q_b = \frac{-q_a \delta}{2(a-b)} \left\{ [2b \log 2a - (a+b) \log (a+b) + (a-b) \log (a-b)] \right\} + 0(\delta^2)$$

will reside on the inner shell. (Adapted from Jeans.)

9. Show that the static potential $\phi(x, y, z)$, correct to third order in a, is equal to the mean of the potentials at the points

$$x \pm a, y, z; \qquad x, y \pm a, z; \qquad x, y, z \pm a.$$

10. For a finite spherically symmetric charge distribution the potential as calculated by $\phi = \int_r^\infty \mathbf{E} \cdot d\mathbf{r}$ is

$$\phi(r) = \frac{1}{\epsilon_0} \int_r^\infty \frac{dr''}{r''^2} \int_0^{r''} \rho r'^2 \, dr'.$$

By dividing the distribution into thin shells, each of which contributes constant potential at all points inside it, obtain an expression for ϕ that involves only single integrals. Prove the equality of the two expressions.

11. Consider two coplanar electric dipoles with their centers a fixed distance apart. Show that if the angles the dipoles make with the line joining their centers are θ and θ' respectively, and if θ is held fixed,

$$\tan \theta = -\tfrac{1}{2} \tan \theta'$$

for equilibrium.

12. The differential equations of the "lines of force" are

$$\frac{dx}{E_x} = \frac{dy}{E_y} = \frac{dz}{E_z}.$$

For a dipole of moment $\mathbf{p}$ directed along the x-axis and located at the origin, find the equation $f(x, y) = $ constant that gives the lines of force in the plane $z = 0$.

13. Calculate the quadrupole moment of two concentric coplanar ring charges q and $-q$, having radii a and b.

14. Two uniform line charges, each of length $2a$, cross each other at the origin in such a way that their ends are at the points $(\pm a, 0, 0)$ and $(0, \pm a, 0)$. Determine ϕ for points $r > a$, up to but not including terms in r^{-4}.

15. Show that the potential of a symmetrical 2^n pole generated by differentiating the point potential n times along successive directions making an angle $2\pi/n$ with one another is given by

$$\phi(r, \theta, \varphi) = \frac{\text{const.}}{r^{n+1}} P_n^n (\cos \theta) \cos n(\varphi - \varphi_0).$$

16. Show that Eq. (1–58) changes sign if $\mathbf{p}_1$ and $\mathbf{p}_2$ are interchanged.

CHAPTER 2

BOUNDARY CONDITIONS AND RELATION OF MICROSCOPIC
TO MACROSCOPIC FIELDS

The dipole moments per unit volume considered in the foregoing chapter are special examples of sources which give rise to electrostatic fields and can therefore be treated as special types of charge densities in Poisson's equation. Since such volume distributions are produced in material media by electric fields, the behavior of a medium in a field can be described in terms of its polarization, i.e., its dipole moment per unit volume. It is customary, in order to clarify the understanding of polarization, to separate the total charge that produces an electrostatic field into two parts: a true, free, movable, net charge ρ, and a bound, zero-net, polarization charge ρ_P. This division is to a certain extent arbitrary, in the sense that the polarization charge ρ_P simply represents separated charges which on the scale of observation being considered in a particular experiment are essentially inaccessible, but which would be treated as free charges on a smaller scale. If, for example, we place a piece of metal between the plates of a capacitor, we can describe the resultant field between the plates either in terms of the true charges produced on the metal or in terms of an equivalent polarization of the piece of metal, depending on whether we consider the charges individually measurable. If, instead of the metal, we introduce a piece of dielectric between the capacitor plates, we are forced to describe the phenomena by a polarization charge, rather than by a true charge, since it is assumed in the theory that observation shall not be made on an atomic scale. An atomic scale observation would be necessary in order to "resolve" the volume polarization into individual charges.

2-1 The displacement vector. It is seen that the distinction between ρ and ρ_P is an arbitrary one, but this arbitrariness will in no way disturb the formalism used to describe the fields produced by polarization charges. Since we have divided the sources of electric fields into these two types, the Poisson source equation becomes

$$\nabla^2 \phi = -\boldsymbol{\nabla} \cdot \mathbf{E} = -\left(\frac{\rho + \rho_P}{\epsilon_0}\right). \qquad (2\text{-}1)$$

The symbol ρ now denotes only the true free charge at the point where the divergence is taken. If ρ_P is expressed in terms of the divergence of the polarization **P**, as given by Eq. (1–71), we obtain from Eq. (2–1)

$$\nabla \cdot \left(\mathbf{E} + \frac{\mathbf{P}}{\epsilon_0} \right) = \frac{\rho}{\epsilon_0}. \tag{2–2}$$

It is thus convenient to define an electric displacement vector **D** (measured in coulombs per square meter in the mks system) by

$$\mathbf{D} = \epsilon_0 \mathbf{E} + \mathbf{P}. \tag{2–3}$$

The source equations then become

$$\nabla \cdot \mathbf{D} = \rho \tag{2–4}$$

and

$$\nabla \cdot \mathbf{E} = \frac{\rho_t}{\epsilon_0}, \tag{2–5}$$

with the total charge density ρ_t written for the sum of the true and polarization charge densities. The corresponding integral relations, secured by means of Gauss's divergence theorem on integration over the volume containing all the charges, are

$$\int \mathbf{D} \cdot d\mathbf{S} = q, \tag{2–6}$$

$$\int \mathbf{E} \cdot d\mathbf{S} = \frac{q_t}{\epsilon_0}, \tag{2–7}$$

where q_t is the total charge, the sum of q (the "true" charge) and the integral of $-(\nabla \cdot \mathbf{P})$ over the volume.

It is clear that **D** represents a partial field in the sense that its sources are the true charges. Note that the relation (2–3) between **D** and **E** is basically an additive one, the difference between **D** and $\epsilon_0 \mathbf{E}$ being the polarization **P**. Note also that the polarization, although defined in a purely geometrical fashion as the dipole moment per unit volume, has the properties of an electric field. The polarization field **P** is that field whose flux arises only from the polarization charges ρ_P. These remarks describe **P** and **D** in terms of their flux and thus their divergence, although $\nabla \times \mathbf{P}$ and $\nabla \times \mathbf{D}$ need not be zero. In fact, $\nabla \times \mathbf{D}$ has a circulation density $\mathbf{c} = \nabla \times \mathbf{P}$; thus mathematically **D** can be derived from **P** by the methods given in Chapter 1, Section 1, since both its circulation density and source density are given.

The solution of an actual field problem involving polarized bodies will depend on the manner in which the polarization depends on the external field. In most cases the polarization is proportional to the field, and can be expressed by an equation of the type

$$\mathbf{P} = \epsilon_0 \chi \mathbf{E}, \tag{2-8}$$

where χ is called the electric susceptibility. Such a description excludes the consideration of electrets (materials possessing a permanent dipole moment), but electrets are not ordinarily of importance. In case we do have a simple medium whose polarization depends linearly on the imposed electric field as expressed by Eq. (2-8), then all three vectors $\mathbf{D}$, $\mathbf{E}$, and $\mathbf{P}$ will be related by constants of proportionality:

$$\mathbf{D} = \epsilon_0 \mathbf{E} + \mathbf{P} = \epsilon_0 (1 + \chi) \mathbf{E}. \tag{2-9}$$

We may define the specific inductive capacity (often called the dielectric constant) by

$$\kappa = 1 + \chi, \tag{2-10}$$

so that

$$\mathbf{D} = \kappa \epsilon_0 \mathbf{E} \tag{2-11}$$

and

$$\mathbf{P} = \epsilon_0 (\kappa - 1) \mathbf{E}. \tag{2-12}$$

As it stands, Eq. (2-8) presupposes that the medium polarizes isotropically, or that the polarization properties of the medium do not depend on the direction of the polarization. This is not the general case and, in fact, the scalar proportionality is valid only for liquids, gases, amorphous solids, and cubic crystals. In crystals of symmetry lower than cubic the relation between each of the components of the polarization vector and of the electric field vector is still linear but the constants of proportionality in the various directions may be different. This means that the relation between the components of the polarization vector and the components of the electric field vector are given by a tensor,

$$P_\alpha = \epsilon_0 \chi_{\alpha\beta} E_\beta, \tag{2-13}$$

and $\mathbf{P}$ is no longer in the same direction as $\mathbf{E}$. It will be shown in Chapter 6 by energy considerations that $\chi_{\alpha\beta}$ is symmetric, i.e., $\chi_{\alpha\beta} = \chi_{\beta\alpha}$. It is therefore possible to express $\chi_{\alpha\beta}$ in terms of principal coordinates by a set of only three constants, and there are at least three directions in which $\mathbf{P}$ and $\mathbf{E}$ are parallel.

The case where $\mathbf{E}$ and $\mathbf{P}$ are not proportional and there is no linear relation between them will not be treated here, but the analogous case will be discussed in connection with magnetic media where nonlinearity is of more practical importance. It should be pointed out, however, that the relation we have here assumed between $\mathbf{P}$ and $\mathbf{E}$ is only a special simplification, and not a fundamental equation of the theory.

2–2 Boundary conditions. Maxwell's field equations, to be discussed later, are a set of equations whose sources are divided into accessible and inaccessible charges. To obtain a solution of Maxwell's equations the inaccessible charges must be related to the accessible charges, or to the fields produced by the accessible charges, by additional equations. The relations that evaluate the inaccessible charge sources in terms of the external fields which produce them are called the constitutive equations. Equation (2–8) is an example of such an equation. The constitutive equations must, of course, depend on the properties of the material in which the inaccessible charges arise. While Eqs. (2–8) to (2–13) are not entirely general, they depend only on linearity and do not imply homogeneity. The susceptibility and specific inductive capacity may therefore be arbitrary functions of the coordinates. A case of much interest is one where the specific inductive capacity varies discontinuously, as at the boundary between two dielectrics.

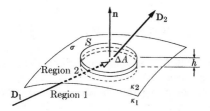

FIG. 2–1. Volume considered for determining the boundary conditions on the normal component of **D**.

To determine the behavior of the fields at a boundary, let us first imagine a small volume, as in Fig. 2–1, whose dimension normal to the interface, h, is smaller than its other dimensions by an order of magnitude, to be placed so that one of its large surfaces ΔA lies in medium 1, the other in medium 2, and both are parallel to the interface. This small volume can be used to derive the behavior of the normal components of the fields.

Let us take the surface integral of **D** over this small volume.

$$\int \mathbf{D} \cdot d\mathbf{S} = q. \tag{2–6}$$

In the limit as the dimension h approaches zero, q approaches $\sigma \, \Delta A$, where σ is the true surface charge density on the interface. The contribution of the sides of the volume normal to the surface vanishes, so that Eq. (2–6) becomes

$$\mathbf{n} \cdot (\mathbf{D}_2 - \mathbf{D}_1) = \sigma, \tag{2–14}$$

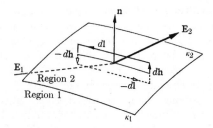

Fig. 2-2. Loop considered in determining the boundary conditions on the tangential components of **E**.

where $\mathbf{n} = d\mathbf{S}/dS$ is the unit vector normal to ΔA, the top of the cylinder of volume $h\,\Delta A$. On the assumption that Eq. (2–11) is valid, we obtain

$$\mathbf{n} \cdot (\kappa_2 \mathbf{E}_2 - \kappa_1 \mathbf{E}_1) = \frac{\sigma}{\epsilon_0},$$

$$\mathbf{n} \cdot (\kappa_2 \boldsymbol{\nabla} \phi_2 - \kappa_1 \boldsymbol{\nabla} \phi_1) = -\frac{\sigma}{\epsilon_0}. \qquad (2\text{–}15)$$

We have assumed that ΔA is small enough so that the fields are essentially uniform over this area.

Equation (2–15) can be written in the more convenient form

$$\kappa_2 \frac{\partial \phi_2}{\partial n} - \kappa_1 \frac{\partial \phi_1}{\partial n} = -\frac{\sigma}{\epsilon_0}, \qquad (2\text{–}15')$$

where $\partial/\partial n$ denotes the derivative taken along the normal to the boundary. If a linear medium [i.e., the validity of Eq. (2–8)] is *not* assumed, but if the polarization **P** is a given function of position, it can be shown that the boundary condition corresponding to Eq. (2–15) is

$$\frac{\partial \phi_2}{\partial n} - \frac{\partial \phi_1}{\partial n} = -\frac{(\sigma + \delta P_n)}{\epsilon_0}, \qquad (2\text{–}15'')$$

where δP_n is the discontinuity of the component of **P** along the normal taken as positive from medium 1 to medium 2.

The behavior of the tangential components of the fields as they cross the interface can be determined from the consideration of a small loop, as in Fig. 2–2, its major extent lying parallel to the surface, one side in medium 1 and the other in medium 2. From Eq. (1–34) we have for the closed line integral of the electric field

$$\oint \mathbf{E} \cdot d\mathbf{l} = 0. \qquad (1\text{–}34)$$

If we apply this to the path indicated in the figure we obtain

$$\mathbf{E}_2 \cdot d\mathbf{l} - \mathbf{E}_1 \cdot d\mathbf{l} + \text{terms of order } \mathbf{E} \cdot d\mathbf{h} = 0. \tag{2-16}$$

Actually, Eq. (2–16) remains valid even if Eq. (1–34) is not assumed, so long as $\mathbf{\nabla} \times \mathbf{E}$ is finite on the interface. In that case

$$\mathbf{E}_2 \cdot d\mathbf{l} - \mathbf{E}_1 \cdot d\mathbf{l} + \text{terms of order } \mathbf{E} \cdot d\mathbf{h} = (\mathbf{\nabla} \times \mathbf{E}) \cdot (d\mathbf{h} \times d\mathbf{l}). \tag{2-16'}$$

If we let $d\mathbf{h} \to 0$ we obtain

$$E_{t2} = E_{t1}, \tag{2-17}$$

where $\mathbf{E}_t$ is the electric field component in the plane of the interface. In terms of the potential,

$$-\mathbf{n} \times (\mathbf{E}_2 - \mathbf{E}_1) = \mathbf{n} \times (\mathbf{\nabla}\phi_2 - \mathbf{\nabla}\phi_1) = 0, \tag{2-18}$$

where $\mathbf{n}$ is the unit normal to the interface of Fig. 2–2. We have assumed that the loop is sufficiently short that the fields are essentially constant over its length. Equation (2–18) can be integrated along the interface to give

$$\phi_1 = \phi_2 + \text{constant}; \tag{2-19}$$

it is usually convenient to omit the constant. Equation (2–15'), or (2–15'') if applicable, together with Eq. (2–19), form the sufficient set of boundary conditions across any interface for a static potential problem.

2–3 The electric field in a material medium. We first defined the electrostatic field produced by free charges in a vacuum, and then we introduced material media containing charges that are inaccessible to measurement. The behavior of these media has been described in terms of their dipole moment per unit volume. Mathematically the fields were defined by Eqs. (2–4), (2–5), and the connecting equation (2–3). Certain difficulties arise in the definition of the electric field within material media if one attempts to maintain a strictly phenomenological point of view. An operational definition of the field might be made by any of the following three methods, which will not necessarily yield answers in agreement with each other:

(A) We may define the field on an atomic electron scale, where the question of the polarizability of material media would presumably not arise. Then for our macroscopic definition of the field we would take the space-time average of these atomic fields. A very fast electron would experience such an average field.

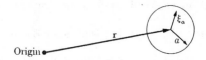

FIG. 2–3. Coordinates for averaging atomic fields. $\xi_\alpha = (\xi, \eta, \zeta)$.

(B) We might consider a hole cut in the dielectric material and define the field as that measured in this hole in terms of a unit charge such as was used in the vacuum definition. This cavity definition of the field will make the field strength depend on the geometry of the cavity and on its orientation with respect to the direction of the field in the medium. This will lead to a unique definition only if the shape and orientation of the cavity are standardized in an arbitrary way.

(C) We may define the field as that acting on an individual molecule of the dielectric.

Let us examine these methods separately.

(A) *Space-time average definition.* Consider a function $f(x, y, z; t)$ defined in a certain region of space during a certain time interval, as indicated in Fig. 2–3. The space-time average of $f(x, y, z; t)$ over a time interval $2T$ and a spherical region of space of radius a is given by

$$\overline{f(x, y, z; t)} =$$

$$\frac{1}{2T \frac{4}{3}\pi a^3} \int_{-T}^{T} \underset{(\xi^2 + \eta^2 + \zeta^2) \le a^2}{\iiint} f[(x + \xi), (y + \eta), (z + \zeta); (t + \theta)]$$

$$\times d\xi\, d\eta\, d\zeta\, d\theta. \qquad (2\text{--}20)$$

Performing this integral is a linear operation and may therefore be commuted with linear differential operators, as, for example,

$$\overline{\boldsymbol{\nabla}f} = \boldsymbol{\nabla}\bar{f}. \qquad (2\text{--}21)$$

On an atomic scale an equation corresponding to Eq. (2–5) holds:

$$\boldsymbol{\nabla} \cdot \boldsymbol{\varepsilon} = \frac{\rho_a}{\epsilon_0}, \qquad (2\text{--}22)$$

where $\boldsymbol{\varepsilon}$ is the atomic electric field and ρ_a is the charge density in the atomic distribution. On taking the space-time average of ρ_a, we obtain

$$\boldsymbol{\nabla} \cdot \bar{\boldsymbol{\varepsilon}} = \overline{\boldsymbol{\nabla} \cdot \boldsymbol{\varepsilon}} = \frac{\overline{\rho_a}}{\epsilon_0} = \frac{\rho_t}{\epsilon_0}, \qquad (2\text{--}23)$$

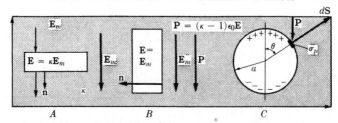

FIG. 2–4. Cavities for defining fields in a dielectric: slot A with short dimension parallel to the field; slot B with long dimension parallel to the field; and sphere C.

and from Eqs. (2–21) and (2–5),

$$\nabla \cdot \bar{\mathcal{E}} = \nabla \cdot \mathbf{E}. \tag{2–24}$$

Hence the macroscopic field $\mathbf{E}$ is actually the space-time average of the atomic field $\mathcal{E}$, even in the presence of dielectrics.

(B) *Cavity definitions.* Consider the three shapes of cavities shown in Fig. 2–4. There are no true charges on the boundaries. From the boundary condition of Eq. (2–15) the field measured in slot A, whose major extent is oriented normal to the field, is $\kappa \mathbf{E}_m$, where $\mathbf{E}_m$ is the field in the medium. (This is rigorously true only in the limit of a very flat "pill box" cavity.) The field in slot B, whose major extent is oriented parallel to the field, is just $\mathbf{E}_m$, by Eq. (2–18), if the width of the slot is sufficiently small. The field measured in the spherical cavity C can be shown to be

$$\mathbf{E} = \frac{3\kappa \mathbf{E}_m}{2\kappa + 1} = \mathbf{E}_m + \frac{\mathbf{P}}{\epsilon_0} \frac{1}{(2\kappa + 1)} \tag{2–25}$$

by methods to be discussed in Chapter 5 for the solution of boundary value problems. It is introduced here only to indicate how cavity definitions depend on the geometry of the cavity. For large values of κ Eq. (2–25) reduces to

$$\mathbf{E} = \tfrac{3}{2}\mathbf{E}_m. \tag{2–26}$$

The three types of fields existing in these cavities are shown graphically in Fig. 2–5.

The measured field in any cavity can in principle be related to the field in the medium and thus used to define the magnitude of that field, provided that the geometry of the cavity and the specific inductive capacity of the medium are known.

(C) *Molecular fields.* A particular molecule is located at a *nonrandom* position in a solid, and hence the field acting on it is not the space-time average discussed above. We now approximate the field acting on a mole-

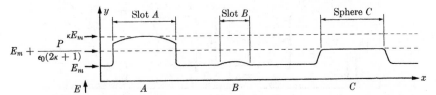

FIG. 2–5. Field profile measured by the cavity technique on a horizontal line passing through the centers of Fig. 2–4 A, B, and C.

cule which itself is one of the molecules of a polarized medium. Consider a dielectric placed between the plates of a parallel plate capacitor, as shown in Fig. 2–6, the dielectric and capacitor being sufficiently large in the directions parallel to the plates so that end effects may be neglected. Consider one of the molecules constituting this dielectric. Let us draw a sphere of radius a about this particular molecule, intended to respresent schematically the boundary between the microscopic and the macroscopic range of phenomena concerning the molecule. The molecule is thus influenced by the fields arising from the following charges:

(1) The charges on the surfaces of the capacitor plates.

(2) The surface charge on the dielectric facing the capacitor plates.

(3) The surface charge on the interior of the spherical boundary of radius a.

(4) The charges of the individual molecules, other than the molecule under consideration, contained within the sphere of radius a.

The fields due to these sources may be computed separately:

(1) The charge on the capacitor plates produces a field at the molecule in question equal to

$$\frac{\mathbf{D}}{\epsilon_0} = \mathbf{E} + \frac{\mathbf{P}}{\epsilon_0}. \tag{2–27}$$

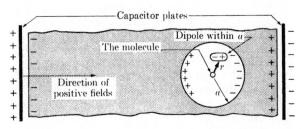

FIG. 2–6. Indicating the contributions of a dielectric to the field on one of its molecules.

(2) The polarization charge on the surface of the dielectric facing the capacitor plates, $\sigma_P = P_n$, produces a field at the molecule given by

$$- \frac{\mathbf{P}}{\epsilon_0}. \tag{2-28}$$

(3) The polarization charge present on the inside of the sphere produces a field that may be calculated as follows. The magnitude of the electric field at the center of the cavity, $\mathbf{E}_P$, due to the polarization on the surface of the cavity, is given by

$$E_P = \frac{1}{4\pi\epsilon_0} \int \frac{\sigma_P \cos\theta \, dS}{a^2}, \tag{2-29}$$

where θ is the angle between $\mathbf{P}$ and the radius vector from the molecule to the surface element dS. The differential element of surface charge between θ and $\theta + d\theta$ is

$$\sigma_P \, dS = |\mathbf{P}| 2\pi a^2 \sin\theta \cos\theta \, d\theta. \tag{2-30}$$

Equation (2–29) thus becomes

$$E_P = \frac{|\mathbf{P}| 2\pi a^2}{4\pi\epsilon_0 a^2} \int_0^\pi \sin\theta \cos^2\theta \, d\theta, \tag{2-31}$$

and on integrating, we obtain $E_P = |\mathbf{P}|/3\epsilon_0$, or

$$\mathbf{E}_P = \frac{\mathbf{P}}{3\epsilon_0}. \tag{2-32}$$

Note that $\mathbf{E}_P$ is *not* the solution of the boundary value problem of a spherical cavity within a dielectric, as was the field given by Eq. (2–25), but is the solution of the problem of a spherical cavity within a dielectric if the polarization is considered to be unaffected by the presence of the cavity.

(4) The field due to the individual molecules within the sphere must be obtained by summing over the fields due to the dipoles within the sphere. We have the potential of an individual dipole from Chapter 1:

$$\phi = \frac{1}{4\pi\epsilon_0} \frac{\mathbf{p} \cdot \mathbf{r}}{r^3}. \tag{1-42}$$

The field at a distance r from a dipole is

$$\mathbf{E} = -\boldsymbol{\nabla}\phi = \frac{-1}{4\pi\epsilon_0} \left[\frac{\mathbf{p}}{r^3} - \frac{3(\mathbf{p} \cdot \mathbf{r})\mathbf{r}}{r^5} \right]. \tag{2-33}$$

Summing over all the dipoles within the sphere is equivalent to taking the spatial average of the x-component of the field:

$$\sum E_x = \frac{-1}{4\pi\epsilon_0} \sum \left[\frac{p_x}{r^3} - \frac{3(p_x x^2 + p_y xy + p_z xz)}{r^5} \right]. \qquad (2\text{--}34)$$

Since according to our assumption the dielectric is isotropic, the x-, y-, and z-directions are equivalent, and

$$\overline{x^2} = \overline{y^2} = \overline{z^2} = \overline{r^2}/3, \qquad \overline{xy} = \overline{yz} = \overline{zx} = 0. \qquad (2\text{--}35)$$

Hence the field due to the dipoles within the sphere vanishes.

On adding the partial fields of expressions (2–27), (2–28), (2–32), and the zero field of Eq. (2–34), we have for the total field acting on one molecule

$$\mathbf{E}_{\text{eff}} = \mathbf{E} + \frac{\mathbf{P}}{\epsilon_0} - \frac{\mathbf{P}}{\epsilon_0} + \frac{\mathbf{P}}{3\epsilon_0} = \mathbf{E} + \frac{\mathbf{P}}{3\epsilon_0}. \qquad (2\text{--}36)$$

This expression, derived for isotropic substances, is also valid for a lattice point within a cubic crystal, but is not valid for crystals of lower symmetry. Note that we have considered only dipole-dipole interactions between neighbors. Clearly, this will be inaccurate for substances having large oriented molecular groups.

The difference between Eq. (2–36) and method (A) is that here we consider what happens to an actual molecule of the medium, rather than take an average of the field at a random point. The physical significance of the space-time average of *all* the atomic fields would be the average field on a fast moving charge traversing the medium.

2–4 Polarizability. The field definition (C) is useful for describing the large-scale behavior of a dielectric in terms of the constants of its molecules. In order that such a description be made, the specific inductive capacity must be associated with the polarizability of a single molecule. This connection may be made by means of Eq. (2–36), which gives the field and thus the force acting on a single molecule within the body of a dielectric in terms of the external field. The quantity α, called the polarizability, is defined by the equation

$$\mathbf{p} = \alpha\epsilon_0 \mathbf{E}_{\text{eff}}, \qquad (2\text{--}37)$$

where $\mathbf{p}$ is the dipole moment induced in a single molecule. If N is the number of molecules per unit volume, the total polarization is

$$\mathbf{P} = N\mathbf{p} = \alpha N \epsilon_0 \mathbf{E}_{\text{eff}}$$

$$\mathbf{P} = \frac{N_0 \rho_m \alpha}{M} \epsilon_0 \mathbf{E}_{\text{eff}}, \qquad (2\text{--}38)$$

where ρ_m is the density and M the molecular weight of the material, and N_0 is Avogadro's number, so that $N = N_0\rho_m/M$. Therefore **P** may be found if α is given for a particular material. Furthermore, by combining Eq. (2–38) with Eqs. (2–12) and (2–36), we may eliminate the fields and obtain the relation sought:

$$\frac{\kappa - 1}{\kappa + 2} = \frac{\alpha N}{3} = \frac{N_0\rho_m\alpha}{3M}. \qquad (2\text{–}39)$$

This formula, known as the Clausius-Mossotti relation, gives the correct dependence of the specific inductive capacity on density for a wide class of solids and liquids. For dilute gases, where κ is not very different from unity, Eq. (2–39) becomes

$$\kappa - 1 = N\alpha = \frac{N_0\rho_m\alpha}{M}, \qquad (2\text{–}40)$$

just as would be expected for an approximation corresponding to the neglect of the interaction between each molecule and its neighbors. The molecular polarizability in general arises from two basic physical causes: (1) the lengthening of the bonds between atoms, and (2) the preferred orientation of molecules along the direction of the field as opposed to the random orientations brought about by thermal motions.

It is this second effect which is responsible for the temperature dependence of the specific inductive capacity. In statistical mechanics it is shown that under conditions of thermal equilibrium the probability that any one molecule has energy U is proportional to $e^{-U/kT}$, where k is Boltzmann's constant and T is the absolute temperature. If we have a molecule of intrinsic moment $\mathbf{p}_0$ in a field **E**, then according to Section 1–8,

$$\begin{aligned} U &= -\mathbf{p}_0 \cdot \mathbf{E} \\ &= -p_0 E \cos\theta, \end{aligned} \qquad (2\text{–}41)$$

which depends on the orientation of the molecule with respect to the field. The contribution of each molecule to the total dipole moment would be $p_0 \cos\theta$, to be summed over all molecules. If there are N molecules per unit volume, the effective polarization would be given by

$$P = \frac{\displaystyle\int Np_0 \cos\theta\, e^{x\,\cos\theta}\, d\Omega}{\displaystyle\int e^{x\,\cos\theta}\, d\Omega}, \qquad (2\text{–}42)$$

where for convenience we have written $x = p_0E/kT$. The denominator is easily integrated, and apart from the constant factor Np_0 the numerator is just the partial derivative of the denominator with respect to x. The final evaluation gives

$$P = Np_0 \left(\coth x - \frac{1}{x} \right), \qquad (2\text{-}43)$$

which for small x becomes

$$P \simeq Np_0 \frac{x}{3} = \frac{Np_0^2 E}{3kT}. \qquad (2\text{-}43')$$

This effect is to be added to the polarizability due to distortion of the molecules by the field, which we may call α_0. Note that the distortion polarizability α_0 would add a term

$$-\alpha_0 \epsilon_0 E_{\text{eff}}^2 / 2$$

to the energy expression (2-41) which does not depend on the angular and position coordinates of the molecule and thus does not result in a temperature-dependent polarization. Equation (2-40) thus becomes

$$\kappa - 1 = N\alpha_0 + \frac{Np_0^2}{3\epsilon_0 kT}, \qquad (2\text{-}44)$$

or, if Eq. (2-40) is not applicable, Eq. (2-39) becomes

$$3 \frac{\kappa - 1}{\kappa + 2} = \frac{N_0 \rho_m}{M} \left(\alpha_0 + \frac{p_0^2}{3\epsilon_0 kT} \right). \qquad (2\text{-}45)$$

Hence examination of the density and temperature dependence of the measured value of κ can often be used to isolate α_0 and p_0.

Suggested References

All the references listed at the end of Chapter 1 for the fundamental principles of vacuum electrodynamics contain treatments of dielectrics and boundary conditions. An excellent discussion of electrostatics of dielectrics constitutes Chapter II of

L. D. Landau and E. M. Lifshitz, *Electrodynamics of Continuous Media.*

For the derivation of Eq. (2-43), called the Langevin formula, see almost any textbook on statistical mechanics, for example:

R. H. Fowler, *Statistical Mechanics.* Uses the elegant methods involving the partition function.

J. E. Mayer and M. G. Mayer, *Statistical Mechanics.* Mathematically simpler than Fowler, and more than adequate for our purpose.

J. H. Van Vleck, *The Theory of Electric and Magnetic Susceptibilities.*

EXERCISES

1. Consider a simple cubic lattice of point dipoles, all of equal moment and like orientation. Show that the field at the position of one dipole due to all others in a sphere of arbitrary radius about this point is zero.

2. A long very thin rod of dielectric constant κ is oriented parallel to a uniform external field $\mathbf{E}$. What are $\mathbf{E}$ and $\mathbf{D}$ within the rod? What are the fields in a very thin disk of dielectric oriented perpendicular to the field?

3. Show that for an electret (fixed $\mathbf{P}$) the integral $\int \mathbf{E} \cdot \mathbf{D} \, dv$ over the entire field volume vanishes.

4. Consider an electron of charge $-e$ moving in a circular orbit of radius a_0 about a charge $+e$ in a field directed at right angles to the plane of the orbit. Show that the polarizability α is approximately $4\pi a_0^3$.

CHAPTER 3

GENERAL METHODS FOR THE SOLUTION
OF POTENTIAL PROBLEMS

We have seen that, in principle, potential problems are solvable if all charge distributions are known, and we shall prove the uniqueness of any solution which reduces to given values of the potentials or their normal derivatives on the boundary of a region. Actually only a few of the most idealized problems can be solved with any degree of simplicity. For practical applications experimental and numerical methods of mapping fields have been devised, as well as graphical and semigraphical procedures involving some calculations. Analytic methods rarely lead to solutions in closed form, while infinite series must converge fairly rapidly to be useful. Nevertheless, analytic solutions of geometrical boundary configurations approximating the actual situation furnish a valuable check even when experimental mapping is finally necessary, and graphical methods depend on prior knowledge of the general behavior of the potential. The solution of problems with relatively simple boundaries and charge distributions is therefore of value, even for more complicated engineering applications.

Unfortunately, no general methods of solution are available which will apply to all types of geometrically simple problems, and therefore each individual case demands, to some extent, special treatment. Certain methods apply to general classes of problems, however, and can be discussed as individually characteristic of these classes.

3–1 Uniqueness theorem. This theorem states that if within a given boundary a solution of a potential problem is found which reduces to the given potential distribution on that boundary, or to the given normal derivative of the potential on that boundary, then this solution is the only correct solution of the potential equations within the boundary. The theorem provides justification for attempting any method of solution so long as the resulting solution can be shown to obey Laplace's equation in a charge-free region. No matter how the solution is obtained, if it satisfies these conditions the problem is considered solved.

The proof of the theorem is very similar to that indicated (Section 1–1) for the unique definition of vector fields from their source and circulation densities, except that here the integration does not extend to infinity. If we put $\phi \boldsymbol{\nabla} \phi$ into Gauss's divergence theorem as the vector field, we obtain

$$\int \phi \boldsymbol{\nabla} \phi \cdot d\mathbf{S} = \int \boldsymbol{\nabla} \cdot (\phi \boldsymbol{\nabla} \phi) \, dv = \int [(\boldsymbol{\nabla} \phi)^2 + \phi \nabla^2 \phi] \, dv. \qquad (3\text{–}1)$$

The last term vanishes, from Laplace's equation, if we choose the surface of integration in such a way as to exclude all charged regions from the interior of the region of integration. It may be necessary to employ surfaces internal to the outer boundary in order to exclude the charges from v.

Let us suppose that two different potentials, ϕ_1 and ϕ_2, are solutions of a given potential problem. Both ϕ_1 and ϕ_2 are to satisfy the boundary condition, and hence on the boundary either $\phi_1 = \phi_2$ or $(\nabla\phi_1)_n = (\nabla\phi_2)_n$. (The component of $\nabla\phi$ normal to a surface is often called the "normal derivative" and may be designated by $\partial\phi/\partial n$.) If we substitute the difference $\phi_1 - \phi_2$ for ϕ in Eq. (3–1), we have

$$\int(\phi_1 - \phi_2)\nabla(\phi_1 - \phi_2) \cdot d\mathbf{S} = \int[\nabla(\phi_1 - \phi_2)]^2 \, dv. \qquad (3\text{–}2)$$

Either boundary condition (equality of the potentials or their normal derivatives) assures the vanishing of the left side of Eq. (3–2). Since the integrand of the right side of Eq. (3–2) is positive definite, it must be zero in order for the integral to vanish; hence throughout the volume v

$$\nabla\phi_1 = \nabla\phi_2, \quad \phi_1 = \phi_2 + \text{constant}. \qquad (3\text{–}3)$$

Thus the two potentials that were assumed to be different yet satisfying the same boundary condition can differ at most by an additive constant which makes no contribution to the gradient; therefore these potentials will give the same electric field distributions.

If linear dielectrics are involved, Eq. (3–1) may be replaced by

$$\int\phi\kappa\nabla\phi \cdot d\mathbf{S} = \int[\kappa(\nabla\phi)^2 + \phi\nabla \cdot (\kappa\nabla\phi)] \, dv. \qquad (3\text{–}1')$$

Laplace's equation for dielectrics is

$$\nabla \cdot (\kappa\nabla\phi) = 0,$$

and hence the proof for uniqueness remains valid. If nonlinear dielectrics are involved, the region may be divided into subregions having uniform polarization densities, and for which the theorem holds separately.

3–2 Green's reciprocation theorem. A large number of theorems that are useful for the solution of electrostatic problems serve to transform the solution of a known, presumably simpler, problem into the solution of another problem whose solution is desired. Of such theorems one of the most useful is Green's reciprocity theorem. Let us consider a set of n point charges q_j, at positions where the potentials due to the other charges are given by a set of numbers ϕ_j. The potential at the point j is related to the charges at the other points (designated by i for purposes of summation) by

$$\phi_j = \frac{1}{4\pi\epsilon_0} \sum_{i=1}^{n}{}' \frac{q_i}{r_{ij}}. \qquad (3\text{-}4)$$

The prime on the summation sign means that the term $i = j$ is to be omitted from the summation. If, on the other hand, a different set of charges q_i' is placed at the same points, giving rise to the corresponding potentials ϕ_j', a similar relation holds:

$$\phi_j' = \frac{1}{4\pi\epsilon_0} \sum_{i=1}^{n}{}' \frac{q_i'}{r_{ij}}. \qquad (3\text{-}5)$$

Let us now multiply Eq. (3–4) by q_j' and Eq. (3–5) by q_j, then sum each expression over the index j:

$$\sum_{j=1}^{n} \phi_j q_j' = \sum_{j=1}^{n} \sum_{i=1}^{n}{}' \frac{q_i q_j'}{r_{ij}} \frac{1}{4\pi\epsilon_0},$$
$$\sum_{j=1}^{n} \phi_j' q_j = \sum_{j=1}^{n} \sum_{i=1}^{n}{}' \frac{q_i' q_j}{r_{ij}} \frac{1}{4\pi\epsilon_0}. \qquad (3\text{-}6)$$

Since i and j are summation indices, we may interchange them in one product of the q's; thus

$$\sum_{j=1}^{n} \phi_j q_j' = \sum_{j=1}^{n} \phi_j' q_j, \qquad (3\text{-}7)$$

which is the desired theorem.

This theorem can be generalized from a set of point charges to a set of n conductors of potentials ϕ_j carrying charges q_j: the generalization follows if we combine the points of equal ϕ_j in Eq. (3–7) into a single term. Equation (3–7) thus applies directly to such a system of conductors. If all but two conductors i and j are grounded, Eq. (3–7) implies that the potential to which the uncharged conductor i is raised by putting a charge q on conductor j is equal to the potential of j, when uncharged, produced by a charge q on i. An application of Eq. (3–7) to the solution of a potential problem is indicated at the close of Section 3–3.

3–3 Solution by Green's function. A great variety of solutions of potential problems can be generated from what is known as a Green's function. The Green's function for a particular geometrical arrangement is the solution of the potential problem for this given geometrical arrangement of grounded conducting boundaries when the only charge present is a unit point charge at point x'_α. It may be shown with the aid of Green's reciprocity theorem that the Green's function for a particular geometry is a symmetrical function of the coordinates of a unit charge located at the point x'_α and the coordinates of the point of observation x_α.

Two general types of problems can be solved by the use of Green's function. One type is that in which the potential distribution ϕ_s over a certain boundary is given, and the other is one in which the charge distribution is given in a region within a conducting boundary. The derivation of the solution of both these problems can be given together by means of Green's theorem,

$$\int (\phi\nabla^2\psi - \psi\nabla^2\phi)\, dv = \int (\phi\nabla\psi - \psi\nabla\phi)\cdot d\mathbf{S}, \qquad (1\text{–}37)$$

where ψ and ϕ are arbitrary functions of position which are required to be nonsingular throughout the volume v. Let ϕ be the desired solution of a particular potential problem and let $\psi = G$ be the Green's function for the geometry of the problem, i.e., the solution of the problem of a unit point charge located at $r = 0$ with the surface S grounded. Then G will be of the form

$$G = \frac{1}{4\pi\epsilon_0 r} + \chi, \qquad (3\text{–}8)$$

where χ represents the potential due to the induced charge on S. Here χ is harmonic in v, i.e., it is a solution of Laplace's equation. The distance r is a symmetric function of coordinates x_α and x'_α, and the volume and surface integrals are to be carried out over x_α. Therefore, G has a singularity only at $r = 0$, which we may handle by means of the δ-function. On substituting Eq. (3–8) into Green's theorem, we have

$$\int (G\nabla^2\phi - \phi\nabla^2 G)\, dv = \int \left[G\nabla^2\phi + \frac{\phi\,\delta(\mathbf{r})}{\epsilon_0} \right] dv$$

$$= \int (G\nabla\phi - \phi\nabla G)\cdot d\mathbf{S}. \qquad (3\text{–}9)$$

Also, by definition, $G = 0$ on S. Hence, on collecting the nonvanishing terms, we find

$$\phi = -\epsilon_0 \left(\int G\nabla^2\phi\, dv + \int \phi_s\nabla G\cdot d\mathbf{S} \right) \qquad (3\text{–}10)$$

for the value of ϕ at x'_α.

Let us now consider the two cases mentioned earlier:

(1) The surface surrounding the point x'_α is grounded, making $\phi_s = 0$, and $\nabla^2\phi = -\rho/\epsilon_0$ due to the charge distribution ρ throughout v. Equa-

tion (3–10) then reduces to

$$\phi(x'_\alpha) = -\epsilon_0 \int G\nabla^2\phi \, dv = \int G\rho \, dv. \qquad (3\text{--}11)$$

This expression is fairly obvious, since it merely represents the principle of superposition applied to the density of point sources within the volume v, with each unit source of which the density ρ consists contributing its share to the potential $\phi(x'_\alpha)$ by the superposition indicated by the integral.

(2) Let there be no sources of ϕ throughout the volume v, so that $\nabla^2\phi = 0$, but let us assume that ϕ is a given function ϕ_s on the surface S. In this case, Eq. (3–10) reduces to

$$\phi(x'_\alpha) = -\epsilon_0 \int \phi_s \nabla G \cdot d\mathbf{S}. \qquad (3\text{--}12)$$

Equation (3–12) gives the potential within a given region enclosed by a boundary where different parts of the boundary are raised to a given set of potentials. This solution expresses the potential within this boundary in terms of the surface integral of the potential on the boundary multiplied by the normal derivative of the Green's function. Physically, the normal derivative of the Green's function represents the surface charge density that would be induced on the boundary by a unit charge at the point x'_α *if* the boundary were a grounded conductor. Equation (3–12) thus gives the solution of the potential problem corresponding to a given potential on the boundary in terms of the integral of this potential multiplied by the charge induced on the grounded boundary by a unit charge placed at the field point. If we wish to express Eq. (3–12) explicitly in terms of the charge σ_{1s} induced on the grounded boundary we note from Eq. (2–15) that

$$\nabla G \cdot \frac{d\mathbf{S}}{dS} = + \frac{\sigma_{1s}}{\epsilon_0}.$$

Thus Eq. (3–12) becomes simply

$$\phi(x'_\alpha) = - \int \phi_s \sigma_{1s} \, dS. \qquad (3\text{--}13)$$

Theorem (3–13) may also be derived directly by the use of Green's reciprocation theorem:

(1) Let the surface S be grounded and let a charge q_0 be placed on a small conductor surrounding the point x'_α. The charge induced by q_0 on the jth region of the boundary S will be designated by q_{1j}.

(2) Let the charge at x'_α be removed, but let the surface S be divided into sections, each at a constant potential, the potential of the jth section of S being ϕ_{js}. In this case $\phi(x'_\alpha)$ represents the potential at x'_α.

If we relate these two cases by means of Eq. (3–7), we obtain

$$q_0\phi(x'_\alpha) + \sum_{j=1}^{n} q_{1j}\phi_{js} = 0 + 0.$$

The two zeros on the right arise from the fact that the potential is zero over the entire boundary in case (1), and that the charge at x'_α is zero in case (2). Remembering that q_0 is unity in our consideration, we have

$$\phi(x'_\alpha) = - \sum_{j=1}^{n} \phi_{js}q_{1j}. \qquad (3\text{–}14)$$

This expression is identical with Eq. (3–13), but it has been obtained directly in terms of the induced charge in a way that is more obvious physically.

In the consideration of more specific problems, we shall derive Green's functions for various conducting boundaries. The solutions for the problems, both for grounded conductors enclosing charge distributions and for charge-free regions surrounded by conductors whose potentials are given, can then be written down immediately.

3–4 Solution by inversion. There are various kinds of transformations by which a set of solutions of one potential problem can be transformed into the solutions of another problem. The process of inversion is a special case, important because it is valid in three dimensions as well as in two dimensions. In two dimensions classes of such transformations more general than inversion can be found.

One of the simplest and most useful methods by which the solution of a problem can often be transformed into the solution of a simpler problem is the inversion in a sphere, as shown in Fig. 3–1. It can be shown by direct

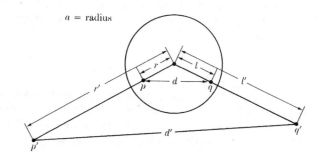

Fig. 3–1. Solution by inversion.

differentiation that if $\phi_p = \phi(r, \theta, \varphi)$ is a solution of Laplace's equation, then

$$\psi(r', \theta, \varphi) = \frac{a}{r'} \phi \left(\frac{a^2}{r'}, \theta, \varphi \right) \tag{3-15}$$

is also a solution of Laplace's equation. In relation to a sphere of radius a this transformation of the point r into the point r', by the relation $rr' = a^2$, maps the point $p(r, \theta, \varphi)$ into its inversion point $p'(a^2/r, \theta, \varphi)$, moving the point along the radius vector from a position inside the sphere to a point outside, or vice versa. Let a charge q be placed at distance l and a charge q' at distance l' from the center such that $ll' = a^2$. The relations $rr' = a^2$ and $ll' = a^2$ imply that $r/l' = l/r'$, and therefore the triangles with sides rld and $r'l'd'$ are similar. Thus we have $r/l' = l/r' = d/d'$. The potential at p before inversion is $\phi_p = q/4\pi\epsilon_0 d$ and the potential at p' after inversion is $\phi_p' = q'/4\pi\epsilon_0 d'$, so that

$$\frac{\phi_p'}{\phi_p} = \frac{q'}{q} \frac{d}{d'} = \frac{q'}{q} \frac{l}{r'} = \frac{q'}{q} \frac{r}{l'}. \tag{3-16}$$

To formulate a law for the inversion of charges, we make use of the fact that zero potential surfaces must transform into zero potential surfaces. We can make the potential of the inversion sphere zero by taking two charges initially, $q_I = q$ at l, and $q_{II} = -qa/l$ at a distance l' from the center such that $ll' = a^2$. Now the inversion sphere at zero potential under the influence of the two charges is to remain so after inversion. This is assured if the two charges change places thus:

$$q_I = q \text{ at } l \text{ becomes } q_I' = aq/l \text{ at } a^2/l,$$

$$q_{II} = -qa/l \text{ at } a^2/l \text{ becomes } q_{II}' = -q \text{ at } l.$$

In either case the transformed charge is the original charge multiplied by the inversion radius a over the original distance of the charge from the center of the sphere,

$$\frac{q'}{q} = \frac{a}{l} = \frac{l'}{a} = \sqrt{\frac{l'}{l}}. \tag{3-17}$$

It seems more convenient for a charge to retain its original sign and change only its magnitude when it is inverted, although this is not necessary if all charges undergoing inversion are treated in the same way. We now secure, by substituting Eq. (3–17) into Eq. (3–16), the rule for the inversion of potentials,

$$\phi_p' / \phi_p = a/r' = r/a, \tag{3-18}$$

in agreement with

$$\phi_p' = \frac{a}{r'} \phi(r, \theta, \varphi),$$

which corresponds to Eq. (3–15). The transformation equations for such quantities as volume or surface charge densities can be obtained by multiplying the charge transformation Eq. (3–17) by the transformation of the appropriate geometrical quantities.

In an inversion transformation a point charge will often appear at the center of inversion in the transformed geometry. This point charge arises from the fact that the net charge in the original geometry had electric field lines that terminated on equal and opposite charges located at infinity and, in the inversion, infinity is brought in to the origin.

The main utility of the inversion transformation is that it rectifies spherical boundaries if the center of inversion is taken on the spherical boundary. Two freely charged intersecting spheres may be inverted into two intersecting planes, and the plane boundary problem is usually easily soluble by the method of images.

3–5 Solution by electrical images. If we have two equal and opposite point charges, the zero potential surface is the plane of points equally distant from the two charges. The zero potential plane could be replaced by a plane grounded conductor, and the potential and field would remain

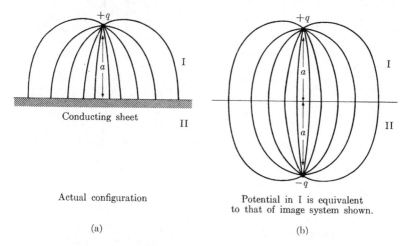

Actual configuration

(a)

Potential in I is equivalent to that of image system shown.

(b)

Fig. 3–2. Point charge and conducting plane. (a) represents the physical system; (b) represents the system of images which would leave the conducting sheet at zero potential and therefore produce the same field distribution in I as the actual physical system.

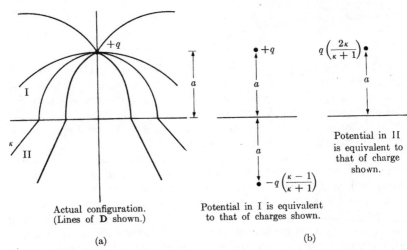

FIG. 3–3. Point charge and "dielectric half-space." (a) represents the actual physical system of a charge $+q$ at a distance a from a dielectric half-space of specific inductive capacity κ. (b) is a system of images representing the configuration: the left-hand side of the distribution (b) is a system of charges which gives the correct field distribution in region I; the right-hand part gives the correct field distribution in region II.

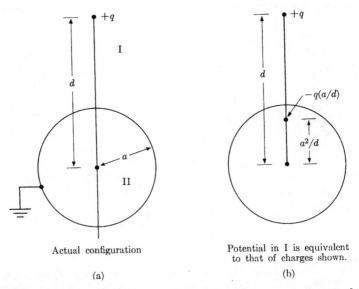

FIG. 3–4. Point charge and grounded sphere. (a) represents the actual physical configuration. (b) represents the system of images which gives the correct distribution in region I. The potential in region II is zero.

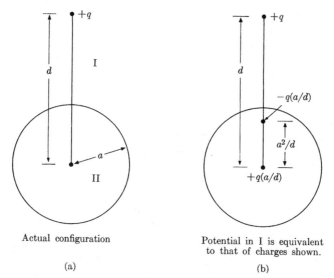

Actual configuration

(a)

Potential in I is equivalent
to that of charges shown.

(b)

FIG. 3–5. Point charge and uncharged isolated conducting sphere. (a) represents the actual physical system. (b) represents the system of images which gives the correct potential distribution in region I. The field, but not the potential, in region II is zero.

unchanged. Thus the solution for a point charge and grounded plane is just that of two point charges throughout region I (see Fig. 3–2) in which the field exists. The fictional charge $-q$ is called the "image" of q, by analogy with reflection in a mirror.

In general, the potentials of the charges induced in infinite plane and spherical conductors and on a plane dielectric boundary can be shown to be equivalent to the potential of a suitable image charge or charges. Typical cases are indicated in Figs. 3–2 through 3–5. Detailed justification of these results is left to the problems.

The method of images leads to geometrically simple expressions for the Green's function involving plane or spherical boundaries. Let us use the theorem of Eq. (3–13) to solve the problem of determining the potential at an arbitrary point p above a plane on which a potential function ϕ_S is given (Fig. 3–6). This figure shows both the physical system described and the "reciprocal" (Green's function) system in which a unit charge is placed at p and S is grounded. The boundary problem in this reciprocal system is solved by images.

The induced charge density σ_{1s} in the reciprocal system is given by $\sigma_{1s} = \epsilon_0 E_n$, where E_n is the electric field at the conductor. On the other hand, E_n (which is normal to the conductor) can be obtained by applying Gauss's theorem, Eq. (1–26'), to the flux passing through a box surrounding the surface element dS, as indicated by dotted lines in Fig. 3–6. It

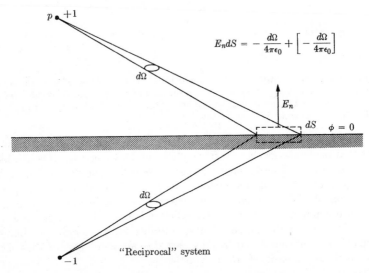

FIG. 3-6. Determination of potential due to plane with arbitrary potential distribution.

is readily seen from the figure that the flux through the upper surface of the box due to both charge and image is

$$\sigma_{1s}\, dS = \epsilon_0 E_n\, dS = -d\Omega/2\pi, \qquad (3\text{--}19)$$

where $d\Omega$ is the element of solid angle subtended by dS at p. Hence from Eq. (3-13) we obtain the simple expression,

$$\phi_p = \frac{1}{2\pi} \int_s \phi_s\, d\Omega. \qquad (3\text{--}20)$$

In general, if the Green's function of a system can be obtained by the method of images, then the problem in which the system boundaries are given an arbitrary potential distribution reduces to a simple sum or integral over solid angles.

3–6 Solution of Laplace's equation by the separation of variables. Except within a distribution of charge, the fundamental problem of potential theory is to find a solution of Laplace's equation which satisfies certain conditions on the boundaries of the region under consideration. If these boundaries correspond to coordinate surfaces in a system of orthogonal coordinates, the solution by separation of variables is often much more convenient than the general Green's function method. For one thing, it is very easy to state the boundary conditions in the appropriate system of coordinates, whether the condition be continuity of ϕ or its derivative or the assignment of some definite value of ϕ or $\partial\phi/\partial n$, when each statement refers to a constant value of a particular coordinate. This would be true of any system of coordinates, but in certain systems we can go further and write the solution as a product of functions of the coordinates separately, so that the boundary conditions can be applied to the separate single-variable factors. It may be added that while there is no direct general method of solving partial differential equations, the separation reduces Laplace's equation to a set of ordinary differential equations which in principle are always solvable.

The essential features of the method can best be demonstrated by means of an example. Consider a pair of parallel grounded conducting plates at $y = 0$ and $y = a$, as shown, with a line charge parallel to the z-axis at the point $(0, d)$. We seek a solution valid between the plates, assuming there are no other charges. The problem is thus a two-dimensional one for which rectangular coordinates are appropriate. Let us assume that $\phi(x, y) = X(x)Y(y)$, where X is a function of x alone and Y is a function of y. Except for the point $(0, d)$ the equation to be satisfied is

$$\nabla^2\phi = YX'' + XY'' = 0, \qquad (3\text{–}21)$$

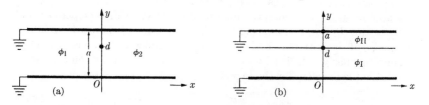

FIG. 3–7. Showing two ways of dividing the space between capacitor plates by planes containing the line of charge at $x = 0$, $y = d$.

where the double prime denotes the second derivative of the function with respect to its argument. If we divide Eq. (3–21) by ϕ, we obtain

$$\frac{X''}{X} + \frac{Y''}{Y} = 0. \tag{3–22}$$

Since x and y can vary independently, both terms of Eq. (3–22) must be independent of either variable, and we can write

$$X''/X = -Y''/Y = K. \tag{3–23}$$

The constant K is called the separation parameter. If there are no restrictions on K the product of the general solutions of the ordinary differential equations is a general solution of the two-dimensional Laplace equation. The boundary conditions of the physical problem, however, will limit both the nature of the solutions and the values of the separation parameter. The solution we seek is a sum (or integral, depending on whether the allowed values of the parameter are discrete or continuous) of allowed product solutions, with coefficients determined so that the boundary conditions are exactly satisfied. In order to determine these coefficients we shall make use of a property of the functions known as *orthogonality*.

There remains a choice in the sign of the separation constant K and thus in the nature of the corresponding solutions. Let us first assume that K of Eq. (3–23) is positive, $K = k^2$, so that the ordinary differential equations become

$$Y'' + k^2 Y = 0,$$
$$X'' - k^2 X = 0, \tag{3–24}$$

having general solutions

$$Y = A \sin ky + B \cos ky,$$
$$X = Ce^{kx} + De^{-kx}. \tag{3–25}$$

The boundary conditions to be satisfied are that $\phi = 0$ at $y = 0$, $y = a$, and $x = \pm\infty$. The potential may be made to vanish at the plates simply by setting $B = 0$ and limiting k to the values $n\pi/a$, where n is an integer. The conditions at plus and minus infinity along x, however, cannot be simultaneously satisfied by either term of X, so that we must write the solutions separately for the regions of positive and negative x:

$$\underset{-\infty < x < 0}{\phi_1} = \sum_{n=1}^{\infty} C_n e^{n\pi x/a} \sin \frac{n\pi y}{a}, \qquad \underset{0 < x < \infty}{\phi_2} = \sum_{n=1}^{\infty} A_n e^{-n\pi x/a} \sin \frac{n\pi y}{a}. \tag{3–26}$$

At $x = 0$ the potential ϕ is continuous, so that the coefficients in the two series are equal term by term, i.e., $C_n = A_n$. We have yet to determine these coefficients, however, and we have not taken account of the flux from the line charge. This physical requirement must exactly correspond to the mathematical determination of the A_n's, since the potential is unique.

Now a line charge could be represented by a two-dimensional δ-function, but since we are writing the solution separately for regions 1 and 2 it is possible to use our knowledge of the boundary conditions at a surface charge and employ a one-dimensional function $\delta(y - d)$ as a special case of an arbitrary charge distribution along the surface $x = 0$. In other words, on $x = 0$,

$$\sigma(y) = q\,\delta(y - d), \tag{3–27}$$

where q is the charge per unit length perpendicular to the xy-plane and $\delta(y - d)$ is defined by the equations

$$\int_0^a \delta(y - d)\,dy = 1,$$
$$\int_0^a f(y)\,\delta(y - d)\,dy = f(d), \qquad 0 < d < a. \tag{3–28}$$

The potentials must then satisfy the conditions

$$\frac{\sigma(y)}{\epsilon_0} = \frac{q\,\delta(y - d)}{\epsilon_0} = \left[\frac{\partial\phi_1}{\partial x} - \frac{\partial\phi_2}{\partial x}\right]_{x=0} = \sum A_n \frac{2n\pi}{a} \sin\frac{n\pi y}{a}. \tag{3–29}$$

The Fourier coefficients are determined in the usual way by multiplying both sides of Eq. (3–29) by $\sin(m\pi y/a)$ and integrating from 0 to a. All terms of the series on the right will vanish except that for which $m = n$, i.e., the sine functions are "orthogonal" over the interval. Therefore

$$\frac{q}{\epsilon_0} \sin\frac{m\pi d}{a} = \frac{2m\pi}{a} A_m \frac{a}{2},$$

or, on writing n for m,

$$A_n = \frac{q}{\epsilon_0 n\pi} \sin\frac{n\pi d}{a}. \tag{3–30}$$

The entire solution is then

$$\phi_1 = \frac{q}{\epsilon_0 \pi} \sum \frac{1}{n} \sin\frac{n\pi d}{a} e^{n\pi x/a} \sin\frac{n\pi y}{a},$$
$$\phi_2 = \frac{q}{\epsilon_0 \pi} \sum \frac{1}{n} \sin\frac{n\pi d}{a} e^{-n\pi x/a} \sin\frac{n\pi y}{a}, \tag{3–31}$$

and the problem is solved.

It is instructive to note that the same potential may look quite different with the opposite choice of sign for the separation parameter in Eq. (3–23). If we put $K = -k^2$, the solution for $X(x)$ is just $\cos kx$, since the potential is obviously an even function of x, but no limitations are imposed on k. No single function Y will vanish on the two conducting plates, however, and we must again divide the region into two parts, this time by the plane $y = d$. It can be easily verified that for any k the two solutions which vanish at $y = 0$ and $y = a$, and are continuous at $y = d$, are

$$Y_{\mathrm{I}} = \frac{\sinh ky}{\sinh kd}, \qquad 0 < y < d,$$

$$Y_{\mathrm{II}} = \frac{\sinh k(a - y)}{\sinh k(a - d)}, \qquad d < y < a.$$

$$(3\text{--}32)$$

The potentials are integrals over k, with coefficients which we may call $A(k)$:

$$\phi_{\mathrm{I}} = \int_{-\infty}^{\infty} A(k) \cos kx \, \frac{\sinh ky}{\sinh kd} \, dk,$$

$$\phi_{\mathrm{II}} = \int_{-\infty}^{\infty} A(k) \cos kx \, \frac{\sinh k(a - y)}{\sinh k(a - d)} \, dk.$$

$$(3\text{--}33)$$

The charge density on the plane $y = d$ is now a function of x, and the condition on the potentials is

$$\frac{q \, \delta(x)}{\epsilon_0} = \left[\frac{\partial \phi_{\mathrm{I}}}{\partial y} - \frac{\partial \phi_{\mathrm{II}}}{\partial y}\right]_{y=d}$$

$$= \int A(k) \cos kx \left[\frac{\cosh kd}{\sinh kd} + \frac{\cosh k(a - d)}{\sinh k(a - d)}\right] k \, dk$$

$$= \int A(k) \, \frac{\cos kx \sinh ka}{\sinh kd \sinh k(a - d)} \, k \, dk. \qquad (3\text{--}34)$$

But it follows from a formal application of the Fourier integral theorem, and has been indicated in a problem at the end of Chapter 1, that

$$\int_{-\infty}^{\infty} \cos kx \, dk = 2\pi \, \delta(x) \qquad (3\text{--}35)$$

in the sense that both behave in the same way as factors in an integrand. Therefore

$$A(k) = \frac{q}{2\pi\epsilon_0} \, \frac{\sinh kd \sinh k(a - d)}{k \sinh ka} \qquad (3\text{--}36)$$

and the potential is given explicitly by

$$\phi_{\mathrm{I}} = \frac{q}{2\pi\epsilon_0} \int \frac{\sinh k(a-d)}{k \sinh ka} \cos kx \, \sinh ky \, dk,$$

$$\phi_{\mathrm{II}} = \frac{q}{2\pi\epsilon_0} \int \frac{\sinh kd}{k \sinh ka} \cos kx \, \sinh k(a-y) \, dk. \tag{3–37}$$

If $q = 1$ this solution [Eq. (3–31) or (3–37)] is the Green's function for the two-dimensional parallel plate geometry, and we shall see that the method is general for determining the Green's function for a set of equi-coordinate planes. But for our immediate purpose the details of this particular problem are less important than the principles they illustrate. Usually one choice of parameters and corresponding functions is more convenient than the other, but a choice must be made.

To achieve separation of variables in three dimensions we follow the same procedure. In Cartesian coordinates $\phi(x, y, z) = X(x)Y(y)Z(z)$, and substitution in Laplace's equation yields

$$\frac{\nabla^2 \phi}{\phi} = \frac{X''}{X} + \frac{Y''}{Y} + \frac{Z''}{Z} = 0. \tag{3–38}$$

We may therefore set

$$\frac{X''}{X} + \frac{Y''}{Y} = -\frac{Z''}{Z} = K_1,$$

$$\frac{X''}{X} = K_1 - \frac{Y''}{Y} = K_2. \tag{3–39}$$

There are now two separation parameters; in general, the number of such parameters corresponding to N independent variables is $N - 1$. The lack of symmetry in the equations for the factor solutions is intrinsic in Laplace's equation, and may be said to correspond to the complete symmetry in sign of the coordinates themselves in the Laplacian operator. In three-dimensional rectangular coordinates two of the factors may be circular functions, but not all three. It is necessary that the functions be orthogonal in order to permit the determination of coefficients at a constant surface of the third variable, but this can be shown to be a general property of solutions of Laplace's equation, independent of the coordinate system for every case in which separation is possible.

The orthogonality of the allowed functions may be demonstrated by means of Green's theorem. Let us apply the proof in the case of spherical polar coordinates, where the geometry is easily visualized, and then see that it is applicable to other coordinate systems as well. Assume that Laplace's equation is separable, so that $\phi(r, \theta, \varphi) = R(r)Y(\theta, \varphi)$, and that whatever the nature of the functions or the parameter involved in the separation, ϕ_1 and ϕ_2 are two allowed solutions. If we put these two potentials into Green's theorem, we obtain

$$\int (\phi_1 \nabla^2 \phi_2 - \phi_2 \nabla^2 \phi_1) \, dv = \int [R_1 Y_1 \nabla(R_2 Y_2) \\ - R_2 Y_2 \nabla(R_1 Y_1)] \cdot d\mathbf{S}. \quad (3\text{--}40)$$

The left side vanishes, since ϕ_1 and ϕ_2 are solutions of Laplace's equation. If we let S be the surface of a sphere, the component of ∇ parallel to $\mathbf{S}$ does not operate on the function Y, and therefore

$$\left(\frac{R_2'}{R_2} - \frac{R_1'}{R_1} \right) \int Y_1 Y_2 \, dS = 0. \quad (3\text{--}41)$$

But if the two radial functions correspond to different values of the separation parameter they have in general different dependence on r, and thus their logarithmic derivatives are unequal. Equation (3–41) is thus a statement that the two functions Y_1 and Y_2 are orthogonal to each other when integrated over the surface of a sphere.

If the coordinate surface is not closed the proof follows in exactly the same way except that use must be made of the fact that the potentials are zero at infinity. For sources confined to a finite region the potentials do approach zero sufficiently fast so that the integral over the infinite surface vanishes. We can therefore conclude that, in general, orthogonal functions are generated in the solution of Laplace's equation, and that if the equation is separable it is possible in principle to complete the solution.

It can be shown that the set of allowed functions is complete as well as orthogonal. We shall not take up the proof of this statement, but it should be remarked that completeness is necessary for the existence of a satisfactory potential theory. Without it the representation of arbitrary potential or charge distributions on surfaces could not be guaranteed.

SUGGESTED REFERENCES

E. WEBER, *Electro-magnetic Fields*, Vol. 1, *Mapping of Fields*. This excellent treatise on practical solutions also presents the principles in a clear and coherent fashion.

O. D. KELLOGG, *Foundations of Potential Theory*. The classical presentation of mathematical foundations.

J. C. MAXWELL, *Electricity and Magnetism*, Vol. 1.

L. D. LANDAU AND E. M. LIFSHITZ, *Electrodynamics of Continuous Media*. Chapter I includes interesting problems, as well as a compact but thorough account of electrostatic fields in the neighborhood of conductors.

Of the works listed at the end of Chapter 1 the most useful for electrostatics problems are those by Jeans, Smythe, and Stratton.

P. Morse and H. Feshbach, *Methods of Mathematical Physics*. This is a very comprehensive work on mathematical methods.

E. T. Whittaker and G. N. Watson, *Modern Analysis*. Much more formal mathematically than Morse and Feshbach, this work is generally useful for the properties of various functions and especially here for a discussion of convergence.

H. and B. S. Jeffreys, *Methods of Mathematical Physics*. Chapter 6 of this excellent work is devoted to potential theory.

Somewhat more elementary treatments of useful mathematical topics are:

W. E. Byerly, *Fourier's Series and Spherical Harmonics*.

R. V. Churchill, *Fourier Series and Boundary Value Problems*.

Exercises

1. If $\phi(x, y, z)$ is a solution of Laplace's equation, show that

$$\frac{1}{r} \, \phi\left(\frac{x}{r^2}, \frac{y}{r^2}, \frac{z}{r^2}\right)$$

is also a solution. [Use spherical coordinates. Compare Eq. (3–15).]

2. Let $\phi(r, z)$ be the electrostatic potential at a point (r, z) in a situation of axial symmetry, with r the two-dimensional radius such that $r^2 = x^2 + y^2$. Let a be a small quantity and let $r = na$. Show that

$$\phi(r, z) = \{2n[\phi(r, z + a) + \phi(r, z - a)] + (2n + 1)\phi(r + a, z)$$
$$+ (2n - 1)\phi(r - a, z)\}/8n,$$

at a charge-free point in space, correct to order a^3. This is one of the typical theorems useful in the "net point" method of field plotting.

3. Two coaxial pipes of the same diameter with a small gap between them are maintained at a potential difference V. Divide the region within the pipe near the gap into a rectangular net and guess the potentials at each point. Check the correctness of your guesses with the result from the theorem of Exercise 2 above, adjust incorrect values, and repeat until a reasonably correct distribution is obtained. This is called the relaxation method.

4. Solve Exercise 3 analytically. Possible methods are:
   (a) Separation of variables and fitting of coefficients.
   (b) Derive a Green's function by methods similar to that used in solving the problem of Fig. 3–7, and then use Eq. (3–14).
   (c) Use the image Green's function as discussed in Section 3–5.

5. Show that the image system of Fig. 3–3 is correct.

6. Show that the image systems of Fig. 3–2 and Fig. 3–4 are related by the inversion transformation.

7. Find the condition that a set of two-dimensional equipotentials $\phi_2 = f(z, y)$ can generate a set of equipotentials when rotated about the z-axis. Show that if

this is possible the potential is

$$\phi = A \int \exp\left[-\int F(\phi_2) d\phi_2\right] d\phi_2 + B$$

where

$$F(\phi_2) = \frac{1}{y} \frac{1}{(\nabla\phi_2)^2} \frac{\partial\phi_2}{\partial y}.$$

(See Jeans or Smythe.)

8. Consider the field due to an electric dipole of moment **p**. What charge distribution would have to be introduced on a sphere with **p** at its center to produce zero field outside the sphere?

9. Find a charge distribution that would produce the Yukawa potential

$$\phi = \frac{q}{4\pi\epsilon_0} \frac{e^{-r/a}}{r}.$$

Why must the total charge be zero?

10. Two closed equipotentials ϕ_1, ϕ_0 are such that ϕ_1 contains ϕ_0; ϕ_p is the potential at any point between them. If a charge q is now put at point p and the equipotentials are replaced by grounded conducting surfaces, show that the charges q_1, q_0 induced on the two conductors satisfy the relation

$$q_1/(\phi_0 - \phi_p) = q_0/(\phi_p - \phi_1) = q/(\phi_1 - \phi_0).$$

11. Show that Eqs. (3–31) and (3–37) converge, and that both correspond to the physical situation.

12. Determine the potential inside an infinitely long rectangular prism with grounded conducting walls at $x = 0$, a, $y = 0$, b, due to a line charge of q per unit length located at the point (c, d) inside the prism.

13. Let the source point in Fig. 3–4 go to infinity, and thus calculate the charge distribution on a grounded spherical conductor in a uniform field **E**. Also calculate the dipole moment of this distribution (which is equal to the dipole moment of the images).

14. Obtain the solution of the problem

$$\frac{d^2\phi(x)}{dx^2} = -\frac{\rho(x)}{\epsilon_0}, \qquad 0 < x < \pi,$$

$$\phi(0) = A, \qquad \phi(\pi) = B,$$

with $\rho(x)$, A, B given arbitrarily, in terms of the Green's function defined by

$$\epsilon_0 \frac{d^2}{dx'^2} G(x', x) = -\delta(x' - x), \qquad 0 < x' < \pi$$

$$G(0, x) = G(\pi, x) = 0.$$

CHAPTER 4

TWO-DIMENSIONAL POTENTIAL PROBLEMS

Potential problems involving geometrical arrangements that may be approximated by a two-dimensional geometry with an infinite uniform extent in the third direction are frequently easier to solve than three-dimensional problems. Some methods are really only simpler in two dimensions, and can be generalized to the study of three-dimensional geometry. On the other hand, certain mathematical techniques which have no genuine counterpart for three dimensions may be applied to two-dimensional problems. The method of complex variable potential description combined with conformal transformation is an especially powerful method of this second kind.

4-1 Conjugate complex functions. We shall show that in two dimensions any analytic function W of a complex variable $z_1 = x_1 + iy_1$ will have real and imaginary parts each of which individually satisfies Laplace's equation in two dimensions. Thus a suitable function $W = W(z_1)$ can completely describe the potential surfaces and field lines of a particular problem. If $\phi + i\psi = W = W(z_1) = W(x_1 + iy_1)$, we may separate real and imaginary parts and obtain $\phi(x_1, y_1)$ and $\psi(x_1, y_1)$. The equations $\phi = \text{constant}$ and $\psi = \text{constant}$ will represent the equipotential and field line surfaces or vice versa. Therefore any transformation from one complex variable z_1 to another z_2 will transform the solution of one potential problem described by the first variable to the solution of another potential problem described by the second variable. In general, a whole class of two-dimensional potential distribution problems can be solved by the following process:

(1) Obtain a transformation $z_2 = f(z_1)$ that will transform the geometric arrangement of the z_1 coordinate system into an arrangement of the z_2 coordinate system so as to bring about a simplification in the problem. This coordinate transformation, $z_2 = f(z_1)$ or $z_1 = g(z_2)$, must be so chosen that it will carry the complex potential geometry $W = W_1(z_1)$ of the original problem into a simpler complex potential geometry $W = W_1(g(z_2)) = W_2(z_2)$.

(2) Express the potential solution ϕ in the transformed (i.e., z_2) plane in such a way that $\phi + i\psi$ is an analytic function of a complex variable.

(3) Transform this solution back into the original z_1 plane.

We shall now discuss the justification for this process. Consider a func-

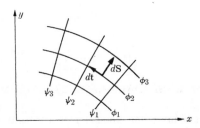

FIG. 4–1. Showing the relation between flux and streamlines.

tion $W = \phi + i\psi = f(z)$ where $z = x + iy$. At all points where the functional relationship is analytic, ϕ and ψ must satisfy the Cauchy-Riemann differential equations:

$$\frac{\partial \psi}{\partial x} = -\frac{\partial \phi}{\partial y}, \tag{4–1}$$

$$\frac{\partial \psi}{\partial y} = \frac{\partial \phi}{\partial x}. \tag{4–2}$$

By partial differentiation of Eq. (4–1) with respect to x and of Eq. (4–2) with respect to y and combination of the two resulting equations,

$$\nabla^2 \psi = 0 = \nabla^2 \phi. \tag{4–3}$$

The last equality follows from a repetition of the partial differentiation with the roles of x and y interchanged. Thus both ϕ and ψ are harmonic functions. The function $W = f(z)$ cannot be analytic everywhere unless it is identically zero; this correlates with the fact that, as discussed in Section 1–1, Laplace's equation cannot be satisfied everywhere if the potential is to be nonvanishing. Singularities of the function $W = f(z)$ represent sources. In general, a singularity of W represents a "divergence-type" source of one real potential function (ϕ or ψ), and a "circulation-type" source of the other.

The functional relationship $W = W(z)$ can be demonstrated graphically (Fig. 4–1) by plotting the lines $\phi = $ constant and the lines $\psi = $ constant in the $z = x + iy$ plane, after the function W has been separated into its real and imaginary parts. Note that the Cauchy-Riemann relations, Eqs. (4–1) and (4–2), ensure that these curves are normal to each other. The curves of constant ϕ obtained by giving a succession of values to W may be taken to represent the potential field of a problem, and the corresponding ψ curves taken to represent the electric field, although the latter are usually referred to as the streamlines.

The flux of the electric field crossing a surface S may be defined by

$$\Phi = \int \mathbf{E} \cdot d\mathbf{S}. \qquad (4\text{--}4)$$

Let us consider a surface lying along one of the equipotential curves, $\phi =$ constant, between two stream curves ψ_1 and ψ_2, and of unit height normal to the xy-plane. For the purposes of this proof let $\mathbf{i}$, $\mathbf{j}$, $\mathbf{k}$ represent unit vectors in the directions of increasing x, y, z and let $\mathbf{t}$ be the length along the ϕ curve, as shown in Fig. 4–1. Since the surface is of unit height, $d\mathbf{t} = \mathbf{k} \times d\mathbf{S}$. Then Eq. (4–4) becomes

$$\Phi = \int_1^2 \mathbf{E} \cdot d\mathbf{S} = -\int_1^2 \boldsymbol{\nabla}\phi \cdot d\mathbf{S} = -\int_1^2 \left(\frac{\partial\phi}{\partial x} \mathbf{i} + \frac{\partial\phi}{\partial y} \mathbf{j} \right) \cdot d\mathbf{S}.$$

By substitution from Eqs. (4–1) and (4–2), we have

$$\Phi = -\int_1^2 \left(\frac{\partial\psi}{\partial y} \mathbf{i} - \frac{\partial\psi}{\partial x} \mathbf{j} \right) \cdot d\mathbf{S} = -\int_1^2 (\boldsymbol{\nabla}\psi \times \mathbf{k}) \cdot d\mathbf{S}$$

$$= -\int_1^2 \boldsymbol{\nabla}\psi \cdot d\mathbf{t} = \psi_1 - \psi_2. \qquad (4\text{--}5)$$

Thus the difference between two stream functions ψ_1 and ψ_2 represents the electric flux passing between the right cylinders of unit height generated by two neighboring lines ψ_1 and ψ_2. This means that no lines of force cross the constant ψ lines. This is the justification for calling ψ the stream function, since in two-dimensional hydrodynamic problems the ψ lines do trace the streamlines of the fluid. In our case, the streamlines traced by giving ψ different constant values will trace the electric field, when ϕ lines are the equipotentials of the field. If, on the other hand, ψ had been assumed to be the potential, then ϕ would have been the stream function. This possibility of exchanging the meaning of ϕ and ψ is frequently useful in the solution of two-dimensional problems.

4–2 Capacity and field strength. The above considerations permit us to obtain immediately the capacity between any two conductors whose boundaries coincide with two equipotential lines ϕ_1 and ϕ_2, and extend between two streamlines ψ_1 and ψ_2. From Eq. (1–26) and the definition of Φ,

$$\Phi = \int \mathbf{E} \cdot d\mathbf{S} = q/\epsilon_0.$$

The capacity is given by

$$C = \frac{q}{\phi_1 - \phi_2} = \frac{\epsilon_0 \Phi}{\phi_1 - \phi_2}. \qquad (4\text{--}6)$$

Since the flux Φ is the change in the stream function ψ between the edges of the conductor surfaces being considered, Eq. (4–6) becomes

$$C = \epsilon_0 \frac{\psi_1 - \psi_2}{\phi_1 - \phi_2}. \tag{4–7}$$

Note that all the charges are assumed to lie on the bounding ϕ surfaces: in general the stream function will be multiple-valued if charges are present in the field. If the conductors are closed, $\psi_1 - \psi_2$ is a measure of the multivaluedness of ψ.

The absolute magnitude of the field strength can also be calculated from a known function of the form $W = W(z)$ representing a particular geometry as in Fig. 4–1. Consider the modulus of the derivative of W:

$$\left| \frac{dW}{dz} \right| = \left| \frac{\partial(\phi + i\psi)}{\partial x} \frac{dx}{dz} + \frac{\partial(\phi + i\psi)}{\partial y} \frac{dy}{dz} \right|,$$

$$\left| \frac{dW}{dz} \right| = \left| \frac{\partial \phi}{\partial x} dx + i \frac{\partial \psi}{\partial y} dy + i \frac{\partial \psi}{\partial x} dx + \frac{\partial \phi}{\partial y} dy \right| \left| \frac{1}{dz} \right|. \tag{4–8}$$

With the aid of the Cauchy-Riemann equations, we obtain

$$\left| \frac{dW}{dz} \right| = \left| \frac{\partial \phi}{\partial x} - i \frac{\partial \phi}{\partial y} \right| = \sqrt{\left(\frac{\partial \phi}{\partial x} \right)^2 + \left(\frac{\partial \phi}{\partial y} \right)^2} = |\boldsymbol{\nabla} \phi| = |\mathbf{E}|. \tag{4–9}$$

The real and imaginary parts of dW/dz are thus respectively the x- and y-components of the gradient of the potential, and therefore the modulus of dW/dz is equal to the magnitude of the electric field strength.

4–3 The potential of a uniform field. Before examining some of the transformations that are useful for simplifying complicated problems, we shall look at two basic potentials from which many more general cases may be generated by transformations and superpositions. The simplest is that of a uniform field $\mathbf{E}$ directed along x, for which $\phi = -|\mathbf{E}|x$. The complex potential can be seen by inspection to be

$$W = -|\mathbf{E}|z = -|\mathbf{E}|(x + iy) = \phi + i\psi, \tag{4–10}$$

and the stream function is $\psi = -|\mathbf{E}|y$. (The potential ϕ has been arbitrarily set equal to zero along the y-axis.)

4–4 The potential of a line charge. The Coulomb field around a line charge with a linear charge density q is found by means of Gauss's electric flux theorem, Eq. (1–26), the surface of integration being that of a circular cylinder of radius r and unit height coaxial with the line charge.

If the charge is located at the origin of coordinates in the xy-plane, this field is given by

$$\mathbf{E} = \frac{q\mathbf{r}}{2\pi\epsilon_0 r^2}. \tag{4–11}$$

The corresponding potential may be secured by substituting this field into Eq. (1–31) and integrating:

$$\phi = -\frac{q}{2\pi\epsilon_0}(\ln r - \ln r_0). \tag{4–12}$$

Again we note that in two-dimensional potential expressions it is not possible to set the potential at infinity equal to zero, since the two-dimensional expression really represents the potentials due to charge distributions of infinite extent perpendicular to the xy-plane. It must be remarked that a two-dimensional problem can be, at most, only an approximation to physical reality, for it implies not only infinite extent but infinite charge. A physical problem can be treated by two-dimensional methods only when it is possible to neglect end effects arising from the finite linear extent of the physical arrangement. In Eq. (4–12) the cylinder surrounding the line charge at a distance r_0 has been arbitrarily set at zero potential.

The complex potential function corresponding to a line charge located at an arbitrary point z_q may be derived by means of the Cauchy-Riemann equations, but it is easily written merely by inspection of Eq. (4–12). We introduce the complex notation $z = re^{i\theta}$ and $z_q = r_q e^{i\theta_q}$. Consider the complex potential

$$W = -\frac{q}{2\pi\epsilon_0}\ln(z - z_q), \tag{4–13}$$

whose complex conjugate is

$$W^* = -\frac{q}{2\pi\epsilon_0}\ln(z^* - z_q^*),$$

and hence the real part ϕ of W is given by

$$\phi = \frac{W + W^*}{2}$$

$$= -\frac{q}{2\pi\epsilon_0}\ln[(z - z_q)(z^* - z_q^*)]^{1/2}$$

$$= -\frac{q}{2\pi\epsilon_0}\ln[r^2 + r_q^2 - 2rr_q\cos(\theta - \theta^*)]^{1/2}. \tag{4–14}$$

This clearly corresponds to Eq. (4–12). By a similar analysis the stream function is seen to be

$$\psi = \left(\frac{W - W^*}{2}\right) = \theta - \theta_q. \tag{4–15}$$

The complex potential function for any system of line charges can be obtained by superposition of appropriately displaced expressions of the form Eq. (4–13), one for each line charge.

4–5 Complex transformations. We now turn to the analysis of the behavior of curves in a small region of the complex potential plane when a transformation of the plane is made. Consider a transformation from the z_1 plane to the z_2 plane given by the equation $z_2 = f(z_1)$ and let the transformation function f be analytic except at a finite number of singularities. At all nonsingular points such a transformation is conformal. This means that the angle between two intersecting lines in the z_1 plane, such as θ_1 in Fig. 4–2(a), transforms into an equal angle between the transformed lines in the z_2 plane, as θ_2 in Fig. 4–2(b). This can be demonstrated as follows. Since all derivatives of an analytic function of a complex variable exist and are continuous, the derivative dz_2/dz_1 will be finite at all points except for the singularities. Let us consider two line elements intersecting at a point P_1, for both of which $dz_2/dz_1 = f'(z_1)$ evaluated at the point P_1. The argument of a product is equal to the sum of the arguments of the factors, so that we have for the argument of the differential line element P_2Q_2:

$$\arg(dz_2) = [\arg f'(z_1)]_{P_1} + \arg(dz_1), \qquad (4\text{–}16)$$

and for the argument of P_2R_2:

$$\arg(dz_2') = [\arg f'(z_1)]_{P_1} + \arg(dz_1'). \qquad (4\text{–}17)$$

Subtracting Eq. (4–17) from Eq. (4–16), and noting that the angles θ_1 and θ_2 are the differences in the arguments of the respective dz's, we obtain

$$\theta_1 = \theta_2. \qquad (4\text{–}18)$$

The modulus of the derivative $dz_2/dz_1 = f'(z_1)$ represents the scale factor by which all spatial intervals in the neighborhood of a point are mul-

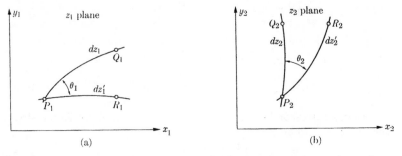

FIG. 4–2. To show that angles are preserved in a complex transformation.

tiplied. This follows from the fact that the modulus of a product is equal to the product of the moduli of the factors. An infinitesimal triangle will thus transform into a similar infinitesimal triangle in the new system, and

$$|dz_2| = |f'(z_1)_{P_1}| \cdot |dz_1|. \tag{4–19}$$

The similarity of this transformed infinitesimal triangle and the original one provides an alternate way of seeing that angles are preserved in analytic complex transformations. This means that the orthogonality between stream functions and equipotentials is invariant under a complex variable transformation.

4–6 General Schwarz transformation. A transformation that will reduce any number of rectilinear boundaries in the z_1 plane to a single straight line boundary in the z_2 plane is due to Schwarz. The Schwarz transformation will map the inside of a polygon in the z_1 plane (although the polygon need not be closed) into the upper half of the z_2 plane. It is based on the special transformation, useful in itself, which changes the size of an angle whose vertex is at the origin in the z_1 plane, as shown in Fig. 4–3.

Consider the simple transformation

$$z_1 = z_2^{\beta}, \tag{4–20}$$

where β is real but not necessarily an integral or a rational number. By this transformation points on the positive real axis are mapped on the positive real axis, although the scale along the axis is changed by raising x_1 to the $1/\beta$ power, or at least a branch of the transformation can be chosen where this is so. On the other hand, for points lying on the negative real axis in the z_2 plane ($z_2 = r_2 e^{i\pi}$), z_1 is complex, since by the transformation z_1 is equal to $r_2^{\beta} e^{i\pi\beta}$. Hence the negative real axis of the z_2 plane is the mapping of a straight line in the z_1 plane, as required by the conformal properties of the transformation, but this line will make an angle $\pi\beta$ with the positive real z_1 axis. The transformation of Eq. (4–20) with $\beta < 1$ therefore maps the area of the upper half of the z_1 plane lying between $\theta_1 = 0$ and $\theta_1 = \pi\beta$ into the entire upper half of the z_2 plane.

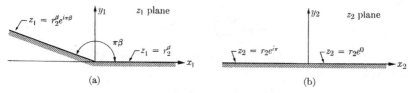

FIG. 4–3. Schwarz transformation for a single angle.

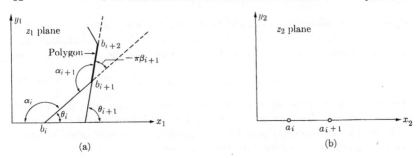

FIG. 4–4. The general Schwarz transformation.

Of course, β may be greater than one, in which case the angle is obtuse. The transformation has a branch point at $z_2 = 0$, at which angles are obviously not preserved, but it is analytic everywhere else.

In the more general case of Fig. 4–4 there are a number of points b_i in the z_1 plane which are the corners of a polygon whose interior angles are α_i. We wish to map the interior of the polygon into the upper half of the z_2 plane. Consider a transformation defined by the differential equation

$$\frac{dz_1}{dz_2} = C_1 \prod_{i=1}^{n} (z_2 - a_i)^{\beta_i}. \tag{4–21}$$

Here C_1 is a constant, possibly complex. This transformation is analytic everywhere except at the points $z_2 = a_i$, which are real but otherwise as yet undetermined. Hence by the conformal properties of such a transformation the real z_2 axis, $z_2 = x_2$, will consist of mappings of straight line segments in the z_1 plane. The angles which each of these straight line segments make with the real axis will be given by the argument of dz_1/dz_2 evaluated in the segment in question. We may take the argument of Eq. (4–21):

$$\arg\left(\frac{dz_1}{dz_2}\right) = \arg C_1 + \beta_1 \arg(z_2 - a_1)$$

$$+ \beta_2 \arg(z_2 - a_2) + \cdots + \beta_n \arg(z_2 - a_n). \tag{4–22}$$

But

$$\arg\left(\frac{dz_1}{dz_2}\right) = \arg dz_1. \tag{4–23}$$

since dz_2 is real. Therefore Eq. (4–23) becomes, when evaluated in the ith interval,

$$\arg\left(\frac{dz_1}{dz_2}\right) = \tan^{-1}\left(\frac{dy_1}{dx_1}\right) = \theta_i. \tag{4–24}$$

Now when z_2 lies on the real axis between a_i and a_{i+1}, the argument of z_2 minus any point to the left is zero and the argument of z_2 minus any point to the right is π. Therefore a combination of Eqs. (4–22) and (4–24) leads to

$$\theta_i = \arg C_1 + \pi(\beta_{i+1} + \beta_{i+2} + \cdots \beta_n). \qquad (4\text{–}25)$$

Thus all points of the real axis segment $a_{i+1} - a_i$ are mappings of a line segment with slope θ_i in the z_1 plane. If we subtract Eq. (4–25) from a similar expression for θ_{i+1}, we obtain

$$\theta_{i+1} - \theta_i = -\pi\beta_{i+1}. \qquad (4\text{–}26)$$

From the geometry of Fig. 4–4(a) we see that this angle difference of $-\pi\beta_{i+1}$ at the point b_{i+1} is related to the interior angle α_{i+1} at each corner by the relation

$$\alpha_{i+1} = \pi + \pi\beta_{i+1}. \qquad (4\text{–}27)$$

For convenience, we may change the subscript to i and solve for β_i,

$$\beta_i = \frac{\alpha_i}{\pi} - 1. \qquad (4\text{–}28)$$

Hence Eq. (4–21) becomes

$$\frac{dz_1}{dz_2} = C_1 \prod_{i=1}^{n} (z_2 - a_i)^{(\alpha_i/\pi - 1)}, \qquad (4\text{–}29)$$

where the scale factor C_1 gives both the relative scale and the relative angular orientation of the two geometries. This is the required transformation for a polygon with internal angles α_i.

In general, the Schwarz transformation is a useful one provided that Eq. (4–29) is integrable in terms of elementary functions. This is possible, with the exception of special cases, only when the angles are multiples of 90° and not more than two corners are involved. One further difficulty in the practical application of the Schwarz transformation is that the resultant transformation is given in terms of $z_1 = f(z_2)$ with the coordinate along the real axis in the z_2 plane as the independent variable, rather than in terms of the coordinates of the z_1 plane, or the problem polygon, as the independent variables. Therefore considerable computational labor is often necessary to find out what the coordinates a_i in the z_2 plane actually are in terms of the geometry of the given problem. Once the a_i are determined, the remainder of the solution of the potential problem is usually simple. The consideration of some special simple cases will serve to indicate the power of the method and to illustrate its use.

4–7 Single-angle transformations. A single angle can always be transformed into the origin of the z_2 plane, so that $a_i = 0$ in each case. For some special angles Eq. (4–29) is immediately integrable:

(a) $\alpha = \pi$. The integration of Eq. (4–29) gives

$$z_1 = C_1 z_2 + C_2. \tag{4–30}$$

This is simply a uniform translation and rotation, and is of no physical interest.

(b) $\alpha = \pi/2$. The integration of Eq. (4–29) gives

$$z_1 = C_3 z_2^{1/2} + C_2. \tag{4–31}$$

If we assume that the constant of translation C_2 is zero, Eq. (4–31) will map the first quadrant of the z_1 plane into the upper half of the z_2 plane. If, for example, the complex potential in the z_2 plane is given by the complex potential solution corresponding to a uniform field **E**,

$$W = -|\mathbf{E}|z_2 = \phi + i\psi,$$

$$\phi = -|\mathbf{E}|x_2, \quad \psi = -|\mathbf{E}|y_2, \tag{4–10}$$

then W in the z_1 plane, according to Eq. (4–31), is

$$W = -C_4|\mathbf{E}|z_1^2$$

$$= -C_4|\mathbf{E}|(x_1^2 - y_1^2 + 2ix_1y_1),$$

$$\phi = -C_4|\mathbf{E}|(x_1^2 - y_1^2),$$

$$\psi = -C_4|\mathbf{E}|(2x_1y_1).$$

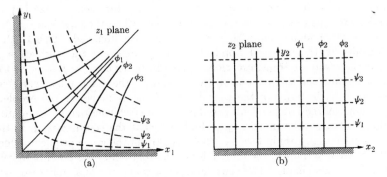

Fig. 4–5. Schwarz transformation for $\alpha = \pi/2$.

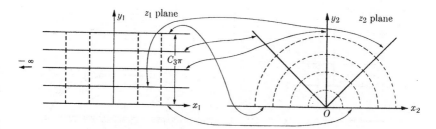

Fig. 4-6. Schwarz transformation for $\alpha = 0$.

This will solve the problem of the charged rectangular boundary (see Fig. 4-5) or, if the interpretations of ϕ and ψ are interchanged, problems involving charged hyperbolic cylinders.

If the complex potential in the z_2 plane is taken to be the logarithmic potential corresponding to a line charge, Eq. (4-12), and if the transformation of Eq. (4-31) is then applied, we obtain the two-dimensional Green's function for an inside rectangular corner, provided that we have translated the line charge into the upper half of the z_2 plane. This same transformation will give the Green's function for a problem having hyperbolic cylindrical boundaries, and thus problems involving such geometries are amenable to solution.

(c) $\alpha = 0$. In this case, the integration yields

$$z_1 = C_3 \ln z_2 = C_3 \ln r_2 + C_3 i \theta_2, \tag{4-32}$$

if we omit the translation constant. If C_3 is real the real part of z_1 is $C_3 \ln r_2$, the positive real z_2 axis is the mapping of the whole real z_1 axis, and the upper half z_2 plane maps into a strip of width $C_3 \pi$, as in Fig. 4-6. The transformation can be visualized by considering the origin in the z_2 plane to be pushed to minus infinity, and the negative real axis of the z_2 plane to be revolved clockwise to a position parallel to the positive real axis but located above it a distance $C_3 \pi$, as seen in Fig. 4-6. This transformation thus results in a periodic configuration in the z_1 plane; of this the strip $C_3 \pi$ wide is the first repeat. The upper half of the z_2 plane is the mapping of the first strip of this configuration in the z_1 plane. The lower half of the z_2 plane is the mapping of the strip lying between $y_1 = C_3 \pi$ and $y_1 = 2 C_3 \pi$ in the z_1 plane, and so on. This transformation is a very useful one in the solution of potential problems involving grids, repeating capacitor plates, and other geometries that repeat in one direction.

4-8 Multiple-angle transformations. A simple example is one of the most useful. If two vertical lines in the z_1 plane are rotated into the positive and negative real axes, respectively, then the upper half of the

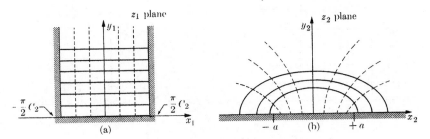

FIG. 4–7. Two-angle transformation.

z_2 plane will be the mapping of the vertically oriented, semi-infinite strip seen in Fig. 4–7(a). The two transformed corners may be taken at $z_2 = \pm a$, and the differential equation becomes

$$\frac{dz_1}{dz_2} = C_1(z_2 - a)^{-1/2}(z_2 + a)^{-1/2} = \frac{C_1}{\sqrt{z_2^2 - a^2}} = \frac{C_2}{\sqrt{a^2 - z_2^2}}. \quad (4\text{-}33)$$

The relation $C_1 = iC_2$ has been introduced to rotate the figure 90° to the orientation shown. Integration yields

$$z_1 = C_2 \sin^{-1}\left(\frac{z_2}{a}\right), \qquad z_2 = a \sin\left(\frac{z_1}{C_2}\right). \quad (4\text{-}34)$$

In practice this transformation is most often used to transform a solution in the z_1 plane into the z_2 plane. If we consider a uniform complex potential field W in the z_1 plane,

$$W = -|\mathbf{E}|z_1, \quad (4\text{-}10)$$

mapped into the z_2 plane by Eq. (4–34), we see that we have obtained the cross section of the potential around a charged conducting strip of width $2a$, or, exchanging potential and stream functions, the potential due to a slot in a conducting sheet. The major axis of the slot or strip is normal to the plane of Fig. 4–7. If the real and imaginary parts of the transformed W are given sets of constant values they will characterize the field of the arrangement. These equations turn out to be the equations of confocal elliptic and hyperbolic cylinders, as indicated in Fig. 4–7.

Many other practical examples of double-angle transformations are readily integrable, and still others may be most easily found by simple successive transformations, as indicated in the references listed at the end of this chapter. Frequently the solution appears in a form that is deceptively simple, since, as we have noted earlier, it is sometimes very difficult to solve for the z_2 coordinates as a function of z_1. Despite these difficulties, the method is obviously a powerful one.

4–9 Direct solution of Laplace's equation by the method of harmonics. Many two-dimensional problems are at least as conveniently solvable by methods for which analogs do exist in three dimensions as by complex potential methods. Two-dimensional inversions and images are useful special cases of general methods already treated. The solution of the two-dimensional Laplace equation by separation of variables for plane polar coordinates is particularly useful, since it has general validity whenever circular or radial boundaries are encountered. Let us consider the application of this method in some detail.

Laplace's equation in the plane polar coordinates r and θ is

$$r \frac{\partial}{\partial r}\left(r \frac{\partial \phi}{\partial r}\right) + \frac{\partial^2 \phi}{\partial \theta^2} = 0. \tag{4–35}$$

To achieve separation we let $\phi(r, \theta) = R(r)\Theta(\theta)$, substitute in Eq. (4–35) and divide by ϕ. The result is

$$\frac{r}{R} \frac{\partial}{\partial r}\left(r \frac{\partial R}{\partial r}\right) + \frac{1}{\Theta} \frac{\partial^2 \Theta}{\partial \theta^2} = 0. \tag{4–36}$$

The two terms must be individually constant, and we may choose the sign of the separation parameter so as to give circular functions in the angle θ. Since the range of θ is always limited, the boundaries of θ are always "closed," so to speak, and only certain values of the parameter will be allowed, just as in the first treatment of the example of Section 3–6. In other words, we may set the first term of Eq. (4–36) equal to k_n^2, to obtain

$$\frac{\partial^2 \Theta}{\partial \theta^2} + k_n^2 \Theta = 0,$$

$$r \frac{\partial}{\partial r}\left(r \frac{\partial R}{\partial r}\right) - k_n^2 R = 0, \tag{4–37}$$

where the separation parameter k_n^2 is in general restricted to discrete values. For $k_n \neq 0$ the solutions are

$$\Theta_n = A_n \cos k_n \theta + B_n \sin k_n \theta,$$

$$R_n = C_n r^{k_n} + D_n r^{-k_n},$$

and if $k_n = 0$,

$$\Theta_0 = E + F\theta,$$

$$R_0 = G + H \ln r.$$

The general solution is obtained by a linear superposition of the individual ("harmonic") solutions:

$$\phi = \sum_{n=0}^{\infty} R_n \Theta_n$$

$$= \sum_{n=1}^{\infty} (A_n \cos k_n\theta + B_n \sin k_n\theta)(C_n r^{k_n} + D_n r^{-k_n})$$

$$+ (E + F\theta)(G + H \ln r). \tag{4–38}$$

In order to apply Eq. (4–38) to the solution of a practical problem, we shall first express certain already known potentials in this form, and then superpose additional potentials, with undetermined coefficients, of the same general form. The coefficients are determined by the boundary conditions of the given problem, just as in Section 3–6.

4–10 Illustration: Line charge and dielectric cylinder. To solve the problem of a line charge located at a distance r_0 from the axis of a dielectric cylinder of radius a and specific inductive capacity κ, as seen in Fig. 4–9, let us first express the logarithmic potential of the line charge alone in the same form as Eq. (4–38). This amounts to shifting the origin of the potential, as indicated in Fig. 4–8. We may omit the arbitrary potential base, so that Eq. (4–12) reduces to

$$\phi = - \frac{q}{2\pi\epsilon_0} \ln R. \tag{4–39}$$

Since Eq. (4–38) is in general nonsingular except at the origin $r = 0$, Eq. (4–39) cannot in general be expressed by a single expansion about the new origin, but must be written as two different solutions, one valid in the region where $r < r_0$ and one valid where $r > r_0$. These two solutions must fit together at $r = r_0$ in such a way that the derivative shall be discontinuous only at the point where the line charge is located, and con-

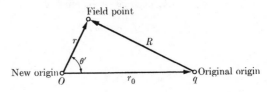

Fig. 4–8. Shifting the origin of coordinates for the potential.

tinuous at all other points. The discontinuity is such that the total flux emerging from that point corresponds to the value of the line charge per unit length.

The logarithmic potential Eq. (4–39) of a line charge at the origin can be put into the form of Eq. (4–38) of an isolated line charge located at $\theta = 0$ and $r = r_0$, as in Fig. 4–8, if we express the radial distance R by the law of cosines, $R = (r_0^2 + r^2 - 2rr_0 \cos \theta)^{1/2}$, and then expand in a power series in r/r_0 for use where $r < r_0$, and in a power series in r_0/r to use where $r > r_0$. Both series converge within their respective ranges of validity. The result gives us the potential due to a line charge:

$$\phi_{0<r<r_0} = \frac{q}{2\pi\epsilon_0} \left\{ \sum_{n=1}^{\infty} \frac{1}{n} \left(\frac{r}{r_0} \right)^n \cos n\theta - \ln r_0 \right\}, \qquad (4\text{–}40)$$

$$\phi_{r_0<r<\infty} = \frac{q}{2\pi\epsilon_0} \left\{ \sum_{n=1}^{\infty} \frac{1}{n} \left(\frac{r_0}{r} \right)^n \cos n\theta - \ln r \right\}. \qquad (4\text{–}41)$$

For the problem of the line charge and dielectric cylinder we shall choose the origin of the polar coordinate system at the center of the dielectric cylinder, with the radius vector corresponding to $\theta = 0$ passing through the charge, as shown in Fig. 4–9. This is the same coordinate system as that in which Eqs. (4–40) and (4–41) are written. To satisfy the boundary condition at the surface of the cylinder $r = a$ we shall superpose on the line charge solution (4–40) a general solution of the type of Eq. (4–38) with $k_n = n$ and with undetermined coefficients A_n, B_n, E, and F, and make a separation of the potential into two parts ϕ_1 and ϕ_2 to be valid outside and inside the cylinder respectively. The coefficients are to be determined so as to account for the polarization of the dielectric cylinder,

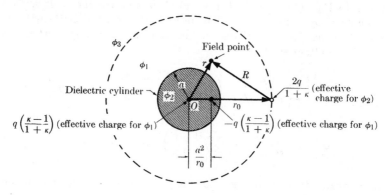

FIG. 4–9. Line charge and dielectric cylinder.

and the separation is made to ensure a finite value for the potential at the origin and convergence of the second series in each expression.

$$\underset{a<r<r_0}{\phi_1} = \frac{q}{2\pi\epsilon_0} \left\{ \sum_{n=1}^{\infty} \frac{1}{n} \left(\frac{r}{r_0}\right)^n \cos n\theta - \ln r_0 \right\}$$

$$+ \sum_{n=1}^{\infty} B_n r^{-n} \cos n\theta + F, \qquad (4\text{-}42)$$

$$\underset{0<r<a}{\phi_2} = \frac{q}{2\pi\epsilon_0} \left\{ \sum_{n=1}^{\infty} \frac{1}{n} \left(\frac{r}{r_0}\right)^n \cos n\theta - \ln r_0 \right\}$$

$$+ \sum_{n=1}^{\infty} A_n r^n \cos n\theta + E. \qquad (4\text{-}43)$$

Since the effect of the induced polarization charges in the cylinder is non-singular both at the origin and at infinity, and the solution is obviously symmetric about the line $\theta = 0$, it has already been possible to simplify the expression (4-38) by omitting the logarithm, the angle and its sine, and by using only positive or negative powers of r as necessary to assure convergence.

For any angle θ the boundary conditions of Section 2-2 for the surface of the dielectric $r = a$ are just

$$\phi_1 = \phi_2, \qquad \frac{\partial\phi_1}{\partial r} = \kappa \frac{\partial\phi_2}{\partial r}. \qquad (4\text{-}44)$$

We can evaluate the coefficients A_n, B_n, E, and F by applying these conditions to Eqs. (4-42) and (4-43) and then equating the coefficients of $\cos n\theta$, term by term, to zero. This procedure is justified mathematically by the fact that these Fourier series functions form a complete orthogonal set. The resultant solution is

$$\underset{a<r<r_0}{\phi_1} = \frac{q}{2\pi\epsilon_0} \left\{ \sum_{n=1}^{\infty} \frac{1}{n} \left[\left(\frac{r}{r_0}\right)^n + \left(\frac{1-\kappa}{1+\kappa}\right)\left(\frac{a^2}{r_0}\right)^n \frac{1}{r^n} \right] \cos n\theta - \ln r_0 \right\}, \quad (4\text{-}45)$$

$$\underset{0<r<a}{\phi_2} = \frac{q}{\pi\epsilon_0(1+\kappa)} \sum_{n=1}^{\infty} \frac{1}{n} \left(\frac{r}{r_0}\right)^n \cos n\theta - \frac{q}{2\pi\epsilon_0} \ln r_0. \quad (4\text{-}46)$$

For r greater than r_0 the solution may be written down immediately, analogously to Eqs. (4-40) and (4-41).

If we add zero to Eq. (4–45) by adding and subtracting

$$q \, \frac{(\kappa - 1)}{(1 + \kappa)} \, \frac{\ln r}{2\pi\epsilon_0},$$

we see that the potential ϕ_1 outside the cylinder corresponds to an effective line charge arrangement, with the role of the cylinder taken by two line charges. This arrangement consists of an effective charge $q(\kappa - 1)/(1 + \kappa)$ located at the origin, an effective charge $-q(\kappa - 1)/(1 + \kappa)$ located at the inversion point of the actual external charge, and the actual charge. The inversion point lies on the vector r_0 at a distance a^2/r_0 from the origin. This problem could therefore have been solved by the method of images, a fact which can be verified directly by the use of the logarithmic potentials. On the other hand, the potential ϕ_2 inside the cylinder is seen to correspond to a single effective charge at the position of the actual charge but of strength $2q/(1 + \kappa)$, except for an additive constant. The potential everywhere, then, is equivalent to some line charge arrangement and the dielectric cylinder absent.

4–11 Line charge in an angle between two conductors. As another example of the solution of a problem in terms of circular harmonics, let us consider a wedge-shaped region bounded by grounded conducting surfaces intersecting at the origin with an interior angle α, as in Fig. 4–10, together with a line charge of strength q per unit length located at the point (r_0, β) within the wedge. The solution of this problem will give the Green's function for the region bounded by the intersecting conducting planes. It is again clear that we cannot hope to express a solution by means of a single equation valid throughout the region from $r = 0$ to $r = \infty$. The solution must be written in two parts, one valid for $r < r_0$ and the other valid for $r > r_0$, joined together at the cylindrical surface $r = r_0$ by the flux condition corresponding to the charge q.

Since the potential vanishes on the boundaries $\theta = 0$ and $\theta = \alpha$, the angular part of the solution must be of the form $\sin (n\pi\theta/\alpha)$. Thus, in

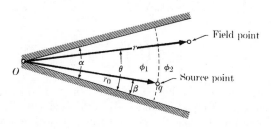

Fig. 4–10. Line charge parallel to two intersecting plane conductors.

Eq. (4–38), $k_n = n\pi/\alpha$. The potentials are then formally:

$$\phi_1 \atop r<r_0 = \sum_{n=1}^{\infty} C_n \left(\frac{r}{r_0}\right)^{n\pi/\alpha} \sin \frac{n\pi}{\alpha}\, \theta,$$

$$\phi_2 \atop r>r_0 = \sum_{n=1}^{\infty} D_n \left(\frac{r_0}{r}\right)^{n\pi/\alpha} \sin \frac{n\pi}{\alpha}\, \theta. \tag{4–47}$$

The coefficients C_n and D_n must be equal in order to assure continuity of the potentials across the cylindrical surface $r = r_0$, so that we shall have solved the problem if we determine the C_n.

The line charge is equivalent to a surface charge $\sigma(\theta)$ on the cylinder $r = r_0$, where $\sigma(\theta) = q\,\delta(\theta - \beta)$. The angular δ-function can be defined by the equations

$$r_0 \int_0^{\alpha} \delta(\theta - \beta)\, d\theta = 1,$$

$$\int_0^{\alpha} f(\theta)\,\delta(\theta - \beta)\, d\theta = f(\beta)/r_0, \qquad 0 < \beta < \alpha. \tag{4–48}$$

The flux condition is thus

$$q\,\delta(\theta - \beta) = -\epsilon_0 \left[\frac{\partial \phi_2}{\partial r} - \frac{\partial \phi_1}{\partial r}\right]_{r=r_0}. \tag{4–49}$$

The determination of the coefficients is almost identical with that of Section 3–6, and leads to

$$C_n = \frac{q}{n\pi\epsilon_0} \sin\left(\frac{n\pi\beta}{\alpha}\right). \tag{4–50}$$

The complete solution is therefore

$$\phi_1 \atop r<r_0 = \frac{q}{\pi\epsilon_0} \sum_{n=1}^{\infty} \frac{1}{n} \left(\frac{r}{r_0}\right)^{n\pi/\alpha} \sin \frac{n\pi\beta}{\alpha} \sin \frac{n\pi\theta}{\alpha},$$

$$\phi_2 \atop r>r_0 = \frac{q}{\pi\epsilon_0} \sum_{n=1}^{\infty} \frac{1}{n} \left(\frac{r_0}{r}\right)^{n\pi/\alpha} \sin \frac{n\pi\beta}{\alpha} \sin \frac{n\pi\theta}{\alpha}. \tag{4–51}$$

Just as in the case of parallel plates, this solution is the desired Green's function when q is set equal to unity. It is thus evident that the method is a general one for deriving the Green's function within a set of boundaries corresponding to equi-coordinate planes.

Suggested References

W. R. Smythe, *Static and Dynamic Electricity*. Chapter IV is devoted to two-dimensional potential problems, with many examples.

J. C. Maxwell, *Electricity and Magnetism*, Vol. 1. There have been many elaborations of potential theory since Maxwell's time, but this remains one of the best fundamental texts.

J. H. Jeans, *The Mathematical Theory of Electricity and Magnetism*. Follows the general presentation of Maxwell, and gives many examples.

E. Weber, *Electro-Magnetic Fields*, Vol. 1. A more modern treatment.

L. A. Pipes, *Applied Mathematics for Engineers and Physicists*. Chapter XX is a good example of the clear and simple treatment of conjugate functions found in several books on applied mathematics.

H. and B. W. Jeffreys, *Methods of Mathematical Physics*. Chapter 13 on conformal representation is particularly useful.

R. Courant, *Dirichlet's Principle*. Treats conformal mapping and relevant existence theorems.

Exercises

1. By inversion, find the law of image formation for a line charge parallel to a conducting circular cylinder. Apply this method to the case of a large cylinder of radius b containing a smaller cylinder of radius a, the distance between their axes being $c < b - a$. Find the capacity per unit length, and show that for $c = 0$ the result reduces to that for coaxial cylinders.

2. Derive the two-dimensional form of Green's boundary value theorem: if $\phi(x, y)$ is the two-dimensional potential, show that

$$\phi(x', y') = \frac{1}{2\pi} \left\{ \int_S (\log r) \, \nabla^2 \phi \, dS - \oint_c \left(\log r \, \frac{\partial \phi}{\partial n} - \phi \, \frac{\partial (\log r)}{\partial n} \right) dl \right\},$$

where S is the area bounded by the contour C and r is the distance between the point x, y and the point x', y'. n is the outward normal.

3. Find the field surrounding a charged conducting cylinder whose cross section is an ellipse.

4. Consider two planes intersecting at right angles raised to potentials $V/2$ and $-V/2$, respectively. Calculate the electrostatic field.

5. The transformation

$$z_2 = \frac{a}{2} \left(\frac{z_1}{a_1} + \frac{a_1}{z_1} \right)$$

transforms the region outside the cylinder $r_1 = a_1$ in the z_1 plane into the entire z_2 plane with a cut along the real axis for $-a < x < a$. By transforming the complex potential function corresponding to a conducting cylinder in a uniform field, find the complex potential function corresponding to a strip of width $2a$ with its plane (a) in the direction of an applied field $\mathbf{E}$, and (b) perpendicular to an applied field $\mathbf{E}$.

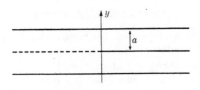

FIGURE 4–11

6. Consider a parallel plate capacitor, $y = \pm a$, of infinite extent, and $y = 0$ semi-infinite, as indicated in Fig. 4–11. The two outside plates are at the same potential. Calculate the capacity per unit length along z contributed by the edge effect, i.e., the difference between the actual capacity per unit length and that estimated by assuming zero field for $x < 0$ and uniform fields for $x > 0$.

7. An infinite circular cylindrical sheet of radius a is divided longitudinally into quarters which are raised to potentials $V, 0, -V, 0$, respectively. Show that the potential inside the cylinder is given by

$$\phi = \frac{V}{\pi} \left\{ \tan^{-1}\left(\frac{2ay}{a^2 - r^2}\right) + \tan^{-1}\left(\frac{2ax}{a^2 - r^2}\right) \right\}.$$

What is the potential outside?

8. Consider the region of space between the cylinder $x^2 + y^2 = b^2$ and the xz-plane. All the curved boundary and that portion of the plane boundary for which $a < |x| < b$ are at zero potential. That portion of the plane boundary for which $-a < |x| < a$ is at potential V. Show that the equation of the lines of force in the region for which $a < r < b$ is

$$\sum_n \frac{a^n}{n}\left[\frac{1}{r^n} + \left(\frac{r}{b^2}\right)^n\right]\cos n\theta = \text{constant},$$

where only odd values of n are taken.

9. Let $\phi(x, y)$ be the potential in a two-dimensional field. Let $F(x) = \phi(x, 0)$ and F_n be the nth derivative of F with respect to x. Show that if $\phi(x, y) = \phi(x, -y)$,

$$\phi(x, y) = \Sigma A_n y^{2n} F_{2n},$$

where $A_n = (-1)^n/(2n)!$.

CHAPTER 5

THREE-DIMENSIONAL POTENTIAL PROBLEMS

Laplace's equation may readily be written in any orthogonal coordinate system for which the infinitesimal line elements are known: the operator ∇^2 is just the divergence of the gradient, and the application of the divergence theorem to the gradient over the surface of an infinitesimal volume element yields the required expression.* In a number of these systems the equation is separable, so that the methods of Section 3–6 can be applied. The two coordinate systems treated in this chapter further illustrate the general method, and correspond to geometrical configurations often encountered in practice, namely, spheres and circular cylinders, or parts thereof. Spherical coordinates also furnish a particularly useful representation of the potential due to an arbitrary distribution of charge confined to a region near the origin of coordinates.

5–1 The solution of Laplace's equation in spherical coordinates. Expressed in spherical polar coordinates, Laplace's equation becomes

$$\nabla^2 \phi = \frac{1}{r^2} \frac{\partial}{\partial r}\left(r^2 \frac{\partial \phi}{\partial r}\right) + \frac{1}{r^2 \sin\theta} \frac{\partial}{\partial\theta}\left(\sin\theta \frac{\partial\phi}{\partial\theta}\right) + \frac{1}{r^2 \sin^2\theta} \frac{\partial^2\phi}{\partial\varphi^2} = 0.$$

$$(5\text{–}1)$$

In order to separate the radial and angular parts of this equation, we put

$$\phi = R(r)Y(\theta, \varphi) \tag{5–2}$$

and proceed in the usual way. The separated equations are

$$\frac{\partial}{\partial r}\left(r^2 \frac{\partial R}{\partial r}\right) - n(n+1)R = 0, \tag{5–3}$$

$$\frac{1}{\sin\theta} \frac{\partial}{\partial\theta}\left(\sin\theta \frac{\partial Y}{\partial\theta}\right) + \frac{1}{\sin^2\theta} \frac{\partial^2 Y}{\partial\varphi^2} + n(n+1)Y = 0. \tag{5–4}$$

The form of the separation constant $n(n+1)$, where n is a real integer, is dictated by the necessity† that there be a regular solution at the singularities of the equation for Y, which occur at $\theta = 0$ and $\theta = \pi$. In general, $Y(\theta, \varphi)$ is known as a spherical harmonic. It has already been proved that

---

* See Appendix III.

† If boundaries are such that $\theta = 0$ and $\theta = \pi$ (i.e., the polar axis) are excluded, then n need not be an integer.

the set of functions $Y_n(\theta, \varphi)$ have orthogonality properties similar to those of the Fourier functions we have considered in connection with two-dimensional solutions. The general solution of the differential equation for the radial part of ϕ, Eq. (5–3), is simply

$$R(r) = A_n r^n + B_n r^{-n-1}. \qquad (5\text{–}5)$$

The spherical surface harmonics can be further separated by means of the substitution

$$Y(\theta, \varphi) = \Theta(\theta)\Phi(\varphi). \qquad (5\text{–}6)$$

If the new separation constant is called m^2, then, in terms of the more convenient polar angle variable, $\mu = \cos\theta$, the two resulting equations are

$$\frac{d}{d\mu}\left[(1 - \mu^2)\frac{d\Theta}{d\mu}\right] + \left[n(n+1) - \frac{m^2}{1 - \mu^2}\right]\Theta = 0, \qquad (5\text{–}7)$$

$$\frac{d^2\Phi}{d\varphi^2} + m^2\Phi = 0. \qquad (5\text{–}8)$$

The solutions of these equations are

$$\Theta = C_n P_n^m(\mu) + D_n Q_n^m(\mu). \qquad (5\text{–}9)$$

$$\begin{aligned}
\Phi &= E_m \cos m\varphi + F_m \sin m\varphi, & m \neq 0, \\
\Phi &= G\varphi + H, & m = 0.
\end{aligned} \right\} \qquad (5\text{–}10)$$

The functions $P_n^m(\cos\theta)$ and $Q_n^m(\cos\theta)$ are the associated Legendre functions of the first and second kind, respectively. Their mathematical properties can be found in numerous references, some of which are listed at the end of the chapter. We need note here only that P_n^m is the solution that is finite for $\mu = \pm 1$, and thus the only solution allowed when the space involved in the problem includes the polar axis.

5–2 The potential of a point charge. For problems having azimuthal symmetry, so that the solution does not depend on the value of the coordinate φ, i.e., $\Phi(\varphi) = $ constant, the separation parameter m is equal to zero. The potential of a point charge at a distance r_0 from the origin of coordinates has such symmetry if the radius vector of the charge is taken as the polar axis. We may obtain the potential of a point charge, expressed in terms of a series expansion in the radial and angular functions obtained in the above separation of coordinates, by expanding the cosine law expression for $1/R$ (see Fig. 5–1) in powers of r/r_0 and r_0/r.

$$\frac{1}{R} = \frac{1}{r_0}\left[\left(\frac{r}{r_0}\right)^2 + 1 - 2\frac{r}{r_0}\cos\theta\right]^{-1/2} = \frac{1}{r}\left[\left(\frac{r_0}{r}\right)^2 + 1 - 2\frac{r_0}{r}\cos\theta\right]^{-1/2}$$

$$(5\text{–}11)$$

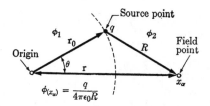

FIG. 5–1. New polar coordinates for the potential due to a point charge.

becomes, according to the region of convergence,

$$\frac{1}{R} = \frac{1}{r_0} \sum_{n=0}^{\infty} \left(\frac{r}{r_0}\right)^n P_n(\mu) = \frac{1}{r} \sum_{n=0}^{\infty} \left(\frac{r_0}{r}\right)^n P_n(\mu). \qquad (5\text{--}12)$$

Equation (5–12) is often taken as the definition of $P_n(\mu)$. It is obvious from carrying out the expansion that these functions are polynomials of degree n in the variable μ. By explicit differentiation of $1/R$ with respect to μ and with respect to the expansion variable, it can be readily shown that $P_n(\mu)$ satisfies Eq. (5–7) with $m = 0$.

For the same reasons as in the harmonic expansions of Chapter 4, we must use two potentials, one valid where r is less than r_0 and one for r greater than r_0. The two potentials are

$$\phi_1 = \frac{q}{4\pi\epsilon_0 r_0} \sum_{n=0}^{\infty} \left(\frac{r}{r_0}\right)^n P_n(\mu), \qquad \phi_2 = \frac{q}{4\pi\epsilon_0 r} \sum_{n=0}^{\infty} \left(\frac{r_0}{r}\right)^n P_n(\mu).$$
$$\scriptstyle r<r_0 \qquad\qquad\qquad\qquad\qquad\qquad\qquad r>r_0$$
$$(5\text{--}13)$$

The resulting potential of the point charge is therefore just a Taylor-Laurent series in r whose coefficients are the Legendre polynomials in $\cos \theta$.

5–3 The potential of a dielectric sphere and a point charge. Problems involving point charges and boundaries which have spherical symmetry can be solved in terms of the functions of the foregoing section. As an example, we shall consider the simple problem of a point charge and a dielectric sphere of radius a, as shown in Fig. 5–2, with r_0 being the dis-

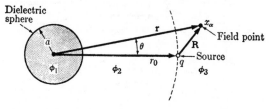

FIG. 5–2. Dielectric sphere and point charge.

tance from the center of the sphere to the point charge. We shall need three expressions for ϕ:

$$\phi_1 \atop 0<r<a = \sum_{n=0}^{\infty} A_n r^n P_n(\mu),$$

$$\phi_2 \atop a<r<r_0 = \frac{q}{4\pi\epsilon_0 r_0} \sum_{n=0}^{\infty} \left(\frac{r}{r_0}\right)^n P_n(\mu) + \sum_{n=0}^{\infty} B_n r^{-n-1} P_n(\mu), \quad (5\text{--}14)$$

$$\phi_3 \atop r_0<r<\infty = \frac{q}{4\pi\epsilon_0 r} \sum_{n=0}^{\infty} \left(\frac{r_0}{r}\right)^n P_n(\mu) + \sum_{n=0}^{\infty} B_n r^{-n-1} P_n(\mu).$$

The fit of ϕ_1 to ϕ_2 at $r = a$ can be made in the same way as for the two-dimensional case. The boundary conditions of Eqs. (2–15) and (2–19) must be satisfied for all values of the angle θ, and the fact that the angular functions are orthogonal makes it possible to equate the terms of the series separately to determine the coefficients A_n and B_n. The fit of ϕ_2 to ϕ_3 is inherent from the nature of the solutions in Eq. (5–13). For $\kappa = \infty$ the solution reduces to that obtained by the inversion process outlined in Chapter 3.

5–4 The potential of a dielectric sphere in a uniform field. As a second example of a problem with spherical geometry, let us consider a dielectric sphere of specific inductive capacity κ in the presence of a uniform field whose force lines are parallel to the x-axis, as shown in Fig. 5–3. The lines of electric displacement **D** are shown.

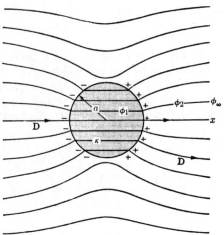

Fig. 5–3. Dielectric sphere in a uniform field.

Since the field at infinity is uniform, the potential is given by

$$\phi_\infty = -E_0 x = -E_0 r \cos\theta = -E_0 r\mu = -E_0 r P_1(\mu). \quad (5\text{-}15)$$

By inspection of Eqs. (5-5) and (5-9), we may write

$$\phi_1 = \sum A_n r^n P_n(\mu),$$

$$\phi_2 = \sum_{n=0}^{\infty} B_n r^{-n-1} P_n(\mu) = E_0 r P_1(\mu), \quad (5\text{-}16)$$

for the potentials inside and outside the sphere. The boundary conditions, $\phi_1 = \phi_2$ and $\kappa(\partial\phi_1/\partial r) = (\partial\phi_2/\partial r)$ at $r = a$, must hold for all values of the angle θ. We therefore evaluate the constants A_n and B_n by equating the coefficients of $P_n(\mu)$ for the same n, and find that

$$A_0 = B_0 = 0 = A_n = B_n \quad \text{for } n \text{ greater than 1,}$$

$$A_1 = \frac{-3E_0}{\kappa + 2}, \quad (5\text{-}17)$$

$$B_1 = \frac{(\kappa - 1)E_0 a^3}{\kappa + 2}.$$

The potentials are therefore

$$\phi_1 = -\frac{3E_0 r}{\kappa + 2} \cos\theta,$$

$$\phi_2 = \frac{(\kappa - 1)}{(\kappa + 2)} \frac{E_0 a^3 \cos\theta}{r^2} - E_0 r \cos\theta. \quad (5\text{-}18)$$

These equations correspond to Eqs. (5-14) in the limit $r_0 \to \infty$. Note that the field $\mathbf{E}$ inside the sphere is uniform, but is smaller than the field at infinity by the ratio $3/(\kappa + 2)$. It is also seen that the induced field of the sphere in the region outside the sphere is that of a dipole whose moment is

$$\mathbf{p} = 4\pi\epsilon_0 a^3 \left(\frac{\kappa - 1}{\kappa + 2}\right) \mathbf{E}_0. \quad (5\text{-}19)$$

Let a quantity L be what is known as the depolarization factor for a dielectric body, defined as

$$\frac{L}{\epsilon_0} = \frac{|\mathbf{E}_0| - |\mathbf{E}_{\text{inside}}|}{|\mathbf{P}_{\text{inside}}|}. \quad (5\text{-}20)$$

It will be remembered that $\mathbf{P} = \epsilon_0(\kappa - 1)\mathbf{E}$, from Eq. (2-12). Then for a sphere $L = \frac{1}{3}$, while for a thin rod oriented parallel to the field $L \simeq 0$, and for a thin disk oriented normal to the field $L = 1$. Thus the electric field within a dielectric body in a uniform filed is always smaller than the field at a large distance, while the dielectric displacement is always larger.

5–5 The potential of an arbitrary axially-symmetric spherical potential distribution. The potential at any point in space corresponding to a given potential distribution, $\phi(a, \theta)$, over a sphere of radius a, can also be written in terms of Legendre polynomials. From Eqs. (5–5) and (5–9), we shall have two potentials, one valid inside and one valid outside the surface:

$$\phi_{r<a} = \sum_{n=0}^{\infty} A_n r^n P_n(\mu),$$

$$\phi_{r>a} = \sum_{n=0}^{\infty} B_n r^{-n-1} P_n(\mu). \tag{5–21}$$

We can determine A_n and B_n by taking advantage of the orthogonality of the functions $P_n(\mu)$ if we know the normalizing factor, i.e., the integral of P_n^2 over the range of its variable. Stated in a general way, so as to include orthogonality, the integral needed is

$$\int_{-1}^{+1} P_n(\mu) P_m(\mu) \, d\mu = \frac{2}{2n+1} \delta_{mn}, \tag{5–22}$$

where $\delta_{mn} = 1$ if $m = n$, and is zero otherwise.

Either of Eqs. (5–21) may then be equated to $\phi(a, \theta)$ for $r = a$, the resulting equation multiplied by $P_m(\mu)$ and integrated from $\mu = -1$ to $\mu = +1$; only one term of the series survives, namely, that for which $n = m$. As a result of solving for A_n,

$$A_n = \frac{2n+1}{2a^n} \int_{-1}^{+1} \phi(a, \theta) P_n(\mu) \, d\mu \tag{5–23}$$

and

$$B_n = a^{2n+1} A_n. \tag{5–24}$$

The potentials therefore become

$$\phi_{r<a} = \sum_{n=0}^{\infty} \frac{2n+1}{2a^n} r^n P_n(\mu) \int_{-1}^{+1} \phi(a, \theta') P_n(\mu') \, d\mu',$$

$$\phi_{r>a} = \sum_{n=0}^{\infty} \frac{2n+1}{2} a^{n+1} r^{-n-1} P_n(\mu) \int_{-1}^{+1} \phi(a, \theta') P_n(\mu') \, d\mu'. \tag{5–25}$$

5-6 The potential of a charged ring. Let us consider the potential of a charged ring, of total charge q, located at a distance r_0 from the origin, as seen in Fig. 5-4, with r_0 making an angle θ_0 with the axis of symmetry. If the distance from the origin along the polar axis is called z, the potential along the polar axis, found by expanding $1/R$ in the Coulomb potential by Eq. (5-13), is

$$\underset{r<r_0}{\phi}\,(z, 0) = \frac{q}{4\pi\epsilon_0 r_0} \sum_{n=0}^{\infty} \left(\frac{z}{r_0}\right)^n P_n\,(\cos\theta_0),$$

$$\underset{r>r_0}{\phi}\,(z, 0) = \frac{q}{4\pi\epsilon_0 r_0} \sum_{n=0}^{\infty} \left(\frac{r_0}{z}\right)^{n+1} P_n\,(\cos\theta_0).$$

$$(5\text{-}26)$$

It is evident from the series of Eq. (5-12) that $P_n(1) = 1$. Therefore the potential at a general point, not lying on the polar axis, may be found by multiplying the nth term in the series by $P_n(\mu)$ and writing r for z:

$$\underset{r<r_0}{\phi}\,(r, \theta) = \frac{q}{4\pi\epsilon_0 r_0} \sum_{n=0}^{\infty} \left(\frac{r}{r_0}\right)^n P_n\,(\cos\theta_0)P_n\,(\cos\theta),$$

$$\underset{r>r_0}{\phi}\,(r, \theta) = \frac{q}{4\pi\epsilon_0 r_0} \sum_{n=0}^{\infty} \left(\frac{r_0}{r}\right)^{n+1} P_n\,(\cos\theta_0)P_n\,(\cos\theta).$$

$$(5\text{-}27)$$

The uniqueness theorem is essential to justify the argument that led to the above result.

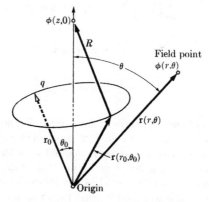

FIG. 5-4. Coordinates for finding the potential due to a charged ring.

5–7 Problems not having axial symmetry. If the geometry of a problem does not have axial symmetry, but does have spherical boundaries, the potential solutions for the problem may be written down formally just as in the examples above. The constant m must be allowed to have values different from zero in order to represent the asymmetry, the spherical harmonics $Y_n^m(\theta, \varphi)$ are characterized by two parameters, and the series solution is a double sum: for each n there is a set of values for m, with the condition that $m \leq n$. The functions Φ are simply the Fourier functions, and the functions Θ are $P_n^m(\mu)$ and $Q_n^m(\mu)$, the associated Legendre functions. Thus the form of the separated functions is different, but no principles are involved which have not been exemplified in problems for which Φ is constant. We shall therefore omit a detailed discussion of such problems. We shall also omit the consideration of problems with conical boundaries that exclude the polar axis from the range of potentials, and which require the use of $Q_n^m(\mu)$ in order that the boundary conditions be satisfied. References for the treatment of many such problems are listed at the end of the chapter.

5–8 The solution of Laplace's equation in cylindrical coordinates. Laplace's equation in the cylindrical coordinates r, φ, z is

$$\frac{1}{r} \frac{\partial}{\partial r} \left(r \frac{\partial \phi}{\partial r} \right) + \frac{1}{r^2} \frac{\partial^2 \phi}{\partial \varphi^2} + \frac{\partial^2 \phi}{\partial z^2} = 0. \tag{5–28}$$

Separating by means of the product functions,

$$\phi(r, \varphi, z) = R(r)\Phi(\varphi)Z(z), \tag{5–29}$$

we obtain

$$r \frac{d}{dr} \left(r \frac{dR}{dr} \right) + (k^2 r^2 - n^2) R = 0, \tag{5–30}$$

$$\frac{d^2 \Phi}{d\varphi^2} + n^2 \Phi = 0, \tag{5–31}$$

$$\frac{d^2 Z}{dz^2} - k^2 Z = 0, \tag{5–32}$$

where k and n are the separation parameters. Equation (5–30) is known as Bessel's equation, and its solutions are called Bessel functions. The character of the solution will depend markedly on the sign of the separation constant, i.e., on whether n and k are real or imaginary. If solutions are desired which are single valued in the azimuth angle φ, then the solutions must be periodic in φ and n must be a real integer. If k is real, the solution $Z(z)$ is a real exponential, and the radial solutions are combinations of the Bessel functions designated by $J_n(kr)$ and $N_n(kr)$. Excellent treatments of Bessel functions and their properties are listed at the end of the chapter. We need remark here only that J_n and N_n are oscillatory functions of their arguments, that both go to zero as $kr \to \infty$, and that J_n is the solution which is regular at $r = 0$, a point for which Eq. (5–30) has a singularity. For real n and k, then, the integrals are of the form:

$$R(r) = A_n J_n(kr) + B_n N_n(kr), \quad k \neq 0,$$

$$R(r) = Ar^n + Br^{-n}, \qquad\qquad k = 0,$$

$$\Phi(\varphi) = C_n \cos n\varphi + D_n \sin n\varphi, \quad n \neq 0,$$

$$\Phi(\varphi) = C\varphi + D, \qquad\qquad\quad n = 0, \qquad\qquad (5\text{–}33)$$

$$Z(z) = E_k e^{kz} + F_k e^{-kz}, \qquad k \neq 0,$$

$$Z(z) = Ez + F, \qquad\qquad\quad k = 0.$$

If n and k are both zero,

$$\phi = (A \ln r + B)(C\varphi + D)(Ez + F).$$

We shall illustrate the use of these cylindrical functions by outlining the solution for a sample problem. Consider a case which has azimuthal symmetry, so that $n = 0$. If the solution extends to infinity radially, so that there are no cylindrical boundaries that might impose restrictions on the radial function $R(r)$, there are correspondingly no restrictions on k and the solution involves an integral over all k. The integral will have the form

$$\int_0^\infty e^{\pm kz} f(k) J_0(kr) \, dk. \qquad\qquad (5\text{–}34)$$

Instead of determining a set of coefficients in a sum, we have now to determine the value of the function $f(k)$. The potential of a point charge may be expressed by an integral of this kind and, in fact, is given in terms of a mathematical identity

$$\frac{1}{R} = \frac{1}{\sqrt{r^2 + z^2}} = \int_0^\infty e^{\pm kz} J_0(kr) \, dk, \qquad\qquad (5\text{–}35)$$

where the exponential is positive for $z < 0$ and negative for $z > 0$. The Coulomb potential $\phi = q/4\pi\epsilon_0 R$ is

$$\phi = \frac{q}{4\pi\epsilon_0} \int_0^\infty e^{\pm kz} J_0(kr) \, dk. \qquad\qquad (5\text{–}36)$$

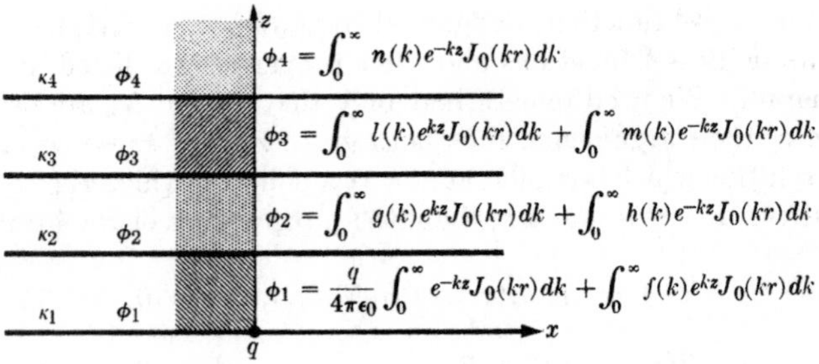

Fig. 5–5. Parallel layers of dielectric materials under the influence of a point charge q.

The potential of Eq. (5–36) can then be used in combination with the induced potential of the form Eq. (5–34) to write down the solution of a problem corresponding to plane boundaries normal to the z-axis and under the influence of a point charge located at the origin. This layer structure, shown in Fig. 5–5, composed of several layers of different inductive capacities, κ_1, κ_2, etc., has the potentials shown in the figure. If we apply the boundary conditions at all of the interfaces and equate the functions under the integral sign, there will be a sufficient number of equations to determine the functions of k, and therefore the solution in every layer.

If the cylindrical solution is required to be periodic in the z direction, k must be imaginary, and the appropriate solutions of the radial equation are Bessel functions of an imaginary argument. It is customary to substitute ik for k, so that Eq. (5–30) becomes what is called the modified Bessel's equation,

$$r \frac{d}{dr}\left(r \frac{dR}{dr}\right) - (k^2 r^2 + n^2) R = 0. \tag{5–37}$$

The required solutions are designated by $I_n(kr)$ and $K_n(kr)$, which are related to the Bessel functions of imaginary argument tabulated by Jahnke and Emde in *Tables of Functions:*

$$I_n(x) = i^{-n} J_n(ix), \qquad (2/\pi) K_n(x) = i^{n+1} H_n^{(1)}(ix) = (-i)^{n+1} H_n^{(2)}(-ix). \tag{5–38}$$

The nature of the functions is indicated by the fact that for large x,

$$I_n(x) \to \left(\frac{1}{2\pi x}\right)^{1/2} e^x, \qquad K_n(x) \to \left(\frac{\pi}{2x}\right)^{1/2} e^{-x}. \tag{5–39}$$

Just as in rectangular and spherical coordinates, we note that it is impossible for all three factors of Eq. (5–29) to be oscillatory functions of their respective variables.

5–9 Application of cylindrical solutions to potential problems. Lists of the mathematical properties of Bessel functions necessary for the solution of actual physical problems are found in the references. To solve problems involving conducting cylindrical boundaries, we must make use of tables, much as we make use of trigonometric tables for circular functions. Inside a grounded cylindrical conductor of radius r_0, for example, each solution must satisfy the condition $J_n(kr_0) = 0$, so that k is limited to a discrete set of values k_l, where l is an index numbering the zeros of J_n from 1 to infinity. The entire solution is, in general, a double sum over n and l, with coefficients determined so as to satisfy the boundary conditions. The orthogonality of the functions follows from the general proof of Section 3–6, or it may be shown from the differential equation (5–30) that

$$\int_0^{r_0} J_n(k_l r) J_n(k_{l'} r) r \, dr = 0$$

if $l \neq l'$. The orthogonality of the partial potentials corresponding to different n is obvious from the form of Φ.

The method can be adequately illustrated even with loss of complete generality by the problem of a point charge at the origin of coordinates within a grounded conducting cylinder about the z-axis. The simplification here arises from the azimuthal symmetry, so that the entire potential is independent of φ, and the radial solutions are confined to $J_0(k_l r)$. The desired solution has the form

$$\phi_1 = \sum_{l=1}^{\infty} A_l e^{k_l z} J_0(k_l r), \qquad \text{for} \quad z < 0,$$

$$\phi_2 = \sum_{l=1}^{\infty} A_l e^{-k_l z} J_0(k_l r), \qquad \text{for} \quad z > 0,$$

(5–40)

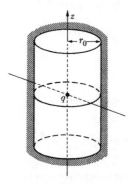

Fig. 5–6. Point charge inside a grounded conducting cylinder.

where k_l is such that at the radius of the cylinder $r = r_0$, $J_0(k_l r_0) = 0$. The flux condition at the plane $z = 0$ can be stated by means of a two-dimensional δ-function defined by the relation

$$2\pi \int_0^{r_0} \delta(r) r \, dr = 1, \qquad (5\text{--}41)$$

so that

$$\frac{q \, \delta(r)}{e_0} = \left[\frac{\partial \phi_1}{\partial z} - \frac{\partial \phi_2}{\partial z} \right]_{z=0} \qquad (5\text{--}42)$$

$$= 2 \sum k_l A_l J_0(k_l r).$$

If we multiply both sides of Eq. (5–42) by $J_0(k_{l'} r) r \, dr$ and integrate over r, we obtain, using the orthogonality properties of Bessel functions,

$$\frac{q J_0(0)}{2\pi \epsilon_0} = 2 k_l A_l \int_0^{r_0} [J_0(k_l r)]^2 r \, dr,$$

or

$$A_l = \frac{q/\epsilon_0}{4\pi k_l} \frac{J_0(0)}{\int_0^{r_0} [J_0(k_l r)]^2 r \, dr}. \qquad (5\text{--}43)$$

Now the Bessel functions are defined in such a way that $J_0(0) = 1$, and it may be shown by multiplying Eq. (5–30) by $(dJ_n/dr) r^2 \, dr$ and integrating by parts that

$$\int_0^{r_0} [J_0(k_l r)]^2 r \, dr = \frac{r_0^2}{2} [J_1(k_l r_0)]^2$$

if we take account of the boundary condition at r_0 and the mathematical identity

$$\frac{dJ_n}{d(kr)} = -J_{n+1} + \frac{n}{kr} J_n. \qquad (5\text{--}44)$$

Therefore

$$A_l = \frac{q/\epsilon_0}{2\pi k_l} \frac{1}{r_0^2 [J_1(k_l r_0)]^2} \qquad (5\text{--}45)$$

and

$$\phi_1 = \frac{q}{2\pi r_0^2 \epsilon_0} \sum_{l=1}^{\infty} e^{k_l z} \frac{J_0(k_l r)}{k_l [J_1(k_l r_0)]^2},$$

$$\phi_2 = \frac{q}{2\pi r_0^2 \epsilon_0} \sum_{l=1}^{\infty} e^{-k_l z} \frac{J_0(k_l r)}{k_l [J_1(k_l r_0)]^2}. \qquad (5\text{--}46)$$

A generalization of this procedure can be used to determine the Green's function for the interior of a conducting cylinder if the point charge is located at an arbitrary position $r = a$, $\varphi = \alpha$, $z = z_0$. In that case the potentials are written separately for $z < z_0$, $z > z_0$, and the flux condition is expressed in terms of the generalized two-dimensional δ-function for which

$$\int \delta(r - a) \, \delta(\varphi - \alpha) r \, dr \, d\varphi = 1. \tag{5-47}$$

The resulting Green's function is a double sum over n and l which reduces to Eq. (5-46) with $q = 1$ for the unit charge at the origin.

SUGGESTED REFERENCES

Of the references listed in earlier chapters those by Smythe, Jeans, Stratton, Weber, Jeffreys and Jeffreys, and Morse and Feshbach will probably be most useful for solving more elaborate potential problems than we have included. The properties of Legendre and Bessel functions are developed in an elementary way by Byerly in *Fourier's Series and Spherical Harmonics*, and more complete treatments are to be found in:

T. M. MacRobert, *Spherical Harmonics*.

G. N. Watson, *Theory of Bessel Functions*.

Tables and graphs of Legendre and Bessel functions, together with summaries of useful functional relations, will be found in:

E. Jahnke and F. Emde, *Tables of Functions*.

EXERCISES

1. What is the potential distribution inside a spherical region bounded by two conducting hemispheres at potentials $\pm V/2$, respectively?

2. Find the potential at points outside a spherical volume distribution of charge in which the charge density is proportional to the distance from a diametral plane.

3. The equation of the surface of a conductor is $r = a[1 + \delta P\nu \, (\cos \theta)]$, where $\delta \ll 1$. Show that if the conductor is placed in a uniform field $\mathbf{E}$ parallel to the polar axis the surface charge density is given by

$$\sigma = \sigma_0 + \delta \left\{ \frac{3n\epsilon_0 E}{2n + 1} [(n + 1)P_{n+1} \, (\cos \theta) + (n - 2)P_{n-1} \, (\cos \theta)] \right\},$$

where σ_0 is the induced charge density for $\delta = 0$.

4. A uniform field E_m is set up in a medium of specific inductive capacity κ. Prove that the field inside a spherical cavity in the medium is given by

$$E = \frac{3\kappa E_m}{2\kappa + 1}.$$

5. A conducting sphere of radius a carrying charge q is placed in a uniform field of strength E_0. Find the potential everywhere. What is the dipole moment of the induced charge on the sphere?

6. What is the capacity of a thin hemispherical shell? (*Hint:* Invert a charged circular disk. The solution of this problem was originally due to Kelvin.)

7. A solid dielectric sphere of radius a has a sector removed so that it fits the edge of an infinite conducting wedge of external angle α, its surface meeting the wedge faces orthogonally. If the wedge is charged, show that the potentials inside and outside the sphere are

$$\phi_{\text{in}} = \text{const.}\,\frac{\alpha + 2\pi}{\alpha(1 + \kappa)}\,(r \sin \theta)^{\pi/\alpha}\cos\frac{\pi\varphi}{\alpha}\,,$$

$$\phi_{\text{out}} = \text{const.}\left[r^{\pi/\alpha} - \frac{\pi(\kappa - 1)a^{\pi/\alpha}}{\alpha + \pi(\kappa + 1)}\left(\frac{a}{r}\right)^{(\pi/\alpha)+1}\right](\sin \theta)^{\pi/\alpha}\cos\frac{\pi\varphi}{\alpha}\,.$$

(*Hint:* Note that the geometry of the problem permits only a single surface harmonic. Which is it?) (Smythe)

8. Show that the potential due to a conducting disk of radius a carrying a charge q is

$$\phi(r, z) = \frac{q}{4\pi\epsilon_0 a}\int_0^\infty e^{-k|z|}J_0(kr)\,\frac{\sin ka}{k}\,dk$$

in cylindrical coordinates. Can you suggest other methods for the solution of this problem?

CHAPTER 6

ENERGY RELATIONS AND FORCES
IN THE ELECTROSTATIC FIELD

Our discussion of the electrostatic field has thus far been based entirely on a single experimental law, namely Coulomb's law of Eq. (1–24) for the action-at-a-distance force between two point charges. The electric field has been introduced as an intermediate agent whose purpose is to simplify the description of the interaction between charges. The question of the reality of the field as an independent physical entity therefore does not arise in these considerations. Maxwell attempted to ascribe a larger degree of physical reality in a direct mechanical sense to the electric field than will be necessary for our purpose, since the fundamental reason for attributing physical reality to the field will not become apparent until nonstatic effects are discussed. However, if the field is a truly valid representation of the experimental facts it is necessary that all of the mechanical properties of an electrically interacting system can be described either in terms of the sources which partake of the interaction or in terms of the fields themselves which are produced by the sources.

This means that the detailed nature of the sources should not influence the action of a field on a given system of charges. The description of the electric field alone must be a sufficient description to determine what interaction occurs if a number of charges are introduced at given points in the field. This interaction must be formulated in terms of the field itself, and not depend explicitly on the configuration of the charges that produce the field.

It should therefore be possible to develop a field theory in which we can describe the over-all mechanical properties, such as energy and force, equivalently in terms of the charges which are the sources of the field, or in terms of integrals over the field produced by the charges. The only criterion for the correctness of such over-all relations when expressed in terms of the electrostatic field shall be that the results are equivalent to those which are obtained from a direct consideration of the action-at-a-distance interaction of the charges responsible for the field.

6–1 Field energy in free space. Let us consider a set of charges q_i in free space. The work done in the course of physical assembly of these already-created charges, which are initially located an infinite distance apart, is given by

$$W = \tfrac{1}{2} \sum_{i=1}^{n} q_i \phi_i^0, \qquad (6\text{--}1)$$

where ϕ_i^0 is the potential produced at the position of q_i by all the other charges. By assembling these charges we have changed the energy of the system, and since all of the forces are conservative, we can identify this expression for the work of assembly with the energy of the system. This energy must be stored somewhere. Just as in mechanics, however, the location that one selects as the place of energy storage depends on one's point of view.

If, for example, we consider two masses at the ends of a compressed spring, we have a system that possesses potential energy which will be released if the spring is allowed to expand. In the expansion the masses will acquire kinetic energy. The physical location of the energy in this mechanical system in its initial condition is not necessarily in the spring. Phenomenologically the masses may be considered to be initially in regions of higher potential energy than they are after the expansion of the spring. Equation (6–1) corresponds to the latter point of view. We shall now try to transform Eq. (6–1) to an expression which would make it appear as if the electrical energy resides in the (figuratively) "elastic" quality of the electric field, as would be required in order to correspond to the point of view that the energy of the mass-spring system resides in the spring.

The expression obtained by Maxwell for the energy in an electric field, expressed as a volume integral over the field, is

$$U = \frac{\epsilon_0}{2} \int E^2 \, dv. \qquad (6\text{--}2)$$

The integral extends over all space. We shall now show that the field energy U is, in fact, the same as the assembly work W. To show this, let us introduce at any arbitrary field point the partial fields $\mathbf{E}_i$, each being the Coulomb field of only one of the point charges q_i. $\mathbf{E}$ and E^2 are then given by

$$\mathbf{E} = \sum_{i=1}^{n} \mathbf{E}_i,$$

$$E^2 = \sum_{i=1}^{n} E_i^2 + \sum_{i=1}^{n} \sum_{j=1}^{n}{}' \mathbf{E}_i \cdot \mathbf{E}_j, \qquad (6\text{--}3)$$

where the prime on the summation symbol indicates that the term for which $i = j$ is omitted, since such terms are grouped separately in the first summation. For point charges, the first sum makes an infinite contribution to the integral over E^2 in Eq. (6–2). However, this infinite term is independent of the relative position of the charges and therefore it must represent the work necessary to create the charges from an arbitrary zero point of energy. We shall therefore designate

$$U_s = \frac{\epsilon_0}{2} \int \sum_{i=1}^{n} E_i^2 \, dv \tag{6–4}$$

as the self-energy of the system. The introduction of a finite radius for the elementary charges enables us to avoid infinite self-energy so long as the charges are stationary, as indicated in a problem at the end of this chapter. Later we shall find that a finite radius involves certain difficulties when moving charges are considered.

With Eq. (6–4) the Maxwell field energy expression, Eq. (6–2), then becomes

$$U = U_s + \frac{\epsilon_0}{2} \sum_{i=1}^{n} \int \mathbf{E}_i \cdot (-\boldsymbol{\nabla} {\textstyle\sum_j'} \phi_j) \, dv, \tag{6–5}$$

where $\sum_j' \phi_j$ denotes the potential at any point due to all of the charges except the ith charge. Using the vector identity

$$\boldsymbol{\nabla} \cdot (\mathbf{A}\phi) = \phi \boldsymbol{\nabla} \cdot \mathbf{A} + \mathbf{A} \cdot \boldsymbol{\nabla}\phi \tag{6–6}$$

in order to perform an integration by parts, we obtain

$$U = U_s - \frac{\epsilon_0}{2} \sum_{i=1}^{n} \int [\boldsymbol{\nabla} \cdot (\mathbf{E}_i {\textstyle\sum_j'} \phi_j) - {\textstyle\sum_j'} \phi_j \boldsymbol{\nabla} \cdot \mathbf{E}_i] \, dv. \tag{6–7}$$

Except at the position of the ith charge, $\boldsymbol{\nabla} \cdot \mathbf{E}_i$ vanishes everywhere. Hence we may write $\boldsymbol{\nabla} \cdot \mathbf{E}_i = q_i \, \delta(\mathbf{r}_i)/\epsilon_0$, and hence

$$U = U_s - \frac{\epsilon_0}{2} \sum_{i=1}^{n} \int \mathbf{E}_i ({\textstyle\sum_j'} \phi_j) \cdot d\mathbf{S} + \frac{\epsilon_0}{2} \sum_{i=1}^{n} \phi_i^0 \frac{q_i}{\epsilon_0}, \tag{6–8}$$

where ϕ_i^0 is $\sum_j' \phi_j$ evaluated at the position of the ith charge, i.e., the potential at q_i due to the other charges. The surface term can be made arbitrarily small by letting the boundary surface go to infinity, since the fields decrease at least as the inverse second power, the potential at least as the inverse first power, and the differential area of integration increases only as the square of the distance. If we consider that the integral in Eq. (6–2) extends over all of space where there is a field, then this integral, as a result of Eq. (6–8), reduces to

$$U = U_s + \tfrac{1}{2} \sum_{i=1}^{n} \phi_i^0 q_i. \tag{6–9}$$

The second term of this final equation is identical with the expression (6–1) for the work of assembly of the charges from infinity, while the first term U_s is the self-energy corresponding to the energy used in the creation of the charges themselves. The analysis shows that Eqs (6–2) and (6–9) correspond to the same energy, but Eq. (6–2) expresses this energy as a volume integral of an energy density $\epsilon_0 E^2/2$ extending over all of space. It is not possible to ascertain experimentally whether the energy resides in the field or is possessed by the charges which produce the field.

6–2 Energy density within a dielectric. In the presence of a dielectric Eq. (6–2) can under certain conditions be replaced by

$$U = \tfrac{1}{2} \int \mathbf{E} \cdot \mathbf{D} \, dv. \tag{6–10}$$

To show this, let us consider the change of energy when a small increment of true charge $\delta\rho$ is added to the field. We shall assume rigid boundaries and rigid constraints on the medium, so that no work is done on mechanical constraints. In the case of continuous charge density the self-energy problem disappears, and the work done is given by

$$\delta W = \int \phi \, \delta\rho \, dv = \int \phi \, \delta(\boldsymbol{\nabla} \cdot \mathbf{D}) \, dv = \int \phi \boldsymbol{\nabla} \cdot \delta\mathbf{D} \, dv. \tag{6–11}$$

The vector theorem of Eq. (6–6) and Gauss's theorem enable us to write

$$\delta W = \int \boldsymbol{\nabla} \cdot (\phi \, \delta\mathbf{D}) \, dv - \int \delta\mathbf{D} \cdot \boldsymbol{\nabla}\phi \, dv$$

$$= \int (\phi \, \delta\mathbf{D}) \cdot d\mathbf{S} - \int \delta\mathbf{D} \cdot \boldsymbol{\nabla}\phi \, dv. \tag{6–12}$$

If we drop the surface term, as we did in the derivation of Eq. (6–9), we obtain

$$\delta W = -\int \delta\mathbf{D} \cdot \boldsymbol{\nabla}\phi \, dv = \int \mathbf{E} \cdot \delta\mathbf{D} \, dv. \tag{6–13}$$

This increment of work usually cannot be integrated unless $\mathbf{E}$ is a given function of $\mathbf{D}$. If, for example, $\mathbf{E}$ and $\mathbf{D}$ are related by a dielectric constant κ, which may be a function of position but not of $\mathbf{E}$, then the energy resulting from the integration of the work increment from $\mathbf{D} = 0$ to $\mathbf{D} = \mathbf{D}$ is

$$U = \int_0^D \delta W = \int_0^D \int \mathbf{E} \cdot \delta \mathbf{D} \, dv = \iint_0^E \frac{\kappa \epsilon_0 \, \delta(E^2)}{2} \, dv$$

$$= \tfrac{1}{2} \int \kappa \epsilon_0 E^2 \, dv = \tfrac{1}{2} \int \mathbf{E} \cdot \mathbf{D} \, dv. \tag{6–14}$$

At least for the simple case, therefore, in which $\mathbf{E}$ and $\mathbf{D}$ are proportional, Eq. (6–10) is justified.

Note that this proof of Eq. (6–10) has involved a particular virtual process: the addition of a true charge increment to a system under rigid constraints. Since we are dealing with conservative forces, the validity of these relations is independent of the particular virtual process chosen.

If the medium is not isotropic, the coefficient which relates $\mathbf{D}$ to $\mathbf{E}$ is a tensor, as already discussed in Section 2–1. That is,

$$D_\alpha = \epsilon_0 \kappa_{\alpha\beta} E_\beta. \tag{6–15}$$

Hence Eq. (6–13) must be replaced by

$$\delta W = \epsilon_0 \int \kappa_{\alpha\beta} E_\beta \, \delta E_\alpha \, dv. \tag{6–16}$$

If $\kappa_{\alpha\beta}$ is independent of the field magnitude, this equation can be integrated to the final field strength, giving

$$W = \frac{\epsilon_0}{2} \int \kappa_{\alpha\beta} E_\beta E_\alpha \, dv, \tag{6–17}$$

since all field components increase proportionally. Hence Eq. (6–14) is also valid for linear nonisotropic media.

We can deduce the symmetry of the dielectric constant tensor $\kappa_{\alpha\beta}$ from Eq. (6–17). The integrand of Eq. (6–17) is

$$\kappa_{\alpha\beta} E_\beta E_\alpha = \kappa_{\alpha\beta} E_\alpha E_\beta = \kappa_{\beta\alpha} E_\beta E_\alpha, \tag{6–18}$$

where we have simply interchanged indices in the last step. Hence

$$\int (\kappa_{\alpha\beta} - \kappa_{\beta\alpha}) E_\alpha E_\beta \, dv = 0 \tag{6–19}$$

for arbitrary E_α and E_β, which shows that $\kappa_{\alpha\beta} = \kappa_{\beta\alpha}$ unless the field energy vanishes identically. The symmetry of the susceptibility tensor assumed in Section 2–1 is thus proved. This proof corresponds to the general theorem that the contracted product

$$A_{\alpha\beta} S_{\alpha\beta} \tag{6–20}$$

of an antisymmetric tensor $A_{\alpha\beta} = -A_{\beta\alpha}$ and a symmetric tensor $S_{\alpha\beta} = S_{\beta\alpha}$ vanishes identically; an arbitrary tensor $T_{\alpha\beta}$ can always be written as

$$T_{\alpha\beta} = S_{\alpha\beta} + A_{\alpha\beta}, \tag{6-21}$$

where

$$S_{\alpha\beta} = \frac{T_{\alpha\beta} + T_{\beta\alpha}}{2}, \qquad A_{\alpha\beta} = \frac{T_{\alpha\beta} - T_{\beta\alpha}}{2}. \tag{6-22}$$

6–3 Thermodynamic interpretation of U. The assumption of a dielectric constant κ that does not change with the field and is a function only of position [as implied in the integrations in Eq. (6–14)] implies that the process of change of field is an isothermal process, since the dielectric constant is usually a function of temperature. To assure isothermal behavior, the dielectric material in question must be in contact with a heat bath which can exchange heat with it to maintain a constant temperature. Thus we cannot equate the increment of work done, δW of Eq. (6–13), to the increase in total energy; since heat transfers are also involved. Equation (6–14) does, however, represent the maximum work which can be extracted at a later time from the total electrical field energy if the temperature is assumed constant.

Thermodynamically, the maximum work which can be obtained from a system under isothermal conditions is the free energy F of the system, not the total energy. This means that in the presence of dielectrics the expression $U = \frac{1}{2}\int \mathbf{E} \cdot \mathbf{D}\, dv$ cannot be identified with the total energy of the system, but can only be identified with the thermodynamic free energy. Of course, the distinction vanishes when no materials with temperature-dependent dielectric properties are present in the field. Thus Eq. (6–13) can be written in terms of the free-energy increment δF,

$$\delta F = \left[\int (\mathbf{E} \cdot \delta \mathbf{D})\, dv\right]_{\text{constant temperature}} \tag{6-23}$$

and

$$F = U - TS = \frac{1}{2}\int \mathbf{E} \cdot \mathbf{D}\, dv, \tag{6-24}$$

where U is the total internal energy, S the entropy, and T the absolute temperature. Formally, in the thermodynamic sense, the electric field $\mathbf{E}$ is analogous to gas pressure and the displacement $\mathbf{D}$ to volume; i.e., $\mathbf{E}$ is the "intensive" and $\mathbf{D}$ the "extensive" variable.

The correct expression for the total energy can be easily derived. If we take the differential of Eq. (6–24),

$$\delta F = \delta U - T\, \delta S - S\, \delta T. \tag{6-25}$$

But according to the first law of thermodynamics, $\delta U - T\,\delta S = \delta W$, where δW is given by Eq. (6–13). Hence, for *arbitrary* changes in field and temperature,

$$\delta F = -S\,\delta T + \int (\mathbf{E} \cdot \delta \mathbf{D})\,dv. \qquad (6\text{–}26)$$

From Eqs. (6–24) and (6–26)

$$S = -\left.\frac{\partial F}{\partial T}\right|_{D=\text{const}} = \int \frac{\epsilon_0}{2} E^2 \frac{d\kappa}{dT}\,dv = \int \frac{\mathbf{E} \cdot \mathbf{D}}{2} \frac{1}{\kappa} \frac{d\kappa}{dT}\,dv,$$

and

$$U = F + TS = \int \frac{\epsilon_0}{2} E^2 \frac{d(T\kappa)}{dT}\,dv = \int \frac{\mathbf{E} \cdot \mathbf{D}}{2} \frac{1}{\kappa} \frac{d(T\kappa)}{dT}\,dv. \quad (6\text{–}27)$$

The heat absorbed by the dielectric during application of a field change at constant temperature is thus

$$\delta Q = T\,\delta S = \int \mathbf{E} \cdot \delta \mathbf{D} \frac{T}{\kappa} \frac{d\kappa}{dT}\,dv. \qquad (6\text{–}28)$$

If, for example, the specific inductive capacity has the form [see Eq. (2–44)]

$$\kappa = 1 + X = 1 + (A/T), \qquad (6\text{–}29)$$

with A constant, as is the case for gases composed of molecules having permanent dipole moments, then $d\kappa/dT$ is less than zero, and hence $\delta Q < 0$ if $\delta \mathbf{D} > 0$. Thus heat is given off when the field is applied, and conversely. If a field is applied to a heat-insulated dielectric, the temperature of the dielectric will therefore increase if $d\kappa/dT < 0$.

6–4 Thomson's theorem. From now on we shall use only the free-energy density in our considerations, although we shall follow the conventions of electromagnetic nomenclature and designate it by U instead of F. This will enable us (in constant temperature processes) to equate changes in the free energy directly to the mechanical work quantities responsible for them, without making it necessary to include thermal quantities in the energy balance. The free-energy expression of Eq. (6–10), applicable even in the presence of dielectrics, behaves in electrical problems in the same manner as does the chemical free energy in chemical kinetics: a reaction will proceed until the free energy reaches a minimum value. In the electric case, charges on a conductor will redistribute themselves in such a way that the entire free field energy will be minimized.

We can show this directly. Let us consider a virtual process in which charges in equilibrium on a number of conductors are displaced by an infinitesimal amount along the constant potential conductor surfaces in such a way that the total charge remains unchanged. The variation of free energy (at constant temperature) is given by

$$\delta U = \int \mathbf{E} \cdot \delta \mathbf{D} \, dv. \tag{6-30}$$

If we apply Gauss's theorem, we obtain as before

$$\delta U = \int \mathbf{E} \cdot \delta \mathbf{D} \, dv = \int (-\boldsymbol{\nabla}\phi \cdot \delta \mathbf{D}) \, dv = \int [\phi \boldsymbol{\nabla} \cdot \delta \mathbf{D} - \boldsymbol{\nabla} \cdot \delta(\mathbf{D}\phi)] \, dv$$
$$= \int \phi \, \delta\rho \, dv - \int \phi \, \delta \mathbf{D} \cdot d\mathbf{S}. \tag{6-31}$$

The surface term vanishes and the integral can be broken into a sum of individual integrals over the ith conductor; thus,

$$\delta U = \sum_i \int_i \phi_i \, \delta\rho \, dv_i. \tag{6-32}$$

If ϕ_i is constant over each conductor δU will vanish, since

$$\int_i \delta\rho_i \, dv_i = \delta q_i = 0 \tag{6-33}$$

because the charge on each conductor is constant. Thus the elementary equilibrium condition that each conductor be an equipotential is equivalent to making the free energy an extremum. This requirement is usually known as "Thomson's theorem."

The term

$$U_v = \frac{\mathbf{E} \cdot \mathbf{D}}{2} \tag{6-34}$$

is known as the energy density (more accurately, of course, as the free-energy density) of the electrostatic field. It is a density in the sense that its volume integral gives the over-all energy of the field. On the other hand, in the same sense as it was impossible to localize the energy either in the field or in the source charges, it is also impossible to associate energy in a definite way with each specific volume of field in a manner which can be verified by experiment.

In deriving the energy expression it is assumed that the medium is held at rest and hence no work is done in motion against forces. This implies that the virtual process of assembling the charges in the dielectrics is a process with particular constraints. The resultant energy expression is nevertheless general, since no nonconservative forces are involved. We shall later consider the more general virtual process permitting mass motion; Eq. (6–10) will still be applicable, however, since the final field energy is independent of history.

6–5 Maxwell stress tensor. In a pure field theory it should be possible to calculate the net force on a given volume element within a dielectric in terms of the field conditions on the surface of the volume element. This implies that the field is the stress-transmitting medium in the same sense that a string tying two weights together is the medium that transmits a force from one weight to the other. This was a point that was emphasized by Maxwell to bring out the importance and the physical reality of field quantities. But, as we have seen before, we can give only an alternate description of the way in which the forces act, and cannot give a definite physical proof of the validity of the field concept as compared with the concept of action at a distance. We must remember that the only physical fact underlying this discussion is Coulomb's law, the remainder of the discussion being mathematical, and therefore we cannot expect to obtain any physical concept regarding the mechanical interaction of charges which will add any physical facts beyond Coulomb's law. New physical facts based on the field concept will arise only when time-dependent effects in the present theory are further investigated in later chapters.

If we consider that the force acting on a given volume is transmitted across the elements of surface bounding that volume, then this transmitting force can be formulated in terms of a quantity known as the stress tensor T. (This stress is an abstract concept, and does not depend on the presence of matter.) The $\alpha\beta$th component $T_{\alpha\beta}$ of the stress tensor T is so constituted that the αth component dF_α of the force $d\mathbf{F}$ transmitted across a surface element $d\mathbf{S}$ whose component in the βth direction is dS_β, is given by

$$dF_\alpha = \sum_{\beta=1}^{3} T_{\alpha\beta}\, dS_\beta. \tag{6–35}$$

It can be shown by the consideration of the rotational equilibrium of a rectangular solid under surface stresses that the stress tensor T must be symmetric.

By means of the summation convention, according to which a summation is to be carried out over indices that are repeated in any single term, Eq. (6–35) may be written

$$dF_\alpha = T_{\alpha\beta}\, dS_\beta. \tag{6–36}$$

Equation (6–36) may be integrated to give the αth component of the total force acting on a given volume:

$$F_\alpha = \int T_{\alpha\beta}\, dS_\beta. \tag{6–37}$$

If this force is to be expressible in terms of a volume force, whose αth component is F_{v_α}, then

$$F_\alpha = \int T_{\alpha\beta}\, dS_\beta = \int F_{v_\alpha}\, dv, \tag{6-38}$$

and hence, by Gauss's divergence theorem expressed in tensor notation,

$$\int \frac{\partial T_{\alpha\beta}}{\partial x_\beta}\, dv = \int T_{\alpha\beta}\, dS_\beta, \tag{6-39}$$

the relation between the stress tensor and the volume force is

$$F_{v_\alpha} = \frac{\partial T_{\alpha\beta}}{\partial x_\beta}. \tag{6-40}$$

Thus if we express a volume force as the tensor divergence of a certain quantity T, then the quantity T can be identified with the surface stress tensor T that gave the stress transmitted by the field across the surface of the volume in Eq. (6–38).

The tensor corresponding to a given volume force $\mathbf{F}_v$ is not unique, since any additional tensor of vanishing divergence can always be added without affecting the value of $\mathbf{F}_v$ in Eq. (6–40).

Let us first consider the problem of identifying such a tensor in the absence of dielectrics, i.e., let the volume force be given by

$$\mathbf{F}_v = \rho\mathbf{E} = (\mathbf{\nabla} \cdot \mathbf{D})\mathbf{E}. \tag{6-41}$$

In tensor notation Eq. (6–41) becomes

$$\begin{aligned}
F_{v_\alpha} &= \epsilon_0 E_\alpha \frac{\partial E_\beta}{\partial x_\beta} \\
&= \epsilon_0\left[\frac{\partial}{\partial x_\beta}(E_\alpha E_\beta) - E_\beta\frac{\partial E_\alpha}{\partial x_\beta}\right]. \tag{6-42}
\end{aligned}$$

The first term of Eq. (6–42) is already in the form (6–40). We can transform the second term in the bracket of Eq. (6–42) by noting that for the electrostatic field $\mathbf{\nabla} \times \mathbf{E} = 0$, i.e., $\partial E_\alpha/\partial x_\beta = \partial E_\beta/\partial x_\alpha$. Hence

$$E_\beta\frac{\partial E_\alpha}{\partial x_\beta} = E_\beta\frac{\partial E_\beta}{\partial x_\alpha} = \frac{1}{2}\frac{\partial}{\partial x_\alpha}(E_\beta E_\beta) = \frac{1}{2}\frac{\partial(E^2)}{\partial x_\alpha}, \tag{6-43}$$

where E is the magnitude of $\mathbf{E}$. Equation (6–43) can also be written as

$$E_\beta\frac{\partial E_\alpha}{\partial x_\beta} = \frac{1}{2}\delta_{\alpha\beta}\frac{\partial}{\partial x_\beta}(E^2) = \frac{\partial}{\partial x_\beta}\left(\frac{1}{2}\delta_{\alpha\beta}E^2\right). \tag{6-44}$$

Hence Eq. (6–42) is of the form (6–40) with

$$T_{\alpha\beta} = \epsilon_0(E_\alpha E_\beta - \tfrac{1}{2}\,\delta_{\alpha\beta}E^2).\tag{6–45}$$

The matrix corresponding to this tensor may be written explicitly,

$$T = \epsilon_0 \begin{bmatrix} \tfrac{1}{2}(E_x^2 - E_y^2 - E_z^2) & E_x E_y & E_x E_z \\ E_x E_y & \tfrac{1}{2}(E_y^2 - E_z^2 - E_x^2) & E_y E_z \\ E_x E_z & E_y E_z & \tfrac{1}{2}(E_z^2 - E_x^2 - E_y^2) \end{bmatrix},$$

$$\tag{6–46}$$

This is the Maxwell electric stress tensor in the absence of dielectrics. Note that additional terms will appear in the stress tensor if the field is not irrotational as assumed above.

The Maxwell tensor is a symmetric tensor of the second rank and can therefore be reduced to three components by transformation to principal axes. The principal values of the matrix can be obtained by solving the secular determinant:

$$|T_{\alpha\beta} - \delta_{\alpha\beta}\lambda| = 0.\tag{6–47}$$

The principal values of the tensor are

$$\lambda_1 = \frac{\epsilon_0}{2}\,E^2, \qquad \lambda_{2,3} = -\frac{\epsilon_0}{2}\,E^2.\tag{6–48}$$

When expressed in terms of principal coordinates the stress tensor therefore takes the simple form:

$$T = \frac{\epsilon_0}{2} \begin{bmatrix} E^2 & 0 & 0 \\ 0 & -E^2 & 0 \\ 0 & 0 & -E^2 \end{bmatrix}.\tag{6–49}$$

The principal axes are so oriented that the coordinate axis corresponding to the single root of the secular determinant, λ_1, is parallel to **E**, while the two axes corresponding to the double roots λ_2 and λ_3 are perpendicular to **E**. This fact is often expressed qualitatively by stating that the electric field transmits a tension $\epsilon_0 E^2/2$ parallel to the direction of the field and a transverse pressure of magnitude $\epsilon_0 E^2/2$ transverse to the direction of the field.

Let us choose a coordinate system in which the x-axis is parallel to the direction of the field, so that $E_y = E_z = 0$, and consider the stress across a surface element, as shown in Fig. 6–1, whose normal makes an angle θ with the x-axis. The force will then have two components, one parallel

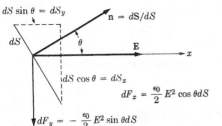

FIG. 6-1. Geometry for considering stresses at a surface element.

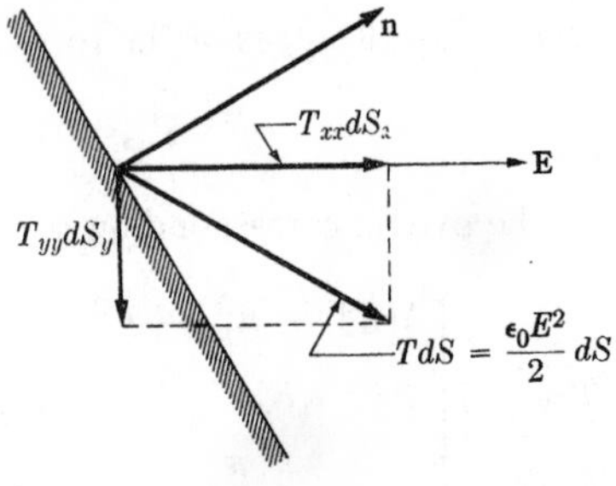

FIG. 6-2. Stresses at a surface.

to $\mathbf{E}$ and the other perpendicular to $\mathbf{E}$ in the plane defined by $\mathbf{E}$ and the normal to the surface element. The magnitudes of these forces are the stress components given by the matrix in Eq. (6-49) multiplied by the surface element components as indicated in the figure. The resultant force on dS is the vector sum of the two force components, as shown in Fig. 6-2. It is seen that the electric field bisects the angle between the normal to the surface and the direction of the resultant force acting on the surface. This construction is frequently a useful one in the graphical evaluation of the forces on a charged region if an experimental field plot is available or, in the analogous magnetic case to be discussed later, it is useful for the computation of forces on magnetized materials, or on current-carrying conductors.

In the special case of stress transmitted across surfaces either parallel or normal to the electric field, we have the simple situation indicated in the first two parts of Fig. 6-3, where the field transmits a pull of magnitude $\frac{1}{2}\epsilon_0 E^2$ across a surface that is normal to the field and a push of magnitude $\frac{1}{2}\epsilon_0 E^2$ across a surface that is tangential to the field. A surface that is oriented at 45° to the direction of the field, as also shown in Fig. 6-3, will receive a force that acts parallel to the surface, also of magnitude $\frac{1}{2}\epsilon_0 E^2$. These relations can be demonstrated for simple cases such as the

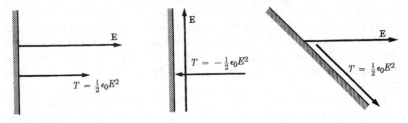

FIG. 6-3. Stresses for fields perpendicular, parallel, and at an angle of 45° to a surface.

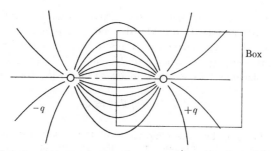

Fig. 6–4. Lines of force for equal and opposite charges.

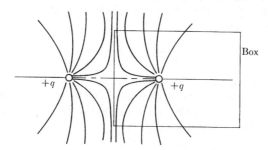

Fig. 6–5. Lines of force for equal charges of the same sign.

attraction and repulsion between two charges of opposite or equal sign. If we consider, for example, two charges of equal magnitude but of opposite sign, then the lines of force are distributed as in Fig. 6–4. If the stress tensor is integrated over the surfaces of a box, one of whose faces is the plane of symmetry between the two charges and the other faces are at infinity, the resulting expression is in agreement with the Coulomb attraction of Eq. (1–24). If we consider the two equal like charges, as in Fig. 6–5, and the same box as before, then the lines of force are parallel at the plane of symmetry, resulting in a repulsion whose magnitude can similarly be shown to be also in accord with the Coulomb repulsion.

6–6 Volume forces in the electrostatic field in the presence of dielectrics. The force per unit volume that acts on a dielectric body when it is under the influence of an external electrostatic field may be derived from the (free) energy as given by Eq. (6–14).

Equation (6–14) was derived by considering a virtual process in which true charges were added to a system of charges and dielectrics under rigid constraints, so that no work was done against physical displacements. We shall now consider a different virtual process, one in which the physical coordinates of charges and dielectrics are given a virtual displacement δx

at each point in space, but where no external charges are added. Since we are dealing with conservative fields, the energy expression (6–14) can still be used although it was derived in terms of another virtual process. The variation in free energy δU when a unit volume of the dielectric undergoes a virtual displacement $\delta \mathbf{x}$ is given in terms of $\mathbf{F}_v$, the force per unit volume, by*

$$\delta U = -\int \mathbf{F}_v \cdot \delta \mathbf{x} \, dv. \tag{6–50}$$

In view of the fact that the magnitude of the virtual displacement $\delta \mathbf{x}$ is quite arbitrary, we can identify $\mathbf{F}_v$ in Eq. (6–50) with the true volume force. This information may be put in a different way: if $\mathbf{u}$ is an arbitrary velocity field within a dielectric, then the rate at which energy is lost by the field is given by

$$\frac{dU}{dt} = -\int \mathbf{F}_v \cdot \mathbf{u} \, dv, \tag{6–51}$$

where $\mathbf{F}_v$ again represents the volume force.

Let us now consider the energy change due to both a change $\delta \rho$ in true charge distribution and a change $\delta \kappa$ in specific inductive capacity caused by the displacements. From Eq. (6–14),

$$\delta U = \tfrac{1}{2}\delta \int \mathbf{E} \cdot \mathbf{D} \, dv$$

$$= \frac{1}{2\epsilon_0} \int D^2 \, \delta(1/\kappa) \, dv + \int \mathbf{E} \cdot \delta \mathbf{D} \, dv \tag{6–52}$$

$$= -\frac{\epsilon_0}{2} \int E^2 \, \delta\kappa \, dv + \int \mathbf{E} \cdot \delta \mathbf{D} \, dv. \tag{6–53}$$

Here the first term represents the energy change caused by the change in dielectric constant due to the virtual displacements, and the second term corresponds to the energy change caused by the displacement of sources. The second term can be reduced in the usual way by partial integration:

$$\int \mathbf{E} \cdot \delta \mathbf{D} \, dv = -\int \boldsymbol{\nabla}\phi \cdot \delta \mathbf{D} \, dv \tag{6–54}$$

$$= \int \phi \boldsymbol{\nabla} \cdot (\delta \mathbf{D}) \, dv = \int \phi \delta \boldsymbol{\nabla} \cdot \mathbf{D} \, dv$$

$$= \int \phi \, \delta\rho \, dv. \tag{6–55}$$

---

* It is assumed here that the virtual velocities $\mathbf{u}$ corresponding to the virtual displacements $\delta \mathbf{x}$ are sufficiently slow so that the process is both reversible and isothermal. Under these conditions the change in free energy can be equated to the mechanical work done.

Hence Eq. (6–53) becomes

$$\frac{dU}{dt} = \int \left(\phi \frac{\partial \rho}{\partial t} - \frac{\epsilon_0}{2} E^2 \frac{\partial \kappa}{\partial t} \right) dv. \tag{6–56}$$

To arrive at an expression for $\mathbf{F}_v$, we must express the time derivatives $\partial \rho / \partial t$ and $\partial \kappa / \partial t$ of Eq. (6–56) in terms of the arbitrary velocity field $\mathbf{u}$. This can be done by means of the hydrodynamic equations of continuity,

$$\boldsymbol{\nabla} \cdot (\rho \mathbf{u}) + \frac{\partial \rho}{\partial t} = 0, \tag{6–57}$$

$$\boldsymbol{\nabla} \cdot (\rho_m \mathbf{u}) + \frac{\partial \rho_m}{\partial t} = 0, \tag{6–58}$$

which represent respectively the conservation of charge and mass if ρ_m is the mass density just as ρ is the charge density. To calculate $\partial \kappa / \partial t$ we must associate the change in dielectric constant with the velocity flow. Since there is net transport of material in a velocity field, the change in dielectric constant can be associated with changes in geometry only if we consider the time history of a volume element that is moving with the velocity $\mathbf{u}$. The total derivative of a particular quantity, such as κ or ρ_m, when evaluated so that the observation point for this derivative moves with a chosen volume element in a velocity field, is known as the substantial derivative, and is related to the partial derivatives and to the velocity $\mathbf{u}$ by

$$\frac{D\kappa}{Dt} = \frac{\partial \kappa}{\partial x} \frac{\partial x}{\partial t} + \frac{\partial \kappa}{\partial y} \frac{\partial y}{\partial t} + \frac{\partial \kappa}{\partial z} \frac{\partial z}{\partial t} + \frac{\partial \kappa}{\partial t} = (\boldsymbol{\nabla}\kappa) \cdot \mathbf{u} + \frac{\partial \kappa}{\partial t}. \tag{6–59}$$

Hence the desired partial derivatives are

$$\frac{\partial \kappa}{\partial t} = -\boldsymbol{\nabla}\kappa \cdot \mathbf{u} + \frac{D\kappa}{Dt},$$

$$\frac{\partial \rho_m}{\partial t} = -\boldsymbol{\nabla}\rho_m \cdot \mathbf{u} + \frac{D\rho_m}{Dt}. \tag{6–60}$$

If there is a dielectric equation of state, i.e., a relation which gives the dependence of the dielectric constant on the density, such as the Clausius-Mossotti relation of Eq. (2–39), then the substantial derivative of the dielectric constant can be expressed in terms of the substantial derivative of the density by

$$\frac{D\kappa}{Dt} = \frac{d\kappa}{d\rho_m} \frac{D\rho_m}{Dt}, \tag{6–61}$$

where $d\kappa / d\rho_m$ is a known property of the dielectric. The assumption that the dielectric constant depends on the mass density alone includes, of course, the assumption that the virtual processes are isothermal. The factor $D\rho_m/Dt$ of Eq. (6–61) may be evaluated with the aid of Eq. (6–60) and the equation of continuity, so that

$$\frac{D\kappa}{Dt} = \frac{d\kappa}{d\rho_m}\left(\frac{\partial\rho_m}{\partial t} + \nabla\rho_m \cdot \mathbf{u}\right)$$

$$= \frac{d\kappa}{d\rho_m}[\nabla\rho_m \cdot \mathbf{u} - \nabla \cdot (\rho_m\mathbf{u})] = -\frac{d\kappa}{d\rho_m}\rho_m(\nabla \cdot \mathbf{u}). \quad (6\text{–}62)$$

Therefore

$$\frac{\partial\kappa}{\partial t} = -\frac{d\kappa}{d\rho_m}\rho_m(\nabla \cdot \mathbf{u}) - (\nabla\kappa) \cdot \mathbf{u}. \quad (6\text{–}63)$$

Equation (6–56) may now be written

$$\frac{dU}{dt} = \int\left[-\phi\nabla \cdot (\rho\mathbf{u}) + \frac{\epsilon_0}{2}E^2\frac{d\kappa}{d\rho_m}\rho_m(\nabla \cdot \mathbf{u}) + \left(\frac{\epsilon_0}{2}E^2\nabla\kappa\right) \cdot \mathbf{u}\right]dv. \quad (6\text{–}64)$$

The integrand must be brought into the form of the dot product of an expression and the velocity $\mathbf{u}$ if we are to identify the volume force $\mathbf{F}_v$. The first term may be written in this form if we use Eq. (6–6) to integrate by parts and assume that the surface boundary is outside of the dielectric and therefore outside the region of charge density. The result is

$$-\int\phi\nabla \cdot (\rho\mathbf{u})\, dv = \int\rho\nabla\phi \cdot \mathbf{u}\, dv. \quad (6\text{–}65)$$

Similarly, the second term can be put into the desired form if we drop a surface term:

$$\frac{\epsilon_0}{2}\int E^2\rho_m\frac{d\kappa}{d\rho_m}\nabla \cdot \mathbf{u}\, dv = -\frac{\epsilon_0}{2}\int\nabla\left(E^2\frac{d\kappa}{d\rho_m}\rho_m\right) \cdot \mathbf{u}\, dv. \quad (6\text{–}66)$$

Equation (6–64) therefore becomes

$$\frac{dU}{dt} = \int\left[-\rho\mathbf{E} + \frac{\epsilon_0}{2}E^2\nabla\kappa - \frac{\epsilon_0}{2}\nabla\left(E^2\frac{d\kappa}{d\rho_m}\rho_m\right)\right] \cdot \mathbf{u}\, dv, \quad (6\text{–}67)$$

and by a comparison of Eqs. (6–67) and (6–51) we see that

$$\mathbf{F}_v = \rho\mathbf{E} - \frac{\epsilon_0}{2}E^2\nabla\kappa + \frac{\epsilon_0}{2}\nabla\left(E^2\frac{d\kappa}{d\rho_m}\rho_m\right) \quad (6\text{–}68)$$

is the volume force.

The first term in Eq. (6–68) is the ordinary electrostatic volume force in agreement with Eqs. (1–23) and (6–41). The second term represents a force which will appear whenever an inhomogeneous dielectric is in an electric field. The last term, known as the electrostriction term, gives a volume force on a dielectric in an inhomogeneous electric field. Note that the magnitude of the electrostriction term depends explicitly through $d\kappa/d\rho_m$ on the electrical equation of state of the material. It is interesting to note that this term will never give a net force on a finite region of dielectric if we integrate it over a large enough portion of dielectric so that its extremities are in a field-free region. Under this condition the electrostriction term, being a pure gradient, will integrate out. It is for this reason that the term is frequently omitted, since in the calculation of total forces on dielectric bodies it usually does not contribute. In cases where it can be omitted, however, an incorrect pressure variation within the dielectric is obtained, even though the total force is given correctly.

6–7 The behavior of dielectric liquids in an electrostatic field. Let us consider the behavior of an uncharged dielectric liquid resulting from the volume force produced by an electrostatic field. If p is the mechanical pressure in the liquid when in equilibrium with the electric volume force $\mathbf{F}_v$, then the mechanical volume force $-\nabla p$ which is set up as a result of the pressure gradient is equal and opposite to $\mathbf{F}_v$. In other words, the equilibrium condition is that $\mathbf{F}_v + (-\nabla p) = 0$, so that $\mathbf{F}_v = \nabla p$. Then by Eq. (6–68) the pressure gradient at any point within the liquid is given by

$$\nabla p = \mathbf{F}_v = -\frac{\epsilon_0 E^2}{2}\nabla\kappa + \frac{\epsilon_0}{2}\nabla\left(E^2\rho_m\frac{d\kappa}{d\rho_m}\right). \qquad (6\text{–}69)$$

This equation can be written as

$$\nabla p = \frac{\epsilon_0\rho_m}{2}\nabla\left(E^2\frac{d\kappa}{d\rho_m}\right). \qquad (6\text{–}70)$$

On integrating this, assuming that there is a definite equation of state for the liquid, we obtain

$$\int_{p_1}^{p_2}\frac{dp}{\rho_m} = \frac{\epsilon_0}{2}\left\{\left[E^2\frac{d\kappa}{d\rho_m}\right]_2 - \left[E^2\frac{d\kappa}{d\rho_m}\right]_1\right\}. \qquad (6\text{–}71)$$

This equation denotes the important fact that the pressure within the dielectric liquid is a unique function of the electric field at a given point, the function depending on the electrical and mechanical equation of state of the liquid. Equation (6–71) also indicates that the net pressure difference (resulting from electrical forces) between two points outside the region

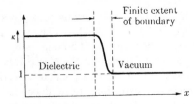

FIG. 6–6. Assumed transition of κ from a dielectric to vacuum.

of the electric field in a dielectric liquid will vanish. A situation that involves boundaries will be analyzed later.

If the liquid is incompressible, Eq. (6–71) reduces to

$$p_2 - p_1 = \frac{\rho_m \epsilon_0}{2} \left[E^2 \frac{d\kappa}{d\rho_m} \right]_1^2, \qquad (6\text{–}72)$$

from which the magnitude of the pressure difference can be estimated numerically in terms of the Clausius-Mossotti relation or a similar equation of state. If the Clausius-Mossotti relation [Eq. (2–39)] is valid, Eq. (6–72) for an incompressible fluid becomes

$$p_2 - p_1 = \frac{N_0 \epsilon_0 \rho_m \alpha}{18M} [E^2 (\kappa + 2)^2]_1^2 = \left[\frac{\epsilon_0 E^2}{2} \frac{(\kappa - 1)(\kappa + 2)}{3} \right]_1^2. \qquad (6\text{–}73)$$

As an example of the stresses that act across the boundary between two dielectrics, let us take the case of a boundary between a dielectric of specific inductive capacity κ and a vacuum. We assume that the transition from dielectric to vacuum takes place in a continuous manner, as indicated by Fig. 6–6. For simplicity, let the problem be a two-dimensional one involving a pair of capacitor plates that dip into a dielectric liquid, such as those shown in Fig. 6–7. If the *net pressure difference* from A to D is

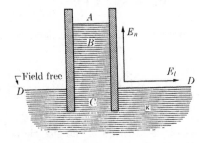

FIG. 6–7. Two capacitor plates dipping into a liquid dielectric. The field free point D may be anywhere in the dielectric, well away from the capacitor. It is indicated as a surface point to facilitate comparison with the hydrostatic pressure.

d can be understood only by considering the effect of the pressure of
liquid on the charges themselves. In accordance with the philosophy
the action-at-a-distance theory, no change in the purely electrical inter-
on between the charges takes place.

SUGGESTED REFERENCES

M. ABRAHAM AND R. BECKER, *The Classical Theory of Electricity and Magnetism.* Much of our treatment parallels that found in Chapter V of this work.

J. A. STRATTON, *Electromagnetic Theory.* The first part of Chapter II is on electrostatic stress and energy.

J. H. JEANS, *Electricity and Magnetism.* Chapter VI, on "the state of the
medium in the electrostatic field," begins with a discussion of field *versus* action-
at-a-distance, but the detailed treatment is rather brief.

M. MASON AND W. WEAVER, *The Electromagnetic Field,* especially Sections 2
and 37.

L. D. LANDAU AND E. M. LIFSHITZ, *Electrodynamics of Continuous Media.*
See especially Chapter II for an excellent discussion of the thermodynamics of
electric fields.

EXERCISES

1. Two charges of q at distance d repel each other with a force given by Coulomb's law. Choose a suitable surface surrounding one of the charges and calculate $\mathbf{F}$ by integration of the Maxwell stress tensor.

2. A conducting spherical shell of radius a is placed in a uniform field $\mathbf{E}$. Show
that the force tending to separate two halves of the sphere across a diametral
plane perpendicular to $\mathbf{E}$ is given by

$$F = \tfrac{9}{4}\pi\epsilon_0 a^2 E^2. \tag{6-81}$$

3. Consider a long flat conducting strip of width $2a$ with its axis perpendicular
to a uniform field $\mathbf{E}$ and its plane inclined at an angle θ to the field. What is the
torque per unit length acting on the strip?

4. A dipole of moment $\mathbf{p}$ is placed in a uniform field $\mathbf{E}$. Show by direct integration of the moment of the Maxwell tensor over a sphere centered at the dipole
that the torque is equal to the conventional result $\mathbf{p} \times \mathbf{E}$.

5. Find the electrostatic energy of a charge q distributed uniformly over the
surface of a sphere of radius r_0. What is the energy if the same charge is distributed uniformly throughout the sphere? (In both cases the energy is of
the order $q^2/4\pi\epsilon_0 r_0$.)

6. Find the interaction energy of two interpenetrating spheres of uniform charge
density ρ_1 and ρ_2. Let the two spheres have equal radii a and let the separation

---

all that is desired, it suffices to integrate the one term of Eq. (6–68) that
is proportional to the gradient of the dielectric constant. The resultant
electrical pressure difference, which has to be balanced by hydrostatic
effects, is

$$p_A - p_D = \frac{\epsilon_0}{2}\int_A^D E^2\boldsymbol{\nabla}\kappa\cdot d\mathbf{x}, \tag{6-74}$$

$$= \frac{\epsilon_0}{2}\int_A^B E^2\boldsymbol{\nabla}\kappa\cdot d\mathbf{x}, \tag{6-75}$$

since the integrand is different from zero only on the boundary $A \to B$.
Hence

$$p_A - p_D = \frac{\epsilon_0}{2}\int_A^D (E_t^2 + E_n^2)\frac{d\kappa}{dx}\,dx. \tag{6-76}$$

If the boundary conditions, Eqs. (2–15) and (2–18), on the normal and
tangential components of the electric field are introduced, Eq. (6–76) becomes

$$p_A - p_D = \frac{\epsilon_0}{2}\left[E_{t_B}^2(\kappa - 1) + \kappa^2 E_{n_B}^2\int_A^B\frac{d\kappa}{\kappa^2}\right]$$

$$= \frac{\epsilon_0(\kappa - 1)}{2}[E_t^2 + \kappa E_n^2]_B. \tag{6-77}$$

Note that the field quantities in Eq. (6–77) refer to the fields inside the
liquid. This formula can be used directly to find the rise of the liquid between the capacitor plates. It does *not* describe the detailed pressure behavior of the liquid from A to D, however, since it does not include the
electrostriction term. In fact, the pressure change from A to B is actually
of opposite sign from the pressure change from A to D. The detailed behavior of the pressure is shown in Fig. 6–8. As the field decreases from B
to D the pressure decreases below the outside value at A by an amount
which is larger than the pressure rise at the surface A–B. The net difference computed in Eq. (6–77) gives only the difference in pressure between
A and D. The pressure that forces the liquid up is actually exerted at

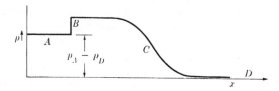

FIG. 6–8. Detailed variation in pressure from A to D. x is simply a path of
integration from A to the field free point D, and $p_A - p_D$ is given by Eq. (6–77).

the region C where the field is inhomogeneous, and not at the surface of the liquid. The physical reason for this is that the energy of dipoles in an electric field is lower than their energy in field-free space, and therefore the dipoles in the liquid are drawn into the regions of higher field in order to satisfy the condition that the potential energy be a minimum. This action on the dipoles takes place in the region C where the field begins to build up. The electrostriction drop at A–B partially counterbalances this minimum energy effect, resulting in the net pressure difference given by Eq. (6–77). The example shows that considerable care is necessary in applying the force equation in dielectrics.

Another example seems extremely simple, but actually leads to an apparent paradox. If a set of charged conductors are so arranged that they may be immersed in a dielectric liquid filling all space, and if the true charges on these conductors are kept constant as a liquid is introduced between them, then the free energy of the system as given by Eq. (6–14) will drop by a factor $1/\kappa$, since $\mathbf{D}$ remains constant but $\mathbf{E}$ is reduced. If, on the other hand, the voltages were maintained at their initial values as the liquid was introduced, then the free energy would be increased by a factor κ, since in this case $\mathbf{E}$ remains constant while $\mathbf{D}$ increases. Of course, these arguments pertain only if all of the space between the conductors and outside of them is filled with a dielectric liquid. At least all space where there are electric fields must be so filled, because otherwise we cannot assume that the distribution of $\mathbf{E}$ and $\mathbf{D}$ remains constant as the dielectric material is introduced between the plates.

This means that if a system maintained at constant charge is totally surrounded by a dielectric liquid all mechanical forces will drop in the ratio $1/\kappa$. A factor $1/\kappa$ is frequently included in the expression for Coulomb's law to indicate this decrease in force. The physical significance of this reduction of force, which is required by energy considerations, is often somewhat mysterious. It is difficult to see on the basis of a field theory why the interaction between two charges should be dependent upon the nature or condition of the intervening material, and therefore the inclusion of an extra factor $1/\kappa$ in Coulomb's law lacks a physical explanation.

To investigate this, let us consider a pair of parallel plate conductors between which is inserted a slab of *solid* dielectric, as shown in Fig. 6–9. Let $\pm\sigma$ be the surface charge per unit area on each capacitor plate and $\pm\sigma_p$ be the polarization charge on the outer surface of the intervening dielectric. The two layers of polarization charge will produce equal and opposite fields on each plate and their effects will therefore cancel each other. From the point of view of electrical interaction alone it is not obvious why any change in force at all is obtained when the dielectric layer is introduced, since the only direct interaction between the charges σ, which are assumed to remain constant, is unaffected by the introduction of the

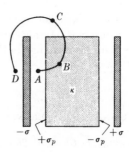

FIG. 6–9. Parallel plate capacitor with a slab of solid dielectri

dielectric slab. That is, the force per unit area remains

$$F_s = \frac{\sigma^2}{2\epsilon_0}$$

as long as the dielectric does not touch the plates of the capacitor. T fore the decrease in force to

$$F_s = \frac{\sigma^2}{2\kappa\epsilon_0},$$

which is experienced when the experiment is performed with a *liquid* th wets the plates and also completely surrounds them, cannot be explaine by electrical forces alone.

This apparent paradox can be explained by taking into account the difference in pressure in the liquid between the field-filled space and the field-free space outside the capacitor plates. This pressure difference is balanced by the internal elastic forces in the case of the solid dielectric in Fig. 6–9, but is transmitted to the plates in the case of the liquid. Let us imagine that the space A in Fig. 6–9 is filled by a nonpolarizable fluid of infinitesimal thickness which can transmit the mechanical pressure. We can then compute the pressure difference $p_A - p_D$ according to Eq. (6–77); moreover, the pressure variation along the path $ABCD$ follows qualitatively the curve of Fig. 6–8. By Eq. (6–77), using the fact that $\mathbf{E}$ has a normal component on the AB interface only, we find that the pressure difference is

$$p_A - p_D = \frac{1}{2\epsilon_0}\left(1 - \frac{1}{\kappa}\right) D_n^2 = \frac{\sigma^2}{2\epsilon_0}\left(1 - \frac{1}{\kappa}\right). \qquad (6\text{–}80)$$

The sum of the force resulting from this pressure and the pure electrical force given by Eq. (6–78) gives the total force of Eq. (6–79) which was derived from energy considerations. Thus the decrease in force that is experienced between two charges when they are immersed in a dielectric between their centers be $d < a$. Show that your answer gives the value expected for the limiting cases (a) $d = 0$, and (b) $d = a$.

7. Show that if the Maxwell tensor of Eq. (6–44) is replaced by

$$T_{\alpha\beta} = E_\alpha D_\beta - \frac{\delta_{\alpha\beta}}{2} (\mathbf{E} \cdot \mathbf{D})(1 - b), \qquad (6\text{–}82)$$

where

$$b = \frac{\rho_m}{\kappa} \left(\frac{d\kappa}{d\rho_m} \right), \qquad (6\text{–}83)$$

then Eq. (6–40) gives the volume force Eq. (6–68) as derived by energy considerations. Hence Eq. (6–82) can be used as the electric Maxwell stress tensor in the presence of dielectrics.

8. From Eq. (6–77) derive the height h to which the dielectric liquid of Fig. 6–7 will rise if the liquid-vacuum boundary is flat and perpendicular to the vertical plates. Also find h from elementary energy considerations.

CHAPTER 7

STEADY CURRENTS AND THEIR INTERACTION

In the discussion of electrostatics only stationary configurations of charge were under consideration. We shall now take into account the flow of charge, or current. According to Oersted's discovery, every current is accompanied by a magnetic field, and it is thus impossible to treat currents fully by means of electric fields alone. Before investigating magnetic interactions, however, we shall consider currents that vary slowly with the time, and which can be assumed to depend entirely on the electric fields present. In the absence of numerical estimates of the relative magnitude of magnetic interactions and the interaction of currents with the lattice structure of the resistive medium that gives rise to electrical resistance, it is not obvious that a situation ever exists in which currents depend only on electric fields. But it turns out that the magnetic effects can be neglected when the fields vary at low frequencies, provided that the dimensions of the conductor are small compared with the so-called "skin depth" of the currents in the particular conductor. We shall also neglect an additional effect, known as the Hall effect, which is present even at zero frequency, and gives rise to redistribution of the equipotential surfaces in a current-carrying conductor. In all but very special substances, however, this effect is extremely small.

7–1 Ohm's law. The conservation of charge in a medium is expressed by the equation of continuity. If $\mathbf{j}$ is the current density within the medium, the rate at which charge leaves a volume V bounded by a surface S will be given by $\int_S \mathbf{j} \cdot d\mathbf{S}$. Since charge is conserved this integral must equal

$$-\frac{dq}{dt} = -\int_V \frac{\partial \rho}{\partial t}\, dV,$$

where q is the charge contained in V. From Gauss's theorem we then obtain the equation of continuity:

$$\mathbf{\nabla} \cdot \mathbf{j} + \frac{\partial \rho}{\partial t} = 0. \tag{7–1}$$

The current is called stationary if there is no accumulation of charge at any point. The criterion for stationary flow is

$$\mathbf{\nabla} \cdot \mathbf{j} = 0. \tag{7–2}$$

To relate the theory of current to the theory of the electric field, an equation is necessary which relates the current and field at any particular point in the conducting material. In many cases they are simply proportional, so that

$$\mathbf{j} = \sigma \mathbf{E}. \tag{7–3}$$

Here σ is the electrical conductivity of the material, measured in mho per meter in the mks system. Equation (7–3) is equivalent to Ohm's law. This relation is a phenomenological characteristic, and is by no means universally valid. The range of current densities over which Eq. (7–3) holds is called the linear range of the particular material, and can be very large, as in metals, or very small, as in a semiconductor. Equation (7–3) implies that the conduction is isotropic. In crystals of low symmetry it must be replaced by a tensor equation.

7–2 Electromotive force. Stationary current is impossible in a purely irrotational electric field, since in a stationary current energy is expended at a rate $\mathbf{j} \cdot \mathbf{E}$ per unit volume, and this energy cannot be provided by an irrotational field. Stationary currents are possible only if there are present sources of electric field known as electromotive forces, which produce non-irrotational fields. If we assume that such electromotive fields exist, and denote them by $\mathbf{E}'$, while $\mathbf{E}$ is the field derivable from a potential, the conduction equation becomes*

$$\mathbf{j} = \sigma(\mathbf{E} + \mathbf{E}'). \tag{7–4}$$

We may define the electromotive force ε by

$$\varepsilon = \oint (\mathbf{E} + \mathbf{E}') \cdot d\mathbf{l} = \oint \mathbf{E}' \cdot d\mathbf{l} = \oint \frac{\mathbf{j} \cdot d\mathbf{l}}{\sigma}. \tag{7–5}$$

The conservative part of the field, $\mathbf{E}$, drops out of the closed line integration, which means that the current is entirely due to the nonconservative forces, although it is influenced by the conductivity and the geometry.

In case the current density is nearly constant over major portions of the path of integration, as frequently happens, Eq. (7–5) can be written as

$$\varepsilon = J \oint \frac{dl}{\sigma S} = JR. \tag{7–6}$$

Here $J = |\mathbf{j}|S$ is the total current, a constant for the circuit (amperes), S is the cross-sectional area of the conductor where the current density is $\mathbf{j}$, and R is the resistance of the conductor (ohms); ε is measured in volts. Equation (7–6) is the form in which Ohm's law is usually stated.

---

* Here $\mathbf{E}'$ consists of all *nonelectrostatic* fields, including those produced by magnetic induction which contribute to the electric field itself, as we shall see in Chapter 9.

Note that in a case where there is no current, we obtain, by integrating Eq. (7–4) along a line between two points 1 and 2 which traverses all of the region in which there is a nonconservative field,

$$-\int_1^2 \mathbf{E} \cdot d\mathbf{l} = \int_1^2 \mathbf{E}' \cdot d\mathbf{l} = \oint \mathbf{E}' \cdot d\mathbf{l} = \varepsilon. \tag{7–7}$$

This indicates that the open circuit electrostatic voltage between two points is equal to the total electromotive force in the circuit. It also follows from Eq. (7–4) that, in a region where $\sigma \neq 0$ and there are nonconservative fields, $\mathbf{E}' = -\mathbf{E}$ in the absence of current flow. Thus, for example, within a given boundary the nonconservative fields (due, e.g., to the chemical potentials) are exactly equal to the electrostatic field which is set up by the charges on the boundaries, if there is no current.

7–3 The solution of stationary current problems. Formally, the current distribution and the field distribution are entirely defined by the nonconservative field and by the conductivity of the medium. By making use of Eqs. (1–28), (7–2), and (7–4), we can write the expressions for $\mathbf{E}$ and $\mathbf{j}$ in terms of $\mathbf{E}'$:

$$\nabla \cdot \mathbf{j} = 0,$$

$$\nabla \times \left(\frac{\mathbf{j}}{\sigma}\right) = \nabla \times \mathbf{E}',$$

$$\nabla \cdot (\sigma \mathbf{E}) = -\nabla \cdot (\sigma \mathbf{E}'), \tag{7–8}$$

$$\nabla \times \mathbf{E} = 0.$$

In the region where there are no nonconservative fields, $\mathbf{E}$ is derivable from a potential, and hence in the case of stationary flow the potential still obeys Laplace's equation:

$$\mathbf{E} = -\nabla \phi,$$

$$\nabla \cdot (\sigma \nabla \phi) = 0,$$

or $\nabla^2 \phi = 0$ if σ is constant. The boundary conditions are different from those of electrostatics, however, since now the conductivities rather than the dielectric constants define the flux relation across a boundary. From Eqs. (7–2) and (7–3) we have, in the absence of nonconservative fields,

$$\nabla \cdot (\sigma \mathbf{E}) = 0. \tag{7–9}$$

Just as in the analogous case for dielectrics in Chapter 2, the divergence equation yields the boundary condition on the normal component of the field at the surface between two media designated by the subscripts 1 and 2:

$$\mathbf{n} \cdot (\sigma_2 \mathbf{E}_2 - \sigma_1 \mathbf{E}_1) = 0,$$
$$\mathbf{n} \cdot (\sigma_2 \boldsymbol{\nabla}\phi_2 - \sigma_1 \boldsymbol{\nabla}\phi_1) = 0. \tag{7–10}$$

The curl equation leads to the condition on the tangential component:

$$\mathbf{n} \times (\mathbf{E}_2 - \mathbf{E}_1) = 0,$$
$$\mathbf{n} \times (\boldsymbol{\nabla}\phi_2 - \boldsymbol{\nabla}\phi_1) = 0. \tag{7–11}$$

It follows that the solution of stationary current distribution problems is mathematically identical to the solution of electrostatic potential distribution problems that have the same geometry, with the conductivity σ substituted for the dielectric constant κ. Thus all of the methods developed in Chapters 3, 4, and 5 are applicable to these problems. The only difference between the static current problems and the electrostatic problems is that the conductivity in a given region may be zero, while the specific inductive capacity is in general not less than unity. This means that the type of boundary value problems which arise in stationary current flow may, under certain conditions, be quite different from any that can exist in electrostatic cases. As an example: if we consider that the region between a set of parallel capacitor plates is filled with a medium of conductivity σ, then the current in the stationary current range will be exactly uniform over the entire area within the conducting medium, while in the analogous electrostatic case the field distribution will be only approximately uniform and will be disturbed by the fringing field at the edges of the plates.

In general, if electrostatic methods permit the calculation of the capacity between two electrodes, then one can conclude immediately what the resistance will be between those electrodes if all of the space in which they are located is filled with a homogeneous medium. The capacity between two electrodes 1 and 2 is given by

$$C = \frac{\kappa \epsilon_0 \int \mathbf{E} \cdot d\mathbf{S}}{\int_1^2 \mathbf{E} \cdot d\mathbf{l}}. \tag{7–12}$$

The numerator is the charge on each electrode by Gauss's flux theorem, and the denominator is the potential difference between the electrodes. The resistance between the two electrodes is

$$R = \frac{\int_1^2 \mathbf{E} \cdot d\mathbf{l}}{\sigma \int \mathbf{E} \cdot d\mathbf{S}}, \tag{7–13}$$

where the denominator gives the net flux of current and the numerator is the potential difference. On comparing Eqs. (7–12) and (7–13), we see that

$$\frac{1}{R\sigma} = \frac{C}{\kappa\epsilon_0} \qquad (7\text{–}14)$$

or

$$RC = \frac{\kappa\epsilon_0}{\sigma} . \qquad (7\text{–}15)$$

We note that the product of the resistance and the capacity is a constant that depends only on the conductivity of the conductor and on the specific inductive capacity of the material between the capacitor plates, and not on the geometry. However, it is not always possible to find an electrostatic problem which will be fully analogous to the corresponding stationary current problem, since the difference in the range of dielectric constants and conductivities mentioned above causes different field patterns in the two cases. The formula for the capacity of a parallel plate capacitor, when edge effects are neglected, is

$$C = \frac{\kappa\epsilon_0 S}{l} , \qquad (7\text{–}16)$$

where S is the area of the plates and l is the distance between them. With this substitution for C, Eq. (7–15) gives

$$R = \frac{l}{\sigma S} \qquad (7\text{–}17)$$

for the resistance of a wire of known length l and cross-sectional area S. The formula for the resistance of a wire is, of course, applicable for large values of l, since the zero conductivity of the surrounding medium prevents fringing of the current flow lines, in strong contrast to the shape of the field lines that would arise from capacitor plates that coincided with the ends of a wire of ordinary length.

7–4 Time of relaxation in a homogeneous medium. If the currents $\mathbf{j}$ are nonstationary but are varying at a sufficiently small rate that magnetic effects are negligible, the equation of continuity (7–1) can be used to determine the rate at which stationary conditions are established in a conducting medium.

From Eqs. (2–4) and (7–1),

$$\nabla \cdot \left(\mathbf{j} + \frac{\partial \mathbf{D}}{\partial t} \right) = 0, \qquad (7\text{–}18)$$

and hence

$$\mathbf{j} + \frac{\partial \mathbf{D}}{\partial t} = \mathbf{j}_S, \qquad (7\text{–}19)$$

where $\mathbf{j}_S$ is the appropriate solution of the equation of stationary current flow (7–2), i.e., $\nabla \cdot \mathbf{j}_S = 0$. If the medium is linear, $\mathbf{D} = (\epsilon_0 \kappa / \sigma)\mathbf{j}$ at all points where no nonconservative fields $\mathbf{E}'$ are present; therefore

$$\left(1 + \frac{\epsilon_0 \kappa}{\sigma} \frac{\partial}{\partial t}\right) \mathbf{j} = \mathbf{j}_S. \tag{7–20}$$

Hence the current will "relax" exponentially with a characteristic time constant

$$\tau = \epsilon_0 \kappa / \sigma, \tag{7–21}$$

such that

$$\mathbf{j} = \mathbf{j}_S + \mathbf{j}_0 e^{-t/\tau}, \tag{7–22}$$

and

$$\frac{\partial \rho}{\partial t} = (-\nabla \cdot \mathbf{j}_0) e^{-t/\tau}, \tag{7–23}$$

where $\mathbf{j}_0$ is the initial nonstationary part of the current distribution.

The relaxation time is a characteristic time for a medium in that it gives an indication of the time in which essentially stationary conditions will be reached after the initiation of a particular flow of charge. The criterion that must be used to determine whether the stationary current equations will be applicable in a particular case is whether the time of observation following the inception of such currents exceeds the relaxation time τ by a sufficiently large amount.

7–5 The magnetic interaction of steady line currents. The magnetic interaction of currents is best described in terms of an experimentally established interaction in vacuum that is analogous to the electrostatic Coulomb's law. The mathematical generalization of the results of Ampère's experiments, which gave the force between two current-carrying elements, as shown in Fig. 7–1, is

$$\mathbf{F}_2 = \frac{\mu_0}{4\pi} J_1 J_2 \oint_1 \oint_2 \frac{d\mathbf{l}_2 \times (d\mathbf{l}_1 \times \mathbf{r}_{12})}{r_{12}^3}. \tag{7–24}$$

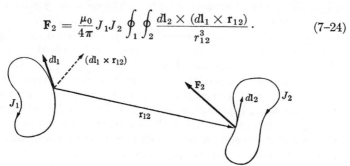

Fig. 7–1. Illustrating Ampère's law.

Here μ_0 is a constant characteristic of the system of units, and is equal to $4\pi \times 10^{-7}$ henry/meter in the mks system; $\mathbf{F}_2$ is the force on the circuit that carries the current J_2 and whose line element is dl_2. Due to the geometry that is involved in expressing the relative directions of $\mathbf{F}_2$, dl_1, dl_2, and $\mathbf{r}_{12}$, this force equation appears to be more complicated than the Coulomb force equation, Eq. (1–24). Also it appears, superficially, to violate Newton's third law of the equality of action and reaction. The integrand of Eq. (7–24) is, in fact, asymmetrical as it stands; when the integral is carried out over two closed circuits, however, the resulting force is symmetrical in terms of the geometry of the two interacting current loops.

The symmetry of the force between the two circuits can be shown as follows. If the integrand is expanded by the vector product rule,

$$\mathbf{A} \times (\mathbf{B} \times \mathbf{C}) = (\mathbf{A} \cdot \mathbf{C})\mathbf{B} - (\mathbf{A} \cdot \mathbf{B})\mathbf{C}, \qquad (7\text{–}25)$$

$\mathbf{F}_2$ becomes

$$\mathbf{F}_2 = \frac{\mu_0}{4\pi} J_1 J_2 \oint_1 \oint_2 \left\{ \frac{(dl_2 \cdot \mathbf{r}_{12})\,dl_1}{r_{12}^3} - \frac{(dl_1 \cdot dl_2)\mathbf{r}_{12}}{r_{12}^3} \right\}. \qquad (7\text{–}26)$$

In the first term the integrand with respect to dl_2 is an exact differential, since it is the line integral of a gradient, $\oint \boldsymbol{\nabla}(1/r_{12}) \cdot dl_2$. Therefore this term vanishes as the integration is carried out over the closed loop. The other integral,

$$\mathbf{F}_2 = -\frac{\mu_0}{4\pi} \left[\oint_1 \oint_2 \frac{(dl_1 \cdot dl_2)\mathbf{r}_{12}}{r_{12}^3} \right] J_1 J_2, \qquad (7\text{–}27)$$

is symmetric in terms of loops 1 and 2.

We have here shown that the basic law of interaction, Eq. (7–24), is, in fact, symmetrical in the current elements of a circuit and is therefore not in violation of Newton's third law. The question is often raised as to how contradiction with the third law can be avoided if each current element is considered, not as part of a closed loop, but as a charge moving with velocity $\mathbf{u}$. In that case the first term in Eq. (7–26) would not vanish, and therefore Newton's third law is not satisfied. This situation, however, represents nonstationary conditions of electromagnetic field, and one cannot exclude the possibility that momentum will be carried away by the change in electromagnetic field. In fact, we shall show later that electromagnetic momentum has to be associated with the electromagnetic field, and that therefore in a nonstationary problem involving electromagnetic interactions action need not be directly equal to reaction between the two interacting sources. A complete discussion of this point must be postponed until nonstationary problems can be treated consistently.

7–6 The magnetic induction field. The reason for taking Eq. (7–24) as the starting point for the discussion of magnetic interactions is that Eq. (7–24) is in such a form that the interaction expression can be separated into a field produced by loop 1 and a force exerted by this field on loop 2. Equation (7–27), on the other hand, which implicitly contains the cosine of the angle between the circuit elements, does not permit such a separation and therefore does not lead directly to a vector field formulation of magnetic interactions. The separation of Eq. (7–24) into a field and a field force can be carried out by putting

$$\mathbf{F}_2 = J_2 \oint dl_2 \times \mathbf{B}_2, \tag{7–28}$$

where $\mathbf{B}_2$ is the magnetic field of induction produced by circuit 1 at the position of circuit 2, and is

$$\mathbf{B}_2 = \frac{\mu_0}{4\pi} J_1 \oint_1 \frac{dl_1 \times \mathbf{r}_{12}}{r_{12}^3}$$

$$= -\frac{\mu_0}{4\pi} J_1 \oint dl_1 \times \mathbf{\nabla}_2 \left(\frac{1}{r_{12}}\right). \tag{7–29}$$

(The mks unit of $\mathbf{B}$ is the weber/meter².)

The magnetic induction field $\mathbf{B}$ is analogous to $\mathbf{E}$ in electrostatic theory in that it determines the force that acts on a circuit element. Equation (7–29) is a generalization of the law of Biot and Savart. It should be noted that thus far we have no differential form of this law. The Biot and Savart law when expressed in terms of volume currents becomes

$$\mathbf{F} = \int (\mathbf{j} \times \mathbf{B}) \, dv, \tag{7–30}$$

$$\mathbf{B} = \frac{\mu_0}{4\pi} \int \frac{\mathbf{j} \times \mathbf{r}}{r^3} \, dv'. \tag{7–31}$$

Note that $\mathbf{r}$ in Eq. (7–31) is directed from the point of integration or source point (where $\mathbf{j}$ is located) toward the field point where $\mathbf{B}$ is being determined. The functional relations are analogous to those discussed in Section 1–1.

7–7 The magnetic scalar potential. Let us inquire under what conditions the magnetic induction field $\mathbf{B}$ can be derived from a scalar potential by the relation

$$\mathbf{B} = -\mu_0 \mathbf{\nabla} \phi_m. \tag{7–32}$$

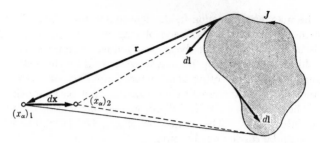

FIG. 7-2. Derivation of magnetic scalar potential.

Consider a closed loop carrying a current J, as in Fig. 7–2, together with a field point $(x_\alpha)_1$. If the field $\mathbf{B}$ were derivable from a scalar magnetic potential ϕ_m, and if the point of observation were moved through a distance $d\mathbf{x}$ from $(x_\alpha)_1$ to $(x_\alpha)_2$, then the increment in the scalar magnetic potential ϕ_m would be given by

$$d\phi_m = -\frac{d\mathbf{x} \cdot \mathbf{B}}{\mu_0}. \qquad (7\text{--}33)$$

The Biot and Savart expression, Eq. (7–29), may be used for $\mathbf{B}$, so that

$$d\phi_m = -\frac{1}{4\pi} J \oint \frac{d\mathbf{x} \cdot (d\mathbf{l} \times \mathbf{r})}{r^3} = -\frac{1}{4\pi} J \oint \frac{\mathbf{r} \cdot (d\mathbf{x} \times d\mathbf{l})}{r^3}. \qquad (7\text{--}34)$$

(The mixed vector-scalar product permits cyclic permutation.) Equation (7–34) is, of course, equal to the change in the scalar magnetic potential which is obtained if the point of observation were held stationary and the loop were moved by an amount $-d\mathbf{x}$, as was discussed in the derivation of Eq. (1–66) for the potential of a dipole sheet. By means of the relation given in Eq. (1–64), the change in the scalar magnetic potential can be written as

$$d\phi_m = \frac{J}{4\pi} d\Omega, \qquad (7\text{--}35)$$

where $d\Omega$ is the change in the solid angle subtended by the loop at the point of observation brought about by an infinitesimal displacement $-d\mathbf{x}$ of all points of the loop.

This scalar potential has the same mathematical properties as the solution of the electrostatic potential of the surface dipole layer discussed earlier, as seen from the similarity of Eq. (1–61) for the static potential of a dipole layer and Eq. (7–35) for the magnetic induction of a current loop. In addition, this means that the scalar potential of a current loop is multiple valued in the sense that it appears to undergo a discontinuity of magnitude J when a surface bounded by the loop is crossed. In the case of the electric dipole sheet, this surface has a physical significance. In the magnetic case, however, the surface can be chosen in any arbitrary fashion. Since the choice of the surface is arbitrary, the magnetic field derived from such a potential outside the current-carrying region is unambiguous, but line integrals of the magnetic field of a current loop will be correct only if the path of integration does not pass through the arbitrary surface. The line integral of the magnetic field of induction $\mathbf{B}$ around a closed path threading the current J is exactly equal to the magnitude of the discontinuity in the magnetic scalar potential ϕ_m across the arbitrary reference surface, and hence

$$\oint \mathbf{B} \cdot d\mathbf{l} = \mu_0 J_{\text{total}}. \tag{7–36}$$

For a graphical representation of the magnetic scalar potential of a current loop, see Fig. 1–7. The dipole layer of Fig. 1–7 corresponds to the arbitrarily located surface of discontinuity of the current loop.

Equation (7–36) is the integral representation of the differential relation that gives the total circulation of the magnetic field vector in terms of the current that causes the magnetic field. Since Eq. (7–36) is valid for any arbitrary closed path of integration, we can convert it into a differential expression by substituting $\mathbf{B}$ into Stokes' theorem and reducing the size of the surface of integration to a differential, thus securing

$$\boldsymbol{\nabla} \times \mathbf{B} = \mu_0 \mathbf{j}. \tag{7–37}$$

We conclude that $\mathbf{B}$ cannot *in general* be derived from a single-valued magnetic scalar potential. The concept of the magnetic scalar potential is of practical utility only to derive magnetic fields in the absence of continuous current distributions. The scalar potential ϕ_m cannot be used if line integrals encircling any currents are considered, or if the fields within current-carrying media are desired.

7–8 The magnetic vector potential. From the defining equation for $\mathbf{B}$, Eq. (7–31), we may determine the divergence as well as the curl by direct differentiation:

$$\boldsymbol{\nabla} \cdot \mathbf{B} = \boldsymbol{\nabla} \cdot \left(\frac{\mu_0}{4\pi}\right) \int \frac{\mathbf{j} \times \mathbf{r}}{r^3} \, dv' = \boldsymbol{\nabla} \cdot \left\{\left(\frac{\mu_0}{4\pi}\right)\left[-\int \mathbf{j} \times \boldsymbol{\nabla}\left(\frac{1}{r}\right) dv'\right]\right\}$$

$$= \frac{\mu_0}{4\pi} \int \mathbf{j} \cdot \left[\boldsymbol{\nabla} \times \boldsymbol{\nabla}\left(\frac{1}{r}\right)\right] dv' = 0. \tag{7–38}$$

Similarly,

$$\mathbf{\nabla} \times \mathbf{B} = -\frac{\mu_0}{4\pi} \int \mathbf{\nabla} \times \left[\mathbf{j} \times \mathbf{\nabla} \left(\frac{1}{r} \right) \right] dv'$$

$$= -\frac{\mu_0}{4\pi} \int \mathbf{j} \nabla^2 \left(\frac{1}{r} \right) dv' + \frac{\mu_0}{4\pi} \int [(\mathbf{j} \cdot \mathbf{\nabla}') \mathbf{\nabla}'] \frac{1}{r} dv'.$$

Integrating the second term by parts and dropping a surface term, we have

$$\mathbf{\nabla} \times \mathbf{B} = -\frac{\mu_0}{4\pi} \int \mathbf{j} \nabla^2 \left(\frac{1}{r} \right) dv' - \frac{\mu_0}{4\pi} \int (\mathbf{\nabla}' \cdot \mathbf{j}) \mathbf{\nabla}' \left(\frac{1}{r} \right) dv'.$$

The last term vanishes [see Eq. (1–16)]; hence

$$\mathbf{\nabla} \times \mathbf{B} = \mu_0 \int \mathbf{j} \, \delta(\mathbf{r}) \, dv' = \mu_0 \mathbf{j}, \tag{7-39}$$

in agreement with Eq. (7–37).

We have in the above implicitly assumed that currents are fundamentally the only sources of magnetic field, and that the field of such currents is entirely given by the law of Biot and Savart. Since the divergence of **B** is zero, it follows that **B** may be derived from a vector potential:

$$\mathbf{B} = \mathbf{\nabla} \times \mathbf{A}. \tag{7-40}$$

The vector potential **A** may be identified if we rewrite the expression for **B**, remembering that the operator **∇** acts only on the variable of field position and can therefore be taken outside of the integration sign:

$$\mathbf{B} = -\frac{\mu_0}{4\pi} \int \mathbf{j} \times \mathbf{\nabla} \left(\frac{1}{r} \right) dv' = \frac{\mu_0}{4\pi} \mathbf{\nabla} \times \int \frac{\mathbf{j}}{r} dv'. \tag{7-41}$$

The explicit expression for **A** in terms of the current is therefore given by

$$\mathbf{A} = \frac{\mu_0}{4\pi} \oint \frac{\mathbf{j}}{r} dv', \tag{7-42}$$

except for an arbitrary function with vanishing curl. For convenience, we may here choose this additional function to be zero, although later this choice will need modification. With this assumption, however, Eq. (7–42) is correct, and the corresponding expression for the vector potential of a linear current distribution is

$$\mathbf{A} = \frac{\mu_0}{4\pi} J \oint \frac{d\mathbf{l}}{r}. \tag{7-43}$$

The fields produced by currents can therefore be computed by first determining the vector potential **A**, using Eq. (7–42) or Eq. (7–43), and then obtaining the magnetic field by means of the relation (7–40). We shall postpone treating the problem of ascertaining vector potentials corresponding to specific current distributions until after we have examined the types of currents encountered and their effect on the fields in material media.

7–9 Types of currents. The original direct experimental observations of the magnetic interactions of currents were made on steady linear currents. Modifications must be introduced into the theory if nonstationary currents are to be treated, and before we can derive the magnetic effects that occur within an arbitrary medium.

We shall here discuss currents in material media in a manner similar to that used to treat charges in material media. Currents may be classified in two categories: true currents that may be identified with the motion of true charges, and other currents which are associated with the medium itself. This separation, which is analogous to the separation that was made in the electrostatic theory between the potentials of true charges and the potentials of polarization charges, will lead us to consider two types of magnetic fields, one derived from true currents and the other derived from the combined effects of all the currents whatever may be their origin. It is this latter field, namely, the field of magnetic induction **B**, that can be considered to be the space-time average of the interatomic fields.

Let us classify the types of currents that must be considered:

1. *True currents*, **j**, identical with the physical transport of true charges.

2. *Polarization currents*, later shown to be $\partial \mathbf{P}/\partial t$, currents that arise from the change of the polarization with time.

3. *Magnetization currents*, $\mathbf{j}_m$, stationary currents that flow within regions that are inaccessible to observation but which might give rise to net boundary or volume currents, due to imperfect orbit cancellation on an atomic scale.

4. *Convective currents.* If a material medium in motion contains charges of various types, additional currents will be obtained which arise from convective effects. These convective currents will be derived from the motion of both true and polarization charges contained in the medium. The convective currents will be discussed in a later chapter. For our present discussion we need a more detailed investigation of polarization and magnetization currents.

7–10 Polarization currents. If ρ represents the charge density within a molecule and the molecular coordinates are designated by $\boldsymbol{\xi}$, then the electrical moment $\mathbf{p}$ of a polarized molecule is defined by

$$\mathbf{p} = \int \mathbf{P} \, dv = \int \rho \boldsymbol{\xi} \, dv, \qquad (7\text{--}44)$$

where $\mathbf{P} = \rho \boldsymbol{\xi}$. If the charge density within the molecule is changing in time,

$$\frac{\partial \mathbf{p}}{\partial t} = \int \frac{\partial \rho}{\partial t} \boldsymbol{\xi} \, dv. \qquad (7\text{--}45)$$

By means of the equation of continuity, Eq. (7–45) may be written

$$\frac{\partial \mathbf{p}}{\partial t} = - \int [\boldsymbol{\nabla} \cdot (\rho \mathbf{u})] \boldsymbol{\xi} \, dv. \qquad (7\text{--}46)$$

Integrating by parts and dropping the surface term, which is justified if we choose the surface of integration so that it lies outside the region where there are molecular charges, we obtain

$$\frac{\partial \mathbf{p}}{\partial t} = \int \rho \mathbf{u} \, dv, \qquad (7\text{--}47)$$

or, on a large scale,

$$\frac{\partial \mathbf{P}}{\partial t} = \frac{\int \rho \mathbf{u} \, dv}{\int dv} = \overline{\rho \mathbf{u}}. \qquad (7\text{--}48)$$

Hence the quantity $\partial \mathbf{P}/\partial t$ does represent the space-time average value of the molecular currents caused by a varying polarization.

7–11 Magnetic moments. The magnetic moment of a particular volume containing currents $\mathbf{j}_m$ is defined as

$$\mathbf{m} = \tfrac{1}{2} \int (\boldsymbol{\xi} \times \mathbf{j}_m) \, dv. \qquad (7\text{--}49)$$

If these currents can be considered as charge densities ρ moving with velocity $\mathbf{u}$, Eq. (7–49) becomes

$$\mathbf{m} = \tfrac{1}{2} \int \rho (\boldsymbol{\xi} \times \mathbf{u}) \, dv, \qquad (7\text{--}50)$$

which is analogous to the expression for the mechanical angular momentum $\mathbf{s}$ in terms of the velocity of a distribution of mass densities ρ_m:

$$\mathbf{s} = \int \rho_m (\boldsymbol{\xi} \times \mathbf{u}) \, dv. \qquad (7\text{--}51)$$

The ratio $\mathbf{m}/\mathbf{s}$ is known as the gyromagnetic ratio Γ of the system.* If the system is composed of particles of mass m and charge e, then clearly,

$$\Gamma = e/2m. \tag{7–52}$$

Hence any classical distribution of charged objects of a single kind will have a gyromagnetic ratio given by (7–52). If the distribution has a more complex structure we can write

$$\Gamma = ge/2m, \tag{7–53}$$

where g (the "g-factor") describes the magnetic structure. The g-factor can be calculated for any classical structure, but it is customary to retain the definition of g as given by (7–53) even for objects for which such a calculation is not possible.

Note that the definition of $\mathbf{m}$ given by Eq. (7–49) is a purely kinematic definition in the sense that it does not involve any mention of an actual interaction, magnetic or otherwise. This corresponds to the definition of the electric moment of a region given in Eq. (7–44), which is also only a kinematic description of a specific alignment of charges, although the net charge of the volume is zero. The magnetic moment definition is a description of a system of currents which need not produce any net flow across a surface large enough to be accessible to macroscopic observation. In the special case of a single "stationary current" loop that encloses a given area $\mathbf{S}$, $\mathbf{m}$ becomes simply the product of the current in the loop and its area, directed normal to the loop in such a way as to agree with the right-hand rule for the current circulation:

$$\mathbf{m} = J\mathbf{S}. \tag{7–54}$$

This is in agreement with the elementary definition of the magnetic moment of a current loop.

Let us now relate the concept of magnetic moment to magnetic interactions by expanding the vector potential, Eq. (7–42), as a "multipole expansion," analogous to that of the scalar potential in Chapter 1. If the current distribution is confined to a volume whose dimensions are small in comparison with the distance to the field point, we may set $\mathbf{r} = \mathbf{R} - \boldsymbol{\xi}$, and expand $1/r$ as in Eq. (1–46). Since the expansion is valid

---

* For a classical system there is no basic reason why the vectors $\mathbf{m}$ and $\mathbf{s}$ should be parallel, and therefore the gyromagnetic ratio Γ should be a 3×3 matrix. We are restricting ourselves here to a system having only a single preferred axis along which both $\mathbf{m}$ and $\mathbf{s}$ are directed.

for any current so confined in space, we shall omit the subscript:

$$\frac{4\pi}{\mu_0}\mathbf{A} = \int \frac{\mathbf{j}\,dv'}{r} = \frac{1}{R}\int \mathbf{j}\,dv' - \left[\frac{\partial}{\partial x_\alpha}\left(\frac{1}{r}\right)\right]_R \int \xi_\alpha \mathbf{j}\,dv'$$

$$+ \frac{1}{2}\left[\frac{\partial}{\partial x_\alpha}\frac{\partial}{\partial x_\beta}\left(\frac{1}{r}\right)\right]_R \int \xi_\alpha \xi_\beta \mathbf{j}\,dv' - \cdots. \quad (7\text{-}55)$$

The first term, $\int \mathbf{j}\,dv'$, vanishes for stationary currents, as can be seen physically by dividing the system into current loops, i.e., by writing

$$\int \mathbf{j}\,dv' = \sum_{\text{loops}} J \oint d\mathbf{l} = 0. \quad (7\text{-}56)$$

Formally we can write, if $\boldsymbol{\nabla} \cdot \mathbf{j} = 0$,

$$0 = \int (\boldsymbol{\nabla} \cdot \mathbf{j})\xi_\alpha\,dv' = \int \boldsymbol{\nabla} \cdot (\xi_\alpha \mathbf{j})\,dv' - \int (\mathbf{j} \cdot \boldsymbol{\nabla}\xi_\alpha)\,dv'$$

$$= \int_S (\xi_\alpha \mathbf{j}) \cdot d\mathbf{S} - \int j_\beta \frac{\partial \xi_\alpha}{\partial \xi_\beta}\,dv' = \int j_\alpha\,dv', \quad (7\text{-}57)$$

where the surface integral has been omitted because we assume the source distribution to be bounded. Hence $\int j_\alpha\,dv'$ and thus $\int \mathbf{j}\,dv'$ vanish, and the vector potential will be given, even in the limit of small distributions of current, by the second and further terms in the expansion. Since $(\partial/\partial x_\alpha)(1/r)$ evaluated at R equals $-R_\alpha/R^3$, we are concerned with

$$\frac{4\pi}{\mu_0} A_\beta = \frac{R_\alpha \int \xi_\alpha j_\beta\,dv'}{R^3}, \quad (7\text{-}58)$$

where the integration is, of course, over $\boldsymbol{\xi}$. The integrand of Eq. (7-58) can be transformed by adding and subtracting $\xi_\beta j_\alpha/2$ into

$$\xi_\alpha j_\beta = \tfrac{1}{2}(\xi_\alpha j_\beta + \xi_\beta j_\alpha) + \tfrac{1}{2}(\xi_\alpha j_\beta - \xi_\beta j_\alpha). \quad (7\text{-}59)$$

The integral of the first part of this expression vanishes for a steady current, as may be seen by writing, as in Eq. (7-57),

$$\xi_\alpha \xi_\beta (\boldsymbol{\nabla} \cdot \mathbf{j}) = 0 = \boldsymbol{\nabla} \cdot (\xi_\alpha \xi_\beta \mathbf{j}) - \mathbf{j} \cdot \boldsymbol{\nabla}(\xi_\alpha \xi_\beta). \quad (7\text{-}60)$$

If we apply Gauss's theorem to the complete divergence, we obtain

$$0 = \int \mathbf{j} \cdot \boldsymbol{\nabla}(\xi_\alpha \xi_\beta)\,dv' = \int (\xi_\alpha j_\beta + \xi_\beta j_\alpha)\,dv'. \quad (7\text{-}61)$$

But

$$R_\alpha(\xi_\alpha j_\beta - \xi_\beta j_\alpha) = [(\boldsymbol{\xi} \times \mathbf{j}) \times \mathbf{R}]_\beta,$$

so that

$$\frac{4\pi}{\mu_0} \mathbf{A} = \int \frac{(\boldsymbol{\xi} \times \mathbf{j}) \times \mathbf{R}}{2R^3} \, dv' = \frac{\mathbf{m} \times \mathbf{R}}{R^3} = -\mathbf{m} \times \boldsymbol{\nabla}\left(\frac{1}{R}\right). \quad (7\text{–}62)$$

Hence the leading term of the multipole expansion yields the vector potential of a magnetic dipole **m**.

Let us consider the torque **L** on a magnetic dipole. By definition the torque **L** acting on an arbitrary current system in an external uniform magnetic field **B** is given by

$$\mathbf{L} = \int \boldsymbol{\xi} \times (\mathbf{j} \times \mathbf{B}) \, dv' \quad (7\text{–}63)$$

$$= \int \mathbf{j}(\boldsymbol{\xi} \cdot \mathbf{B}) \, dv' - \mathbf{B} \int (\mathbf{j} \cdot \boldsymbol{\xi}) \, dv'. \quad (7\text{–}64)$$

The second term vanishes for a bounded current distribution, as may be shown by a procedure similar to Eqs. (7–57) and (7–59):

$$0 = \int \xi^2(\boldsymbol{\nabla} \cdot \mathbf{j}) \, dv' = \int \boldsymbol{\nabla} \cdot (\xi^2 \mathbf{j}) \, dv' - \int \mathbf{j} \cdot \boldsymbol{\nabla}(\xi^2) \, dv'$$
$$= \int_S (\xi^2 \mathbf{j}) \cdot d\mathbf{S} - 2\int (\mathbf{j} \cdot \boldsymbol{\xi}) \, dv'. \quad (7\text{–}65)$$

The surface term may be omitted, and thus we find that the second term of Eq. (7–64) vanishes. The first term has an α-component which, by using Eq. (7–60), we may write:

$$L_\alpha = B_\beta \int j_\alpha \xi_\beta \, dv' = B_\beta \int \left(\frac{\xi_\beta j_\alpha - \xi_\alpha j_\beta}{2}\right) dv'. \quad (7\text{–}66)$$

Hence

$$\mathbf{L} = \mathbf{m} \times \mathbf{B} \quad (7\text{–}67)$$

is the simple expression for the torque, in complete analogy with the torque on an electric dipole **p** in an electrostatic field **E** [Eq. (1–57)].

If the angular momentum is **s**, the classical equations of motion are

$$d\mathbf{s}/dt = \mathbf{m} \times \mathbf{B} = \Gamma(\mathbf{s} \times \mathbf{B}), \quad (7\text{–}68)$$

where Γ is the gyromagnetic ratio, as before. Hence if **s** makes an angle with the direction of **B**, the direction of **s** will "precess" about the direction of **B** with an angular velocity

$$\boldsymbol{\omega} = \Gamma\mathbf{B}. \quad (7\text{–}69)$$

7-12 Magnetization and magnetization currents. Let us consider a region in which there is a distribution of magnetic moments such that each volume element dv' has a magnetic moment

$$d\mathbf{m} = \mathbf{M} \, dv'. \qquad (7\text{-}70)$$

From Eq. (7–62) the vector potential of such a distribution becomes

$$\frac{4\pi}{\mu_0} \mathbf{A} = + \int \mathbf{M} \times \mathbf{\nabla}' \left(\frac{1}{r}\right) dv' = \int \frac{\mathbf{\nabla}' \times \mathbf{M}}{r} \, dv' - \int \mathbf{\nabla}' \times \left(\frac{\mathbf{M}}{r}\right) dv.' \qquad (7\text{-}71)$$

But

$$\int \mathbf{\nabla}' \times \frac{\mathbf{M}}{r} \, dv' = - \int \frac{\mathbf{M} \times d\mathbf{S}}{r}. \qquad (7\text{-}72)$$

If the surface may be taken outside the region where there is current, then the second term of Eq. (7–71) vanishes. From the general form of the vector potential $\mathbf{\nabla} \times \mathbf{M}$ can therefore be identified as the current density and, in particular,

$$\mathbf{j}_m = \mathbf{\nabla} \times \mathbf{M} \qquad (7\text{-}73)$$

enables us to express a volume magnetization in terms of an equivalent current.

The equivalence of magnetized materials and currents in their effect on outside points was first noted by Ampère. The concept is necessary for treating the inaccessible currents of atomic origin in a macroscopic theory. According to Eq. (7–73), the microscopic currents cancel within a region of homogeneous magnetization; $\mathbf{j}_m$ is the net current density produced where there is inhomogeneous magnetization.

Physically, Eq. (7–73) can be understood as follows. Consider the z-component of the magnetic moment of neighboring current loops in a rectangular network in the xy-plane, as seen in Fig. 7–3. If the magnetization is inhomogeneous, there will not be complete cancellation between the boundaries of the loops and there will be a net current. This net current will bring about whatever net effects are ascribable to the currents. The moment of a single loop in Fig. 7–3 is

$$d\mathbf{m} = \mathbf{M} \, dx \, dy \, dz. \qquad (7\text{-}74)$$

Then, from Eq. (7–54), the current in rectangle 1 is

$$J_1 = \frac{M_z \, dx \, dy \, dz}{dx \, dy},$$

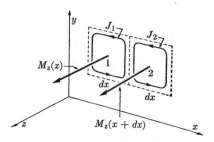

FIG. 7–3. Equivalent Amperian current loops in a magnetized medium, showing cancellation effect on internal boundary.

and by means of Taylor's expansion we may write the current in the neighboring rectangle 2:

$$J_2 = \frac{[M_z + (\partial M_z/\partial x)\ dx]\ dx\ dy\ dz}{dy\ dx}.$$

The difference between J_1 and J_2 results in a net current in the y-direction along the mutual boundary of rectangles 1 and 2:

$$J_y = -\frac{\partial M_z}{\partial x}\ dx\ dz. \qquad (7\text{–}75)$$

This form will be recognized as one of the six components of the curl and, in general,

$$\mathbf{j}_m = \boldsymbol{\nabla} \times \mathbf{M}, \qquad (7\text{–}76)$$

as before in Eq. (7–73). In a region of discontinuous magnetization it is easily seen that a surface current equal to the change in the tangential component of the magnetization will result at such a discontinuity. This follows when Eq. (7–73) is applied to a limiting transverse surface bounding such a discontinuity.

7–13 Vacuum displacement current. In a stationary medium the total current is given by the sum of the first three types of current enumerated in Section 7–9:

$$\mathbf{j}_{\text{total}} = \mathbf{j}_{\text{true}} + \frac{\partial \mathbf{P}}{\partial t} + \boldsymbol{\nabla} \times \mathbf{M}. \qquad (7\text{–}77)$$

To assure the conservation of charge, it is necessary that this total current obey the equation of continuity, $\boldsymbol{\nabla} \cdot \mathbf{j} + (\partial \rho/\partial t) = 0$. The divergence of the total current is

$$\nabla \cdot \mathbf{j}_{total} = \nabla \cdot \mathbf{j}_{true} + \nabla \cdot \left(\frac{\partial \mathbf{P}}{\partial t}\right) + \nabla \cdot (\nabla \times \mathbf{M}), \qquad (7\text{--}78)$$

so that the equation of continuity as applied to the total charge and current densities is

$$\nabla \cdot \mathbf{j}_{true} + \nabla \cdot \left(\frac{\partial \mathbf{P}}{\partial t}\right) + \nabla \cdot (\nabla \times \mathbf{M}) + \left(\frac{\partial \rho}{\partial t}\right)_{total} = 0. \quad (7\text{--}79)$$

Since

$$\rho_{total} = \epsilon_0 \nabla \cdot \mathbf{E},$$
$$\left(\frac{\partial \rho}{\partial t}\right)_{total} = \epsilon_0 \nabla \cdot \frac{\partial \mathbf{E}}{\partial t}, \qquad (7\text{--}80)$$

and thus

$$\nabla \cdot \mathbf{j}_{true} + \nabla \cdot \left(\frac{\partial \mathbf{P}}{\partial t}\right) + \nabla \cdot (\nabla \times \mathbf{M}) + \epsilon_0 \nabla \cdot \left(\frac{\partial \mathbf{E}}{\partial t}\right) = 0, \quad (7\text{--}81)$$

or

$$\nabla \cdot \left(\mathbf{j}_{true} + \frac{\partial \mathbf{D}}{\partial t} + \nabla \times \mathbf{M}\right) = 0, \qquad (7\text{--}82)$$

if we make use of the relation $\mathbf{D} = \epsilon_0 \mathbf{E} + \mathbf{P}$.

The divergence of the total current, Eq. (7–78) is not zero, and thus the total current is not solenoidal. However, the quantity

$$\mathbf{c} = \mathbf{j}_{true} + \frac{\partial \mathbf{D}}{\partial t} + \nabla \times \mathbf{M} = \mathbf{j}_{total} + \epsilon_0 \frac{\partial \mathbf{E}}{\partial t}, \qquad (7\text{--}83)$$

generated by adding the term $\epsilon_0 (\partial \mathbf{E}/\partial t)$ to the total current, is a solenoidal current. The need for the addition of this term to produce a solenoidal net current vector was recognized by Maxwell. The "vacuum displacement current" $\epsilon_0 (\partial \mathbf{E}/\partial t)$ does not have the significance of a current in the sense of being the motion of charges. We shall see later that the magnetic effects of currents can be formulated only in terms of solenoidal currents, and therefore that the vacuum displacement current term must be introduced in order to be able to apply the formulas which will be developed for the magnetic interaction to cases involving nonstationary currents.

The geometrical significance of the solenoidal current $\mathbf{c}$ is that at points where there is an accumulation of charge the current is assumed to be continuous across the discontinuity in the form of the rate of change of the field resulting from the accumulation of the charges on the boundaries of the discontinuity. As an example, a battery charging a capacitor produces a closed current loop in terms of $\mathbf{c}$.

Suggested References

M. Abraham and R. Becker, *The Classical Theory of Electricity and Magnetism*. Applications are not emphasized, but the approach to steady currents in Chapter VI is much the same as ours.

L. D. Landau and E. M. Lifshitz, *Electrodynamics of Continuous Media*. The treatment of constant current in Chapter III also includes the Hall effect and thermoelectric phenomena.

G. P. Harnwell, *Principles of Electricity and Electromagnetism*. Stresses linear circuit theory.

W. R. Smythe, *Static and Dynamic Electricity*. Contains many useful and interesting applications.

J. A. Stratton, *Electromagnetic Theory*. Discusses conduction problems along with the electrostatic field.

J. H. Jeans, *Electricity and Magnetism*. The order of material is so different from ours that Jeans may be of limited use in the matter of physical principles, but there are interesting applications.

Exercises

1. Calculate the change in resistance of a wire of diameter D (expressed as a change in length of the wire) due to an internal fissure of diameter d. Let $D - d \ll D$, and treat as a two-dimensional problem. (See Fig. 7–4.)

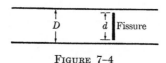

FIGURE 7–4

2. The core (radius a) of a coaxial cylindrical cable is surrounded by an insulating sheath of conductivity σ_1, outer radius b, and a second layer of conductivity σ_2 extending to the outer metallic conductor of radius c. Find the resistance per meter of cable between the core and the outer conductor.

3. Current enters an infinite plane conducting sheet at some point p and leaves at infinity. A circular hole, exclusive of p, is cut anywhere in the sheet. Show that the potential difference between any two points on the edge of the hole is twice that between the same two points before the hole was cut.

4. Two small spherical electrodes of radius a are embedded in a semi-infinite medium of conductivity σ, each at distance $d > a$ from the plane face of the medium and at distance b from each other. Find the resistance between the electrodes.

5. A spherical electrode of radius a is surrounded by a concentric spherical electrode of radius b, while the intervening space is filled with a medium whose conductivity is inversely proportional to the distance from the center of the system. If the outer sphere is maintained at potential ϕ_0 and a total current J flows between the electrodes, find the potential at a distance $r > a$ from the center.

6. The surface of a circular disk of radius a is covered with a continuous and uniform spiral winding of N turns of fine wire, starting at the center and continuing to the edge, through which flows a steady current J. Find the magnetic dipole moment.

7. A circular loop, $z = b$, $x^2 + y^2 = a^2$, carries current J. Show that for $r < R = \sqrt{a^2 + b^2}$ the magnetic scalar potential is proportional to

$$\sum_{n=1}^{\infty} \frac{n+1}{2n+1} \left[P_{n+1} \left(\frac{b}{R} \right) - P_{n-1} \left(\frac{b}{R} \right) \right] \left(\frac{r}{R} \right)^n P_n(\cos \theta).$$

8. In the equations (7–8) of stationary current flow, derive the current density $\mathbf{j}$ from a vector potential and the electric field $\mathbf{E}$ from a scalar potential. Write differential equations for these potentials, assuming σ and $\mathbf{E}'$ to be given functions of position.

CHAPTER 8

MAGNETIC MATERIALS AND BOUNDARY VALUE PROBLEMS

Thus far only magnetic fields in a vacuum have been treated, although we have analyzed the currents which will have to be taken into account if material media are introduced into magnetic fields. In general the current $\mathbf{j}$ in Eq. (7–37) must be replaced by the total current including the magnetization and polarization currents, as given in Eq. (7–78). We have seen, however, that *in vacuo* the curl of $\mathbf{B}$ is proportional to the total stationary current density, while $\mathbf{j}_{\text{total}}$ is not solenoidal if the polarization changes with the time. The relation $\boldsymbol{\nabla} \times \mathbf{B} = \mu_0 \mathbf{j}$ can be generalized to nonstationary cases in one of two ways: either the current remains solenoidal or the relation used in deriving the magnetic field from the current is modified. The choice between these alternatives made by Maxwell was to retain the relations that derive the magnetic field from the current, Eqs. (7–29), (7–31), or (7–48), but to use the general current $\mathbf{c}$ of Eq. (7–83), which includes the displacement current and which remains solenoidal. This choice has been amply justified by its further consequences.

8–1 Magnetic field intensity. If $\mathbf{c}$ is used as the total current the defining equations for the vector field $\mathbf{B}$ are

$$\boldsymbol{\nabla} \cdot \mathbf{B} = 0, \qquad (8\text{–}1)$$

$$\boldsymbol{\nabla} \times \mathbf{B} = \mu_0 \mathbf{c} = \mu_0 \left(\mathbf{j}_{\text{true}} + \boldsymbol{\nabla} \times \mathbf{M} + \frac{\partial \mathbf{D}}{\partial t} \right). \qquad (8\text{–}2)$$

In the treatment of the polarization of dielectrics in Chapter 2 it was found mathematically convenient to separate the field whose sources were true charges from the total field whose sources were the true charges plus the polarization charges. In a similar manner, it is convenient to separate from the total field that part whose circulation density arises from atomic magnetization currents. Therefore if we write Eq. (8–2) in the form

$$\boldsymbol{\nabla} \times (\mathbf{B} - \mu_0 \mathbf{M}) = \mu_0 \left(\mathbf{j}_{\text{true}} + \frac{\partial \mathbf{D}}{\partial t} \right) \qquad (8\text{–}3)$$

and define a new field $\mathbf{H}$ by

$$\mathbf{H} = \frac{1}{\mu_0} (\mathbf{B} - \mu_0 \mathbf{M}) = \frac{\mathbf{B}}{\mu_0} - \mathbf{M}, \qquad (8\text{–}4)$$

this definition reduces to

$$\nabla \times \mathbf{H} = \mathbf{j}_{\text{true}} + \frac{\partial \mathbf{D}}{\partial t}. \qquad (8\text{--}5)$$

The quantity $\mathbf{H}$, the magnetic field intensity, is measured in amp-turns/m in the mks system of units. Equation (8–5) then means that the circulation density of $\mathbf{H}$ arises from the true current plus the total displacement current, the latter including both the polarization current $\partial \mathbf{P}/\partial t$ and the vacuum displacement current $\epsilon_0 \, \partial \mathbf{E}/\partial t$. Under stationary conditions, or quasi-stationary conditions in which the magnetic effect of the displacement current is negligible compared with the magnetic effect of the true current,

$$\nabla \times \mathbf{H} = \mathbf{j}_{\text{true}}, \qquad (8\text{--}6)$$

and in integral form,

$$\oint \mathbf{H} \cdot d\mathbf{l} = J_{\text{true}}. \qquad (8\text{--}7)$$

Note that in the sense of the separation between the effect that is produced by the total and the true charges and the total and true currents, respectively, $\mathbf{B}$ plays a role that corresponds to $\mathbf{E}$, while $\mathbf{H}$ plays a role that corresponds to $\mathbf{D}$, as may be seen by comparing the above equations with

$$\int \mathbf{E} \cdot d\mathbf{S} = \frac{q_{\text{total}}}{\epsilon_0},$$

$$\int \mathbf{D} \cdot d\mathbf{S} = q_{\text{true}}$$

of Chapter 2.

8–2 Magnetic sources. The discussion of dielectrics in Chapter 2 was limited to the case of linear media, namely, media in which the polarization was proportional to the applied electric field. In ferromagnetic substances, however, the case of nonlinear behavior is most common, and therefore we must discuss some of the properties of the magnetic field which arise in cases when the magnetization $\mathbf{M}$ is not a linear and often not even a unique function of the external fields. At first we shall assume $\mathbf{M}$ to be a given function of the material medium independent of external fields. In the most extreme case, that of a permanent magnet, there will be a magnetic moment $\mathbf{M}$ per unit volume, even in the absence of any true currents. In this case, as we see from Eq. (8–6), $\mathbf{H}$ is irrotational and will therefore behave mathematically like an electrostatic field, while $\mathbf{B}$ remains, of course, solenoidal:

$$\nabla \times \mathbf{H} = 0,$$

$$\nabla \times \mathbf{B} = \mu_0(\nabla \times \mathbf{M}) \neq 0. \qquad (8\text{--}8)$$

However, the magnetic field will have sources; if we call the magnetic source density ρ_M,

$$\mathbf{\nabla} \cdot \mathbf{H} = -\mathbf{\nabla} \cdot \mathbf{M} = \rho_M. \tag{8–9}$$

The "magnetostatic field" $\mathbf{H}$ can therefore be derived from a magnetic source density which is equal to the negative divergence of the magnetization. One unit of this equivalent magnetic charge density is usually known as a unit magnetic pole. In terms of this description, a magnetic pole has no physical reality other than that it makes the mathematical description of the resultant magnetic field of a permanent magnet formally the same as that of the electric field of charges. Since the magnetic field $\mathbf{H}$ of a permanent magnet is irrotational, it can be derived from a magnetic scalar potential ϕ_m in the same way that $\mathbf{E}$ may be derived from the electrostatic potential ϕ. If we put

$$\mathbf{H} = -\mathbf{\nabla}\phi_m, \tag{8–10}$$

then the resultant scalar potential, in terms of the equivalent volume and surface pole densities, is given by

$$\phi_m = \frac{1}{4\pi}\left[\int \frac{\mathbf{M} \cdot d\mathbf{S}}{r} - \int \frac{\mathbf{\nabla} \cdot \mathbf{M}}{r}\, dv'\right]. \tag{8–11}$$

The field of a permanent magnet with a given magnetization can also be described by the vector potential $\mathbf{A}$ which is derived from the equivalent surface *currents* and volume *currents* within the magnetized body. Equation (7–71) becomes

$$\mathbf{A} = \frac{\mu_0}{4\pi}\left[\int \frac{\mathbf{\nabla} \times \mathbf{M}}{r}\, dv' - \int \frac{\mathbf{n} \times \mathbf{M}}{r}\, dS\right], \tag{8–12}$$

where the surface now coincides with the boundary of the magnet. The surface current is equivalent to $-\mathbf{n} \times \mathbf{M}$, where $\mathbf{n}$ is a unit vector normal to the surface. The magnetic induction field is then derived from Eq. (8–12) by the use of Eq. (7–40).

In the case of a uniformly magnetized medium all internal currents cancel and hence the equivalent surface currents are the only ones present. A cylindrical magnet magnetized in a direction parallel to the axis of the cylinder therefore has a magnetic induction field equivalent to the field of a solenoidal coil, carrying current on the *cylindrical face* of the magnet, with the current flow lines lying in planes normal to the axis of the cylinder. In this case it is clear that contributions to the scalar potential of Eq. (8–11) come only from the *ends* of the cylinder. The situation can be described by noting that for a permanent magnet $\mathbf{H}$ can be thought of as arising from a layer of equivalent pole charges located on the magnet

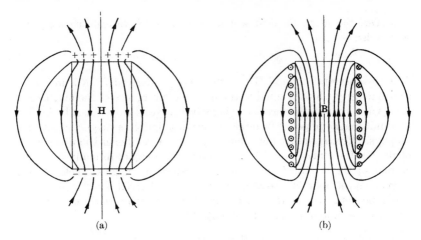

Fig. 8–1. A permanent magnet of uniform magnetization: (a) showing equivalent magnetic charges from which **H** is derived, and (b) showing equivalent solenoid with resultant **B**.

pole faces in the same manner as an electrostatic field would be formed by charges so placed. On the other hand, **B** arises from an equivalent solenoid which can be thought of as being wound on the cylindrical surface of the magnet in the same manner as a vacuum steady current field arises. $\mathbf{B}/\mu_0$ and **H** are identical outside of the region where **M** has a finite value, but they differ by **M** inside the magnet. Note that **B** and **H** in Fig. 8–1 are actually in opposite directions inside the magnet, as must be true if the line integral of **H** is to be zero over any closed path.

Equations (8–11) and (8–12) give scalar and vector potentials in terms of the equivalent pole or current distributions. The potentials can, of course, be described in terms of the integral over the potentials of the individual magnetic moments themselves. We have already seen in Eq. (7–71) that

$$\mathbf{A} = -\frac{\mu_0}{4\pi} \int \mathbf{M} \times \nabla \left(\frac{1}{r}\right) dv'. \tag{8–13}$$

Similarly, the scalar potential can be written in a form analogous to Eq. (1–68):

$$\phi_m = -\frac{1}{4\pi} \int \mathbf{M} \cdot \nabla \left(\frac{1}{r}\right) dv'. \tag{8–14}$$

The equivalence of these expressions with Eqs. (8–11) and (8–12) is evident from an integration by parts. The fields derived from either set of equations must, of course, be the same.

By the use of vector identities it can be shown that the vacuum field due to a magnetic dipole of moment $\mathbf{m} = \int \mathbf{M} \, dv'$ is

$$\mathbf{B} = \boldsymbol{\nabla} \times \mathbf{A} = -\frac{\mu_0}{4\pi} \left\{ \boldsymbol{\nabla} \times \left[\mathbf{m} \times \boldsymbol{\nabla} \left(\frac{1}{r} \right) \right] \right\}$$
$$= \frac{\mu_0}{4\pi} \left[(\mathbf{m} \cdot \boldsymbol{\nabla}) \boldsymbol{\nabla} \left(\frac{1}{r} \right) - \mathbf{m} \nabla^2 \left(\frac{1}{r} \right) \right]. \tag{8-15}$$

The last term in Eq. (8-15) vanishes except at $r = 0$. From Eq. (8-14),

$$\mathbf{H} = -\boldsymbol{\nabla} \phi_m = \frac{1}{4\pi} \left\{ \boldsymbol{\nabla} \left[\mathbf{m} \cdot \boldsymbol{\nabla} \left(\frac{1}{r} \right) \right] \right\} = \frac{(\mathbf{m} \cdot \boldsymbol{\nabla}) \boldsymbol{\nabla} (1/r)}{4\pi}. \tag{8-16}$$

Thus the two fields differ, except for a factor μ_0, only at $r = 0$. This formal distinction in the behavior of the two fields at the origin reflects a difference in the nature of the source singularity implied by the two otherwise equivalent descriptions.

We have now seen that the fields of permanent magnets (outside the magnets) may be described either in terms of "equivalent currents" or "equivalent poles." Since the entire description of magnetic fields has been based on the premise that they are produced by moving charges,[*] we are led to believe that the interpretation of the field of a permanent magnet in terms of the circulation of atomic currents is a more fundamental one than the concept of magnetic charges, and that therefore $\mathbf{B}$, which arises from currents, is a more fundamental field than $\mathbf{H}$, which arises from "magnetic charges." The description in terms of $\mathbf{H}$ is more attractive from a practical point of view, however, since it reduces problems that involve permanent magnets or problems involving magnetized pieces of iron whose magnetization can be determined by other means, to problems in electrostatics.

The question as to whether $\mathbf{B}$ or $\mathbf{H}$ is the more fundamental field can be formulated in a different way. Let us pose the problem: "Consider a charge q moving with a velocity $\mathbf{u}$ in a magnetized medium and let us suppose that the force acting on it is of the form $\mathbf{F} = q(\mathbf{u} \times \mathbf{X})$. Should we use $\mathbf{B}$ for $\mathbf{X}$, or should we use $\mathbf{H}$ or even a combination of the two?" This question was first tested experimentally by Rasetti[†] by measuring

---

[*] There is no basic objection to the existence of magnetic poles; their fields are simply not considered here since there is no experimental evidence as to their existence. If single magnetic poles did exist all the above equations would have to be supplemented. It has been shown quantum mechanically that if magnetic poles did exist the magnitude of the "elementary" unit pole would have to be related to the inverse of the elementary charge by a constant factor.

[†] F. Rasetti, *Phys. Rev.* **66**, 1 (1944).

the deflection of cosmic rays in magnetized iron, and has been studied theoretically by Wannier* by analyzing in detail the motion of charged particles in magnetized media. The answer is essentially this: If the motion of the charged particle is truly random relative to the magnetized material—that is, if it is not affected by the presence of the magnetized medium, to a first approximation—then the force that is exerted on it corresponds to the use of **B** as the magnetic field in the force equation. If, on the other hand, the particle is moving slowly and its motion is substantially affected by the magnetized medium, then it is effectively prevented from passing through the equivalent atomic current loops, and in this case, since the individual current loops act like impenetrable dipoles, the averaging process favors a deflection that corresponds to the use of **H** in the force equation. Rasetti's experimental results actually indicate that the deflection for very high speed particles corresponds approximately to the use of **B** in the force equation. In order for the results to correspond to the use of **H** it would have been necessary for the deflection to have been in the opposite direction. The answer to the above question can be given precisely in the limit of high velocities: $X_{u \to c} = B$.

Further evidence for the basic importance of **B** has been obtained from experiments on the reflection and diffraction of neutrons by magnetic materials performed by Hughes and others. It has been shown that the effective field for neutrons is **B**, not **H**; that is to say, the neutrons and the magnetic domains interact as if both were Amperian currents, not impenetrable dipoles. A discussion of the relevant experiments and their interpretation is to be found in *Pile Neutron Research*, by D. J. Hughes.†

8–3 Permeable media : magnetic susceptibility and boundary conditions.
Thus far we have considered the magnetization **M** as a given function of position, as in a permanent magnet. We now turn to the case of an ideally permeable medium, i.e., a medium that has no magnetic moment in the absence of external true currents, and in which there is a magnetic moment proportional to the field produced by any external true currents. The field equations are then

$$\nabla \cdot \mathbf{B} = 0 \qquad (8\text{--}1)$$

and

$$\oint \mathbf{H} \cdot d\mathbf{l} = J_{\text{true}}. \qquad (8\text{--}7)$$

If we assume that $\mathbf{M} = \chi_m \mathbf{H}$, we get a relation that corresponds to Eq.

---

* G. H. Wannier, *Phys. Rev.* **72**, 304 (1947).
† D. J. Hughes, *Pile Neutron Research*, Addison-Wesley, 1953, particularly Sections 11–4 and 10–6.

(2–9) in the discussion of electrostatics:

$$\mathbf{B} = \mu_0(\chi_m + 1)\mathbf{H} = \kappa_m\mu_0\mathbf{H} = \mu\mathbf{H}, \qquad (8\text{–}17)$$

where χ_m is the magnetic susceptibility, $\kappa_m = \chi_m + 1$ is the relative permeability of the medium, and μ is its absolute permeability.

By means of a derivation which is completely analogous to the one used in Chapter 2 to derive the boundary conditions for $\mathbf{E}$ and $\mathbf{D}$, the boundary conditions for $\mathbf{B}$ and $\mathbf{H}$ may be found. At a boundary between two linear media 1 and 2 the relation between the normal components is

$$\mathbf{n}_1 \cdot (\mathbf{B}_2 - \mathbf{B}_1) = \mathbf{n}_1 \cdot (\mu_2\mathbf{H}_2 - \mu_1\mathbf{H}_1) = 0, \qquad (8\text{–}18)$$

while for the tangential components,

$$\mathbf{n}_1 \times (\mathbf{H}_2 - \mathbf{H}_1) = \mathbf{n}_1 \times (\nabla\phi_{m_1} - \nabla\phi_{m_2}) = \mathbf{n}_1 \times \left(\frac{\mathbf{B}_2}{\mu_2} - \frac{\mathbf{B}_1}{\mu_1}\right) = \mathbf{K},$$
$$(8\text{–}19)$$

where $\mathbf{K}$ is the true surface current on the boundary between the two media. Equation (8–18) is analogous to Eq. (2–14) and Eq. (8–19) to Eq. (2–18), but there are differences in each case. The normal component of $\mathbf{B}$ is strictly continuous across the boundary, while the normal component of $\mathbf{D}$ is continuous only if there is no surface charge. On the other hand, the tangential component of $\mathbf{E}$ is strictly continuous, while that of $\mathbf{H}$ is continuous across the boundary only if there is no surface current. These differences correspond, of course, to the existence of true electric charge and the absence of true magnetic charge.

8–4 Magnetic circuits. It may be noted that Eqs. (8–1), (8–17), and (8–7) are mathematically identical with the equations governing stationary flow in a continuous medium in the presence of a nonconservative electromotive force, which were

$$\nabla \cdot \mathbf{j} = 0, \qquad (7\text{–}2)$$

$$\mathbf{j} = \sigma\mathbf{E}, \qquad (7\text{–}3)$$

$$\oint \mathbf{E} \cdot d\mathbf{l} = \mathcal{E}. \qquad (7\text{–}5)$$

These last three equations led to the expression

$$R = \frac{l}{\sigma S} = \sum_i \frac{l_i}{\sigma S_i} \qquad (7\text{–}17)$$

for the "resistance" of linear conductors in series. This analogy gives rise to the concept of the magnetic circuit, and the solution of linear magnetic media problems given by the expression for the magnetic flux, Φ_m:

$$\Phi_m = \int \mathbf{B} \cdot d\mathbf{S} = \frac{J}{R_m}, \tag{8-20}$$

where

$$R_m = \sum_i \frac{l_i}{\mu S_i} \tag{8-21}$$

is the magnetic "reluctance" (ampere/weber) of the circuit.

In Eq. (8–20), J plays the role of a "magnetomotive force," in analogy with the "electromotive force" $\mathcal{E}$ in Eq. (7–5). If a given conductor carrying the current J links the closed flux n times, then J must be replaced by nJ (measured in "ampere-turns") to give the correct magnetomotive force. The magnetic circuit solution is based solely on the correspondence of the differential equations for linear magnetization problems with those for steady current problems, and the solutions themselves will actually correspond only in case the boundary conditions for the magnetic and current problems are identical. This cannot be true in general. In fact, it can never be completely accurate, since the conductivity of free space is zero, while the permeability of free space is unity. This means that the magnetic circuit solution will be valid only if the permeability of the media being considered is large compared with unity, or if the regions of space accessible to the magnetic field which have $\kappa_m \simeq 1$ are small compared with those regions in which the permeability is much larger. The set of solutions (8–20) and (8–21) do, however, form the basis of industrial magnetic machinery design, since they permit an approximate treatment in cases where direct boundary value solutions are impractical.

8–5 Solution of boundary value problems by magnetic scalar potentials. In general, boundary value problems involving magnetic media can be attacked by the use of either the magnetic scalar or the vector potential. Problems involving magnetic media located in external fields, where true currents do not enter the region of interest, are best treated by means of magnetic scalar potentials. In this case the magnetic boundary conditions, Eqs. (8–18) and (8–19), may be expressed in terms of the scalar potentials, and are completely analogous to the electrostatic boundary conditions of Chapter 2 if the relative permeability replaces the specific inductive capacity.

For example, problems involving magnetic shields can be treated by electrostatic boundary value methods. The only additional complication which enters in the magnetic problem is the fact that in practical cases the permeability μ for a particular material is not constant for all values of the external field. This is especially true for high flux densities within the permeable medium. In such cases of high flux density or saturation the method of successive approximations may be used if the permeability is known as a function of the flux density. A solution based on the assumption that μ is constant is first obtained. The resultant flux density is computed, after which the problem is repeated, using the corresponding permeability. This method will give an accurate solution if the field inside the permeable medium turns out to be uniform. The problem of an ellipsoid of magnetic material situated in a uniform magnetic field is such a problem. For this reason, the torque acting on an ellipsoid suspended in a uniform field can be used as a measurement of the permeability of the material of the ellipsoid as a function of the external field. Problems of the behavior of permeable media in high fields where the resultant magnetization is appreciably nonuniform are essentially impossible to treat by purely analytical methods.

8–6 Uniqueness theorem for the vector potential. Problems in which currents are present must be treated by the use of the vector potential unless it is possible to introduce an equivalent dipole sheet in place of the current. We have seen in Section 1–1 that the vector potential is unique if the integration over sources extends to infinity and the sources are confined to a finite region of space. In order to use it with confidence in practical problems we must investigate its uniqueness for a region with definite boundaries. The proof is very similar to the proof used in Chapter 3 for the uniqueness of the scalar potential. We shall need the vector form of Green's theorem. Let $\mathbf{U}$ and $\mathbf{W}$ be arbitrary vector functions, and substitute

$$\mathbf{V} = \mathbf{U} \times (\boldsymbol{\nabla} \times \mathbf{W})$$

into Gauss's divergence theorem, $\int \boldsymbol{\nabla} \cdot \mathbf{V}\, dv = \int \mathbf{V} \cdot d\mathbf{S}$. We obtain

$$\int [(\boldsymbol{\nabla} \times \mathbf{W}) \cdot (\boldsymbol{\nabla} \times \mathbf{U}) - \mathbf{U} \cdot \boldsymbol{\nabla} \times (\boldsymbol{\nabla} \times \mathbf{W})]\, dv = \int \mathbf{U} \times (\boldsymbol{\nabla} \times \mathbf{W}) \cdot d\mathbf{S},$$

$$(8\text{–}22)$$

which is the required theorem. Let us now consider a region in space that is bounded by the surface S, on which the vector potential is specified, and which has a volume v within which there is no current. In order to accomplish this it may be necessary to choose subsurfaces which will exclude the regions of surface flow. If we now let $\mathbf{U} = \mathbf{W} = \mathbf{A}$ for all points within v, and remember that if $\mathbf{j}_{\text{total}} = 0$ then

$$\boldsymbol{\nabla} \times \mathbf{B} = \boldsymbol{\nabla} \times (\boldsymbol{\nabla} \times \mathbf{A}) = 0,$$

we obtain, on substitution in Eq. (8–22),

$$\int (\mathbf{\nabla} \times \mathbf{A})^2 \, dv = \int (\mathbf{A} \times \mathbf{B}) \cdot d\mathbf{S} = \int (\mathbf{A} \times \mathbf{B}) \cdot \mathbf{n} \, dS. \qquad (8\text{–}23)$$

Equation (8–23) may be written in the form

$$\int (\mathbf{\nabla} \times \mathbf{A})^2 \, dv = \int (\mathbf{B} \times \mathbf{n}) \cdot \mathbf{A} \, dS = \int \mathbf{B}_t \cdot \mathbf{A} \, dS, \qquad (8\text{–}24)$$

where $\mathbf{B}_t$ is the tangential component of $\mathbf{B}$, parallel to the surface and perpendicular to $\mathbf{S}$. Now let us assume that $\mathbf{A}$ in Eqs. (8–23) and (8–24) represents the difference between alternative solutions corresponding to the same boundary values of either the tangential component of the magnetic induction field $\mathbf{B}$ or the tangential component of vector potential $\mathbf{A}$. The right side of Eq. (8–24) then vanishes, since it is evaluated on the boundary where the alternate solutions are equal. On the other hand, the left side of Eq. (8–24) is positive definite and hence the integrand must vanish. Therefore $\mathbf{\nabla} \times \mathbf{A}$, where $\mathbf{A}$ is the difference of the two vector potentials, must be zero throughout the volume v and hence, in general, the field $\mathbf{B} = \mathbf{\nabla} \times \mathbf{A}$ is unique. Since it is the curl of $\mathbf{A}$ that is unique, we have defined $\mathbf{A}$ itself only within an additive irrotational vector.

8–7 The use of the vector potential in the solution of problems. We have proved that the tangential component of the magnetic induction field or the tangential component of the vector potential on the surface S uniquely defines the magnetic field within the volume bounded by this surface. This is equivalent to the analogous electrostatic consideration in which the value of the scalar potential on the bounding surface or the value of the normal electrostatic field defines the electrostatic field in the volume bounded by this surface. It is possible to carry the analogy of this procedure still further by writing the vector potential within v explicitly in terms of the currents $\mathbf{j}$ within v and the boundary values of the field over S which are chosen such as to make the field outside S equal to zero. The surface terms will then correspond to the complementary solution of the differential equation, while the volume integral of the currents will correspond to the particular integral. We shall not carry out the details of this process.* The particular integral is just that which becomes equal to the general solution given by Eq. (1–5) or, more specifically, by Eq. (7–42), in case we let the boundary be at infinity, i.e., the solution that corresponds to knowing the sources over all space.

---

* *Cf.* J. A. Stratton, *Electromagnetic Theory*, pp. 245 ff.

The differential equation for **A** must be derived from the field equations by use of a vector identity for the double curl. This gives

$$\boldsymbol{\nabla} \times (\boldsymbol{\nabla} \times \mathbf{A}) = \boldsymbol{\nabla}(\boldsymbol{\nabla} \cdot \mathbf{A}) - \nabla^2 \mathbf{A} = \mu_0 \mathbf{j}. \qquad (8\text{-}25)$$

Some care must be used in the interpretation of the operation of the symbol ∇^2 when it is applied to a vector. In the Cartesian coordinate system $\nabla^2 \mathbf{A}$ means a vector whose αth component is $\nabla^2 A_\alpha$. In a non-Cartesian coordinate system it is a vector whose αth component must be evaluated by means of the identity

$$\nabla^2 \mathbf{A} = -\boldsymbol{\nabla} \times (\boldsymbol{\nabla} \times \mathbf{A}) + \boldsymbol{\nabla}(\boldsymbol{\nabla} \cdot \mathbf{A}). \qquad (8\text{-}26)$$

The choice of $\boldsymbol{\nabla} \cdot \mathbf{A}$ has thus far been left arbitrary, since **A** was defined only in terms of the equation $\mathbf{B} = \boldsymbol{\nabla} \times \mathbf{A}$. It is convenient here to take

$$\boldsymbol{\nabla} \cdot \mathbf{A} = 0, \qquad (8\text{-}27)$$

which involves no new physical assumptions. We shall find later in considering nonstationary currents that a more complicated expression must be substituted for Eq. (8-27) in order to preserve symmetry between the electric and magnetic cases, and in the more general application to obtain relativistic covariance of the resulting equations. There will be no conflict with Eq. (8-27) for stationary currents, however.

With the simplifying assumption of Eq. (8-27) the differential equation for **A** reduces to

$$\nabla^2 \mathbf{A} = -\mu_0 \mathbf{j}. \qquad (8\text{-}28)$$

This is the vector form of Poisson's equation, of which the particular integral is

$$\mathbf{A} = \frac{\mu_0}{4\pi} \int \frac{\mathbf{j}}{r} \, dv', \qquad (8\text{-}29)$$

which becomes the general solution if the integral extends over all of the currents that contribute to the field. This is identical to Eq. (7-42); the assumption $\boldsymbol{\nabla} \cdot \mathbf{A} = 0$ corresponds to setting the arbitrary function generated by integrating Eq. (7-41) equal to zero.

The solution of Eq. (8-28) subject to arbitrary boundary conditions is usually considerably more complicated than that of the corresponding scalar potential equation. The reason for this is that, because of the restriction of Eq. (8-27), **A** does not actually have three independent components. This means that we cannot expect to expand the components of **A** separately in normal orthogonal functions and have a sufficient number of boundary conditions to determine the coefficients. In other words, each "harmonic" of the separated scalar solution involving a single coordinate includes two constants of integration, and thus there are six constants in all for each term of the series. Here there are not eighteen constants, six for each component of the vector, but, because of the condition $\nabla \cdot \mathbf{A} = 0$, only twelve constants altogether.

We therefore seek an expression for $\mathbf{A}$ which involves two scalar potential functions, say U and V, and which reduces the equation $\nabla \cdot \mathbf{A} = 0$ to an identity. Let us examine a solution of the type

$$\mathbf{A} = \nabla V + \nabla \times (\mathbf{a}U), \qquad (8\text{–}30)$$

which obviously satisfies $\nabla \cdot \mathbf{A} = 0$. Here $\mathbf{a}$ is the vector of the (in general curvilinear) coordinate system which is to be chosen to fulfill certain conditions. The expression $\mathbf{B} = \nabla \times \mathbf{A}$ will not depend on V; hence U is the only function necessary to specify the field. This is as it should be, because in a source-free region there is no distinction between a field derivable from a scalar or vector potential. In general, however, V is needed to meet boundary value requirements on $\mathbf{A}$.

The function $\mathbf{A}$ as given by Eq. (8–30) satisfies the condition $\nabla \times (\nabla \times \mathbf{A}) = 0$ if $\mathbf{a}$ fulfills the condition that (in tensor notation) the function $\partial a_\beta / \partial x_\alpha$ is a diagonal matrix with equal constant coefficients, that is, $\partial a_\beta / \partial x_\alpha = \text{const } \delta_{\alpha\beta}$. Note that this condition implies that $\nabla \times \mathbf{a} = 0$. This condition is satisfied, for example, when $\mathbf{a} = \hat{\mathbf{x}}, \hat{\mathbf{y}},$ or $\hat{\mathbf{z}}$, the unit vectors along Cartesian coordinate axes, or by the vector $\mathbf{r}$ in spherical coordinates.

To prove this, we note that from Eq. (8–30),

$$\mathbf{B} = \nabla \times \mathbf{A} = -\nabla \times (\mathbf{a} \times \nabla U). \qquad (8\text{–}31)$$

In tensor notation this becomes

$$-\nabla \times \mathbf{A}|_\alpha = \frac{\partial}{\partial x_\beta}\left[a_\alpha \frac{\partial U}{\partial x_\beta} \right] - \frac{\partial}{\partial x_\beta}\left[a_\beta \frac{\partial U}{\partial x_\alpha} \right], \qquad (8\text{–}32)$$

which can be reduced to

$$-\nabla \times \mathbf{A}|_\alpha = -\frac{\partial}{\partial x_\alpha}\left(\alpha_\beta \frac{\partial U}{\partial x_\beta} \right) + \frac{\partial U}{\partial x_\beta}\left(\frac{\partial a_\alpha}{\partial x_\beta} + \frac{\partial a_\beta}{\partial x_\alpha} \right) - \frac{\partial U}{\partial x_\alpha}\left(\frac{\partial a_\beta}{\partial x_\beta} \right), \qquad (8\text{–}33)$$

using the condition $\nabla^2 U = 0$. The first term of this relation is a pure gradient; the remaining terms are also pure gradients if $\mathbf{a}$ satisfies the condition above. Hence $\nabla \times (\nabla \times \mathbf{A})$ vanishes.

The boundary conditions on $\mathbf{A}$ at a magnetic interface are derived from the conditions on $\mathbf{B}$. Conservation of magnetic flux across the boundary demands conservation of the line integral of $\mathbf{A}$ along any arbitrary curve in the boundary surface. We thus require that on the interface between regions 1 and 2

$$\mathbf{A}_{t_1} = \mathbf{A}_{t_2},$$

where the subscript t denotes the tangential component. In the absence of true surface currents the tangential component of $\mathbf{H}$ is conserved; if the permeabilities are linear this is equivalent to

$$\frac{1}{\mu_1} (\nabla \times \mathbf{A})_{t_1} = \frac{1}{\mu_2} (\nabla \times \mathbf{A})_{t_2}.$$

Both of these boundary conditions have two components, and hence we have four boundary equations. Note that in the analogous scalar case we had two conditions, Eqs. (2–15) and (2–19), on the scalar potential ϕ. Here we have a sufficient number of conditions to join V and U of Eq. (8–30) across a boundary.

In practical cases not involving boundaries it is conventional to write down solutions of Eq. (8–30) with $V = 0$; then, for a given choice of $\mathbf{a}$, orthogonal expansions for $\mathbf{A}$ can be made by using the same number of functions as in the scalar case.

The derivation of $\mathbf{A}$ from scalar solutions of Laplace's equation also serves to circumvent an additional complication arising from the fact that only in rectangular coordinates is $\nabla^2 \mathbf{A}$ simply the vector sum of the Laplacians of the separate components of $\mathbf{A}$. We shall meet this problem again in connection with the vector wave equation.

8–8 The vector potential in two dimensions. In two-dimensional problems, where it can be assumed that the fields are not functions of the z-coordinate, a simple use can be made of the vector potential. In the case in which all lines of flow are parallel to the z-axis, it follows from Eq. (8–29) that the vector potential has only a z-component. The magnetic fields are then derived from the potential by

$$B_x = \frac{\partial A_z}{\partial y}, \qquad B_y = -\frac{\partial A_z}{\partial x}, \qquad (8\text{–}34)$$

and Laplace's equation becomes

$$\nabla^2 A_z = 0. \qquad (8\text{–}35)$$

Since A_z is not a function of z, the divergence of $\mathbf{A}$ is obviously equal to zero. Equation (8–35) is the two-dimensional Laplace's equation, and thus we can make the vector potential either the real or the imaginary part of a complex potential. Equations (8–34) are mathematically the same as the equations that relate the stream function to the corresponding electrostatic field. Since the form of the potential for each rectangular component of the vector potential is the same as the scalar Coulomb potential, the vector potential of the line current in two dimensions has the form, by analogy with Eq. (4–12),

$$- A_z = \frac{\mu_0}{2\pi} J \ln \left(\frac{r}{r_0} \right), \tag{8–36}$$

and the corresponding complex potential is given by

$$W = \phi + i\psi = \frac{\mu_0}{2\pi i} J \ln \left(\frac{x + iy}{x_0 + iy_0} \right). \tag{8–37}$$

The imaginary part of W, the stream function ψ, is the vector potential. All the methods developed for finding the stream function in the solution of electrostatic problems in two dimensions can be used for the solution of two-dimensional magnetic boundary value problems.

Complex transformations, such as the Schwarz transformation, can be used to transform simple problems involving currents in the neighborhood of permeable media having rectilinear boundaries into more complicated configurations. For example, the problem of a line current located at a given distance from the surface of a semi-infinite permeable medium is soluble by the method of images (the image current here is of the same magnitude and sign as the current in the original conductor) and therefore the solution of various problems concerning slots or gaps in permeable materials under the influence of magnetizing windings can be derived from this simple image solution by means of a suitable Schwarz transformation.

The method of two-dimensional harmonics is also useful, particularly in problems involving cylinders of permeable material. Consider, for example, a permeable conducting cylinder of radius a, carrying a steady current J in the presence of an external field $\mathbf{B}_0$ at right angles to its axis. As usual, we may divide the plane into two regions. For $r > a$ the vector potential A_z is a solution of Laplace's equation, which for large r must give the constant field as well as that produced by the current J. For $r < a$ we must solve the inhomogeneous equation

$$\nabla^2 A_z = -\mu_i j, \tag{8–38}$$

where $j = J/\pi a^2$ and μ_i is the permeability of the cylinder. A particular solution of Eq. (8–35) is $-\mu_i j r^2/4$, and since negative powers of r are excluded by the condition that the origin be a regular point, the general solution is

$$A_z \atop r<a = - \frac{\mu_i j r^2}{4} + \sum_{n=0}^{\infty} (a_n \cos n\varphi + b_n \sin n\varphi) r^n. \tag{8–39}$$

In order to write a solution valid for the region outside the cylinder, we note that a constant field in the x-direction may be derived from the vector potential $B_0 y = B_0 r \sin \varphi$, and that the effect of the current is that of a line current at the origin. Therefore

$$A_z \atop r>a} = \frac{\mu_0 J}{2\pi} \ln \frac{r_0}{r} + B_0 r \sin \varphi + \sum_{n=0}^{\infty} (c_n \cos n\varphi + d_n \sin n\varphi) r^{-n}, \quad (8\text{–}40)$$

where r_0 is an arbitrary length. The continuity of the normal component of $\mathbf{B}$ at the boundary $r = a$ is equivalent to the continuity of A_z, while the continuity of the tangential component of $\mathbf{H}$ demands that

$$\frac{1}{\mu_i} \frac{\partial}{\partial r} (A_{z<a}) = \frac{1}{\mu_0} \frac{\partial}{\partial r} (A_{z>a}) \qquad (8\text{–}41)$$

at $r = a$. These two conditions serve to determine the coefficients, with the result that

$$
\begin{aligned}
&A_z \atop r<a} = -\frac{\mu_i j r^2}{4} + \frac{2\mu_i}{\mu_i + \mu_0} B_0 r \sin \varphi, \\
&A_z \atop r>a} = \frac{\mu_0 J}{2\pi} \ln \left(\frac{a}{r}\right) - \frac{\mu_i J}{4\pi} + \left(1 + \frac{\mu_i - \mu_0}{\mu_i + \mu_0} \frac{a^2}{r^2}\right) B_0 r \sin \varphi,
\end{aligned}
\qquad (8\text{–}42)
$$

from which B_r and $\mathbf{B}$ may be immediately computed. It is instructive to write the potential valid inside the cylinder in rectangular coordinates:

$$A_z \atop r<a} = -\frac{\mu_i j (x^2 + y^2)}{4} + \frac{2\mu_i}{\mu_i + \mu_0} B_0 y, \qquad (8\text{–}43)$$

from which it is more easily seen that magnetization produced by the external field is uniform, and that in the limit of large permeability the induction field is twice that applied, in addition to that produced by the current.

8–9 The vector potential in cylindrical coordinates. A three-dimensional case of practical importance is that of cylindrical symmetry, i.e., current flow in coaxial circles only. For this case the only component of $\mathbf{A}$ is A_φ, which satisfies the differential equation

$$[\mathbf{\nabla} \times (\mathbf{\nabla} \times \mathbf{A})]_\varphi = \frac{\partial^2 A_\varphi}{\partial r^2} + \frac{1}{r} \frac{\partial A_\varphi}{\partial r} - \frac{A_\varphi}{r^2} + \frac{\partial^2 A_\varphi}{\partial z^2} = -\mu j_\varphi \qquad (8\text{–}44)$$

in cylindrical coordinates r, φ, z; the solution is independent of the coordinate φ.

Equation (8–44) separates into Bessel's equation, Eq. (5–30), with $n = 1$, or the corresponding modified Bessel's equation, Eq. (5–30'), and the accompanying equation in z. As indicated in Section 5–8, the solutions are of the form

$$A_\varphi = \cos (kz - \alpha_k)[A(k)I_1(kr) + B(k)K_1(kr)] \qquad (8\text{–}45)$$

or

$$A_\varphi = e^{\mp kz}[A^\pm(k)J_1(kr) + B^\pm(k)N_1(kr)], \qquad (8\text{–}46)$$

depending on which form of the solution is appropriate to the physical problem. If the current flows in a loop or solenoid of radius r_0 the solution may be written in the form of Eq. (8–45) for region 1, $r < r_0$, and for region 2, $r > r_0$. The constants may be determined by the use of the boundary conditions on the interface, $r = r_0$, between regions 1 and 2:

$$A_\varphi^{(1)} = A_\varphi^{(2)}, \qquad (8\text{–}47)$$

$$\frac{1}{\mu_1} \frac{1}{r} \frac{\partial}{\partial r} (rA_\varphi^{(1)}) - \frac{1}{\mu_0} \frac{1}{r} \frac{\partial}{\partial r} (rA_\varphi^{(2)}) = j_{s\varphi}, \qquad (8\text{–}48)$$

where $j_{s\varphi}$ is the surface current density at the interface. Equations (8–47) and (8–48) correspond to $\boldsymbol{\nabla} \cdot \mathbf{B} = 0$ and $\boldsymbol{\nabla} \times \mathbf{H} = \mathbf{j}$ respectively. (The z-component of $\mathbf{H}$ is discontinuous at $r = r_0$). If the current loops are in a plane, which we may take as $z = 0$, the solution should be written in the form of Eq. (8–46) for the regions of positive and negative z. The boundary condition replacing Eq. (8–48) is that the radial component of $\mathbf{H}$, namely $(\partial A_\varphi / \partial z)/\mu$, suffers a discontinuity corresponding to the surface current in the plane $z = 0$.

Let us apply this method to the determination of the vector potential due to a current J in a single plane loop of radius r_0. In this case the choice of functions is optional. If we choose Eq. (8–45) so as to take the cylinder of radius r_0 as the surface dividing the two regions, we first note that $K_1(kr)$ must be excluded from the inner region and $I_1(kr)$ from the outer region in order that our solutions be regular everywhere. Therefore we may write

$$A_\varphi^{(1)} = \int_0^\infty A(k)I_1(kr) \cos kz \, dk,$$

$$\qquad (8\text{–}49)$$

$$A_\varphi^{(2)} = \int_0^\infty B(k)K_1(kr) \cos kz \, dk.$$

Equation (8–47) demands that

$$A(k)I_1(kr_0) = B(k)K_1(kr_0), \qquad (8\text{–}50)$$

while Eq. (8–48), with $j_\varphi = J\,\delta(z)$, gives

$$\left[\int_0^\infty \left\{ A(k)\,\frac{\partial}{\partial r}\,[rI_1(kr)] - B(k)\,\frac{\partial}{\partial r}\,[rK_1(kr)] \right\} \frac{\cos kz\,dk}{r} \right]_{r=r_0} = \mu J\,\delta(z). \tag{8–51}$$

We can multiply by $\cos k'z$ and apply the Fourier integral theorem or, equivalently, remember the δ-function character of $\int\cos kz\,dk$, to obtain from Eq. (8–51)

$$\frac{\pi}{r_0}\left[A(k)\,\frac{\partial}{\partial r}\,[rI_1(kr)] - B(k)\,\frac{\partial}{\partial r}\,[rK_1(kr)] \right]_{r=r_0} = \mu J. \tag{8–52}$$

But there is a mathematical relation between the Bessel functions and their derivatives, namely,

$$I_n'(kr)K_n(kr) - K_n'(kr)I_n(kr) = \frac{1}{kr}. \tag{8–53}$$

Therefore we may obtain the coefficients at once,

$$A(k) = \frac{r_0\mu J}{\pi}\,K_1(kr_0),$$

$$B(k) = \frac{r_0\mu J}{\pi}\,I_1(kr_0),$$

and the potentials are

$$A_\varphi^{(1)} = \frac{r_0\mu J}{\pi}\int_0^\infty K_1(kr_0)I_1(kr)\cos kz\,dk,$$

$$A_\varphi^{(2)} = \frac{r_0\mu J}{\pi}\int_0^\infty I_1(kr_0)K_1(kr)\cos kz\,dk. \tag{8–54}$$

If, on the other hand, the $z = 0$ plane is used as the division between the two regions, Eq. (8–46) leads to

$$A_\varphi^\pm = \int A^\pm(k)J_1(kr)e^{\mp kz}\,dk, \tag{8–25}$$

since $N_1(kr)$ is not regular at the origin. The surface current is confined to $J\,\delta(r - r_0)$ in the plane, and introduces a discontinuity in $\partial A_\varphi/\partial z$ at $z = 0$. The determination of the coefficients $A(k)$ requires the use of the Fourier-Bessel integral

$$\int_0^\infty k\,dk\int_0^\infty r\,dr\,f(k)J_n(kr)J_n(k'r) = f(k'), \tag{8–56}$$

which leads at once to

$$A_\varphi^\pm = \frac{r_0 \mu J}{2} \int_0^\infty J_1(kr_0) J_1(kr) e^{\mp kz} \, dk \qquad (8\text{-}57)$$

for the required potential.

Suggested References

J. A. Stratton, *Electromagnetic Theory*. Chapter IV includes a derivation of the complete solution for the vector potential including boundary conditions.

P. M. Morse and H. Feshbach, *Methods of Theoretical Physics*. A more complete mathematical treatment than that of Stratton.

W. R. Smythe, *Static and Dynamic Electricity*. Chapter VII contains methods for determining the vector potential and many examples of its use.

Exercises

1. Consider a magnet with pole pieces wide compared with the pole gap, and with the windings far removed from the gap. Find the complex potential function and plot the field in the plane of symmetry from a point well outside the poles to a position in the gap where the field is essentially constant.

2. Find **B** and **H** inside and outside a spherical shell of radii a and b which is magnetized permanently to a constant magnetization **M**. What is the effect of making the spherical cavity not concentric with the outside surface of the shell?

3. A cylindrical hole of radius a is bored parallel to the axis of a long cylindrical conductor of radius b which carries a uniformly distributed current of density **j**. The distance between the center of the conductor and the center of the hole is x_0. Find the magnetic field in the hole. (The generalization of this result to cylinders of elliptical cross section is used in the design of the magnetic field for some high-energy accelerating devices.)

4. A coil is wound on the surface of a sphere such that the field inside the sphere is uniform. What is the winding? (This form of winding is used in the Westinghouse-Goudsmit mass spectrometer.)

5. What is the effective magnetic dipole moment of a sphere at points outside if the sphere has a uniform surface charge density and rotates with angular velocity ω?

6. A spherical shell of material with permeability $\mu \neq \mu_0$, whose inner and outer radii are a and b respectively, is placed in an originally uniform field B_0. What is the field in the spherical cavity of radius a?

7. Consider an infinitely long wire at $x = a$, $y = 0$, carrying current J in vacuum, while all space for which x is negative is filled with a medium of $\mu \neq \mu_0$. Find the field at all points in space.

8. Find the magnetic field of a long wire at $x = a$, $y = b$, carrying current J in vacuum, while all of space for which x is negative and all for which y is negative are filled with magnetic material of very large permeability, $\mu \to \infty$.

9. Calculate the magnetic vector potential due to a ring of radius a coaxial with a permeable circular cylinder of permeability $\mu \neq \mu_0$ and radius $b < a$ (Smythe).

10. A soft iron ring of radius b, cross section radius $a \ll b$, is wound with N turns of wire carrying current J. A small air gap of length δ is cut at one point in the ring. Assuming a large permeability for the iron, compute the reluctance of the magnetic circuit.

11. Show that if $\mathbf{B} = \nabla \times \mathbf{A}$ and $\mathbf{H} = -\nabla \phi_m$, and if $\mathbf{A}$ and ϕ_m are given by Eqs. (8–13) and (8–14) respectively, then $\mathbf{B}$ and $\mathbf{H}$ satisfy the relation $\mathbf{B} = \mu_0(\mathbf{H} + \mathbf{M})$.

12. Put $\mathbf{a} = \mathbf{k}$ in Eq. (8–30) where $\mathbf{k}$ is the unit vector along the z-axis in cylindrical coordinates. Take $V = 0$. Show that the resulting vector potential expansion, if U is expanded as a scalar potential solution in cylindrical coordinates, agrees with Eqs. (8–45) and (8–46), obtained by direct integration.

CHAPTER 9

MAXWELL'S EQUATIONS

9–1 Faraday's law of induction. In electrostatics the electric field is conservative, i.e., curl $\mathbf{E} = 0$. We have seen in Chapter 7 that in order to produce stationary currents there must be electric fields, such as $\mathbf{E}'$ in Eq. (7–5), which violate this condition. And it is found experimentally that a nonconservative electric field actually accompanies varying magnetic fields. The law describing this situation is usually known as the Faraday law of induction, and can be formulated as follows. Consider a circuit of resistance R carrying a current J and containing an electromotive force $\mathcal{E}$. The magnetic flux linking this circuit is defined by

$$\Phi_m = \int \mathbf{B} \cdot d\mathbf{S}, \tag{9-1}$$

where the surface of integration is bounded by the circuit. If this flux changes in time, it is found experimentally that

$$JR - \mathcal{E} = -\frac{d\Phi_m}{dt}. \tag{9-2}$$

This means that the current in the circuit differs from that predicted by Ohm's law by an amount which can be attributed to an additional electromotive force equal to the negative time rate of change of flux through the circuit. Note that Eq. (9–2) is an independent experimental law and is in no way derivable from any of the relations that have been previously used. In particular, contrary to the statement that is sometimes made, Faraday's law of induction is not the consequence of the law of conservation of energy applied to the over-all energy balance of currents in magnetic fields.

Equation (9–2) is formulated in terms of the total flux passing through the given circuit. This flux can change for several reasons: it can change because of changes in the external field with time; it can change because of motion of the circuit itself or parts of the circuit. We shall consider Eq. (9–2) to be an experimental law which holds for all such cases, including that of currents in moving media.

It was recognized by Maxwell that the Faraday law of induction had a very much more general significance than the case actually described by Eq. (9–2) would indicate. The equation can be written in the equivalent form,

$$\oint \mathbf{E} \cdot d\mathbf{l} = -\frac{d\Phi_m}{dt} = JR - \mathcal{E}, \tag{9-3}$$

which indicates that there must be an electric field along the wire that is nonelectrostatic. However, from the boundary condition that requires the tangential components of the electric field across the boundary of a wire to be continuous, we can conclude that Eq. (9–3) is also valid in the region immediately adjacent to the wire. Since the characteristics of the wire, namely, its resistance and its electromotive forces, are not contained in the left side of Eq. (9–3), it appears likely that this relation is, in fact, independent of the presence of a current-carrying conductor, and is a general physical law relating an electric field *in vacuo* to the rate of change of a magnetic field. If this is so, then the equation can be transformed into a differential expression valid for free space or a stationary medium. In either case, the total derivative of the flux integral, Eq. (9–1), can be written as an integral of the partial time derivative of the magnetic field, and Eq. (9–3) becomes

$$\oint \mathbf{E} \cdot d\mathbf{l} = - \frac{d}{dt} \int \mathbf{B} \cdot d\mathbf{S} = - \int \frac{\partial \mathbf{B}}{\partial t} \cdot d\mathbf{S}. \qquad (9\text{--}4)$$

The use of Stokes' theorem leads to

$$\mathbf{\nabla} \times \mathbf{E} = - \frac{\partial \mathbf{B}}{\partial t}. \qquad (9\text{--}5)$$

Equation (9–5) expresses the modification of the electrostatic field irrotationality which is a necessary consequence of Faraday's law of induction.

9–2 Maxwell's equations for stationary media. We now have expressions for both the source densities (divergence) and circulation densities (curl) of the two basic field vectors **E** and **B**. From Eqs. (2–5), (7–38), (8–3), and (9–5), we have

(1) $$\mathbf{\nabla} \cdot \mathbf{E} = \epsilon_0^{-1} \rho_{\text{total}} = \epsilon_0^{-1} (\rho_{\text{true}} - \mathbf{\nabla} \cdot \mathbf{P}),$$

(2) $$\mathbf{\nabla} \cdot \mathbf{B} = 0,$$

(3) $$\mathbf{\nabla} \times \mathbf{E} = - \frac{\partial \mathbf{B}}{\partial t},$$

(4) $$\mathbf{\nabla} \times \mathbf{B} = \mu_0 \left(\mathbf{j}_{\text{true}} + \frac{\partial \mathbf{P}}{\partial t} + \mathbf{\nabla} \times \mathbf{M} + \epsilon_0 \frac{\partial \mathbf{E}}{\partial t} \right).$$

(9–6)

Equations (9–6) are Maxwell's electrodynamic field equations, formulated so as to be valid for media at rest. The restriction to material media at rest arises from the omission of any convective current terms from Eq. (9–6) (4), and from ignoring, in the derivation of Eq. (9–5), flux changes due to motion of the medium. Equations (9–6) are written in terms of the equivalent vacuum charges or currents which give rise to the fields, and contain the equivalent current and charge densities explicitly. If the additional field vectors $\mathbf{D}$ and $\mathbf{H}$ are introduced by the defining equations

$$\mathbf{D} = \epsilon_0\mathbf{E} + \mathbf{P}, \tag{2–3}$$

$$\mathbf{H} = \frac{\mathbf{B}}{\mu_0} - \mathbf{M}, \tag{8–4}$$

Maxwell's field equations become

(1) $\nabla \cdot \mathbf{D} = \rho_{\text{true}},$

(2) $\nabla \cdot \mathbf{B} = 0,$

(3) $\nabla \times \mathbf{E} = -\dfrac{\partial \mathbf{B}}{\partial t},$ (9–7)

(4) $\nabla \times \mathbf{H} = \mathbf{j}_{\text{true}} + \dfrac{\partial \mathbf{D}}{\partial t}.$

The field equations (9–7) appear formally simpler than Eqs. (9–6), but they are actually more complicated physically. The solution of these equations is possible only if "constitutive" equations are available connecting $\mathbf{D}$ to $\mathbf{E}$, $\mathbf{j}$ to $\mathbf{E}$, and $\mathbf{H}$ to $\mathbf{B}$, such as $\mathbf{D} = \kappa\epsilon_0\mathbf{E}$, $\mathbf{j}_{\text{true}} = \sigma\mathbf{E}$, $\mathbf{H} = \mathbf{B}/\mu$, for a linear medium, or whatever forms apply for a nonlinear medium.

9–3 Faraday's law for moving media. It is necessary to use considerable care in extending the law of induction to take account of motion in general. We must first derive the subsidiary theorem which expresses the total time rate of change of the flux across a given surface in terms of a surface integral of a function of the vector $\mathbf{B}$, even when the surface itself across which the flux is evaluated is in motion.

Again let Φ_m be the flux of the vector field $\mathbf{B}$ across the surface S. We are seeking the function $D\mathbf{B}/Dt$, defined by

$$\frac{d}{dt}\Phi_m = \frac{d}{dt}\int \mathbf{B} \cdot d\mathbf{S} = \int \frac{D\mathbf{B}}{Dt} \cdot d\mathbf{S}. \tag{9–8}$$

In order to evaluate $D\mathbf{B}/Dt$, let us consider the surface in Fig. 9–1 in position 1 at time t, and in position 2 at time $t + dt$. By the rules for differentiation,

$$\frac{\Delta}{\Delta t}\int \mathbf{B} \cdot d\mathbf{S} = \frac{1}{\Delta t}\int (\mathbf{B}_{t+dt} \cdot d\mathbf{S}_2 - \mathbf{B}_t \cdot d\mathbf{S}_1). \tag{9–9}$$

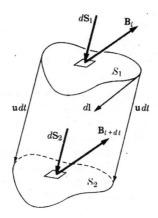

FIG. 9–1. Figure for evaluation of $D\mathbf{B}/Dt$.

If we apply Gauss's theorem at the time t to the volume enclosed by S_1, S_2, and the traces of the edges of S, we have

$$\int \mathbf{\nabla} \cdot \mathbf{B}\, dv = \int (\mathbf{B}_t \cdot d\mathbf{S}_2 - \mathbf{B}_t \cdot d\mathbf{S}_1) - \oint \mathbf{B}_t \cdot (\mathbf{u}\, dt \times d\mathbf{l}). \quad (9\text{–}10)$$

The last term represents the flux change across the side surface generated by the motion of the boundary of S, of which an element is $d\mathbf{l}$.* Note that the flux across the surfaces S_1 and S_2 in Eq. (9–10) is considered at time t, since Gauss's theorem applied only to simultaneous values of the vector field $\mathbf{B}$. The value of $\mathbf{B}$ on S_2 at time $t + dt$ may be found in terms of its value at t by Taylor's theorem,

$$\mathbf{B}_{t+dt} = \mathbf{B}_t + \frac{\partial \mathbf{B}}{\partial t}\, dt + \cdots. \quad (9\text{–}11)$$

With the substitution of Eqs. (9–10) and (9–11), Eq. (9–9) becomes, in the limit,

$$\frac{d}{dt} \int \mathbf{B} \cdot d\mathbf{S} = \int \frac{\partial \mathbf{B}}{\partial t} \cdot d\mathbf{S} + \oint \mathbf{B} \times \mathbf{u} \cdot d\mathbf{l} + \int \frac{\mathbf{\nabla} \cdot \mathbf{B}\, dv}{dt}. \quad (9\text{–}12)$$

Then, by using Stokes' theorem and the fact that

$$dv = \mathbf{u} \cdot d\mathbf{S}\, dt, \quad (9\text{–}13)$$

---

* The direction of $d\mathbf{l}$ is chosen to correspond to $d\mathbf{S}$ according to the right-hand rule.

we obtain the desired relation,

$$\frac{D\mathbf{B}}{Dt} = \frac{\partial \mathbf{B}}{\partial t} + \mathbf{\nabla} \times (\mathbf{B} \times \mathbf{u}) + (\mathbf{\nabla} \cdot \mathbf{B})\mathbf{u}. \tag{9-14}$$

The first term of this expression represents the change in the flux through S that is caused by the time variation of the vector field. The second term represents the flux loss across the boundary of the moving surface. The third term arises from the passage of surface S through an inhomogeneous vector field in which flux lines are generated.

Equation (9–14) can now be used to express Faraday's law, Eq. (9–2), in differential form in a moving medium. Since the magnetic induction $\mathbf{B}$ is always solenoidal, we have from Eqs. (9–3) and (9–14),

$$\oint \mathbf{E}' \cdot d\mathbf{l} = -\frac{d\Phi_m}{dt} = -\int \left[\frac{\partial \mathbf{B}}{\partial t} + \mathbf{\nabla} \times (\mathbf{B} \times \mathbf{u})\right] \cdot d\mathbf{S}. \tag{9-15}$$

We have designated the field by $\mathbf{E}'$ around the circuit, since $\mathbf{E}'$ is to be measured in the moving frame of reference, i.e., Faraday's law applies specifically to the current measured in the wire through which the flux is changing, no matter what might be the cause of the flux change. By applying Stokes' theorem to Eq. (9–15), we arrive at

$$\mathbf{\nabla} \times \mathbf{E}' = -\frac{\partial \mathbf{B}}{\partial t} - \mathbf{\nabla} \times (\mathbf{B} \times \mathbf{u}), \tag{9-16}$$

where $\mathbf{E}'$ still represents the field measured in the moving medium. Equation (9–16) can be written in the form

$$\mathbf{\nabla} \times (\mathbf{E}' - \mathbf{u} \times \mathbf{B}) = -\frac{\partial \mathbf{B}}{\partial t}. \tag{9-17}$$

But the argument of the curl in Eq. (9–17), $\mathbf{E}' - \mathbf{u} \times \mathbf{B}$ actually represents the field that is measured by a stationary observer. The reason for this is that an observer carrying a charge q through a magnetic field $\mathbf{B}$ with a velocity $\mathbf{u}$ will experience a force, $q(\mathbf{u} \times \mathbf{B})$, in addition to the force of the electric field $\mathbf{E}$ which may also be present. Hence the electric field observed by a stationary observer is equal to the electric field $\mathbf{E}'$ seen by the moving observer minus the effective field $\mathbf{u} \times \mathbf{B}$, i.e., $\mathbf{E} = \mathbf{E}' - \mathbf{u} \times \mathbf{B}$. Hence for a stationary observer, Eq. (9–17) becomes

$$\mathbf{\nabla} \times \mathbf{E} = -\frac{\partial \mathbf{B}}{\partial t}. \tag{9-5}$$

This means that the differential formulation of Faraday's law of induction is independent of the motion of the medium inside the field. This is as it should be, since Eq. (9–5) is purely a field relation in terms of the equivalent vacuum fields **B** and **E**, and should therefore be independent of the characteristics of the medium, including its motion. However, the electric field observed by the moving observer does contain two terms, namely, the "induced field" produced by the time rate of change of the external magnetic fields, and the "motional field," **u** × **B**, produced by the motion of the observer in the magnetic field. Note that in this discussion it has been assumed that the electric field proper is not affected by the state of motion of the observer. This assumption is actually justified only in case the motion of the observer is small compared with the velocity of light, as we shall see later. Therefore all the conclusions which we now draw concerning Maxwell's equations in moving media can be applied with confidence only when the velocities of such media are small compared with the velocity of light.

9–4 Maxwell's equations for moving media. We have concluded that the third of Maxwell's four equations in the form of Eq. (9–6) is not affected by the motion of the medium in which the fields are measured. The first and second equations are not affected by this motion either, since nonrelativistically the charge density of the medium is not affected by the state of motion of the observer. The only modification is a correction to the current in the fourth of Eqs. (9–6), a convective term, and correction to the polarization current. The convection current, due to the motion of the charge density and equivalent polarization charge, is simply $\mathbf{u}(\rho_{\text{true}} - \boldsymbol{\nabla} \cdot \mathbf{P})$. The polarization current is properly given by $D\mathbf{P}/Dt$ instead of $\partial\mathbf{P}/\partial t$, to take account of the charges lost due to the change of polarization flux across the moving surface. The total current which gives rise to the magnetic field is thus composed of the following parts:

(1) True currents, $\mathbf{j}_{\text{true}}$.

(2) Convective currents, $\mathbf{u}(\rho_{\text{true}} - \boldsymbol{\nabla} \cdot \mathbf{P})$.

(3) Currents caused by the rate of change of the polarization and the motion of polarized media, in analogy with Eq. (9–14), given by

$$\frac{D\mathbf{P}}{Dt} = \frac{\partial\mathbf{P}}{\partial t} + \boldsymbol{\nabla} \times (\mathbf{P} \times \mathbf{u}) + (\boldsymbol{\nabla} \cdot \mathbf{P})\mathbf{u}.$$

(4) Vacuum displacement current, $\epsilon_0 \, \partial\mathbf{E}/\partial t$.

With these corrections to the fourth of Maxwell's equations, it is now possible to write the entire set of field equations so as to be valid in a nonmagnetized medium moving with a velocity **u** which is small compared with the velocity of light:

(1) $\quad \nabla \cdot \mathbf{D} = \rho_{\text{true}},$

(2) $\quad \nabla \cdot \mathbf{B} = 0,$

(3) $\quad \nabla \times \mathbf{E} = -\dfrac{\partial \mathbf{B}}{\partial t},$

(4) $\quad \nabla \times \mathbf{B} = \mu_0 \left[\mathbf{j}_{\text{true}} + \mathbf{u}(\rho_{\text{true}} - \nabla \cdot \mathbf{P}) + \dfrac{\partial \mathbf{P}}{\partial t} + \nabla \times (\mathbf{P} \times \mathbf{u}) \right.$

$$\left. + (\nabla \cdot \mathbf{P})\mathbf{u} + \epsilon_0 \dfrac{\partial \mathbf{E}}{\partial t} \right] \quad (9\text{--}18)$$

$$= \mu_0 \left[\mathbf{j}_{\text{true}} + \rho_{\text{true}}\mathbf{u} + \dfrac{\partial \mathbf{D}}{\partial t} + \nabla \times (\mathbf{P} \times \mathbf{u}) \right].$$

Note that Maxwell's equations for moving (nonmagnetic) media in the form given by Eq. (9–18) (4) are "mixed," i.e., the sources $\mathbf{j}_{\text{true}}$, **P**, ρ_{true}, are measured in the moving medium, while the fields are given in the stationary frame.

The constitutive equations which give the true currents in the moving medium and the polarization of the moving medium are derived from the fields measured in the moving medium:

$$\mathbf{j}_{\text{true}} = \sigma(\mathbf{E} + \mathbf{u} \times \mathbf{B}) = \sigma \mathbf{E}',$$

$$\mathbf{P} = \epsilon_0(\kappa - 1)(\mathbf{E} + \mathbf{u} \times \mathbf{B}). \quad (9\text{--}19)$$

For a noncharged and current free dielectric, Maxwell's equations can then be written in the form:

(1) $$\nabla \cdot \mathbf{D} = 0,$$

(2) $$\nabla \cdot \mathbf{B} = 0,$$

$$(9\text{--}20)$$

(3) $$\nabla \times \mathbf{E} = -\dfrac{\partial \mathbf{B}}{\partial t},$$

(4) $$\nabla \times [\mathbf{B} - \mu_0 \mathbf{P} \times \mathbf{u}] = \mu_0 \dfrac{\partial \mathbf{D}}{\partial t}.$$

It is seen from (4) of Eqs. (9–20) that from the macroscopic point of view a moving polarized dielectric is equivalent to a magnetized material of magnetic moment

$$\mathbf{M}_{\text{eq}} = \mathbf{P} \times \mathbf{u}. \quad (9\text{--}21)$$

This can be easily understood by considering a polarized slab of material moving at right angles to the direction of polarization. In these circumstances, there is an equivalent positive current in the direction of motion, and another, due to the motion of the negative charges and displaced from the first, in the opposite direction. These currents constitute a net current loop, and thus give rise to a magnetic moment. Hence the moving polarized dielectric will produce a magnetic field which is indistinguishable from that of a magnetized material. This effect has been demonstrated experimentally by Roentgen, Eichenwald, and others.

9–5 Motion of a conductor in a magnetic field. As an example of the application of Maxwell's equations to fields in moving media, let us consider a conducting bar, infinitely long and of rectangular cross section, moving in the direction of its length with velocity **u** relative to a uniform magnetic field **B**. As shown in Fig. 9–2, **B** is at right angles to the direction of motion of the bar, and we shall assume that it is constant in time. Two sliding contacts, terminals of a stationary galvanometer, touch the bar on opposite sides, as shown, so that the distance l between them is the active length of the bar.

From physical considerations, one would expect a current to flow in this composite loop. An electron moving with the bar will experience an effective field given by $\mathbf{E}' = \mathbf{u} \times \mathbf{B}$, so that a current will flow through the contacts to be measured in the external circuit by a stationary observer. Curl **E** must vanish for an observer in any one frame, since **B** is not changing in time. (Whether the source of **B** is stationary or is in motion is entirely irrelevant, since any observed phenomena which depend on a field description must be describable in terms of the behavior of the field quantities alone, independent of the nature of the mechanism which produces the field quantities.) Hence if, as appears logical from the above electron argument, there is an electric field, then such a field must be irrotational, i.e., electrostatic.

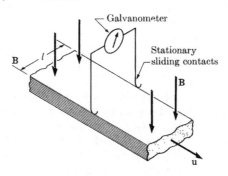

Fig. 9–2. Conducting bar moving along its length in a magnetic field.

TABLE 9-1

ELECTROMOTIVE FORCE RESULTING FROM VARIOUS POSSIBILITIES OF
MOTION FOR THE BAR, SOURCE OF MAGNETIC FIELD, AND
OBSERVER OF FIG. 9-2

Motion of			Electromotive force measured by the observer
Bar	Source of B	Observer (galvanometer)	
u	0	0	uBl
0	u	0	0
0	u	u	uBl
u	u	0	uBl
0	0	u	uBl
u	0	u	0

The effective electric fields within the moving bar will cause a current within the bar that will produce charges on the faces, and these charges will produce the observed external electrostatic field. On the other hand, the same charge displacement will exactly cancel the effective electric field $u \times B$ within the bar, and therefore if we consider an integration path partially contained in the bar and partially outside the bar the integral of the electric field around a closed loop will not vanish. This result is in agreement with the physically observed result that an electromotive force of magnitude uBl is measured by the galvanometer. Note that if the entire circuit were stationary there would be no electromotive force.

If, instead of moving the bar, we move the galvanometer link relative to the bar in the field B, again the electromotive force uBl is observed, since the roles of the link and the bar are simply interchanged in the above integration. The various cases of relative motion are summarized in Table 9-1. Two salient facts characterize the results: (1) the state of motion of the source of B is irrelevant as long as B is uniform; and (2) absolute motion cannot be detected in this arrangement. The latter fact is an indication that Maxwell's equations, if carefully interpreted, are in agreement with relativistic principles. This will be shown later in greater detail and generality.

The situation is more complicated if, in addition to an external magnetic field, there is a field caused by the magnetic moment, either induced or permanent, of the slab. Our conclusion that the electric field observed in a stationary loop will be purely electrostatic remains valid. However, the source of the electrostatic field will not become fully clear until the equations for moving media are modified to include permeable media.

Unfortunately, this modification cannot be made in a reasonable way without introducing relativistic considerations. Nevertheless, the result is physically clear, since the source of the magnetic field, provided it is constant in time, does not affect the force produced by the field. In other words, the force which acts on a moving electron within a moving bar is independent of whether the magnetic field is external or due to the magnetic moment of the bar itself. Therefore we should expect to obtain an electromotive force, given by uBl as before, where $\mathbf{B}$ is the magnetic field in the moving magnetized bar. The only thing that seems paradoxical is that since a moving magnet is essentially an assembly of current loops, we have apparently concluded that the motion of loops carrying steady currents gives rise to an electrostatic field. This, as will be shown later by relativistic considerations, is in fact true. We shall be able to show in general that if an observer moves with a velocity $\mathbf{u}$ relative to a medium of magnetization $\mathbf{M}$ he will observe an equivalent electric moment given by

$$\mathbf{P}_{eq} = \frac{1}{c^2}\mathbf{u} \times \mathbf{M}. \tag{9-22}$$

Therefore,

$$\mathbf{\nabla} \cdot \mathbf{E} = -\frac{\mathbf{\nabla} \cdot \mathbf{P}_{eq}}{\epsilon_0} \tag{9-23}$$

will define the sources of the field. Note that this effect, although it appears deceptively similar to the classical effect of Eq. (9–21), is actually explainable only in terms of the special theory of relativity. It is caused by the fact that to the observer the time spent by the charge traveling in the direction parallel to the relative motion of the circuit and the observer is different from the time that it spends moving in the antiparallel direction. This gives rise to an effective polarization that is perpendicular to the direction of motion and which lies in the plane of the current loop. We shall discuss this effect in detail later.

If the length of the magnetized slab is finite, the field $\mathbf{E}$ will no longer be irrotational, since $\partial\mathbf{B}/\partial t$ is no longer zero in the rest frame. In fact, since $\partial/\partial t = -\mathbf{u} \cdot \mathbf{\nabla}$ for uniform motion $\mathbf{u}$, $\mathbf{\nabla} \times \mathbf{E} = -\partial\mathbf{B}/\partial t = (\mathbf{u} \cdot \mathbf{\nabla})\mathbf{B}$, which because of the vanishing divergence of $\mathbf{B}$ reduces to $-\mathbf{\nabla} \times (\mathbf{u} \times \mathbf{B})$. Hence if $\mathbf{B}$ is no longer uniform, then $\mathbf{E}$ is no longer irrotational; its curl, however, is identical with the curl of $-(\mathbf{u} \times \mathbf{B})$, which is also the effective electric field in the moving medium.

Suggested References

The law of induction and its consequences are treated in all books on electrodynamics. The treatment most nearly parallel to ours is that in *Classical Electricity and Magnetism*, Vol. 1, M. Abraham and R. Becker.

EXERCISES

1. Show that if a flip coil is wound on the surface of a sphere with a winding density proportional to sin θ, where θ is the polar angle, then this coil will measure the axial field component at the center of the coil, independent of the degree of inhomogeneity of the field. (See Exercise 4, Chapter 8.)

2. Approximate the properties of the coil of problem 1 by a simple cylindrical coil. What are the optimum proportions (ratio of height to diameter) of such a coil to minimize gradient effects?

3. A dielectric ($\kappa > 1$) cylinder of radius a is oriented with its axis parallel to a magnetic induction field $\mathbf{B}$. If it is rotated about its axis with an angular velocity ω, find the resultant polarization per unit volume, and the charge per unit length that appears on its surface.

4. A particle of charge q and mass m moves in a circular orbit of radius r_0 under the influence of a central attractive force varying as the inverse square of the distance. Let an external uniform magnetic field $\mathbf{B}(t)$ be applied perpendicular to the orbit such that

$$\mathbf{B}(t) = 0, \qquad\qquad t < 0,$$
$$= \mathbf{B}_0(t/T), \qquad 0 < t < T,$$
$$= \mathbf{B}_0, \qquad\qquad t > T.$$

(a) Calculate the motion.
(b) Calculate the magnetic moment for $t < 0$ and $t > T$.
(c) Calculate the angular velocity ω for $t > T$ assuming the angular velocity to be ω_0 for $t < 0$.

5. The vector potential $\mathbf{A}$ is given in a certain region of space by

$$\mathbf{A}(r, \varphi, z) = a\varphi\hat{r} + (b + cr^2)\hat{z},$$

where $\hat{r}$ and $\hat{z}$ are the unit vectors along the $\mathbf{r}$ and $\mathbf{z}$ directions. What is the current distribution $\mathbf{j}(r, \varphi, z)$ corresponding to this vector potential?

6. Consider a region in space where the magnetic induction field is changing at the rate $\dot{\mathbf{B}}(x_\alpha')$. Show that the electric field $\mathbf{E}$ induced in space is given by

$$\mathbf{E}(x_\alpha) = \frac{1}{4\pi} \int \frac{(\mathbf{r} \times \dot{\mathbf{B}})}{r^3} \, dv', \tag{9-24}$$

where $r(x_\alpha, x_\alpha')$ is here the magnitude of the distance between the field point x_α and the point x_α'.

7. Consider the special case of Exercise 6 in which $\dot{\mathbf{B}}$ has axial symmetry, i.e., $\dot{\mathbf{B}}$ is not a function of the azimuthal angle φ about an axis z. Show that in this case (9–24) reduces to

$$\mathbf{E}(r, z, \varphi) = \frac{1}{2\pi r} \dot{\Phi}_m\hat{\varphi}, \tag{9-25}$$

where Φ_m is the magnetic flux contained in a circle of radius r. Rederive Eq. (9-25) by elementary arguments directly from Faraday's law.

Let $B_z(r, z)$ be the z-component of the axially symmetric magnetic induction field. Consider a particle describing circular orbits of radius r under the influence of this field. Show that if

$$2\pi r^2 \dot{B}_z = \dot{\Phi}_m, \tag{9-26}$$

the radius r of the orbit will remain constant. This equation is the "betatron condition."

CHAPTER 10

ENERGY, FORCE, AND MOMENTUM RELATIONS
IN THE ELECTROMAGNETIC FIELD

In the discussion of the energy relations in electrostatic fields (Chapter 6) we succeeded in associating an energy density with the electric field by considering a specific process. In that case the process was the assembly of charges, during which work was done and changes were produced in the fields. It was possible to obtain a free-energy density of the electric field, in the thermodynamic sense, by balancing the work and energy terms. Let us consider an analogous process to establish a magnetic field before proceeding to obtain a general set of relations for the electromagnetic field.

10–1 Energy relations in quasi-stationary current systems. Consider a process in which a battery with an electromotive field $\mathbf{E}'$ is feeding energy both into heat losses and into a magnetic field. If we take the scalar product of $\mathbf{j}$ and the equation

$$\mathbf{j} = \sigma(\mathbf{E} + \mathbf{E}'), \qquad (7\text{–}4)$$

we obtain

$$\mathbf{E}' \cdot \mathbf{j} = \frac{j^2}{\sigma} - \mathbf{E} \cdot \mathbf{j}. \qquad (10\text{–}1)$$

The left side of Eq. (10–1) represents the time rate at which the battery does work; the first term on the right represents the Joule heat loss in the current-carrying medium, and the last term we tentatively identify as the rate at which energy is fed into the magnetic field. If all fields are quasi-stationary, i.e., slowly varying, the displacement current may be neglected, and the fourth of Maxwell's equations becomes $\nabla \times \mathbf{H} = \mathbf{j}$. When we make this substitution in Eq. (10–1) and integrate over all space, we obtain an expression for the total power expended by the battery in terms of the fields:

$$\int \mathbf{E}' \cdot (\nabla \times \mathbf{H}) \, dv = \int \frac{(\nabla \times \mathbf{H})^2}{\sigma} \, dv - \int \mathbf{E} \cdot (\nabla \times \mathbf{H}) \, dv. \qquad (10\text{–}2)$$

The last term may be integrated by parts by means of the relation

$$\nabla \cdot (\mathbf{E} \times \mathbf{H}) = \mathbf{H} \cdot (\nabla \times \mathbf{E}) - \mathbf{E} \cdot (\nabla \times \mathbf{H}); \qquad (10\text{–}3)$$

and if we use the divergence theorem and the third Maxwell equation, we find that

$$\int \mathbf{E} \cdot (\boldsymbol{\nabla} \times \mathbf{H}) \, dv = - \int \mathbf{H} \cdot \frac{\partial \mathbf{B}}{\partial t} \, dv - \int (\mathbf{E} \times \mathbf{H}) \cdot d\mathbf{S}. \quad (10\text{–}4)$$

Since $\mathbf{E} \times \mathbf{H}$ falls off at least as $1/r^5$ in electrostatic and quasi-stationary magnetic fields, the surface integral in Eq. (10–4) may be dropped. Note that this will not be possible in case $\mathbf{E}$ and $\mathbf{H}$ fall off as $1/r$, as they do in radiation fields. In the absence of the displacement current term no such fields arise: our restriction to slowly varying fields corresponds to neglect of all radiation terms in the magnetic field. Similar limitations apply to our considerations of electrostatic field energy. Thus far we are using, separately, energy relations in electrostatic fields on the one hand, and quasi-stationary current magnetic fields on the other. Later we must see how these concepts can be modified in a consistent way to obtain the general energy expressions.

The total power expended by the battery, Eq. (10–2), may thus be written

$$\int \mathbf{E}' \cdot (\boldsymbol{\nabla} \times \mathbf{H}) \, dv = \int \frac{(\boldsymbol{\nabla} \times \mathbf{H})^2}{\sigma} \, dv + \int \mathbf{H} \cdot \frac{\partial \mathbf{B}}{\partial t} \, dv. \quad (10\text{–}5)$$

The first term on the right has already been identified with the rate of Joule heat loss. The last term is now obviously the rate at which energy is fed into the field. The variation δU_m in magnetic field energy can therefore be given by

$$\delta U_m = \int \mathbf{H} \cdot \delta \mathbf{B} \, dv. \quad (10\text{–}6)$$

This is analogous to the electrostatic energy variation

$$\delta U = \int \mathbf{E} \cdot \delta \mathbf{D} \, dv. \quad (6\text{–}30)$$

As was the case with U, U_m represents the free energy.

In order to make Eq. (10–6) integrable, we must assume a functional relation between $\mathbf{H}$ and $\mathbf{B}$. For a medium which magnetizes linearly the integration can be carried out in the same manner as Eq. (6–14), giving

$$U_m = \tfrac{1}{2} \int \mathbf{H} \cdot \mathbf{B} \, dv. \quad (10\text{–}7)$$

In nonlinear materials such as ferromagnets Eq. (10–6) can be integrated only between definite states, and the answer will, in general, depend on the past history of the sample. For ferromagnets the integral of Eq. (10–6) has a finite value, not zero, when **B** is evaluated around a complete cycle, as in a field produced by an alternating current. The cyclic energy loss is given by

$$\Delta U_m = \iint \oint \mathbf{H} \cdot d\mathbf{B} \, dv. \qquad (10\text{--}8)$$

This means that the energy expended per unit volume when a magnetic material is carried through a magnetization cycle is equal to the area of its hysteresis loop as plotted in a graph of B against H.

Equation (10–7) gives the energy in terms of a volume integral over the fields. If we wish an expression as a volume integral over the sources, we need only write **B** in terms of the vector potential **A**, and **H** in terms of the stationary current **j**. First we obtain

$$U_m = \tfrac{1}{2} \int \mathbf{H} \cdot (\boldsymbol{\nabla} \times \mathbf{A}) \, dv. \qquad (10\text{--}9)$$

If we now integrate by parts and drop the surface terms as before, we find that

$$U_m = \tfrac{1}{2} \int \mathbf{j} \cdot \mathbf{A} \, dv, \qquad (10\text{--}10)$$

which is analogous to the expression for the electrostatic energy in terms of volume charge density and the scalar potential.

Equations (10–6), (10–7), and (10–10) have been derived by a particular "virtual process." The expressions depend only on the final fields, however, and not on the nature of the process. They may thus be taken to represent the magnetic field energy in general, with Eqs. (10–7) and (10–10) subject only to the restrictions of a linear relation between **B** and **H**.

The factor $\tfrac{1}{2}$ in Eq. (10–10) is similar to the factor $\tfrac{1}{2}$ in Eq. (6–1), and is due to the fact that the vector potential **A** includes the fields of the currents **j** themselves. The interaction energy of a system of currents and charges in an *external* field of potentials ϕ and **A** is given by

$$U_{\text{interaction}} = \int (\mathbf{j} \cdot \mathbf{A}_{\text{external}} + \rho \phi_{\text{external}}) \, dv. \qquad (10\text{--}11)$$

10–2 Forces on current systems. We shall now use the energy expressions for two purposes: first, to derive expressions for the forces between currents in terms of the currents themselves and suitable geometrical parameters which depend on their location; and second, to express the variation of the magnetic field energy in terms of the variations of the currents and of the geometrical coordinates.

Let us analyze a system of n geometrically linear circuits carrying currents J_k. For these line circuits Eq. (10–10) reduces to

$$
\begin{aligned}
U_m &= \tfrac{1}{2} \sum_{k=1}^{n} J_k \oint \mathbf{A} \cdot d\mathbf{l}_k = \tfrac{1}{2} \sum_{k=1}^{n} J_k \int (\boldsymbol{\nabla} \times \mathbf{A}) \cdot d\mathbf{S}_k \\
&= \tfrac{1}{2} \sum_{k=1}^{n} J_k \int \mathbf{B} \cdot d\mathbf{S}_k = \tfrac{1}{2} \sum_{k=1}^{n} J_k \Phi_k,
\end{aligned}
\tag{10–12}
$$

where we have made use of Stokes' theorem, and $\Phi_k = \int \mathbf{B} \cdot d\mathbf{S}_k$ is the flux linking the kth circuit, as in Section 8–4 and Eq. (9–1). For a system of *line* currents the flux integrals diverge logarithmically as the radius of the wire goes to zero. Hence, in the subsequent discussion, we shall use $U'_m = U_m - U_s$ as that part of the magnetic field energy which depends on the relative positions of the circuit elements; U_s is the "self-energy" of the elements.

To derive the forces, let us assume that the ith circuit is subjected to a virtual, infinitesimally slow velocity $\mathbf{u}_i$. Then the rate at which a force $\mathbf{F}_i$ acting on the ith circuit is doing work is given by $\mathbf{F}_i \cdot \mathbf{u}_i$. The total rate of energy change in this virtual process is zero, but it is balanced between four quantities:

(1) Rate at which mechanical work is done on the ith circuit, $\mathbf{F}_i \cdot \mathbf{u}_i$.

(2) Rate of change of magnetic field energy, dU'_m/dt.

(3) Rate of Joule heat losses, $\sum J_k^2 R_k$, where R_k is the resistance of the kth circuit.

(4) Rate at which work is being done on the electromotive forces within the circuits, $-\sum J_k \mathcal{E}_k$.

Thus we obtain

$$
\mathbf{F}_i \cdot \mathbf{u}_i + \frac{dU'_m}{dt} + \sum J_k^2 R_k - \sum J_k \mathcal{E}_k = 0.
\tag{10–13}
$$

We are assuming that the magnetic field energy U'_m is expressed explicitly as a function of the coordinates $\mathbf{x}_k$ of the kth current loop and of the current J_k as independent variables. Note that because of the Joule heat and electromotive force work terms we cannot simply equate the force on the ith circuit, $\mathbf{F}_i$, to the negative gradient of the field energy U'_m at constant current, a procedure that would be justified if no other energy terms were present.

Let us now consider a special type of constant current process, namely, that in which the external electromotive forces are adjusted as a function of the virtual velocity $\mathbf{u}_i$ corresponding to the change of a single parameter, $\mathbf{x}_i$, so that the currents within the system remain constant. In this case, if we use Eq. (10–12) for U'_m in Eq. (10–13) and write $\mathcal{E} = JR + d\Phi/dt$ according to Faraday's law [Eq. (9–2)], we obtain

$$\mathbf{F}_i \cdot \mathbf{u}_i + \frac{1}{2} \sum_{k=1}^n J_k \frac{d\Phi_k}{dt} - \sum_{k=1}^n J_k \frac{d\Phi_k}{dt} = 0,$$

or

$$\mathbf{F}_i \cdot \mathbf{u}_i = \frac{\partial U'_m}{\partial t} = \frac{\partial U'_m}{\partial x_{i_\alpha}} u_{i_\alpha}, \tag{10–14}$$

under the conditions that all J's are constant and the positions of all circuits but the ith are held fixed. Hence,

$$F_{i_\alpha} = \frac{\partial U'_m}{\partial x_{i_\alpha}}\bigg|_{J \text{ constant}} \tag{10–15}$$

is the force exerted *by* the field *on* the ith conductor. Note that the sign is opposite to the sign which would be expected if other energy terms were neglected. This means that in order to maintain a constant current in the circuits, as the geometry changes, the batteries must do exactly twice as much work as that done against the external forces, in addition to supplying the Joule heat losses. Equation (10–15) is very useful for computing the forces acting on current-carrying circuits if the magnetic field energy is expressible in terms of the current producing the field.

10–3 Inductance. To express the magnetic field energy

$$U'_m = \tfrac{1}{2} \sum_k J_k \Phi_k \tag{10–12}$$

as a function of current and geometry it is useful to introduce the concept of inductance. The flux through the kth circuit is given by

$$\Phi_k = \int \mathbf{B} \cdot d\mathbf{S}_k = \int (\boldsymbol{\nabla} \times \mathbf{A}) \cdot d\mathbf{S}_k = \oint_k \mathbf{A} \cdot d\mathbf{l}_k. \tag{10–16}$$

The general solution for $\mathbf{A}$,

$$\mathbf{A} = \frac{\mu_0}{4\pi} \int \frac{\mathbf{j}}{r}\, dv \tag{8–29}$$

can be written as a line integral for linear circuits:

$$\mathbf{A}(x_k) = \frac{\mu_0}{4\pi} \sum_i \oint \frac{J_i\, d\mathbf{l}_i}{r_{ik}}. \tag{10–17}$$

With Eq. (10–17) substituted in Eq. (10–16), we may write

$$\Phi_k = \sum_{i \neq k} L_{ik} J_i, \tag{10–18}$$

where

$$L_{ik} = \frac{\mu_0}{4\pi} \oint \oint \frac{d\mathbf{l}_i \cdot d\mathbf{l}_k}{r_{ik}} = L_{ki} \tag{10–19}$$

is a purely geometrical quantity. L_{ik} is called the mutual inductance between the ith and kth circuits, and Eq. (10–19) is known as Neumann's formula.

Equation (10–12) becomes, upon use of Eq. (10–18),

$$U'_m = \frac{1}{2} \sum_i \sum_k{}' L_{ik} J_i J_k, \tag{10–20}$$

where the term $i = k$ is not included in the sum. The force acting on the ith circuit is thus, from Eq. (10–15),

$$F_{i_\alpha} = + \frac{\partial U'_m}{\partial x_{i_\alpha}} \bigg|_{J \text{ constant}} = \frac{1}{2} \sum_j \sum_k{}' J_j J_k \frac{\partial L_{jk}}{\partial x_{i_\alpha}}. \tag{10–21}$$

In the sums of Eqs. (10–20) and (10–21) each term for which the two indices are different occurs twice. The mutual energy of two circuits is therefore

$$U'_m = J_1 J_2 L_{12}. \tag{10–22}$$

We can also formally introduce the self-energy of the two circuits as

$$U_s = \frac{1}{2}(J_1^2 L_{11} + J_2^2 L_{22}), \tag{10–23}$$

where the L_{ii} are called the self-inductances.

The force expression (10–21) is in agreement with the original magnetic interaction of Eq. (7–24). If we substitute Eq. (10–19) in the force equation as applied to two circuits, we obtain

$$\mathbf{F}_1 = \frac{\mu_0}{4\pi} J_1 J_2 \oint \oint (d\mathbf{l}_1 \cdot d\mathbf{l}_2) \boldsymbol{\nabla} \left(\frac{1}{r_{12}} \right)$$

$$= -\frac{\mu_0}{4\pi} J_1 J_2 \oint \oint \mathbf{r}_{12} \left(\frac{d\mathbf{l}_1 \cdot d\mathbf{l}_2}{r_{12}^3} \right),$$

which is identical with Eq. (7–27).

The force equation (10–21) may be written in the simple form

$$F_{i_\alpha} = J_i \sum_k J_k \frac{\partial L_{ik}}{\partial x_{i_\alpha}}, \tag{10–24}$$

since only the terms for which either $j = i$ or $k = i$ depend on x_i and give nonzero derivatives. Equation (10–24) can also be written as

$$F_{i_\alpha} = J_i \left. \frac{\partial \Phi_i}{\partial x_{i_\alpha}} \right|_{J \text{ constant}}, \tag{10–25}$$

where Φ_i is the flux linking the ith circuit. This expression is in evident agreement with the elementary force relation

$$d\mathbf{F} = J \, d\mathbf{l} \times \mathbf{B}. \tag{10–26}$$

These considerations enable us to express the general variation of the field energy as a function, independently, of the geometrical coordinates and of the currents. Since

$$U_m = \tfrac{1}{2} \sum_i \sum_k L_{ik} J_i J_k, \tag{10–20}$$

we have (note that each term occurs twice):

$$\delta U_m = \sum_i \left[\left(\sum_k L_{ik} J_k \right) \delta J_i + \sum_k \frac{\partial L_{ik}}{\partial x_{i_\alpha}} \delta x_{i_\alpha} J_i J_k \right]. \tag{10–27}$$

Hence, from Eqs. (10–18) and (10–24),

$$\delta U_m = \sum_i (\Phi_i \, \delta J_i + F_{i_\alpha} \, \delta x_{i_\alpha}), \tag{10–28}$$

where Φ_i is the total flux linking the ith circuit. It may be seen that positions and currents play the roles of extensive variables in the thermodynamic sense, while the forces and flux play the roles of intensive variables.

The "back emf" terms can be ignored in force calculations if the flux linkages are held constant; in that case the work supplied by the battery exactly balances the Joule heat loss. In these circumstances, directly from Eq. (10–13), we have

$$F_{i_\alpha} = -\left.\frac{\partial U'_m}{\partial x_{i_\alpha}}\right|_{\Phi \text{ constant}}, \tag{10–29}$$

in contrast to Eq. (10–15).

The mutual inductances can be calculated by several means other than Neumann's formula. One method is to use the defining equation, Eq. (10–18). The flux linking the kth circuit due to the current in the ith circuit can be evaluated directly from the known field or vector potential of the ith circuit. A second method, which is particularly useful when continuous current distributions and therefore partial flux linkages are involved, makes explicit use of the two expressions for the magnetic field,

$$U_m = \tfrac{1}{2}\sum_{i=1}^{n}\sum_{k=1}^{n} L_{ik}J_i J_k = \tfrac{1}{2}\int \mathbf{H}\cdot\mathbf{B}\,dv. \tag{10–30}$$

Solutions for $\mathbf{H}$ and $\mathbf{B}$ may be obtained by methods already discussed. The calculation of inductances is then carried out by evaluating the integrals of $\mathbf{H}\cdot\mathbf{B}$. This is the only method available for the calculation of the self-inductances L_{ii}.

In computing the inductances of current-carrying conductors it is usually advantageous to separate the problem into two parts: (1) the inductance associated with the field outside the conductor, and (2) the contribution to the inductance by the field energy inside the wire. It is necessary to make this separation because the inductance due to the external field cannot be computed under the assumption of zero radius for the conductor: such an assumption will generally lead to a logarithmic divergence of the integral involved. We shall see that at high frequencies the contribution to the inductance by the field within the wire becomes negligible, since the currents do not penetrate into the wire. At lower frequencies, and particularly in case the conductors are ferromagnetic, this internal term may make an appreciable contribution.

10–4 Magnetic volume force. We have now calculated the forces between current systems in terms of the currents and the appropriate geometrical quantities. These forces are, as would be expected, re-expressions of the original Ampère interaction law given in Eq. (7–24). We can also, in analogy to the electrostatic case, derive an expression for the magnetic body force per unit volume in terms of the field, the permeability, and the current at a given point. In the electrostatic case we defined $\mathbf{F}_v$ by the relation

$$\frac{dU}{dt} = -\int \mathbf{F}_v \cdot \mathbf{u}\,dv \tag{6–51}$$

and found that $\mathbf{F}_v$ was given by

$$\mathbf{F}_v = \rho \mathbf{E} - \frac{\epsilon_0}{2} E^2 \nabla \kappa + \frac{\epsilon_0}{2} \nabla \left(E^2 \frac{d\kappa}{d\rho_m} \right). \qquad (6\text{--}68)$$

Similarly, we may write

$$\frac{dU_m}{dt} = - \int \mathbf{F}_v \cdot \mathbf{u} \, dv$$

and identify $\mathbf{F}_v$ as the magnetic body force. Instead of writing the general expression, let us make the following restrictions: (1) The medium is linear, i.e., its permeability is not a function of the field. (2) There is no permanent magnetic moment present. (3) There is no magnetostriction, i.e., $d\kappa_m/d\rho_m = 0$. Under these conditions a straightforward calculation yields

$$\mathbf{F}_v = \mathbf{j} \times \mathbf{B} - \frac{\mu_0}{2} H^2 \nabla \kappa_m. \qquad (10\text{--}31)$$

It is again possible, in accord with the requirements of a satisfactory field theory, to derive the total force on a given volume in terms of the value of the field on the boundary of this volume. In other words, it is possible to define a stress tensor from which the volume force is derivable by the tensor divergence relation of Eq. (6–35). The form of the Maxwell tensor in the magnetic case, in the absence of a magnetostriction term, is formally analogous to Eq. (6–82):

$$T_{\alpha\beta} = H_\alpha B_\beta - \frac{\delta_{\alpha\beta}}{2} H_\gamma B_\gamma, \qquad (10\text{--}32)$$

with the customary summation convention. The geometrical interpretation of this tensor leads to the same conclusions about magnetic forces as were reached in Chapter 6 about electrical forces: the direction of the magnetic field bisects the angle between the normal to a surface and the direction of the resultant magnetic stress that acts on the surface. The magnitude of the magnetic stress normal to the field or parallel to the field is $HB/2$.

10–5 General expressions for electromagnetic energy. Thus far we have considered the electrostatic field and the magnetostatic field, or quasi-stationary current fields, separately. Let us now inquire as to which of the energy, force, or momentum relations will need modification for the general case in which no restrictions on the time rate of change of the field quantities are imposed. We can confine our attention to vacuum fields and nonpermeable conductors without omitting results of any great interest.

Maxwell's equations *in vacuo* were shown to be:

$$(1) \qquad\qquad \mathbf{\nabla} \cdot \mathbf{D} = \rho_{\text{true}},$$

$$(2) \qquad\qquad \mathbf{\nabla} \cdot \mathbf{B} = 0,$$

$$(3) \qquad\qquad \mathbf{\nabla} \times \mathbf{E} = - \frac{\partial \mathbf{B}}{\partial t}, \qquad\qquad (9\text{–}7)$$

$$(4) \qquad\qquad \mathbf{\nabla} \times \mathbf{H} = \mathbf{j}_{\text{true}} + \frac{\partial \mathbf{D}}{\partial t}.$$

These equations completely represent the behavior of electromagnetic fields when they are considered in combination with suitable constitutive equations and boundary conditions. This is true even for rapidly varying fields, or at least no internal contradiction is present if an arbitrary rate of change is assumed. Care must be taken in using constitutive equations, however, since the material constants are generally dependent on the frequency of the fields. We shall often restrict ourselves to vacuum conditions in order to avoid complications due to such special properties of the constitutive equations.

An energy integral of Maxwell's equations can be obtained as a result of taking the scalar product of the third and fourth of these equations with $\mathbf{H}$ and $\mathbf{E}$ respectively. If we subtract the two resulting equations and make use of the vector identity for the divergence of a cross product, Eq. (10–3), we obtain

$$\mathbf{\nabla} \cdot (\mathbf{E} \times \mathbf{H}) = - \frac{\partial \mathbf{B}}{\partial t} \cdot \mathbf{H} - \mathbf{E} \cdot \mathbf{j} - \mathbf{E} \cdot \frac{\partial \mathbf{D}}{\partial t}. \qquad (10\text{–}33)$$

Let us now take the volume integral of Eq. (10–33), assuming that the constitutive equations are linear:

$$- \frac{\partial}{\partial t} \int \tfrac{1}{2} (\mathbf{H} \cdot \mathbf{B} + \mathbf{E} \cdot \mathbf{D}) \, dv = \int \mathbf{E} \cdot \mathbf{j} \, dv + \int (\mathbf{E} \times \mathbf{H}) \cdot d\mathbf{S}, \qquad (10\text{–}34)$$

where the surface term arises as a result of using the divergence theorem. The left side of Eq. (10–34) is recognized as the rate of decrease of the sum of the electric and magnetic field energies [Eqs. (6–14) and (10–7)] as derived for the static case. By means of Eq. (10–1) the first term on the right can be written

$$\int \mathbf{E} \cdot \mathbf{j} \, dv = \int \left(\frac{j^2}{\sigma} - \mathbf{E}' \cdot \mathbf{j} \right) dv. \qquad (10\text{–}35)$$

Equation (10–35) therefore represents the sum of the Joule heat losses and the negative rate at which the electromotive forces are doing work.

The last term on the right side of Eq. (10–34) demands careful consideration. It represents an energy which has been hitherto neglected, since for static and quasi-static fields the integral can be made to vanish by taking an arbitrarily large enclosing surface. As we shall see later, however, the electric and magnetic radiation fields of charges in motion and of changing currents fall off in general only as $1/r$ at large distances. In that case $\int (\mathbf{E} \times \mathbf{H}) \cdot d\mathbf{S}$ will approach a constant value for an arbitrarily large surface, and thus the integral may contribute to the energy balance.

The vector

$$\mathbf{N} = \mathbf{E} \times \mathbf{H} \tag{10–36}$$

is known as the Poynting vector. In terms of Eq. (10–34) it can be considered to represent the electromagnetic field energy flow per unit area per unit time across a given surface. It must be noted, however, that only the entire surface integral of $\mathbf{N}$ contributes to the energy balance. Paradoxical results may be obtained if one tries to identify the Poynting vector with the energy flow per unit area at any particular point. The energy term arose as the volume integral of $\boldsymbol{\nabla} \cdot (\mathbf{E} \times \mathbf{H})$, and the net energy flow in the electromagnetic field will always vanish if the divergence of the Poynting vector is zero. For example, in static superposed electric and magnetic fields in the absence of currents we may have nonzero values of the Poynting vector at various points in space, but its divergence vanishes everywhere, as can be seen from Eq. (10–33).

We have now seen that Eq. (10–34) can be taken to represent the overall energy balance between the electric and magnetic energy of the field, the loss due to resistive heating, the work done by sources of electromotive force, and the radiation loss. It appears that the expressions derived for the energy densities of the electrostatic and magnetostatic fields retain their validity when the fields are allowed to vary arbitrarily with the time. The only additional consideration needed in order to conserve energy is to

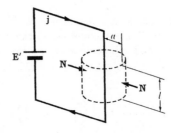

Fig. 10–1. Illustrating the role of the Poynting vector in the conservation of energy over a part of a circuit.

assume that fields may carry energy in or out of the volume of integration at a rate that is given by the surface integral of the Poynting vector.

A new consequence of the introduction of the surface term into the conservation laws is the possibility of balancing energy and momentum over only part of a system. The surface integral over the Poynting vector permits us to conserve energy in one part of a system whether radiative processes are present or not. To illustrate this point, consider the simple process of a battery (electromotive field $\mathbf{E}'$) feeding a current (density $\mathbf{j}$) to a resistor (of conductivity σ), as in Fig. 10–1. Let us take the energy balance over the volume of length l of the resistor, whose radius a has been magnified in the drawing. By elementary considerations,

$$E = \frac{j}{\sigma}; \qquad H = \frac{1}{2}\,ja; \qquad N = \frac{j^2 a}{2\sigma}$$

where $\mathbf{N}$ is directed inward over the lateral surface of the cylinder. Hence

$$\int \mathbf{N} \cdot d\mathbf{S} = -j^2\,\frac{(\pi a^2 l)}{\sigma} = -\int \frac{j^2}{\sigma}\,dv = -\text{Joule heat.}$$

Equation (10–34) is thus satisfied, and energy is balanced by considering the *field* as "feeding" the resistor via the Poynting vector, without the explicit introduction of the source of energy, i.e., the battery field $\mathbf{E}'$.

10–6 Momentum balance. In order to investigate the momentum relations in the electromagnetic fields, we shall inquire as to whether the divergence of the complete Maxwell stress tensor will give a volume force which is in accord with experience. If we neglect the electrostriction and magnetostriction terms the stress tensor is the sum of Eqs. (6–82) and (10–32):

$$T_{\alpha\beta} = E_\alpha D_\beta - \tfrac{1}{2}\,\delta_{\alpha\beta}E_\gamma D_\gamma + H_\alpha B_\beta - \tfrac{1}{2}\,\delta_{\alpha\beta}H_\gamma B_\gamma. \qquad (10\text{–}37)$$

The tensor divergence of Eq. (10–37) is

$$\frac{\partial T_{\alpha\beta}}{\partial x_\beta} = E_\alpha\,\frac{\partial D_\beta}{\partial x_\beta} + D_\beta\,\frac{\partial E_\alpha}{\partial x_\beta} - \frac{1}{2}\,E^2\epsilon_0\,\frac{\partial \kappa}{\partial x_\alpha} - D_\beta\,\frac{\partial E_\beta}{\partial x_\alpha}$$

$$+ H_\alpha\,\frac{\partial B_\beta}{\partial x_\beta} + B_\beta\,\frac{\partial H_\alpha}{\partial x_\beta} - \frac{1}{2}\,H^2\mu_0\,\frac{\partial \kappa_m}{\partial x_\alpha} - B_\beta\,\frac{\partial H_\beta}{\partial x_\alpha}. \qquad (10\text{–}38)$$

Equation (10–38) can be expressed in vector form: by Maxwell's first equation $\boldsymbol{\nabla} \cdot \mathbf{D} = \rho_{\text{true}}$ in the first term; the second and fourth terms become $-\mathbf{D} \times (\boldsymbol{\nabla} \times \mathbf{E})$, which is equal to $\mathbf{D} \times \partial \mathbf{B}/\partial t$; the sixth and eighth terms become $-\mathbf{B} \times (\boldsymbol{\nabla} \times \mathbf{H})$, which is equal to $-\mathbf{B} \times (\mathbf{j}_{\text{true}} + \partial \mathbf{D}/\partial t)$;

the fifth term vanishes due to the fact that $\nabla \cdot \mathbf{B} = 0$; and the third and seventh terms involve merely gradients of κ and κ_m. Therefore

$$\frac{\partial T_{\alpha\beta}}{\partial x_\beta} = \left[\mathbf{E}\rho_{\text{true}} - \frac{\epsilon_0}{2} E^2 \nabla \kappa - \frac{\mu_0}{2} H^2 \nabla \kappa_m - \mathbf{B} \times \mathbf{j}_{\text{true}} + \frac{\partial}{\partial t} (\mathbf{D} \times \mathbf{B}) \right]_\alpha .$$
(10–39)

This entire tensor divergence may be expressed as the sum of two terms:

$$\frac{\partial T_{\alpha\beta}}{\partial x_\beta} = \left[\mathbf{F}_{ev} + \frac{\partial}{\partial t} (\mathbf{D} \times \mathbf{B}) \right]_\alpha ,$$
(10–40)

where the first term,

$$\mathbf{F}_{ev} = \mathbf{E}\rho_{\text{true}} - \frac{\epsilon_0}{2} E^2 \nabla \kappa - \frac{\mu_0}{2} H^2 \nabla \kappa_m - \mathbf{B} \times \mathbf{j}_{\text{true}} \qquad (10–41)$$

is the ordinary volume force acting on material bodies in a quasi-stationary electromagnetic field. Equation (10–41) accounts fully for the volume force on true charges or inhomogeneous dielectrics in an electric field, and for that on true currents or inhomogeneous permeable material in a magnetic field. The second term in Eq. (10–40) is new, and is proportional to the time rate of change of the Poynting vector. Equation (10–40) may be written explicitly

$$\frac{\partial T_{\alpha\beta}}{\partial x_\beta} = \left[\mathbf{F}_{ev} + \mu\epsilon \frac{\partial(\mathbf{E} \times \mathbf{H})}{\partial t} \right]_\alpha$$

$$= \left[\mathbf{F}_{ev} + \frac{\kappa\kappa_m}{c^2} \frac{\partial \mathbf{N}}{\partial t} \right]_\alpha ,$$
(10–42)

where

$$\epsilon_0 \mu_0 = \frac{1}{c^2} .$$
(10–43)

The existence of $\mathbf{F}_{ev}$ depends on the presence of material bodies carrying charges or endowed with dielectric properties. On the other hand, the second term in Eq. (10–42) does not vanish even *in vacuo*, and therefore it would superficially suggest the idea of a volume force on the vacuum. This term has evoked a great deal of speculation. It fits into an ether theory in which vacuum possesses various mechanical properties that enable it to transmit elastic waves and also to sustain body forces. The only way such an ether force could be measured would be by means of the action of the ether on matter.

According to Lorentz' electron theory, however, the only force which has physical significance is a resultant force which arises from the space-time average forces acting on material charges and currents, namely, those obtained by averaging

$$\mathbf{F} = \rho(\mathbf{E} + \mathbf{u} \times \mathbf{B}). \tag{10–44}$$

We shall also find that within the framework of the special theory of relativity no measurement can be devised which can determine the velocity or other properties of the ether. If therefore we adopt the point of view that the only volume force which has a place in physical theory is a force derivable from the Lorentz force, Eq. (10–44), it follows that the second term in Eq. (10–42) must be subtracted out. We then have for the volume force, when $\kappa = \kappa_m = 1$,

$$F_{v\alpha} = \frac{\partial T_{\alpha\beta}}{\partial x_\beta} - \frac{1}{c^2} \frac{\partial N_\alpha}{\partial t}, \tag{10–45}$$

which is equal to the Lorentz force.

If we apply this equation to a volume containing both matter and radiation, and bounded by a finite surface, we obtain*

$$F_\alpha = \int T_{\alpha\beta} \, dS_\beta - \frac{1}{c^2} \frac{\partial}{\partial t} \int N_\alpha \, dv. \tag{10–46}$$

Since the integrated body force F_α represents the total rate of change of mechanical momentum, p_α, of the volume, Eq. (10–46) can be written

$$\frac{d}{dt} \left[p_\alpha + \frac{1}{c^2} \int N_\alpha \, dv \right] = \int T_{\alpha\beta} \, dS_\beta. \tag{10–47}$$

This equation states that the sum of the rate of change of the mechanical momentum, plus a term equal to the volume integral of the Poynting vector divided by c^2, is equal to the surface integral of the total Maxwell stress transmitted across the surface surrounding the volume. If it were possible to choose a surface so large that it is in field-free space, then the sum of the mechanical momentum and the volume integral of the Poynting vector would be constant in time. This implies that the correction term, whose introduction into Eq. (10–45) was required only by our demand for a physical interpretation of the volume force, makes necessary a change in our concept of momentum.

---

* We have omitted a term which, in matter, is given by

$$\frac{(\kappa\kappa_m - 1)}{c^2} \frac{\partial}{\partial t} \int N_\alpha \, dv,$$

and is actually present when an electromagnetic wave travels through matter. Its net impulse due to a finite wave train always vanishes; we shall not discuss its rather complicated interpretation.

In the absence of measurable physical properties for the ether, we are thus forced to modify the law of the conservation of momentum. It must apply not just to matter alone, but must also include a momentum density of the electromagnetic radiation field which is equal to the Poynting vector divided by a constant, c^2. The Poynting vector therefore appears in a dual role, as carrying energy and also as carrying momentum. It will turn out in the special theory of relativity that the transfer of energy corresponds to a transfer of momentum in the proportions that have been derived here. Actually, this is a property of all forms of energy flow, and is not confined to electromagnetic radiation.

SUGGESTED REFERENCES

M. ABRAHAM AND R. BECKER, *Classical Theory of Electricity and Magnetism.* The energy theorems are developed in a clear and elementary way.

L. D. LANDAU AND E. M. LIFSHITZ, *Electrodynamics of Continuous Media.* Chapter IV contains a thorough discussion of the magnetic field energy and the interaction of systems of currents.

W. R. SMYTHE, *Static and Dynamic Electricity.* Many specific problems.

J. H. JEANS, *Electricity and Magnetism.* Again the organization of subject matter is very different from ours, but there are interesting applications.

EXERCISES

1. Use Neumann's formula to compute the mutual inductance per unit length between two equal parallel conductors. If the two wires constitute a single circuit, find the force between them.

2. Consider a coaxial cable consisting of a center wire of radius a and a thin sheath of radius $b \gg a$. Find the self-inductance per unit length of the pair of conductors as a circuit, and the mutual inductance of the core and sheath. Show that the mutual inductance of the core and sheath is the same as the self-inductance of the sheath. Are the answers altered if the core and sheath are not coaxial?

3. Find the torque on a solid conducting cylinder rotating slowly in a uniform magnetic field perpendicular to the axis of the cylinder.

4. A thin spherical shell of conductivity σ, thickness t, radius a, rotates with uniform angular velocity about an axis perpendicular to a uniform magnetic field. Calculate the power needed to maintain the rotation.

5. A steel sphere of radius a is magnetized permanently and uniformly with magnetization **M**. A circular coil of N turns and radius $b > a$, carrying a current J, is mounted in a plane parallel to **M** so that the center of the coil coincides with the center of the sphere. Find the torque on the coil.

6. Find the total field energy, and the field energy density in the iron and air gap of the incomplete iron ring of Exercise 10, Chapter 8. Discuss the energy balance of the process of widening the gap slightly by means of an external mechanical force.

7. Show that the tensor divergence $\partial T_{\alpha\beta}/\partial x_\alpha$, where $T_{\alpha\beta}$ is given by Eq. (10–32), gives the β-component of the volume force, Eq. (10–31).

CHAPTER 11

THE WAVE EQUATION AND PLANE WAVES

11–1 The wave equation. In the energy integral of Maxwell's equations, Eq. (10–34), we interpreted a nonvanishing integral over a distant surface as representing a flow of energy which we called radiation. The equations may be combined to exhibit these propagating fields more explicitly: the four first-order linear partial differential field equations can be reduced to a system of two second-order linear partial differential equations. To make this reduction, let us recall that Maxwell's equations are

$$\text{(1)} \qquad\qquad \mathbf{\nabla} \cdot \mathbf{D} = \rho_{\text{true}},$$

$$\text{(2)} \qquad\qquad \mathbf{\nabla} \cdot \mathbf{B} = 0,$$

$$\text{(3)} \qquad\qquad \mathbf{\nabla} \times \mathbf{E} = -\frac{\partial \mathbf{B}}{\partial t}, \qquad\qquad\qquad \text{(9–7)}$$

$$\text{(4)} \qquad\qquad \mathbf{\nabla} \times \mathbf{H} = \mathbf{j}_{\text{true}} + \frac{\partial \mathbf{D}}{dt}.$$

Consider a region where there are no true charges and no sources of electromotive force, so that $\rho_{\text{true}} = 0$ and $\mathbf{E}' = 0$, and where ϵ and μ are not functions of the coordinates or of the time. Take the curl of Eq. (9–7) (3) and substitute $\mu\mathbf{H}$ for $\mathbf{B}$:

$$\mathbf{\nabla} \times (\mathbf{\nabla} \times \mathbf{E}) = -\frac{\partial}{\partial t}(\mathbf{\nabla} \times \mu\mathbf{H}). \qquad (11\text{–}1)$$

On substituting Eq. (9–7) (4) for curl $\mathbf{H}$, we obtain

$$\mathbf{\nabla} \times (\mathbf{\nabla} \times \mathbf{E}) = -\mu\frac{\partial}{\partial t}\left(\mathbf{j}_{\text{true}} + \epsilon\frac{\partial \mathbf{E}}{\partial t}\right). \qquad (11\text{–}2)$$

Now for any vector it is true that

$$\mathbf{\nabla} \times (\mathbf{\nabla} \times \mathbf{E}) = \mathbf{\nabla}(\mathbf{\nabla} \cdot \mathbf{E}) - \nabla^2\mathbf{E},$$

but $\mathbf{\nabla} \cdot \mathbf{E} = 0$ in the charge-free region, and by Eq. (7–3) $\mathbf{j} = \sigma\mathbf{E}$. Therefore Eq. (11–2) becomes

$$\nabla^2\mathbf{E} - \frac{\kappa\kappa_m}{c^2}\frac{\partial^2 \mathbf{E}}{\partial t^2} - \mu\sigma\frac{\partial \mathbf{E}}{\partial t} = 0, \qquad (11\text{–}3)$$

where we have written $1/c^2$ for the product of free space constants $\mu_0\epsilon_0$. It can be seen from Eq. (11–3) that c must have the dimensions of velocity.

Equation (11–3) is known as the general wave equation. Usually either the second or the third term drops out in any particular application of this equation. In a nonconducting medium the third term vanishes, leaving a propagation equation for waves that travel with the velocity $u = 1/\sqrt{\mu\epsilon}$, as we shall see. In a conducting medium the second term is usually negligible, and we are left with the same differential equation as for heat conduction or diffusion. The relative magnitude of the two terms can be easily estimated for a field that varies with an angular frequency ω:

$$\mathbf{E} = \mathbf{E}(x_a)e^{-i\omega t}. \tag{11–4}$$

When Eq. (11–4) is substituted in Eq. (11–3), we obtain an equation that does not depend on the time:

$$\nabla^2\mathbf{E} + \frac{KK_m}{c^2}\omega^2\mathbf{E} + i\mu\sigma\omega\mathbf{E} = 0. \tag{11–5}$$

Equation (11–5) may be written

$$\nabla^2\mathbf{E} + \left(1 + \frac{i\sigma}{\epsilon\omega}\right)\mu\epsilon\omega^2\mathbf{E} = 0, \tag{11–6}$$

and the relaxation time of the medium, discussed in Section 7–4, is given by

$$\tau = \frac{\epsilon}{\sigma}. \tag{7–20}$$

Hence we see that if the relaxation time τ is long compared with the period $2\pi/\omega$ of the sinusoidal vibration, so that the imaginary term in Eq. (11–6) can be neglected, we are left with a propagation equation. On the other hand, if the relaxation time is short compared with the period, then the imaginary term is large compared with unity and we have essentially a diffusion condition.

For all pure metals the relaxation time is of the order of 10^{-14} sec, and hence the diffusion type of equation is valid for all frequencies lower than those of the optical spectrum. This means that within metallic conductors the propagation term $\mu\epsilon\,\partial^2\mathbf{E}/\partial t^2$ can be neglected even in the ultra-high-frequency radio region. To put it in a different way, the displacement current is negligible relative to the conduction current in metals at the highest frequencies that are theoretically attainable with macroscopic oscillators.

The coefficients that appear in Eq. (11–5) can be expressed in a variety of useful ways. If, for example, we write

$$\frac{\omega}{c} = \frac{1}{\lambda_0}, \tag{11–7}$$

λ_0 is the free space wavelength divided by 2π. Similarly, for a medium characterized by κ and κ_m, we may write

$$\frac{\omega}{u} = \frac{1}{\lambda}, \tag{11-8}$$

where $u = c/\sqrt{\kappa\kappa_m}$. The ratio of the magnitude of the conduction current to the magnitude of the displacement current can be rewritten in terms of these relations as

$$\frac{1}{\omega\tau} = \frac{\sigma}{\omega\epsilon} = \frac{\sigma}{\epsilon}\frac{\lambda}{u} = \sigma\lambda\sqrt{\frac{\mu}{\epsilon}} = \sigma\lambda R_0\sqrt{\frac{\kappa_m}{\kappa}}, \tag{11-9}$$

where

$$R_0 = \sqrt{\frac{\mu_0}{\epsilon_0}} \tag{11-10}$$

is a resistance whose numerical value is 376.7 ohms. This number is sometimes called the characteristic impedance of free space. The analogy between wave propagation and "lumped" transmission line parameters is useful in some practical considerations, but we shall here avoid such analogies in favor of considering the characteristics of the electromagnetic fields themselves. The significance of Eq. (11-9) might be stated by saying that if the resistance of a cube whose edge is λ is larger than R_0, then in such a medium the displacement current is dominant, while if the reverse is true the conduction current governs the behavior of electromagnetic fields in the medium.

Equation (11-3) and the analogous equation for **H** or **B** are homogeneous in the field vectors because they were derived subject to the condition of no true charges and no sources of electromotive force. To investigate the relation between these fields and their sources it is convenient, as in the static case, to introduce potentials as intermediary field quantities. We shall return to this problem in Chapter 14. A number of interesting and important consequences that are independent of the origin of the fields can be considered without reference to the potentials.

11-2 Plane waves. Let us consider the case in which all fields are functions only of the distance of a given plane from the origin, as in Fig. 11-1. If this distance is ζ and if **n** is a unit vector normal to the plane, then for all points spatial derivatives are with respect to ζ only, and the operator ∇ becomes

$$\nabla = \mathbf{n}\frac{\partial}{\partial\zeta}.$$

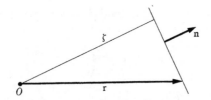

Fig. 11-1. All fields are constant on such plane surfaces as that indicated by the unit normal **n**.

Thus Maxwell's equations reduce to

(1) $$\mathbf{n} \cdot \frac{\partial \mathbf{D}}{\partial \zeta} = 0,$$

(2) $$\mathbf{n} \cdot \frac{\partial \mathbf{B}}{\partial \zeta} = 0,$$ (11-11)

(3) $$\mathbf{n} \times \frac{\partial \mathbf{E}}{\partial \zeta} = -\frac{\partial \mathbf{B}}{\partial t},$$

(4) $$\mathbf{n} \times \frac{\partial \mathbf{H}}{\partial \zeta} = \mathbf{j} + \frac{\partial \mathbf{D}}{\partial t} = \sigma \mathbf{E} + \frac{\partial \mathbf{D}}{\partial t}.$$

By taking the scalar product of **n** and Eq. (11-11) (4), we see that

$$\mathbf{n} \cdot \left(\frac{\sigma}{\epsilon} + \frac{\partial}{\partial t} \right) \mathbf{D} = 0. \qquad (11\text{-}12)$$

Equations (11-12) and (11-11) (1) imply that the longitudinal components of **D** and **E**, i.e., those components that are perpendicular to the plane surface of the figure, are independent of ζ and that their time dependence follows an exponential decay law in accordance with the characteristic relaxation time of the medium. Thus

$$E_n = E_{n_0} e^{-t/\tau} = E_{n_0} e^{-(\sigma t/\epsilon)}. \qquad (11\text{-}13)$$

This means that the only longitudinal solution of the field equations is an electrostatic solution, and that in the presence of finite conductivity the electrostatic solutions will vanish exponentially with time.

Scalar multiplication of **n** and Eq. (11-11) (3) leads to

$$\mathbf{n} \cdot \frac{\partial \mathbf{B}}{\partial t} = 0, \qquad (11\text{-}14)$$

which, together with Eq. (11-11) (2), shows that the only longitudinal component of the magnetic field compatible with the field equations is a stationary uniform magnetic field. Thus if there is a nonstatic solution it must be composed of transverse fields, fields whose vectors lie in the plane indicated by **n** in Fig. 11–1.

Each transverse component of the electric field obeys the differential equation

$$\frac{\partial^2 \mathbf{E}}{\partial \zeta^2} - \mu\epsilon \frac{\partial^2 \mathbf{E}}{\partial t^2} - \mu\sigma \frac{\partial \mathbf{E}}{\partial t} = 0, \tag{11–15}$$

which may be derived by eliminating **H** between Eqs. (11–11) (3) and (4) or may be written down directly from Eq. (11–3). Equation (11–15) is called the "equation of telegraphy." If the medium has zero conductivity, this equation reduces to

$$\frac{\partial^2 \mathbf{E}}{\partial \zeta^2} - \mu\epsilon \frac{\partial^2 \mathbf{E}}{\partial t^2} = 0. \tag{11–16}$$

The general solution of Eq. (11–16) is

$$E = f(\zeta - ut) + g(\zeta + ut), \tag{11–17}$$

where f and g are arbitrary functions of their respective arguments, and $u = c/\sqrt{\kappa\kappa_m} = 1/\sqrt{\mu\epsilon}$, as before. The general solution represents waves traveling along ζ, i.e., a disturbance of form f is propagated in the positive ζ-direction as time progresses, and g is propagated in the opposite direction. u is the velocity with which the form of the disturbance travels, and is called the phase or wave velocity. The ratio of the free-space velocity to that in a medium characterized by κ and κ_m, $c/u = \sqrt{\kappa\kappa_m}$, is called the index of refraction of the medium, and is often designated by n.

If **E** is assumed to have a sinusoidal time variation, given by the factor $e^{-i\omega t}$, with an angular frequency ω, the solution of Eq. (11–16) is

$$\mathbf{E} = \mathbf{E}_0 e^{-i(\omega t \pm k\zeta)}, \tag{11–18}$$

where

$$k = \omega/u = 2\pi/\lambda = 1/\lambdabar.$$

It is often useful to write k as a vector in the direction of propagation. The propagation vector is given by

$$\mathbf{k} = \frac{\omega}{u^2} \mathbf{u} = \frac{1}{\lambdabar} \frac{\mathbf{u}}{u}, \tag{11–19}$$

so that a solution of Eq. (11–16) may be written as

$$\mathbf{E} = \mathbf{E}_0 e^{i(\mathbf{k} \cdot \mathbf{r} - \omega t)}. \tag{11–20}$$

Let us consider the case of the negative sign in the complex exponential of Eq. (11–18), so as to have a wave moving toward greater positive ζ. Equation (11–11) (3) becomes

$$\mathbf{n} \times \frac{\partial \mathbf{E}}{\partial \zeta} = i\omega \mathbf{B} = ik\mathbf{n} \times \mathbf{E},$$

or

$$\mathbf{B} = \sqrt{\mu\epsilon}\,\mathbf{n} \times \mathbf{E} = \frac{\mathbf{k}}{\omega} \times \mathbf{E}. \qquad (11\text{–}21)$$

Thus we see that associated with each transverse component of $\mathbf{E}$ there is a magnetic field. If the direction of $\mathbf{E}$ or $\mathbf{B}$ is constant in time the wave is said to be plane polarized, or linearly polarized. Since the field equations show no correlation between different components of any one field vector, we lose no generality in restricting our discussion to a plane polarized wave. $\mathbf{E}$ and $\mathbf{B}$ are perpendicular to each other as well as to the direction of propagation, and the equations indicate that $\mathbf{E}$, $\mathbf{B}$, and $\mathbf{n}$ (or $\mathbf{k}$) denote a right-handed coordinate system, in that order.

It is clear that the complex exponential time dependence for the fields leads to a particularly simple form for Maxwell's equations, and complex expressions such as that for $\mathbf{E}$ in Eq. (11–20) are, in general, very convenient mathematically, but we must remember that the actual fields are the real part of the complex quantity. The energy density associated with sinusoidally varying fields pulsates in the time, since it involves the factor $E^2 = E_0^2 \cos^2{(k\zeta - \omega t)}$, the square of the real part of Eq. (11–20). The time average of E^2 is just $\frac{1}{2}E_0^2$. In general, the time average of the product of the real parts of two vectors both of which vary as $e^{-i\omega t}$ is one-half the real part of the product of one vector and the complex conjugate of the other. Thus

$$\overline{(\mathrm{Re}\ \mathbf{F}) \cdot (\mathrm{Re}\ \mathbf{G})} = \tfrac{1}{2}\,\mathrm{Re}\ (\mathbf{F} \cdot \mathbf{G}^*) = \tfrac{1}{2}\,\mathrm{Re}\ (\mathbf{F}^* \cdot \mathbf{G}), \qquad (11\text{–}22)$$

where * denotes complex conjugate. The products on the right are independent of the time, so that the line denoting the average would have no significance. The verification of Eq. (11–22) follows readily from writing $\mathbf{F}$ and $\mathbf{G}$ as a sum of real and imaginary parts, and the details are left to a problem.

For the case of the plane wave fields given by Eqs. (11–20) and (11–21), and the related fields $\mathbf{D}$ and $\mathbf{H}$, the time average of the energy density is

$$\overline{U} = \tfrac{1}{2}\overline{(\mathbf{H} \cdot \mathbf{B} + \mathbf{E} \cdot \mathbf{D})} = \tfrac{1}{2}\epsilon E_0^2 \text{ joules/meter}^3, \qquad (11\text{–}23)$$

to which the electric and magnetic fields contribute equally. The Poynting vector, $\mathbf{N} = \mathbf{E} \times \mathbf{H}$, is in the direction of propagation of the wave, and its time average is

$$\overline{\mathbf{E} \times \mathbf{H}} = \tfrac{1}{2}\sqrt{\epsilon/\mu}\, E_0^2\, \mathbf{n} \quad \text{watts/meter}^2$$
$$= \overline{U}\mathbf{u}. \tag{11–24}$$

Thus the energy density associated with a plane wave in a stationary homogeneous nonconducting medium is propagated with the same velocity as that of the fields.

11–3 Radiation pressure. We have seen in Section 10–6 that the ether theory and the Lorentz electron theory lead to different expressions for the volume force. The divergence of the Maxwell stress tensor yields a volume force which does not vanish in the absence of charges and currents, and the correction needed to make this force consistent with the Lorentz force implies the existence of a momentum density in the electromagnetic field. Now we see that a plane wave carries energy. Let us next investigate the consequences of momentum balance applied to a plane wave.

Let us first consider a linearly polarized plane wave incident normally on a slab of material that absorbs it completely—a blackbody. If we take the x-axis as the direction of propagation, we may let y be the direction of the electric field and the magnetic field will be parallel to the z-axis. The Maxwell stress tensor (without electro- or magnetostriction),

$$T_{\alpha\beta} = E_\alpha D_\beta + H_\alpha B_\beta - \frac{\delta_{\alpha\beta}}{2}(E_\gamma D_\gamma + H_\gamma B_\gamma), \tag{10–37}$$

has in this case only the diagonal components

$$T_{xx} = -\tfrac{1}{2}(E_y D_y + H_z B_z),$$
$$T_{yy} = E_y D_y - \tfrac{1}{2}(E_y D_y + H_z B_z), \tag{11–25}$$
$$T_{zz} = H_z B_z - \tfrac{1}{2}(E_y D_y + H_z B_z).$$

The force derived by Eq. (10–45), which makes the volume force agree with the Lorentz force in a material medium, is

$$F_{v\alpha} = \frac{\partial T_{\alpha\beta}}{\partial x_\beta} - \mu\epsilon \frac{\partial}{\partial t}(\mathbf{E} \times \mathbf{H})_\alpha, \tag{11–26}$$

of which the x-component is

$$-\frac{1}{2}\frac{\partial}{\partial x}(E_y D_y + H_z B_z) - \mu\epsilon \frac{\partial}{\partial t}(E_y H_z). \tag{11–27}$$

The y- and z-components of the force vanish, since the fields are not functions of y and z, and since the Poynting vector has no y- and z-components.

Since the fields are absorbed, the x-component of the time average volume force can be integrated to give the total pressure on the slab of material:

$$\int_0^\infty \overline{F_{v_x}}\, dx = \tfrac{1}{2}(\epsilon \overline{E^2} + \mu \overline{H^2}), \qquad (11\text{--}28)$$

which is exactly equal to the energy density of the incoming radiation field. The time average of the time derivative of the Poynting vector vanishes if E_y and H_z vary sinusoidally, and therefore this term has been omitted from Eq. (11–28). If we wish to consider the net impulse transmitted by a wave train of finite length the pressure must be integrated from time $-\infty$ to time $+\infty$, and the time derivative will also integrate out in this case; it can contribute only to the instantaneous value of the pressure during the absorption of a wave train, and not to the over-all effect. The radiation pressure is thus the same as in the ether theory—it is unaffected (except for transients) by the correction term needed to bring the volume force into accord with the Lorentz force. Note that Eq. (11–28) agrees with the pressure calculated by assuming that the momentum of volume $c\, dt$ given by $\mathbf{N}c\, dt/c^2$ is absorbed per unit area of the body.

This result can be summarized by saying that the phenomenon of radiation pressure is consistent with the concept of momentum of electromagnetic waves, and with the more general concept of momentum carried by any energy transmitting process. This pressure was predicted by the classical prerelativity ether theory of electromagnetic radiation, however, and is not changed by modern refinements of electromagnetic theory. The force involved is simply the Lorentz force, and all detailed models of the absorption process lead to the same answer, namely, that the pressure is equal to the energy density of the incident radiation. Measurements of radiation pressure have verified the theoretical predictions of its intensity.

Radiation pressure can also be computed for the case of wholly or partly reflecting surfaces and oblique incidence. Let us take the case of isotropic homogeneous radiation in a cavity with absorptive walls, i.e., the radiation is unpolarized and of equal intensity in all directions. By symmetry it is evident that the only nonvanishing component of the volume force, Eq. (11–26), is that normal to the wall. If we take the x-axis normal to the element of surface under consideration, this force may be written out:

$$F_{v_x} = \epsilon \left[\frac{\partial E_x^2}{\partial x} + \frac{\partial}{\partial y}(E_x E_y) + \frac{\partial}{\partial z}(E_x E_z) \right]$$
$$+ \mu \left[\frac{\partial H_x^2}{\partial x} + \frac{\partial}{\partial y}(H_x H_y) + \frac{\partial}{\partial z}(H_x H_z) \right] - \frac{\epsilon}{2} \frac{\partial E^2}{\partial x} - \frac{\mu}{2} \frac{\partial H^2}{\partial x}. \quad (11\text{--}29)$$

(The time derivative term is omitted for the same reasons as before.)
Because different components of the electric field are uncorrelated, the
time averages of all the cross terms vanish, and we obtain

$$\overline{F_{v_x}} = \epsilon \overline{\frac{\partial E_x^2}{\partial x}} - \frac{\epsilon}{2} \overline{\frac{\partial E^2}{\partial x}} + \mu \overline{\frac{\partial H_x^2}{\partial x}} - \frac{\mu}{2} \overline{\frac{\partial H^2}{\partial x}}. \tag{11–30}$$

Since, according to our assumption, the fields are oriented completely at
random,

$$\overline{E_x^2} = \overline{E_y^2} = \overline{E_z^2} = \tfrac{1}{3}\overline{E^2},$$

and, similarly,

$$\overline{H_x^2} = \tfrac{1}{3}\overline{H^2}.$$

Therefore the normal force becomes

$$\overline{F_{v_x}} = -\frac{1}{6} \frac{\partial}{\partial x} [\epsilon \overline{E^2} + \mu \overline{H^2}] = -\frac{\partial}{\partial x} \left(\frac{\overline{U}}{3} \right),$$

where U is the energy density of the radiation. Integrating from the surface of the medium into field-free space, we obtain the result that the total
radiation pressure is equal to $\tfrac{1}{3}$ of the energy density of the radiation:

$$\int_0^\infty \overline{F_{v_x}} \, dx = \tfrac{1}{3}\overline{U}. \tag{11–31}$$

This equation forms the basis for thermodynamic derivations of the Stefan-
Boltzmann law and the Wien displacement law for blackbody radiation.

It should be noted that for the existence of completely homogeneous
isotropic radiation, and thus of the randomness utilized in simplifying Eq.
(11–29), it is essential that the walls be "black," i.e., perfectly absorbing.
Equation (11–31) must not be assumed valid for other cases, as, for example, cavities with specularly reflecting walls.

11–4 Plane waves in a moving medium.   Unlike radiation pressure,
which was first predicted by electromagnetic theory, the velocity of light
in a moving medium was derived by Fresnel on the assumption of elastic
vibrations in a stationary ether, and confirmed by Fizeau as early as 1853.
Fresnel's prediction, which will later be derived relativistically, is also in
agreement with Maxwell's equations in moving media, provided that we
interpret the velocity of the medium as that relative to the frame of reference in which the free-space velocity of light would be c.

According to Fresnel and Fizeau, the velocity of light in a body moving
with a velocity **v** relative to the observer is given by

$$u = u_0 + \left(1 - \frac{1}{n^2} \right) \mathbf{v} \cdot \mathbf{n} \tag{11–32}$$

where $n = \sqrt{\kappa\kappa_m}$ is the index of refraction of the medium and $u_0 = c/n$. It should not be confused with the unit vector $\mathbf{n}$, which is in the direction of wave propagation, i.e., $\mathbf{u} = u\mathbf{n}$. We may derive the velocity of electromagnetic waves in a nonpermeable medium moving with a velocity small compared with c from the considerations of Section 9–4. Maxwell's equations in a moving medium free from true charges and currents become

$$(1) \qquad\qquad \boldsymbol{\nabla} \cdot \mathbf{D} = 0,$$

$$(2) \qquad\qquad \boldsymbol{\nabla} \cdot \mathbf{B} = 0,$$

$$(3) \qquad\qquad \boldsymbol{\nabla} \times \mathbf{E} = -\frac{\partial \mathbf{B}}{\partial t}, \qquad\qquad (11\text{--}33)$$

$$\cdot(4) \qquad\qquad \boldsymbol{\nabla} \times \mathbf{B}/\mu_0 = \frac{\partial \mathbf{P}}{\partial t} + \epsilon_0 \frac{\partial \mathbf{E}}{\partial t} + \boldsymbol{\nabla} \times (\mathbf{P} \times \mathbf{v}),$$

as may be seen from Eqs. (9–18). The polarization is produced by the effective field in the moving medium:

$$\mathbf{P} = \epsilon_0(\kappa - 1)(\mathbf{E} + \mathbf{v} \times \mathbf{B}). \qquad\qquad (9\text{--}19)$$

Note that all the fields in Eqs. (11–33) and (9–19) are those which would be measured in the stationary frame of the observer. If we substitute the polarization field, Eq. (9–19), into Eq. (11–33) (4), and then make use of the third Maxwell equation, we obtain, correct to terms linear in $\mathbf{v}$:

$$\boldsymbol{\nabla} \times \mathbf{B}$$
$$= \mu_0 \left\{ \epsilon_0 \frac{\partial \mathbf{E}}{\partial t} + \epsilon_0(\kappa - 1) \frac{\partial \mathbf{E}}{\partial t} + \epsilon_0(\kappa - 1) \left[\mathbf{v} \times \frac{\partial \mathbf{B}}{\partial t} + \boldsymbol{\nabla} \times (\mathbf{E} \times \mathbf{v}) \right] \right\}$$
$$= \frac{\kappa}{c^2} \left\{ \frac{\partial \mathbf{E}}{\partial t} + \left(1 - \frac{1}{\kappa} \right) [-\mathbf{v} \times (\boldsymbol{\nabla} \times \mathbf{E}) + \boldsymbol{\nabla} \times (\mathbf{E} \times \mathbf{v})] \right\}. \qquad (11\text{--}34)$$

If we expand the vector triple products, noting that $\boldsymbol{\nabla} \cdot \mathbf{E} = 0$, Eq. (11–34) becomes, for constant $\mathbf{v}$,

$$\boldsymbol{\nabla} \times \mathbf{B} = \frac{\kappa}{c^2} \left\{ \frac{\partial \mathbf{E}}{\partial t} + \left(1 - \frac{1}{\kappa} \right) [2(\mathbf{v} \cdot \boldsymbol{\nabla})\mathbf{E} - \boldsymbol{\nabla}(\mathbf{v} \cdot \mathbf{E})] \right\}. \qquad (11\text{--}35)$$

On taking the curl and making use of Eqs. (11–33) (3) and (2) we obtain

$$\nabla^2 \mathbf{B} = \frac{\kappa}{c^2} \left[\frac{\partial^2 \mathbf{B}}{\partial t^2} + 2 \left(1 - \frac{1}{\kappa} \right) (\mathbf{v} \cdot \boldsymbol{\nabla}) \frac{\partial \mathbf{B}}{\partial t} \right] \qquad\qquad (11\text{--}36)$$

for the wave equation in slowly moving media.

Since we are here interested only in terms linear in the velocity, it is permissible to make the approximation that for plane wave propagation in the direction $\mathbf{n}$ along the coordinate ζ,

$$\mathbf{v} \cdot \nabla = (\mathbf{v} \cdot \mathbf{n}) \frac{\partial}{\partial \zeta} = - \frac{\mathbf{v} \cdot \mathbf{n}}{u_0} \frac{\partial}{\partial t}.$$

Hence,

$$\nabla^2 \mathbf{B} + \frac{1}{u_0^2} \frac{\partial^2 \mathbf{B}}{\partial t^2} \left[1 - 2 \left(1 - \frac{1}{\kappa} \right) \frac{\mathbf{v} \cdot \mathbf{n}}{u_0} \right] = 0. \qquad (11\text{–}37)$$

This is a wave equation that corresponds to the propagation velocity

$$u = u_0 \left[1 - \left(1 - \frac{1}{\kappa} \right) 2 \frac{\mathbf{v} \cdot \mathbf{n}}{u_0} \right]^{-1/2} \simeq u_0 + \left(1 - \frac{1}{\kappa} \right) \mathbf{v} \cdot \mathbf{n}, \quad (11\text{–}38)$$

which agrees to the order v/c with Eq. (11–32) and has been checked not only by Fizeau but also, with greater accuracy, by Michelson and Morley.

The physical interpretation of Eq. (11–38) is that the only part of the propagation velocity of the wave which is affected by the motion of the medium is that proportional to

$$1 - \frac{1}{\kappa} = \frac{\kappa - 1}{\kappa} \propto \frac{\partial \mathbf{P}/\partial t}{\partial \mathbf{D}/\partial t}. \qquad (11\text{–}39)$$

This fraction is thus proportional to the ratio of the polarization current to the displacement current. Since a polarization current does actually correspond to the motion of dipoles, it is quite reasonable to assume that the portion of the wave corresponding to these dipoles will be affected by the velocity of the medium. We shall see that the effect of a medium on a plane wave is simply that the medium is polarized by the incident wave and that the resulting dipoles produce a wave which combines with the incident radiation in such a way as to correspond to the over-all phase velocity. It is this coherent retarded component which is being radiated from a moving source in this case, and which gives rise to the Fresnel-Fizeau coefficient. It may be remarked that the physical assumptions involved in this classical nonrelativistic theory are consistent with those of relativity, and we shall see that all these results are in accord with relativistic principles.

11–5 Reflection and refraction at a plane boundary. The boundary conditions inherent in the field equations determine what happens when a plane wave is incident on a boundary between media of different electric and magnetic properties. Let us consider a plane wave traveling in the direction of $\mathbf{k}$ in medium 1, and incident on the plane boundary between

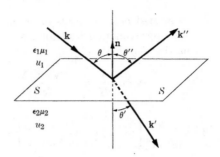

Fig. 11-2. The propagation vectors for refraction and reflection at a plane surface. **k** and **n**, the normal to the reflecting surface, determine the plane of incidence.

media 1 and 2, as shown in Fig. 11-2. The media are characterized by constants μ_1 and μ_2, ϵ_1 and ϵ_2. The propagation vector **k** and the unit vector **n** specifying the plane boundary determine what is called the plane of incidence of the wave. In accord with Eqs. (11-20) and (11-21) the incident fields are represented by

$$\mathbf{E} = \mathbf{E}_0 e^{i(\mathbf{k}\cdot\mathbf{r}-\omega t)},$$

$$\mathbf{H} = \frac{\mathbf{k}\times\mathbf{E}}{\omega\mu_1}. \tag{11-40}$$

The boundary conditions cannot be satisfied by a single progressive wave, since the tangential components of both **E** and **H** must be continuous across the boundary. If we use primed symbols for the wave in medium 2, and double primes for the reflected wave, we may write

$$\mathbf{E}' = \mathbf{E}_0' e^{i(\mathbf{k}'\cdot\mathbf{r}-\omega t)}, \qquad \mathbf{H}' = \frac{\mathbf{k}'\times\mathbf{E}'}{\omega\mu_2},$$

$$\mathbf{E}'' = \mathbf{E}_0'' e^{i(\mathbf{k}''\cdot\mathbf{r}-\omega t)}, \qquad \mathbf{H}'' = \frac{\mathbf{k}''\times\mathbf{E}''}{\omega\mu_1}. \tag{11-41}$$

The continuity of the tangential fields is possible only if the exponentials are the same at the boundary for all three fields. Thus we were justified in assuming throughout Eqs. (11-41) that the frequency is unchanged in the transmitted and reflected waves, and have the further conditions that

$$\mathbf{k}\cdot\mathbf{r} = \mathbf{k}'\cdot\mathbf{r} = \mathbf{k}''\cdot\mathbf{r} \tag{11-42}$$

over the boundary surface. It is evident from Eq. (11-42) that all the propagation vectors are coplanar. If for convenience we put the origin of the coordinate vector $\mathbf{r}$ in the boundary plane specified by the unit vector $\mathbf{n}$, so that $\mathbf{n} \cdot \mathbf{r} = 0$ is the equation of the plane, then

$$\mathbf{n} \times (\mathbf{n} \times \mathbf{r}) = (\mathbf{n} \cdot \mathbf{r})\mathbf{n} - \mathbf{r} = -\mathbf{r}$$

on the boundary, and substitution in Eqs. (11–42) leads to

$$(\mathbf{k} - \mathbf{k}'') \times \mathbf{n} \cdot (\mathbf{n} \times \mathbf{r}) = 0,$$
$$(\mathbf{k} - \mathbf{k}') \times \mathbf{n} \cdot (\mathbf{n} \times \mathbf{r}) = 0. \tag{11–43}$$

Now the propagation vectors $\mathbf{k}$ and $\mathbf{k}''$ in the same medium are equal in magnitude, and $|\mathbf{k}|/|\mathbf{k}'| = u_2/u_1$, the ratio of the phase velocities in the two media. Therefore the laws of reflection and refraction are implicit in Eq. (11–43), and in terms of the angles may be written in the usual forms:

$$\theta = \theta'',$$
$$\frac{\sin \theta}{\sin \theta'} = \frac{u_1}{u_2}. \tag{11–44}$$

The second of these equations is known as Snell's law.

The relations between the various amplitudes can be determined to satisfy the boundary conditions

$$\mathbf{n} \times (\mathbf{E} + \mathbf{E}'') = \mathbf{n} \times \mathbf{E}',$$

$$\mathbf{n} \times (\mathbf{k} \times \mathbf{E} + \mathbf{k}'' \times \mathbf{E}'')/\mu_1 = \mathbf{n} \times (\mathbf{k}' \times \mathbf{E}')/\mu_2,$$

where we have written the magnetic amplitudes in terms of those of the electric fields. It will suffice for us to consider the case in which $\mu_1 = \mu_2$. The orientation of $\mathbf{E}$ is arbitrary, but it may always be written as the sum of two components at right angles, and any difference in phase can be taken care of by making $\mathbf{E}$ complex.

(a) $\mathbf{E}$ *at right angles to the plane of incidence.* If $\mathbf{E}$ is perpendicular to the plane of incidence, all the electric field vectors are tangential to the surface, and

$$E_0 + E_0'' = E_0'. \tag{11–45}$$

Since $\mathbf{n} \cdot \mathbf{E}$ vanishes the expansion of the vector triple product in the second boundary condition leads to

$$E_0(\mathbf{n} \cdot \mathbf{k}) + E_0''(\mathbf{n} \cdot \mathbf{k}'') = E_0'(\mathbf{n} \cdot \mathbf{k}'),$$

or, in terms of the angles of Fig. 11–2,

$$E_0 \cos \theta - E_0'' \cos \theta'' = \frac{u_1}{u_2} E_0' \cos \theta'. \tag{11–46}$$

Equations (11–45) and (11–46) may be combined to give the transmitted and reflected amplitudes in terms of that of the incident wave:

$$E' = E \frac{2 \cos \theta}{\cos \theta + (u_1/u_2) \cos \theta'} ,$$

$$E'' = -E \frac{(u_1/u_2) \cos \theta' - \cos \theta}{(u_1/u_2) \cos \theta' + \cos \theta} . \tag{11–47}$$

Here use has been made of the fact that $\theta'' = \theta$, and the angle θ' may be eliminated by the use of Snell's law of refraction. The relative index of refraction of medium 2 with respect to medium 1 is $u_1/u_2 = n_{21}$, and n without subscripts is usually interpreted as n_{21}.

(b) **E** *in the plane of incidence.* In this case **H** is parallel to the boundary, and the derivation is formally identical with that above except that the relation

$$\mathbf{E} = -\mu\omega \frac{\mathbf{k} \times \mathbf{H}}{k^2} , \tag{11–48}$$

which follows from Maxwell's equations for a plane wave, leads to a slightly different distribution of constants. If we again assume that the two media have the same permeability, the final results are

$$H' = H \frac{2 \cos \theta}{\cos \theta + (u_2/u_1) \cos \theta'} ,$$

$$H'' = -H \frac{(u_2/u_1) \cos \theta' - \cos \theta}{(u_2/u_1) \cos \theta' + \cos \theta} , \tag{11–49}$$

for **E** in the plane of incidence.

Equations (11–47) and (11–49) are known as Fresnel's formulas, and were originally derived on the assumption of an elastic ether. Their transformation to other convenient forms, and the derivation of the resulting reflection and transmission coefficients, are left to the problems.

There are several interesting consequences of the fact that the two components are not transmitted and reflected equally. For the first case treated above, in which the electric vector is normal to the plane of incidence, there is reflection unless $\theta = \theta'$, i.e., unless the two media have the same optical properties. The reflected amplitude of Eq. (11–49), however, vanishes if $\theta + \theta' = 90°$, i.e., if **k′** and **k″** are perpendicular to each other.* The angle of incidence that satisfies this condition is called Brewster's

---

* This is obvious from the formulas of Exercise 2, below.

angle, or the polarizing angle, and is defined in terms of Snell's law by

$$\tan \theta_0 = u_1/u_2. \tag{11-50}$$

Since at this angle the reflected beam consists entirely of the component whose electric vector is at right angles to the plane of incidence, a plate or a stack of plates of dielectric may be used to produce a plane polarized wave from unpolarized light.

On the other hand, there may be a whole range of angles for which there is no transmission. If u_2 is greater than u_1 the law of refraction (Snell's law) leads to values of $\sin \theta'$ that are greater than unity when θ exceeds $\sin^{-1}(u_1/u_2)$. Under these conditions the reflection is said to be total, but since the wave amplitudes may be complex in any case, we may apply Fresnel's formulas to examine the process in detail. If we eliminate θ' by means of Snell's law, using $n = u_1/u_2$, $(n < 1)$, the reflected amplitude for case (a), $\mathbf{E}$ normal to the plane of incidence, becomes

$$E'' = E \frac{\sin \theta \cos \theta - \sin \theta' \cos \theta'}{\sin \theta \cos \theta + \sin \theta' \cos \theta'}$$

$$= E \frac{\cos \theta - i\sqrt{\sin^2 \theta - n^2}}{\cos \theta + i\sqrt{\sin^2 \theta - n^2}}, \tag{11-51}$$

where the fact that $\cos \theta' = i\sqrt{\sin^2 \theta - n^2}/n$ is imaginary has been made explicit. It is evident from Eq. (11–51) that the reflected amplitude is equal in magnitude to that of the incident wave, but the phase is changed:

$$E'' = E e^{-2i\phi}, \tag{11-52}$$

where

$$\phi = \arctan \frac{\sqrt{\sin^2 \theta - n^2}}{\cos \theta}. \tag{11-53}$$

For the component whose electric vector is in the plane of incidence, we have from Eq. (11–49)

$$H'' = H \frac{\cos \theta - (i/n^2)\sqrt{\sin^2 \theta - n^2}}{\cos \theta + (i/n^2)\sqrt{\sin^2 \theta - n^2}}$$

$$= H e^{-2i\psi}, \tag{11-54}$$

where

$$\psi = \arctan \frac{\sqrt{\sin^2 \theta - n^2}}{n^2 \cos \theta}.$$

Thus the components are changed in phase by unequal amounts, and as a result plane polarized radiation is totally reflected with elliptical polarization. The phase difference between the two components, $2(\phi - \psi)$, depends on both the angle of incidence and the relative index of refraction. A simple computation leads to

$$\tan(\phi - \psi) = -\frac{\cos\theta\sqrt{\sin^2\theta - n^2}}{\sin^2\theta}. \tag{11-55}$$

Although the transmission coefficient is zero for total reflection, the instantaneous fields do not vanish in the second medium. The fact that $\cos\theta'$ is a pure imaginary yields an exponential decrease of the field strength within the medium. The field in the second medium can be detected if an additional transparent object of large index is placed very near the surface.

11-6 Waves in conducting media and metallic reflection. The solutions of the general wave equation, Eq. (11-15), can be written in the same form as those of the propagation equation, Eq. (11-16), if we again assume a sinusoidal time dependence. We may write for any transverse component of $\mathbf{E}$

$$\frac{\partial^2 \mathbf{E}}{\partial\zeta^2} = -(\mu\epsilon\omega^2\mathbf{E} + i\omega\mu\sigma\mathbf{E}) = -K^2\mathbf{E}, \tag{11-56}$$

where

$$K = \alpha + i\beta = k\sqrt{1 + \frac{i\sigma}{\epsilon\omega}}. \tag{11-57}$$

When Eq. (11-57) is solved for the real and imaginary parts of K, the results are

$$\alpha = k\sqrt{\frac{\sqrt{1 + (\sigma/\epsilon\omega)^2} + 1}{2}},$$

$$\beta = k\sqrt{\frac{\sqrt{1 + (\sigma/\epsilon\omega)^2} - 1}{2}}. \tag{11-58}$$

The fields are therefore

$$\mathbf{E} = \mathbf{E}_0 e^{-\beta\zeta} e^{i(\alpha\zeta - \omega t)},$$

$$\mathbf{H} = \frac{\mathbf{k} \times \mathbf{E}}{k\mu\omega}(\alpha + i\beta), \tag{11-59}$$

where $\mathbf{k}$ is along ζ, as before. Two features of Eq. (11-59) are immediately evident: the conductivity gives rise to exponential damping of the wave, and the electric and magnetic fields are no longer in phase.

The application of the Fresnel formulas to find the reflection at a plane metallic boundary is straightforward, but leads in general to rather complicated results. It is useful to consider the approximation, valid for all frequencies below the optical region, in which $\epsilon\omega \ll \sigma$. In that case,

$$K = (1 + i)k \sqrt{\frac{\sigma}{2\omega\epsilon}} = (1 + i) \sqrt{\frac{\mu\sigma\omega}{2}}. \tag{11–60}$$

For simplicity, let a plane wave in a dielectric be incident normally on the surface of a conductor in the direction of ζ. The fields inside the conductor are then

$$\mathbf{E}' = \mathbf{E}'_0 e^{-\sqrt{(\mu\sigma\omega/2)}\zeta} e^{i\left(\sqrt{(\mu\sigma\omega/2)}\zeta - \omega t\right)},$$

$$\mathbf{H}' = \frac{\mathbf{B}'}{\mu} = \frac{1 + i}{\sqrt{2}} \sqrt{\frac{\sigma}{\mu\omega}}\, \mathbf{n} \times \mathbf{E}', \tag{11–61}$$

where we have measured ζ from the boundary surface. Both fields decrease with penetration, falling to $1/e$ of their surface values in a distance

$$\delta = \sqrt{\frac{2}{\mu\sigma\omega}}, \tag{11–62}$$

which is known as the skin depth. Note that δ goes to zero as the conductivity approaches infinity, and is smaller for high frequencies. Equation (11–62) cannot be extrapolated to infinite frequencies, however, and an expression derived from Eq. (11–58) must be used where the approximation $\epsilon\omega \ll \sigma$ is not valid.

Inside the metal the energy density is no longer shared equally by the two fields: the time average of the electrical energy density is now $\epsilon\overline{E^2}/2$, while that associated with the magnetic fields is $\sigma\overline{E^2}/2\omega$, so that the ratio is $\epsilon\omega/\sigma$, which we have assumed is a very small number. Within the conductor the electrical energy density is negligible in comparison with the magnetic energy, and good conductors are essentially not penetrated by electric fields satisfying the condition that $\epsilon\omega \ll \sigma$.

The "reflecting power" of the metallic surface is readily found if we remember that the reflected wave travels toward increasing negative ζ, so that

$$\mathbf{E}'' = \mathbf{E}''_0 e^{-i(k\zeta + \omega t)},$$

and then impose the conditions that $\mathbf{E}$ and $\mathbf{H}$ are both continuous across the boundary. The ratio of the reflected energy to the incident energy is

$$\left|\frac{\mathbf{E}''}{\mathbf{E}}\right|^2 = 1 - 2\sqrt{\frac{2\epsilon_0\omega}{\sigma}} \tag{11–63}$$

if higher powers of $\sqrt{2\epsilon_0\omega/\sigma}$ than the first are neglected.

11–7 Group velocity. The velocity u which appears in the wave equation, such that the plane wave solution is given by Eq. (11–17), is the phase velocity of the wave, and the index of refraction is defined by $n = c/u$. If the index of refraction were independent of the frequency this would be the only velocity involved. In actual media, however, the dielectric constant is a function of the frequency, and the passage of light through a medium is modified by the difference in phase velocity among its various Fourier components.

The concept of group velocity applies to a pulse of waves, or "wave packet," with a continuous spectrum confined to a narrow band of frequencies. Let $\psi(x, t)$ stand for a component of **E** or **B** corresponding to such a narrow group, which can be represented by a Fourier integral

$$\psi = \int a(k)e^{i(kx-\omega t)} \, dk. \tag{11–64}$$

If the amplitude $a(k)$ differs from zero only in the neighborhood of a particular frequency designated by ω_1, k_1, we may write

$$\psi = A(x, t)e^{i(k_1 x-\omega_1 t)}, \tag{11–65}$$

where

$$A(x, t) = \int a(k)e^{i[(\Delta k)x-(\Delta\omega)t]}dk, \tag{11–66}$$

with $\Delta\omega = \omega - \omega_1$ and $\Delta k = k - k_1$. If A is slowly varying, the group may be considered as a wave of angular frequency ω_1 propagated with amplitude A. The velocity u_g with which A travels corresponds to having the amplitude be constant to an observer moving with that velocity. Hence

$$\frac{dA}{dt} = \frac{\partial A}{\partial t} + \frac{\partial A}{\partial x}(u_g) = 0, \tag{11–67}$$

from which

$$u_g = \left(\frac{dx}{dt}\right)_{A \text{ constant}} = \frac{-\partial A/\partial t}{\partial A/\partial x} = \frac{\Delta\omega}{\Delta k} \rightarrow \left(\frac{d\omega}{dk}\right)_{\omega=\omega_1} \tag{11–68}$$

may be taken as the velocity of the pulse. This quantity, called the group velocity, may be written in a variety of useful ways:

$$u_g = \frac{d\omega}{dk} = u - \lambda\frac{du}{d\lambda} = \frac{1}{dk/d\omega}. \tag{11–69}$$

Note that u_g is defined for a particular frequency in such a way that in the integral over k, Fourier components of frequencies far from ω_1 tend to cancel each other, whereas in the neighborhood of ω_1 the contributions to the integral reinforce to constitute the pulse. The group velocity is therefore essentially the rate of propagation of energy. If the phase velocity is a slowly varying function of the frequency, a pulse may travel through a dispersive medium with relatively little change; but if this condition is not satisfied, the group itself is highly distorted, and the concept of group velocity is no longer valid.

SUGGESTED REFERENCES

Plane electromagnetic waves are treated in many books on electrodynamics and physical optics, and in general textbooks on theoretical physics. Our list includes some of the better and more easily available accounts.

M. ABRAHAM AND R. BECKER, *Electricity and Magnetism*, Chapter X. Characteristically clear, rigorous, and rather elementary.

M. BORN, *Optik*, Chapter 1. Probably the most complete and rigorous of the physical optics treatments.

L. D. LANDAU AND E. M. LIFSHITZ, *Electrodynamics of Continuous Media*, especially Sections 66 and 67.

J. A. STRATTON, *Electromagnetic Theory*, Chapter V. A very useful account, presented in both detail and generality.

Briefer accounts will be found in:

J. H. JEANS, *The Mathematical Theory of Electricity and Magnetism*.

G. JOOS, *Theoretical Physics*. Chapter XIX includes propagation in anisotropic media.

J. C. SLATER AND N. H. FRANK, *Electromagnetism*.

W. R. SMYTHE, *Static and Dynamic Electricity*.

H. A. LORENTZ, *The Theory of Electrons*, contains an excellent account of radiation pressure.

EXERCISES

1. Prove that the time average of the product of the real parts of two complex vectors, both of which depend on the time as $e^{-i\omega t}$, is given by one-half the real part of the product of one with the complex conjugate of the other.

2. Prove that for the electric vector perpendicular to the plane of incidence Fresnel's formulas are

$$\mathbf{E}' = \frac{2 \cos \theta \sin \theta'}{\sin (\theta + \theta')} \mathbf{E} \quad \text{for the transmitted wave,}$$

and

$$\mathbf{E}'' = \frac{\sin (\theta' - \theta)}{\sin (\theta' + \theta)} \mathbf{E} \quad \text{for the reflected wave.}$$

Prove also that when $\mathbf{E}$ is in the plane of incidence, the reflection is given by

$$\frac{|\mathbf{E}''|}{|\mathbf{E}|} = \frac{\tan (\theta - \theta')}{\tan (\theta + \theta')}.$$

3. By considering the waves with Fresnel amplitudes in the second medium, prove that the transmission coefficient vanishes under the conditions of total reflection.

4. Prove that for normal incidence the reflection coefficient, or percent of light reflected from the surface between two dielectrics, is given by $(n - 1)^2/(n + 1)^2$, where n is the relative index of refraction of medium 2 with respect to medium 1. What is the transmission coefficient? Compute the amount of light transmitted by a transparent glass plate in air at normal incidence, in terms of the index of refraction of the glass.

5. A plane wave is specularly reflected by a surface at an angle θ from the normal. Find the radiation pressure.

6. Estimate the ratio of the resistance of a conducting wire for high-frequency alternating currents to its direct current resistance. The result is known as Rayleigh's resistance formula.

7. Plane monochromatic waves are propagated parallel to the z-axis in both the positive and negative directions. At the origin the field strengths are given by

$$E_x = A \cos \omega t, \qquad E_y = 0,$$

$$H_x = 0, \qquad\qquad H_y = B \cos \omega t$$

Calculate the mean intensity of the radiation in each of the two directions in terms of A, B, and the constants of the medium.

CHAPTER 12

CONDUCTING FLUIDS IN A MAGNETIC FIELD

(MAGNETOHYDRODYNAMICS)

In the previous chapter we have considered wave motions which are basically concerned with an interchange of electric and magnetic energies. Material media and their motions were involved only to the extent of modifying the fields, not in terms of adding significant amounts of mechanical energy to the system. If, however, fluid media of large conductivity σ exist in the presence of magnetic fields, the inertia of the medium carrying currents is no longer negligible, and we meet a wholly new set of phenomena usually described as "magnetohydrodynamics."

12-1 "Frozen-in" lines of force. Consider the field equation

$$\nabla \times \mathbf{E} = -\frac{\partial \mathbf{B}}{\partial t}, \tag{9-5}$$

and the equation giving the current density $\mathbf{j}$ in a medium moving with velocity $\mathbf{u}$ in an electric field $\mathbf{E}$ and magnetic induction field $\mathbf{B}$:

$$\mathbf{j} = \sigma(\mathbf{E} + \mathbf{u} \times \mathbf{B}). \tag{9-19}$$

On combining these equations we obtain

$$\frac{\partial \mathbf{B}}{\partial t} = \nabla \times \left(\mathbf{u} \times \mathbf{B} - \frac{\mathbf{j}}{\sigma}\right). \tag{12-1}$$

If the conductivity is so high that the second term on the right-hand side is negligible, we have

$$\frac{\partial \mathbf{B}}{\partial t} = \nabla \times (\mathbf{u} \times \mathbf{B}). \tag{12-2}$$

The condition that the last term of Eq. (12-1) be negligible is

$$\left|\frac{\mathbf{j}}{\sigma}\right| \ll |\mathbf{u} \times \mathbf{B}|. \tag{12-3}$$

If displacement currents are neglected, $\nabla \times \mathbf{B} = \mu_0 \mathbf{j}$, and we can therefore say that

$$|\mathbf{B}| \sim \mu_0 L |\mathbf{j}|, \tag{12-4}$$

where L is a length characteristic of the dimensions of the system. Hence the inequality (12–3) is equivalent to

$$R_M = \sigma|\mathbf{u}|\mu_0 L \gg 1, \tag{12–5}$$

where R_M is a quantity often called the "magnetic Reynolds number" by analogy with the usual hydrodynamic Reynolds number which also scales linearly with L and $|\mathbf{u}|$. Numerical examples show that it is difficult to satisfy the condition (12–5) in laboratory experiments using ordinary conducting materials; magnetohydrodynamics is important only in the physics of ionized gases (plasma physics), or in very large conducting systems such as astronomical bodies.

The interpretation of Eq. (12–2) is both simple and striking. In Section 9–3 we derived an expression, Eq. (9–12), for the total rate of flux through a surface S moving with a medium having velocity $\mathbf{u}$. With $\mathbf{\nabla} \cdot \mathbf{B} = 0$, Eq. (9–12) becomes

$$\frac{d}{dt} \int_S \mathbf{B} \cdot d\mathbf{S} = \int_S \left[\frac{\partial \mathbf{B}}{\partial t} - \mathbf{\nabla} \times (\mathbf{u} \times \mathbf{B}) \right] \cdot d\mathbf{S}. \tag{12–6}$$

This expression vanishes if Eq. (12–2) is valid. Hence the flux through any surface moving with the material remains constant. This implies that magnetic lines remain "frozen" in a moving medium of sufficiently large magnetic Reynolds number.

If the inequality (12–5) is not satisfied, the material motion does not dominate the situation; $\mathbf{B}$ then obeys the diffusion-type equation as discussed in Section 11–1;

$$\nabla^2 \mathbf{B} - \mu\sigma \frac{\partial \mathbf{B}}{\partial t} = 0. \tag{12–7}$$

The diffusion of lines across the medium predicted by this equation will take place in a time τ_D of order

$$\tau_D \sim \mu\sigma L^2, \tag{12–8}$$

where L is again a length characteristic of the dimensions of the system. The time τ_D is that necessary for the ohmic losses to dissipate the energy change involved in moving the material across the field.

It is instructive to compare the time τ_D with the time necessary for the medium to cross the flux lines if the magnetic Reynolds number is large. The equation of motion is

$$\rho_m \frac{d\mathbf{u}}{dt} \equiv \rho_m \left[\frac{d\mathbf{u}}{dt} + (\mathbf{u} \cdot \mathbf{\nabla})u \right] = -\mathbf{\nabla}p + \mathbf{j} \times \mathbf{B} + \mathbf{F}_v, \tag{12–9}$$

where ρ_m is the density, p is the pressure, and $\mathbf{F}_v$ allows for the consideration of external volume forces, such as gravity or a viscous drag force per unit volume.   From Eq. (9–19),

$$\rho_m \frac{d\mathbf{u}}{dt} = \sigma[\mathbf{E} \times \mathbf{B} - \mathbf{u}B^2 + \mathbf{B}(\mathbf{u} \cdot \mathbf{B})] - \nabla p + \mathbf{F}_v. \qquad (12\text{–}10)$$

The transverse velocity $u_\perp$ across the lines of force, in the absence of electric fields and mechanical pressure gradients and forces, is thus given by the equation

$$\rho_m \frac{du_\perp}{dt} = -\sigma u_\perp B^2. \qquad (12\text{–}11)$$

As a result of the "drag" of the mechanical density ρ_m "attached" to the flux lines, any transverse relative motion between field and matter decays with a time constant τ_i, given by

$$\tau_i = \frac{\rho_m}{\sigma B^2}. \qquad (12\text{–}12)$$

Thus the effect of the field $\mathbf{B}$ is to produce a large magnetic "viscosity" which tends to destroy motion across the flux lines.

If the magnetic lines are completely "frozen" into the fluid, as is the case in the limit $\sigma \to \infty$, the hydrostatic and hydrodynamic forces acting on the fluid must be supplemented by the forces derived from the magnetic stress tensor, Eq. (10–32), i.e., by a tension $B^2/2\mu_0$ along the lines of force and an equal transverse pressure.   This is equivalent to a total tension B^2/μ_0 and a hydrostatic (omnidirectional) pressure $B^2/2\mu_0$. Equation (12–9) becomes

$$\rho_m \frac{du_\alpha}{dt} = \frac{\partial}{\partial x_\beta}\left[\frac{B_\alpha B_\beta}{\mu_0} + \left(-p - \frac{B^2}{2\mu_0}\right)\delta_{\alpha\beta}\right] + F_{v_\alpha}. \qquad (12\text{–}13)$$

We shall not here discuss many solutions of this equation, some of which are of great interest in plasma physics and in the study of astronomical systems.

12–2 Magnetohydrodynamic waves.

If the condition for "frozen-in lines" is satisfied, a new type of wave motion is to be expected.   Qualitatively we can understand this by analogy with the theory of waves on a loaded string: Each tube of force of unit cross section is "loaded" by a mechanical mass density ρ_m per unit length and is under a tension B^2/μ_0. (It is also under a hydrostatic pressure $B^2/2\mu_0$.)   Accordingly we should expect a wave motion propagated with a phase velocity v_M given approximately by

$$v_M = \left[\frac{(B^2/\mu_0)}{\rho_m}\right]^{1/2} = \frac{B}{(\mu_0\rho_m)^{1/2}} \cdot \quad (12\text{-}14)$$

The waves are propagated along the lines of force. The detailed theory of these waves is very complicated, but a simplified treatment shows many characteristic features. We shall confine our attention to incompressible fluids and ignore all dissipative terms, i.e., consider only zero viscosity and infinite conductivity.

Under these assumptions the differential equations are

$$\frac{\partial \mathbf{B}}{\partial t} = \boldsymbol{\nabla} \times (\mathbf{u} \times \mathbf{B}), \quad (12\text{-}15)$$

$$\rho_m \frac{\partial \mathbf{u}}{\partial t} + \rho_m(\mathbf{u} \cdot \boldsymbol{\nabla})\mathbf{u} = -\boldsymbol{\nabla}(p + \psi) + \frac{1}{\mu_0} (\boldsymbol{\nabla} \times \mathbf{B}) \times \mathbf{B}, \quad (12\text{-}16)$$

where any external volume force F_v is represented as derived from a potential energy ψ.

Let us assume that there exists a uniform magnetic field $\mathbf{B}_0$ and let $\mathbf{b}$ be the magnetic disturbance propagated, i.e.,

$$\mathbf{B} = \mathbf{B}_0 + \mathbf{b}. \quad (12\text{-}17)$$

The field $\mathbf{B}_0$ is constant, and in a direction which we may take as that of the z coordinate axis. The differential equation (12–16) is nonlinear; hence, in general, waves traveling in different directions and having different polarization will affect one another. We shall first consider the simple (if not completely realistic) case in which the wave propagates along the positive z-direction only, and in which $\mathbf{b}$ and $\mathbf{u}$ are transverse and plane polarized. Equation (12–15) becomes

$$\frac{\partial \mathbf{B}}{\partial t} = \frac{\partial \mathbf{b}}{\partial t} = B_0 \frac{\partial \mathbf{u}}{\partial z} \cdot \quad (12\text{-}18)$$

Hence $\mathbf{b}$ and $\mathbf{u}$ are polarized in the same plane, as indicated in Fig. 12–1. Since the operator $\boldsymbol{\nabla}$ has a component only along z, Eq. (12–16) reduces to

$$\rho_m \frac{\partial \mathbf{u}}{\partial t} = -\boldsymbol{\nabla}\left(p + \psi + \frac{B^2}{2\mu_0}\right) + \frac{B_0}{\mu_0} \frac{\partial \mathbf{b}}{\partial z} \cdot \quad (12\text{-}19)$$

Note the appearance of $B^2/2\mu_0$ as an additional pressure. The transverse part of Eq. (12–19) is simply

$$\rho_m \frac{\partial \mathbf{u}}{\partial t} = \frac{B_0}{\mu_0} \frac{\partial \mathbf{b}}{\partial z}, \quad (12\text{-}20)$$

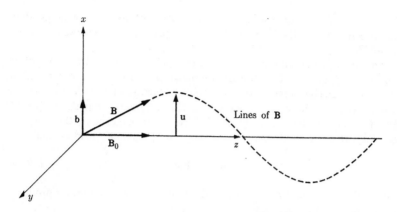

Fig. 12–1. Amplitudes in magnetohydrodynamic waves.

which, when combined with Eq. (12–18), gives

$$\frac{\partial^2 \mathbf{b}}{\partial z^2} - \frac{\rho_m \mu_0}{B_0^2} \frac{\partial^2 \mathbf{b}}{\partial t^2} = 0, \qquad \frac{\partial^2 \mathbf{u}}{\partial z^2} - \frac{\rho_m \mu_0}{B_0^2} \frac{\partial^2 \mathbf{u}}{\partial t^2} = 0. \qquad (12\text{–}21)$$

This appears to be the one-dimensional wave equation corresponding to a phase velocity given by Eq. (12–14) with $\mathbf{B}$ replaced by $\mathbf{B}_0$. It must be remembered, however, that Eq. (12–21) involves partial derivatives and that the basic equation is three-dimensional. In fact, a general solution of Eq. (12–21) corresponding to a single frequency is

$$\mathbf{b} = \mathbf{b}_0 e^{i(k_\alpha x_\alpha - \omega t)}, \qquad (12\text{–}22)$$

where

$$k_z^2 = \frac{\rho_m \mu_0}{B_0^2} \omega^2 = \frac{\omega^2}{v_M^2}. \qquad (12\text{–}23)$$

The phase velocity u_p is thus

$$u_p = \frac{\omega}{(k_x^2 + k_y^2 + k_z^2)^{1/2}} = \frac{v_M}{\{1 + [(k_x^2 + k_y^2)/k_z^2]\}^{1/2}}, \qquad (12\text{–}24)$$

which is not in general equal to v_M; the medium is of course unisotropic. The α-component of the group velocity is, according to Eq. (11–69), given by

$$v_{g\alpha} = \frac{d\omega}{dk_\alpha},$$

so that

$$v_{gz} = v_M, \qquad v_{gy} = v_{gx} = 0. \qquad (12\text{–}25)$$

Hence the wave fronts and the group velocity vector are not perpendicular; the group velocity (and also the energy propagation) is always directed along $\mathbf{B}_0$.

The above discussion assumes a single polarization and a single direction of propagation of the magnetohydrodynamic wave; as noted earlier, the nonlinear character of the equations indicates that such a solution is unrealistic. Without this assumption the differential equations (12–15) and (12–16) can be approximated by the wave equations (12–21) if it is assumed that the perturbation of the guiding magnetic field is small, i.e.,

$$|\mathbf{b}| \ll |\mathbf{B}_0|. \qquad (12\text{–}26)$$

Let us therefore treat $\mathbf{b}$ and $\mathbf{u}$ as small quantities and ignore their products. Equation (12–15) becomes

$$\frac{\partial \mathbf{b}}{\partial t} = (\mathbf{B}_0 \cdot \boldsymbol{\nabla})\mathbf{u}, \qquad (12\text{–}27)$$

since the divergence of $\mathbf{u}$ is zero for an incompressible fluid. Since

$$(\boldsymbol{\nabla} \times \mathbf{B}) \times \mathbf{B} \cong -\boldsymbol{\nabla}(\mathbf{B}_0 \cdot \mathbf{b}) + (\mathbf{B}_0 \cdot \boldsymbol{\nabla})\mathbf{b},$$

Eq. (12–16) becomes

$$\rho_m \frac{\partial \mathbf{u}}{\partial t} = -\boldsymbol{\nabla}\left[p + \psi + \frac{1}{\mu_0}(\mathbf{B}_0 \cdot \mathbf{b}) \right] + \frac{1}{\mu_0}(\mathbf{B}_0 \cdot \boldsymbol{\nabla})\mathbf{b}. \qquad (12\text{–}28)$$

It can be shown that $\boldsymbol{\nabla}[p + \psi + (\mathbf{B}_0 \cdot \mathbf{b})/\mu_0]$ vanishes. The argument runs as follows: The other two terms in Eq. (12–28) have vanishing divergence, since for an incompressible fluid ρ_m is constant and $\boldsymbol{\nabla} \cdot \mathbf{u} = 0$; moreover, $\boldsymbol{\nabla} \cdot \mathbf{b} = 0$. Hence $\nabla^2[p + \psi + (\mathbf{B}_0 \cdot \mathbf{b})/\mu_0]$ vanishes. On the other hand, outside the region of the disturbance $\mathbf{b} = 0$ and $\boldsymbol{\nabla}p = -\boldsymbol{\nabla}\psi$ in order that the fluid be in equilibrium. The function $[p + \psi + (\mathbf{B}_0 \cdot \mathbf{b})/\mu_0]$ thus vanishes over a bounding surface, as well as being harmonic; according to the theorems of Section 1–1 it will have a vanishing gradient everywhere. Therefore the two equations reduce to

$$\mu_0 \rho_m \frac{\partial \mathbf{u}}{\partial t} = (\mathbf{B}_0 \cdot \boldsymbol{\nabla})\mathbf{b}, \qquad (12\text{–}29)$$

and

$$\frac{\partial \mathbf{b}}{\partial t} = (\mathbf{B}_0 \cdot \boldsymbol{\nabla})\mathbf{u}. \qquad (12\text{–}27)$$

Equations (12–29) and (12–27) are identical to Eqs. (12–18) and (12–20) if $\mathbf{B}_0$ is taken along the z-axis. The wave equations (12–21) then follow.

Thus far we have neglected all dissipative effects. Let us now estimate the effect of a finite conductivity, assumed uniform throughout the medium. Since displacement currents can be neglected, the relation

$$\mathbf{\nabla} \times \left(\frac{\mathbf{j}}{\sigma}\right) = \mathbf{\nabla} \times \left(\mathbf{\nabla} \times \frac{\mathbf{B}}{\mu_0 \sigma}\right) = -\frac{1}{\mu_0 \sigma} \nabla^2 \mathbf{B} \qquad (12\text{--}30)$$

follows from the relation $\mathbf{\nabla} \times \mathbf{B} = \mu_0 \mathbf{j}$ and the zero divergence of $\mathbf{B}$. It is then clear from a consideration of Eq. (12–1) that the effect of allowing for finite electrical resistance is an additional term in the wave equation for $\mathbf{b}$ which involves $\nabla^2(\partial \mathbf{B}/\partial t) = \nabla^2(\partial \mathbf{B}/\partial t)$,

$$\frac{\partial^2 \mathbf{b}}{\partial t^2} = v_M^2 \frac{\partial^2 \mathbf{b}}{\partial z^2} + \frac{1}{\mu_0 \sigma} \nabla^2 \left(\frac{\partial \mathbf{b}}{\partial t}\right). \qquad (12\text{--}31)$$

This equation, involving derivatives in x and y as well as z, corresponds to disturbances diffusing out as they travel along the z-direction. The time required for this diffusion is of the order of the time τ_D given by Eq. (12–8); therefore the waves along "frozen-in" lines are not appreciably affected if the frequency is large compared to $1/\tau_D$.

SUGGESTED REFERENCES

The references listed here treat some of the basic points given above as well as many applications to sunspot theory, theory of the earth's magnetism, variable stars, and the behavior of highly ionized gases in a magnetic field.

H. ALFVEN, *Cosmical Electrodynamics.*

T. G. COWLING, *Magnetohydrodynamics.*

W. M. ELSASSER, *Rev. Modern Phys.* **28**, 135 (1956). Review of work on the magnetohydrodynamics of the earth's core and the relation to the earth's magnetism. Also an excellent general discussion of magnetohydrodynamics.

L. SPITZER, JR., *Physics of Fully Ionized Gases.*

S. CHANDRASEKHAR, *Plasma Physics.*

L. D. LANDAU AND E. M. LIFSHITZ, *Electrodynamics of Continuous Media.* Chapter VIII constitutes a thorough account of the fundamentals of "magnetic fluid dynamics," including shock waves and turbulent flow.

EXERCISES

1. Show from the equation of continuity and Eq. (12–2) that

$$\frac{d}{dt}\left(\frac{\mathbf{B}}{\rho_m}\right) = \left(\frac{\mathbf{B}}{\rho_m} \cdot \mathbf{\nabla}\right) \mathbf{u}.$$

2. Derive Eq. (12–13) from Eq. (12–9) and Maxwell's equations.

CHAPTER 13

WAVES IN THE PRESENCE OF METALLIC BOUNDARIES

Before undertaking the theory of radiation, which examines the origin of electromagnetic waves in terms of their sources, we shall consider further the solutions of the homogeneous wave equation in regions containing various geometrical boundaries, particularly in regions bounded by conductors. The boundary value problem is of great theoretical significance and has many practical electromagnetic applications, especially in the microwave region of the spectrum. We shall here be concerned primarily with the underlying principles, not the devices based on these principles.

We may remark that there is usually no essential advantage in introducing conventional potentials in the absence of charges and currents: it is equally convenient to work with the fields directly, remembering, however, that they are not independent, since they must satisfy the field equations. In curvilinear coordinates the fields are in fact most simply expressed in terms of "superpotentials," but these special functions are readily related to particular field components. We shall see that two independent field components play the role of potentials from which the complete fields may be derived.

13–1 The nature of metallic boundary conditions. Let us review the general boundary conditions on the field vectors at a surface between medium 1 and medium 2:

$$\begin{aligned}
\mathbf{n} \cdot (\mathbf{D}_1 - \mathbf{D}_2) &= \tau, \\
\mathbf{n} \times (\mathbf{E}_1 - \mathbf{E}_2) &= 0, \\
\mathbf{n} \cdot (\mathbf{B}_1 - \mathbf{B}_2) &= 0, \\
\mathbf{n} \times (\mathbf{H}_1 - \mathbf{H}_2) &= \mathbf{K},
\end{aligned} \tag{13–1}$$

where τ is used for the surface charge density to avoid confusion with the conductivity, and $\mathbf{K}$ is the surface current density. As usual, $\mathbf{n}$ is a unit vector normal to the surface, directed from medium 2 to medium 1. We have seen in Section 11–6 that for normal incidence an electromagnetic wave falls off very rapidly inside the surface of a good conductor, and that at the surface the tangential component of $\mathbf{H}$ may be large, while that of $\mathbf{E}$ vanishes in the limit of perfect conductivity. Let us examine the behavior of the normal components.

Let medium 2 be a good conductor for which $\sigma/\omega\epsilon \gg 1$, while medium 1 is a perfect insulator. The surface charge τ will be related to the currents in the conductor; in fact, the conservation of charge requires that

$$\mathbf{n} \cdot \mathbf{j} = \partial\tau/\partial t = -i\omega\tau.$$

But $\mathbf{n} \cdot \mathbf{j} = \mathbf{n} \cdot \sigma\mathbf{E}_2$, and therefore we obtain from the first of Eqs. (13–1)

$$\left(1 - \frac{i\omega\epsilon_2}{\sigma}\right)\mathbf{n} \cdot \mathbf{E}_2 = -\frac{i\omega\epsilon_1}{\sigma}\mathbf{n} \cdot \mathbf{E}_1. \tag{13–2}$$

Thus we see that the normal component of $\mathbf{E}$ within the conductor also becomes vanishingly small as the conductivity approaches infinity, and suffers a discontinuity at the surface in the presence of real surface charge. The normal component of $\mathbf{B}$, on the other hand, is continuous across the surface.

If $\mathbf{E}$ vanishes inside a perfect conductor the curl of $\mathbf{E}$ also vanishes, and the time rate of change of $\mathbf{B}$ is correspondingly zero. This means that there are no oscillatory fields whatever inside such a conductor, and that the boundary values of the fields outside are given by

$$\begin{aligned} \mathbf{n} \cdot \mathbf{D} &= \tau, \\ \mathbf{n} \times \mathbf{E} &= 0, \\ \mathbf{n} \cdot \mathbf{B} &= 0, \\ \mathbf{n} \times \mathbf{H} &= \mathbf{K}. \end{aligned} \tag{13–3}$$

Thus the electric field is normal and the magnetic field tangential at the surface of a perfect conductor. For good conductors these conditions lead to excellent representations of the geometrical configurations of external fields, but they correspond to complete neglect of some important features of real fields, such as losses in cavities and attenuation of signal in the case of transmitted waves.

It is useful for the estimation of losses to see how the tangential and normal field components compare with the first approximation in $1/\sigma$. The tangential component of $\mathbf{E}$ is strictly continuous, so that we have, from Section 11–6,

$$\mathbf{H} = \frac{1 + i}{\sqrt{2}} \sqrt{\frac{\sigma}{\mu\omega}} \, \mathbf{n} \times \mathbf{E} \tag{13–4}$$

at the surface of the conductor. Let us assume, without obtaining a complete solution, that a wave with $\mathbf{H}$ very nearly tangential and $\mathbf{E}$ very nearly normal is propagated along the surface of the metal. If k is the component of the propagation vector along the surface, then just outside the surface,

$$|\mathbf{H}_{||}| \simeq \frac{k}{\mu\omega} |\mathbf{E}_\perp|. \tag{13–5}$$

But a tangential component of $\mathbf{H}$ is accompanied by a small tangential component of $\mathbf{E}$, according to Eq. (13–4). By comparing these two expressions, we see that

$$\frac{|\mathbf{E}_{||}|}{|\mathbf{E}_\perp|} \simeq k \sqrt{\frac{2}{\mu\omega\sigma}} = \frac{\delta}{\lambda}, \tag{13–6}$$

and the ratio of the tangential component of $\mathbf{E}$ to its normal component has the same order of magnitude as the skin depth divided by the wavelength. It can be readily shown that the ratio of the normal component of $\mathbf{H}$ to its tangential component is of this same magnitude, but for the estimation of losses it is the tangential component of $\mathbf{E}$ in which we shall be interested. Thus we see that in the limit of high conductivity, which means vanishing skin depth, no fields penetrate the conductor, and the boundary conditions are just those given by Eqs. (13–2). We shall investigate the solution of the homogeneous wave equation subject to such boundaries.

13–2 Eigenfunctions and eigenvalues of the wave equation. The vector wave equation in Cartesian coordinates is simply a set of three scalar equations, one in each of the rectangular components of the vector. In other coordinate systems the situation is more complicated, but we shall see that at least in cylindrical and spherical polar coordinates the fields may also be written in terms of scalar functions. Our problem is to solve an equation of the form

$$\nabla^2\Psi - \frac{1}{c^2}\frac{\partial^2\Psi}{\partial t^2} = 0, \tag{13–7}$$

where Ψ may represent a component of $\mathbf{E}$ or $\mathbf{H}$, in such a way that certain boundary conditions are satisfied.

Equation (13–7), unlike Laplace's equation, is not symmetrical in its independent variables: the second derivative with respect to the time occurs with the sign opposite to that of the space derivatives. The resulting character of possible solutions corresponds to a fundamental physical distinction from the case of static potentials: entirely oscillatory solutions are now allowed in a shielded region free of sources, whereas, for example, the static field inside a grounded conductor is zero throughout.

We may assume simple harmonic time dependence for the function $T(t)$ in the expression

$$\Psi(x, y, z, t) = \psi(x, y, z)T(t) = \psi(x, y, z)e^{-i\omega t}, \qquad (13\text{–}8)$$

where ψ is a function of the space variables alone, and satisfies the differential equation

$$\nabla^2 \psi = -\frac{\omega^2}{c^2}\psi = -k^2\psi. \qquad (13\text{–}9)$$

In rectangular coordinates, we may set $\psi = X(x)Y(y)Z(z)$, whereupon separation yields

$$X'' + k_1^2 X = 0,$$

$$Y'' + k_2^2 Y = 0, \qquad (13\text{–}10)$$

$$Z'' + k_3^2 Z = 0,$$

with

$$k_1^2 + k_2^2 + k_3^2 = k^2 = \omega^2/c^2. \qquad (13\text{–}11)$$

Equation (13–11) indicates that there are no more than the expected number of separation constants, namely three, corresponding to the four independent variables. Equations (13–10), or their counterparts in other coordinate systems, must be solved subject to the boundary conditions on the field component represented by or related to the scalar function ψ.

Let us examine this problem somewhat more generally. Each of Eqs. (13–10) is a simple example of the general form

$$\frac{d}{dx}\left[p(x)\frac{dX}{dx}\right] + [q(x) + \lambda r(x)]X = 0, \qquad (13\text{–}12)$$

where λ is a parameter, in our case the separation constant. The solutions are required to be continuous and bounded throughout an interval $a \leq x \leq b$, and to satisfy certain boundary conditions at $x = a$, $x = b$. [As to the nature of the unspecified functions $p(x)$, $q(x)$, $r(x)$, we need only assume that they are such as to introduce no singularities within the interval, although there may be singularities at the boundaries.] The boundary conditions can be satisfied in general only by certain values of λ, and by particular solutions of Eq. (13–12) corresponding to these values of λ. The allowed values of λ are called eigenvalues, and the allowed solutions are called eigenfunctions. It will be recognized that we have met this one-dimensional problem before, in applying the method of separation of variables to the solution of static potential problems, but here the solution of the entire three-dimensional problem as a whole may be characterized by an eigenvalue related to the frequency.

To facilitate discussion at first, let us consider the case where $q(x) = 0$, $r(x) = 1$, under the conditions that $X(x) = 0$ at the ends of the interval. Equation (13–12) becomes

$$\frac{d}{dx}\left[p(x)\frac{dX}{dx}\right] = -\lambda X. \tag{13–13}$$

From Eq. (13–13) it can be seen that if we have a solution $X(x)$ equal to zero at $x = a$, it can never satisfy the same condition at $x = b$ if λ is negative, for in that case it would continue to depart from its zero value once it began to do so. For positive values of λ, however, the curvature is opposite in sign to that of $X(x)$, and for some value of λ the function will be brought back to the x-axis just at b. Further increase in λ will cause $X(x)$ to change sign short of b, but a value will be reached for which $X(x)$ will again be zero at $x = b$, this time with one zero in the interval $a \leq x \leq b$. Thus we see that there is an absolute lower bound for λ, and the eigenvalues may be numbered λ_0, λ_1, ... , with subscripts corresponding to the number of zeros in the interval.

This whole argument may be repeated for the more general case of Eq. (13–12), except that the sign of the curvature will depend on $q(x)$ and $r(x)$. There is still a lowest allowed value of λ, but the curvature is determined by the sign of the entire coefficient of $X(x)$ in the differential equation. For example, Bessel's equation becomes

$$\frac{d}{dr}\left(r\frac{dJ_n}{dr}\right) + \left(k^2 r - \frac{n^2}{r}\right)J_n = 0$$

when written in this form, and the curvature will not be such as to bring the function back toward the axis unless $k^2 r > n^2/r$.

If the boundary conditions are such that either $X(x)$ or its first derivative must vanish at the ends of the interval, the eigenfunctions of Eq. (13–12) are orthogonal over the interval. Let X_n and X_m be two allowed solutions corresponding to two eigenvalues of the parameter, λ_n and λ_m. To prove orthogonality, let the equation for X_n be multiplied by X_m, that for X_m by X_n, one subtracted from the other, and the whole integrated over the interval. Since either all functions X or their derivatives vanish at the boundaries, the result is

$$(\lambda_m - \lambda_n)\int_a^b r(x)X_m X_n\, dx = 0. \tag{13–14}$$

The function $r(x)$ is often called the "weight function" in mathematical treatments.

If more than one solution is allowed for a particular value of λ the problem is said to be degenerate, and the number of independent solutions corresponding to that value of λ is called the degree of the degeneracy. In this case Eq. (13–14) fails to show orthogonality, i.e., λ_m may be equal to λ_n for $n \neq m$. The solutions may nevertheless be chosen to be orthogonal, due to the fact that any linear combination of the independent solutions belonging to a given λ is also a solution, and these linear combinations may be made in such a way that they are orthogonal to each other. The choice of coefficients in the orthogonal linear combinations is not unique, but is usually made so that the final solutions correspond to some readily identifiable physical property of the system. In any case, we may always speak of the independent solutions as orthogonal, assuming that orthogonalization has been carried out for degenerate λ's.

In any coordinate system in which the wave equation is separable the function ψ is a product of single variable functions such as $X(x)$. The frequency for which there is an allowed solution of Eq. (13–9) depends on three allowed parameters λ, one for each coordinate variable. It is instructive to prove that the wave equation generates orthogonal functions. If we apply Green's theorem to two product eigenfunctions, ψ_n and ψ_m, we obtain

$$\int (\psi_m \nabla^2 \psi_n - \psi_n \nabla^2 \psi_m)\, dv = \int (\psi_m \nabla \psi_n - \psi_n \nabla \psi_m) \cdot d\mathbf{S}.$$

Either boundary condition, the vanishing of the field component or its normal derivative, causes the surface integral to vanish. The volume integral can be written, in view of Eq. (13–9), as

$$(k_n^2 - k_m^2) \int \psi_m \psi_n\, dv = 0, \tag{13–15}$$

which is the desired orthogonality relation. By comparison of Eqs. (13–14) and (13–15) it is seen that the "weight function" $r(x)$ is just the factor required to give the correct volume element in three dimensions.

In closed cavities, then, with boundaries on which the field components or their derivatives vanish, we may expect to find characteristic electromagnetic vibrations, similar to the modes of a vibrating string. The frequencies of these vibrations will correspond to a lowest eigenvalue, the "fundamental," and an infinite discrete set of higher frequencies. These modes of vibration are sharp and independent insofar as the approximation of perfect conductivity is valid; in practice they will not sustain themselves indefinitely, and we shall have to look at the effect of finite conductivity in order to estimate losses. It should be pointed out that the actual configuration of fields in a cavity will depend on the method of excitation.

Although every one of the four single variable factors in $\Psi(x, y, z, t)$ may now be an oscillatory function, only three correspond to closed boundaries. The fourth independent variable, time, is still "open," and the complete solution will depend on the initial conditions, $\Psi(x, y, z, t_0)$ and $(\partial\Psi/\partial t)_{t=t_0}$, at some particular time $t = t_0$. There is no solution corresponding to the specification of both initial and final conditions.

13–3 Cavities with rectangular boundaries. Let us consider a vacuum region totally enclosed by rectangular conducting walls. In this case, all field components are products of the solutions of Eqs. (13–10), with k_1, k_2, and k_3 so chosen that the electric fields are normal to the walls at the boundary and the magnetic fields tangential. If A, B, and C are the dimensions of the cavity, it may be readily verified that the electric field components are

$$E_x = E_1 \cos(k_1 x) \sin(k_2 y) \sin(k_3 z)\, e^{-i\omega t},$$

$$E_y = E_2 \sin(k_1 x) \cos(k_2 y) \sin(k_3 z)\, e^{-i\omega t}, \qquad (13\text{–}16)$$

$$E_z = E_3 \sin(k_1 x) \sin(k_2 y) \cos(k_3 z)\, e^{-i\omega t},$$

where the eigenvalues of the separation parameters needed to satisfy the boundary conditions are

$$k_1 = l\pi/A, \qquad k_2 = m\pi/B, \qquad k_3 = n\pi/C, \qquad (13\text{–}17)$$

with l, m, and n integers. The allowed frequencies are thus given by

$$\frac{\omega^2}{c^2} = k^2 = \pi^2 \left(\frac{l^2}{A^2} + \frac{m^2}{B^2} + \frac{n^2}{C^2} \right). \qquad (13\text{–}18)$$

It is clear from Eqs. (13–16) that at least two of the integers l, m, n must be different from zero in order to have nonvanishing fields. The magnetic fields obtained by the use of $\nabla \times \mathbf{E} = i\omega\mathbf{B}$ automatically satisfy the appropriate boundary conditions, and are seen to be in time quadrature with the electric fields. Thus the sum of the total electric and magnetic energies within the cavity is constant although the two terms fluctuate separately.

The amplitudes of the electric field components are not independent, but are related by the divergence condition $\nabla \cdot \mathbf{E} = 0$, which yields

$$E_1 k_1 + E_2 k_2 + E_3 k_3 = 0. \qquad (13\text{–}19)$$

There are, in general, two linearly independent vectors $\mathbf{E}$ that satisfy this condition, corresponding to two possible polarizations. (The exception is the case that one of the integers l, m, n is zero, so that $\mathbf{E}$ is fixed in direction.) These vectors may be chosen arbitrarily so long as they are not collinear, and each is accompanied by a magnetic field at right angles. Thus again $\mathbf{E}$, $\mathbf{H}$, and $\mathbf{k}$ form an orthogonal set of vectors. The fields corresponding to a given set of integers l, m, n constitute a mode of vibration such as we predicted in the previous section. The total field is a sum over all possible modes of vibration, with amplitude factors that depend on the method of excitation.

Actual conducting walls gradually absorb energy from the cavity at a rate that can be readily estimated from the considerations of Section 13–1. For finite σ the small tangential component of $\mathbf{E}$ may be found from Eq. (13–4):

$$\mathbf{E}_{||} = \frac{1-i}{\sqrt{2}} \sqrt{\frac{\mu\omega}{\sigma}} \, \mathbf{H}_{||} \times \mathbf{n}.$$

The tangential component of $\mathbf{H}$ will be slightly different from the ideal solution, but its relative change will be very small and may be neglected. To find the power loss, we may compute the time average of the Poynting vector into the wall at the surface,

$$\overline{\mathbf{N}} = \frac{\mathrm{Re}(\mathbf{E}_{||} \times \mathbf{H}_{||})}{2} = \frac{\mathbf{n}}{2} \sqrt{\frac{\mu\omega}{2\sigma}} \mathbf{H}_{||0}^{\,2} = \frac{\mathbf{H}_{||0}^{\,2}}{2\sigma\delta} \mathbf{n}. \qquad (13\text{–}20)$$

Here $\mathbf{H}_{||0}$ is the crest value of the tangential magnetic field, which equals the crest value of the surface current $\mathbf{K}$. If we define a surface resistance such that

$$\overline{\mathbf{N}} = \mathbf{n}(\overline{\mathbf{K}^2} R_s),$$

then

$$R_s = \frac{1}{\sigma\delta}.$$

The total power loss in a cavity is obtained if we integrate Eq. (13–20) over the entire area of the walls. Mode losses are often expressed by giving the "Q" for a cavity, defined by

$$Q = \frac{\text{energy stored in the cavity}}{\text{energy lost per cycle to the walls}},$$

which is independent of the amplitude of mode excitation.

13–4 Cylindrical cavities. The rectangular parallelepiped we have considered is a special kind of right cylinder, and has many features in common with cylindrical cavities of arbitrary cross section. In every cavity the allowed values of k, and thus the allowed frequencies, are determined by the geometry of the cavity. But we have seen that for each set of k_1, k_2, k_3 in the rectangular cavity there are, in general, two linearly independent modes, i.e., the polarization remains arbitrary. We may take advantage of this arbitrariness to classify modes into two kinds according to the orientation of the field vectors: let us choose one mode such that the electric field lies in the cross-sectional plane, and the other so that the magnetic vector lies in this plane. This classification into transverse electric (TE) and transverse magnetic (TM) modes turns out to be possible for all cylindrical cavities, although the rectangular parallelepiped with unequal axes is unique in having one mode of each kind correspond to the same allowed frequency. (There is, of course, a higher degree of degeneracy if any axes of the parallelepiped are equal.)

To show that this classification is both possible and complete for the modes of any cylindrical cavity, we first note that the factor depending on the altitude coordinate z may be separated out of the wave equation. The boundary conditions at $z = 0$ and $z = C$ demand that the z dependence be given by either $\sin k_3 z$ or $\cos k_3 z$, where $k_3 = n\pi/C$. In other words, every field component satisfies the equation

$$\left(\frac{\partial^2}{\partial z^2} + k_3^2\right) \begin{Bmatrix} \mathbf{E} \\ \mathbf{H} \end{Bmatrix} = 0 \qquad (13\text{–}21)$$

as well as

$$(\nabla^2 + k^2) \begin{Bmatrix} \mathbf{E} \\ \mathbf{H} \end{Bmatrix} = 0, \qquad (13\text{–}22)$$

with k as yet undetermined. The field equations

$$\boldsymbol{\nabla} \times \mathbf{E} = i\omega\mu\mathbf{H},$$
$$\boldsymbol{\nabla} \times \mathbf{H} = -i\omega\epsilon\mathbf{E} \qquad (13\text{–}23)$$

must also be satisfied. If we write each vector and each operator in Eqs. (13–23) as the sum of a transverse part, designated by the subscript s, and a component along z, we find for the transverse fields

$$i\omega\mu\mathbf{H}_s = \boldsymbol{\nabla}_s \times \mathbf{E}_z + \boldsymbol{\nabla}_z \times \mathbf{E}_s,$$
$$-i\omega\epsilon\mathbf{E}_s = \boldsymbol{\nabla}_s \times \mathbf{H}_z + \boldsymbol{\nabla}_z \times \mathbf{H}_s, \qquad (13\text{–}24)$$

since, for example, $\boldsymbol{\nabla}_s \times \mathbf{E}_s = i\omega\mu\mathbf{H}_z$. When one of Eqs. (13–24) is substituted for the transverse field on the right side of the other, and use is made of Eq. (13–21), there results

$$\mathbf{E}_s = \frac{\nabla_s(\partial E_z/\partial z)}{k^2 - k_3^2} + \frac{i\omega\mu}{k^2 - k_3^2}\, \nabla_s \times \mathbf{H}_z,$$

$$\mathbf{H}_s = \frac{\nabla_s(\partial H_z/\partial z)}{k^2 - k_3^2} - \frac{i\omega\epsilon}{k^2 - k_3^2}\, \nabla_s \times \mathbf{E}_z. \tag{13–25}$$

All transverse fields are thus expressed in terms of the z-components, each of which satisfies the differential equation

$$[\nabla_s^2 + (k^2 - k_3^2)] \begin{Bmatrix} E_z \\ H_z \end{Bmatrix} = 0, \tag{13–26}$$

where ∇_s^2 is the two-dimensional Laplacian operator.

The conditions on E_z and H_z at the boundary of the cylinder cross section differ from each other, however: E_z must vanish on the boundary, while the normal derivative of H_z must vanish to satisfy the conditions on the second of Eqs. (13–25). When the cross section is a rectangle, these two conditions may lead to the same eigenvalues of $(k^2 - k_3^2) = k_s^2 = k_1^2 + k_2^2$, as we have seen. Otherwise they correspond to two *different* frequencies, one for which E_z is permitted but $H_z = 0$, the other with H_z but no E_z. In every case, it is possible to classify the modes as transverse magnetic or transverse electric. The field components E_z and H_z play the role of independent potentials, from which all other field components of the TM and TE modes respectively may be derived by means of Eqs. (13–25).

The frequencies are determined by the eigenvalues of Eqs. (13–21) and (13–26). If we denote the functional dependence of E_z or H_z on the plane cross section coordinates by $f(x, y)$, we may write Eq. (13–26) as

$$\nabla_s^2 f = -(k^2 - k_3^2)f = -k_s^2 f. \tag{13–27}$$

Let us first show that $k_s^2 > 0$, and hence $k > k_3$. From Green's theorem, we note that

$$-k_s^2 \int f^2\, dv + \int (\nabla_s f)^2\, dv = \int f \nabla f \cdot d\mathbf{S}.$$

If either f or its normal derivative is to vanish on S, the cylindrical surface, then

$$k_s^2 = \frac{\displaystyle\int (\nabla_s f)^2\, dv}{\displaystyle\int f^2\, dv} > 0. \tag{13–28}$$

We have already seen that $k_3 = n\pi/C$. The allowed values of k_s depend both on the geometry of the cross section and the nature of the mode.

For TM modes $H_z = 0$, and the z dependence of E_z is given by $\cos n\pi z/C$. Equation (13–27) must be solved subject to the condition that f vanishes on the boundaries of the plane cross section, thus completing the determination of E_z and k. The transverse fields are special cases of Eqs. (13–25):

$$\mathbf{E}_s = \frac{1}{k_s^2} \, \boldsymbol{\nabla}_s \frac{\partial E_z}{\partial z} \, ,$$

$$\mathbf{H} = - \frac{i\omega\epsilon}{k_s^2} \, \boldsymbol{\nabla}_s \times \mathbf{E}_z \tag{13–29}$$

$$= \frac{i\omega\epsilon}{k_s^2} \hat{\mathbf{z}} \times \boldsymbol{\nabla}_s E_z,$$

where $\hat{\mathbf{z}}$ is a unit vector along the axis of the cylinder.

For TE modes, in which $E_z = 0$, the condition that H_z vanish at the ends of the cylinder demands the use of $\sin n\pi z/C$, and k_s must be such that the normal derivative of H_z is zero at the walls. Equations (13–25), giving the transverse fields, then become

$$\mathbf{H}_s = \frac{1}{k_s^2} \, \boldsymbol{\nabla}_s \frac{\partial H_z}{\partial z} \, ,$$

$$\mathbf{E} = \frac{i\omega\mu}{k_s^2} \, \boldsymbol{\nabla}_s \times \mathbf{H}_z = - \frac{i\omega\mu}{k_s^2} \hat{\mathbf{z}} \times \boldsymbol{\nabla}_s H_z, \tag{13–30}$$

and the mode determination is completed.

13–5 Circular cylindrical cavities. Let us apply the methods of the previous section to the TM modes of a right circular cylinder. We may write

$$E_z = A f(r, \varphi) \cos (k_3 z) e^{-i\omega t},$$

where $f(r, \varphi)$ satisfies the equation

$$\frac{1}{r} \, \frac{\partial}{\partial r} \left(r \, \frac{\partial f}{\partial r} \right) + \frac{1}{r^2} \, \frac{\partial^2 f}{\partial \varphi^2} + k_s^2 f = 0. \tag{13–31}$$

Separation of variables leads to Bessel's equation for the radial coordinate, just as in electrostatic problems of cylindrical symmetry. If the axis $r = 0$ is included in the cavity, that solution must be chosen which is regular at the origin. For the TM mode in a cavity of radius r_0 and height C,

$$E_z = A J_m(k_l r) e^{im\varphi} \cos (k_3 z) e^{-i\omega t}. \tag{13-32}$$

The k_l are determined by the condition that $J_m(k_l r_0) = 0$, and are the eigenvalues of k_s in Eq. (13-31). The frequency of any mode is given by

$$\frac{w^2}{c^2} = k^2 = k_l^2 + \frac{n^2 \pi^2}{C^2}.$$

Knowledge of the zeros of the Bessel functions is thus necessary in order for us to ascertain the numerical values of the frequencies. If l is the ordinal number of a zero of a particular Bessel function of order m (l increases with increasing values of the argument) then each mode is characterized by three integers, l, m, n, as in the rectangular case. All transverse components of the fields may be found by taking the appropriate derivatives of E_z, as indicated in Eqs. (13-29).

The frequencies of the transverse electric modes are determined by the fact that $J_m(k_l r_0)$ must have zero slope at the boundary. The eigenvalues k_l are thus different from those of the TM case, and the cavity frequencies are correspondingly different. Again there is an infinite discrete set of modes, each characterized by three integers.

Sectors of cylinders can be treated by means of Bessel functions of fractional order, since m is no longer an integer but is determined by the boundary condition on the $\varphi = $ constant surfaces. Cavities in the form of coaxial cylinders are amenable to the same methods of solution, except that the two boundary surfaces require the use of the general solution of Bessel's equation. Perhaps the most interesting feature of the double-walled cavity, however, is that in addition to TM and TE modes there are also TEM modes for which $E_z = 0 = H_z$. So far as these modes are concerned, the boundaries of the multiply connected region behave like internal sources for the fields. It is easy to see physically that there is even a zero-frequency mode having no electric field components, which would correspond to an upward current in the outside cylinder and a downward current in the inside one, for example.

13-6 Wave guides. Let us consider the possibility of transmitting electromagnetic waves along the axis of a wave guide, which is simply a long cylinder with open ends. To represent a progressive wave along z, we may write the dependence of the fields on the coordinate variables and the time as

$$f(x, y) e^{i(k_g z - \omega t)}. \tag{13-33}$$

The "guide propagation constant," k_g, is just the k_3 of Eq. (13–21) except that it is no longer restricted by boundary conditions to discrete values. The general considerations of Section 13–4 apply, so that we may treat TM and TE modes separately.

One other case, briefly noted in Section 13–5, is of greater importance in this connection. If the boundaries are not singly connected, i.e., if the region is bounded by at least two conducting cylinders, the two-dimensional Laplacian of the fields, or $\nabla_s^2 f(x, y)$, can vanish subject to the boundary conditions. The solution of the two-dimensional electrostatic problem will then define the solution in a transverse plane and $k^2 = k_g^2$, so that the velocity of propagation in vacuum is simply c. Such a wave will have no field components along the axis of the cylinder: E_z vanishes on all boundaries and, being a solution of $\nabla_s^2 E_z = 0$, will vanish everywhere. Similarly, since the normal derivative of H_z vanishes on all boundaries and since $\nabla_s^2 H_z = 0$, H_z vanishes everywhere. Hence such a wave is purely transverse (TEM mode). The ordinary coaxial and bifilar line structures are typical examples of such cylinders.

If the boundary is singly connected, the solutions for f are identical with those for cylindrical cavities already discussed. Although k_g is not restricted in magnitude, we note that for every eigenvalue of the two-dimensional equation, k_s, there is a lowest value of k, namely, $k = k_s$ (often designated by k_c for wave guides), for which k_g is real. This corresponds to a "cutoff frequency" below which waves are not transmitted by that mode, and the fields fall off exponentially with increasing z. An absolute cutoff frequency is associated with the mode of lowest frequency, i.e., with the lowest value of k_s. In general,

$$k_s^2 = k^2 - k_g^2. \tag{13–34}$$

For real k_g it is obvious from the form of Eq. (13–33) that the wave is propagated along the guide with a phase velocity

$$u_p = \frac{\omega}{k_g} = \frac{k}{k_g} c.$$

Since a minimum condition for propagation is that k is greater than k_g, it is evident that the wave or phase velocity is greater than that of electromagnetic waves in free space. This velocity is not constant, however, but depends on the frequency. The wave guide thus behaves like a dispersive medium, and the group velocity of a pulse propagated along the guide is, by Eqs. (11–68) and (13–34),

$$u_g = \frac{d\omega}{dk_g} = \frac{k_g}{k} c. \tag{13–35}$$

It is seen that u_g is always smaller than c and, as a matter of fact,

$$u_p u_g = c^2. \tag{13-36}$$

Only if the pulse consists of a narrow range of frequencies can it be said to be transmitted with a definite velocity, the velocity u_g being evaluated at some median frequency.

The fields may be expressed even more simply in terms of the longitudinal field components than in the case of cylindrical cavities, since here $\partial/\partial z = ik_g$. It may be seen from Eqs. (13-25) that

for TM modes ($H_z = 0$),

$$\mathbf{E}_s = \frac{ik_g}{k_s^2} \, \boldsymbol{\nabla}_s E_z,$$

$$\mathbf{H}_s = \frac{\omega\epsilon}{k_g} \, \hat{z} \times \mathbf{E}_s \qquad \text{or} \qquad c\mathbf{B}_s = \frac{k}{k_g} \, \hat{z} \times \mathbf{E}_s, \tag{13-37}$$

and for TE modes ($E_z = 0$),

$$\mathbf{H}_s = \frac{ik_g}{k_s^2} \, \boldsymbol{\nabla}_s H_z,$$

$$\mathbf{E}_s = -\frac{\omega\mu}{k_g} \, \hat{z} \times \mathbf{H}_s \qquad \text{or} \qquad \mathbf{E}_s = \frac{ck}{k_g} \, \mathbf{B}_s \times \hat{z}. \tag{13-38}$$

The time average z-component of the Poynting vector $\mathbf{N}$ is given by

$$\overline{N_z} = \frac{|\mathbf{E}_s \times \mathbf{H}_s^*|}{2}$$

$$= \sqrt{\frac{\mu_0}{\epsilon_0}} \, \frac{k}{k_g} \, \frac{H_{s0}^2}{2} \qquad \text{for TE modes}$$

$$= \sqrt{\frac{\mu_0}{\epsilon_0}} \, \frac{k_g}{k} \, \frac{H_{s0}^2}{2} \qquad \text{for TM modes,} \tag{13-39}$$

where the subscript 0 denotes the crest value of the field. Wave guide losses may be estimated by integrating Eq. (13-20) over the walls of the guide for any given mode. The energy flow of a progressive wave will be attenuated as e^{-Kz}, where

$$K = \frac{\text{power loss per unit length of guide}}{\text{power transmitted through guide}}$$

$$= \frac{1}{2\sigma\delta} \int (H_s^2 + H_z^2) \, dS \, \bigg/ \int\!\!\int \overline{N_z} \, dS, \tag{13-40}$$

with the numerator integrated over unit length of wall and the denominator integrated over the guide cross section. It is customary to define a "guide impedance" Z_g by writing

$$\int \overline{N_z}\, dS = \tfrac{1}{2} Z_g \int H_{s0}^2\, dS. \tag{13-41}$$

The guide impedance can be written down by inspection of Eqs. (13-39):

$$Z_g = \sqrt{\frac{\mu_0}{\epsilon_0}} \frac{k}{k_g} \qquad \text{for TE modes,}$$

$$Z_g = \sqrt{\frac{\mu_0}{\epsilon_0}} \frac{k_g}{k} \qquad \text{for TM modes.}$$

The attenuation constant K of Eq. (13-40) can be computed by performing the appropriate integrations. The exponential attenuation that results when $k < k_s$ does not involve power loss in the walls, but is generated by a superposition of interfering reflections.

13-7 Scattering by a circular cylinder. The scattering of a plane wave by a cylindrical conductor offers an instructive application of boundary conditions on the homogeneous wave equation solutions. Consider a perfectly conducting cylinder of radius a, its axis coincident with the z-coordinate axis, as shown in Fig. 13-1. Let there be incident normal to the axis of the cylinder a plane wave of angular frequency ω, polarized in such a way that its magnetic vector is in the z-direction. The total field will consist of the incident wave and that scattered (reflected) by the cylinder. This field is essentially a transverse electric mode, with $k_3 = 0$ and no restrictions on the frequency.

The generating function for Bessel functions permits us to write down immediately the appropriate expansion for the magnetic vector of the incoming wave:

$$H_{z0} e^{i(kr \cos \varphi - \omega t)} = H_{z0} \sum_{m=-\infty}^{\infty} J_m(kr) i^m e^{i(m\varphi - \omega t)}. \tag{13-42}$$

In order to represent an outgoing wave from the scattering cylinder, we must choose that combination of Bessel functions which varies as $e^{ikr}/\sqrt{r}$ for large values of r. This combination is just $J_m(kr) + iN_m(kr)$, for which the standard notation is $H_m^{(1)}(kr)$; we may drop the superscript, and note that this function will not be confused with a magnetic field because of the subscript m and the explicit argument kr. Thus the scattered wave is of the form

$$H_{z0} \sum_m a_m H_m(kr) e^{i(m\varphi - \omega t)}.$$

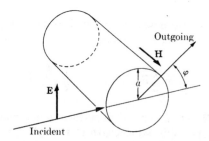

Fig. 13–1. Cross section of a conducting circular cylinder with radiation incident at right angles to the axis. **H** is parallel to the axis of the cylinder.

The coefficients a_m are determined by the condition that the tangential component of the total electric field vanishes on the surface of the cylinder. Since H_z does not depend on z, it remains the only wave component of the magnetic field, a result we should also expect for physical reasons. It therefore follows from the curl relation that

$$E_\varphi = \frac{-i}{\epsilon_0 \omega} \frac{\partial}{\partial r} H_z. \qquad (13\text{–}43)$$

For each term of the series, therefore, the boundary condition requires that

$$i^m J'_m(ka) + a_m H'_m(ka) = 0,$$

where the prime denotes differentiation with respect to the argument. Solving for a_m, we obtain

$$a_m = -i^m \frac{J'_m(ka)}{H'_m(ka)}. \qquad (13\text{–}44)$$

We are interested in the scattered wave at large distances from the cylinder. For large values of its argument,

$$H_m(kr) \rightarrow \sqrt{\frac{2}{\pi kr}} e^{ikr} e^{-(2m+1)(\pi/4)i}, \qquad (13\text{–}45)$$

and the scattered wave fields are thus

$$H_z = H_{z0} \sqrt{\frac{2}{\pi kr}} e^{i(kr-\omega t)} \sum a_m e^{i[m\varphi-(2m+1)(\pi/4)]}$$

$$E_\varphi = \sqrt{\frac{\mu_0}{\epsilon_0}} H_z,$$

if we neglect terms that fall off more rapidly with increasing r.

The Poynting vector associated with these fields is, of course, pointed along positive r. Its time average is

$$\overline{N_r} = \frac{1}{2}\operatorname{Re}(E_\varphi H_z^*) = \frac{1}{2}\sqrt{\frac{\mu_0}{\epsilon_0}}\,|H_z|^2. \qquad (13\text{--}46)$$

If we integrate $\mathbf{N}$ over a cylindrical surface to determine the total rate of energy scattered per unit axial length of the cylinder, the double sum in the product reduces to a single sum because of the orthogonality of the functions $e^{im\varphi}$, and we obtain

$$\frac{dU}{dt} = \int_0^{2\pi} Nr\,d\varphi = \frac{1}{2}\sqrt{\frac{\mu_0}{\epsilon_0}}\,H_{z_0}^2\,\frac{4}{k}\sum_m \left|\frac{J_m'(ka)}{H_m'(ka)}\right|^2 \qquad (13\text{--}47)$$

or, in terms of real Bessel functions,

$$\frac{dU}{dt} = \frac{1}{2}\sqrt{\frac{\mu_0}{\epsilon_0}}\,H_{z_0}^2\,\frac{4}{k}\sum_m \frac{[J_m'(ka)]^2}{[J_m'(ka)]^2 + [N_m'(ka)]^2}. \qquad (13\text{--}47')$$

The time average of the energy flow per unit area in the incident beam is

$$\overline{N_0} = \frac{1}{2}\sqrt{\frac{\mu_0}{\epsilon_0}}\,H_{z_0}^2, \qquad (13\text{--}48)$$

and the coefficient of this factor in Eq. (13–47) is called the scattering cross section of the scatterer. It represents the area from which the energy flow of the incident beam is "removed" by the scatterer. The cross section here appears to have the dimensions of length, but this is because we have throughout treated a unit length of the cylinder. The geometrical area of the cylinder per unit length is just $2a$.

In the limit of long wavelengths, for which $ka \to 0$, the expression for the scattering cross section can be simplified by means of the power series expansions for the Bessel functions. The term $m = 0$ and the two terms $m = \pm 1$ contribute equally in this limit, while all other terms involve only higher powers of ka. The result is

$$\sigma = \tfrac{3}{4}\pi^2(ka)^3 a, \qquad (13\text{--}49)$$

which varies as the inverse cube of the wavelength of the radiation and is smaller than the geometrical cross section of the cylinder. To determine the angular distribution of the scattered radiation we must go back to the time average of the Poynting vector. The leading terms give

$$\sigma(\varphi)\, d\varphi = \frac{\overline{N} r\, d\varphi}{\overline{N_0}},$$

$$\sigma(\varphi) = \frac{2}{\pi k} \left| \frac{J_0'(ka)}{J_0'(ka) + iN_0'(ka)} - \frac{2J_1'(ka)\cos\varphi}{J_1'(ka) + iN_1'(ka)} \right|^2. \quad (13\text{–}50)$$

If $ka \ll 1$, this becomes

$$\sigma(\varphi) = \frac{2}{\pi k} \frac{\pi^2 (ka)^4}{2^4} (1 - 2\cos\varphi)^2 = \frac{\pi (ka)^3}{8} a (1 - 2\cos\varphi)^2.$$

The scattering is thus predominantly backward, with a subsidiary maximum in the forward direction, and zero intensity at $\varphi = 60°$.

13–8 Spherical waves. The scalar wave equation is easily separable in spherical polar coordinates, and the solution of

$$\nabla^2 \psi(r, \theta, \varphi; t) = \frac{1}{c^2} \frac{\partial^2 \psi}{\partial t^2} = -k^2 \psi \quad (13\text{–}51)$$

is

$$\psi = \Pi(r, \theta, \varphi) e^{-i\omega t} = \frac{1}{\sqrt{kr}} Z_{l+\frac{1}{2}}(kr)\, Y_l^m(\theta, \varphi) e^{-i\omega t}. \quad (13\text{–}52)$$

Here $Y_l^m(\theta, \varphi)$ is the spherical harmonic of order l, m, with l a positive integer and $m = l, l - 1, \ldots -l$,

$$Y_l^m(\theta, \varphi)$$

$$= \frac{(-1)^{l+m}}{2^l l!} \left[(2l + 1) \frac{(l - m)!}{(l + m)!} \right]^{1/2} (\sin\theta)^m \frac{d^{l+m}}{(d\cos\theta)^{l+m}} [(\sin\theta)^{2l}] e^{im\varphi}$$

$$= \left[(2l + 1) \frac{(l - m)!}{(l + m)!} \right]^{1/2} P_l^m(\cos\theta) e^{im\varphi}. \quad (13\text{–}53)$$

Two properties of these functions should be noted explicitly, namely,

$$Y_l^{-m}(\theta, \varphi) = (-1)^m Y_l^{m*}(\theta, \varphi),$$

$$\int Y_l^m(\theta, \varphi) Y_{l'}^{m'*}(\theta, \varphi)\, d\Omega = 4\pi\, \delta_{mm'}\, \delta_{ll'}. \quad (13\text{–}53')$$

The Bessel function $Z_{l+1/2}(kr)$ must be chosen as that appropriate to the boundary conditions of the given problem. It is convenient to use "spherical Bessel functions" following a terminology according to which

$$z_l(kr) = \sqrt{\frac{\pi}{2kr}} Z_{l+1/2}(kr). \tag{13–54}$$

The spherical Bessel functions $z_l(kr)$ may be specified as $j_l(kr)$, $n_l(kr)$, or $h_l(kr)$, just as Z_l is used to represent J_l, N_l, or H_l. (Note that again we have dropped the superscript usually found on the Hankel function; H_l and h_l are identical with the usual $H_l^{(1)}$ and $h_l^{(1)}$.) For example

$$j_l(kr) = \sqrt{\frac{\pi}{2kr}} J_{l+1/2}(kr),$$

$$h_l(kr) = \sqrt{\frac{\pi}{2kr}} H_{l+1/2}(kr). \tag{13–54'}$$

Thus we may write the space dependent factor of Eq. (13–52), corresponding to a particular set of separation parameters l and m, as

$$\Pi_l^m(r, \theta, \varphi) = z_l(kr) Y_l^m(\theta, \varphi). \tag{13–52'}$$

But it is not immediately apparent that the scalar solution is at all useful in determining a vector field. In the coordinate systems utilized thus far we have been able to demonstrate simple yet physically interesting results while avoiding the intrinsic difficulty of the vector wave equation very similar to that mentioned in Chapter 8 in connection with the vector Laplace equation: only in Cartesian coordinates, where the coordinate vectors are in the same direction for all points of space, is Laplace's or D'Alembert's equation simply equivalent to three scalar equations for the three components of the vector. Actually, the homogeneous wave equation is in general solvable in terms of two independent scalar functions.* The classification of fields in problems involving cylindrical boundaries as TE and TM modes is a special case of this general theorem. (In Section 13–4 the two scalar functions were taken as E_z and H_z.) In spherical coordinates the two corresponding "modes" are those in which the electric and magnetic fields, respectively, are transverse to the radius vector from the center of coordinates.

Consider a function $\Pi_l^m(r, \theta, \varphi)$ which satisfies the scalar equation (13–51). The gradient of Π_l^m is a solution of the corresponding vector equa-

---

* See A. Nisbet, *Proc. Roy. Soc.* **231 A**, pp. 250–263 (1955) for a general proof of this theorem and its extension to points within a prescribed source distribution.

tion, and so is $\mathbf{r} \times \nabla \Pi_l^m$, if the equation is invariant under an infinitesimal rotation of the coordinates, as indicated in a problem. If we note that

$$\mathbf{r} \times \nabla \Pi = -\nabla \times (\mathbf{r}\Pi)$$

we see that the divergence of this vector vanishes; thus it satisfies a necessary condition on $\mathbf{E}$ or $\mathbf{H}$ in a uniform source-free medium. A vector field having no radial component could therefore be expressed as a sum over l and m of such solutions, with coefficients appropriate to the detailed nature of the problem. If we let

$$\mathbf{H}_1 = \mathbf{r} \times \nabla \Pi_1, \qquad (13\text{--}55)$$

where Π_1 is a solution (or a sum of solutions) of Eq. (13–51), we may immediately determine the corresponding electric field from the relation $-i\omega\mathbf{D}_1 = \nabla \times \mathbf{H}_1$. But, even though the scalar functions Π_l^m form a complete set, $\mathbf{H}_1$ cannot represent a general magnetic field, since it has no component along $\mathbf{r}$. A linearly independent solution of the vector equation can be obtained, however, by interchanging the roles of $\mathbf{H}$ and $\mathbf{E}$, i.e.,

$$\mathbf{E}_2 = \mathbf{r} \times \nabla \Pi_2 \qquad (13\text{--}56)$$

where Π_2 is also a solution of Eq. (13–51); the corresponding $\mathbf{H}$ is then obtained from the relation $i\omega\mathbf{B}_2 = \nabla \times \mathbf{E}_2$. In general, an electromagnetic field can be expressed as the sum of these two kinds of partial fields, and it is evident that the boundary conditions at the surface of a sphere whose center is the origin of coordinates may be readily imposed.

As we shall note again in Chapter 14, the two scalar functions Π_1 and Π_2 are known as Debye potentials. These functions are, however, essentially equivalent to the radial components of $\mathbf{E}_1$ and $\mathbf{H}_2$ respectively, as may be seen from examination of the space dependence of the two varieties of fields. It can be readily shown from the wave equation that if

$$\mathbf{H} = \mathbf{r} \times \nabla \Pi_l^m$$

then

$$i\omega\epsilon_0 E_r = [\nabla \times (\nabla \times \mathbf{r}\, \Pi_l^m)]_r = \frac{l(l+1)}{r}\, \Pi_l^m, \qquad (13\text{--}57)$$

and thus all fields of the kind we have designated by the subscript 1 could be expressed in terms of E_r. Similarly, fields of the kind we have designated by subscript 2 may be expressed in terms of H_r. The fields are not often written in this way, but it is convenient to characterize them by subscripts E and M rather than by 1 and 2; $\mathbf{E}_E$, $\mathbf{H}_E$ is the set of fields for which the electric field, but not the magnetic field, has a radial component, whereas for $\mathbf{E}_M$, $\mathbf{H}_M$ only the magnetic field has a component along $\mathbf{r}$. For each value of the separation parameters l and m we may define a "unit" field of each type:

$$\mathbf{E}_E(l, m) = \boldsymbol{\nabla} \times \boldsymbol{\nabla} \times \mathbf{r}\,\Pi_l^m,$$

$$\mathbf{H}_E(l, m) = -ik\,\sqrt{\frac{\epsilon_0}{\mu_0}}\,\boldsymbol{\nabla} \times \mathbf{r}\,\Pi_l^m,$$

(13–58)

$$\mathbf{E}_M(l, m) = ik\boldsymbol{\nabla} \times \mathbf{r}\,\Pi_l^m,$$

$$\mathbf{H}_M(l, m) = \sqrt{\frac{\epsilon_0}{\mu_0}}\,\boldsymbol{\nabla} \times \boldsymbol{\nabla} \times \mathbf{r}\,\Pi_l^m,$$

(13–59)

where the constants are introduced so that both types of fields, $\mathbf{E}_E$ and $\mathbf{E}_M$, are of the same dimensions. Explicitly, the components are, for type E,

$$E_r = \frac{l(l + 1)}{r}\,z_l(kr)\,Y_l^m(\theta, \varphi), \qquad H_r = 0,$$

$$E_\theta = \frac{1}{r}\,\frac{d}{dr}\,[rz_l(kr)]\,\frac{\partial}{\partial\theta}\,[Y_l^m(\theta, \varphi)], \qquad H_\theta = \sqrt{\frac{\epsilon_0}{\mu_0}}\,\frac{km}{\sin\theta}\,z_l(kr)\,Y_l^m(\theta, \varphi),$$

$$E_\varphi = \frac{im}{r\sin\theta}\,\frac{d}{dr}\,[rz_l(kr)]\,Y_l^m(\theta, \varphi), \qquad H_\varphi = ik\,\sqrt{\frac{\epsilon_0}{\mu_0}}\,z_l(kr)\,\frac{\partial}{\partial\theta}\,[Y_l^m(\theta, \varphi)],$$

(13–60)

and, independently, for type M,

$$E_r = 0, \qquad H_r = \sqrt{\frac{\epsilon_0}{\mu_0}}\,\frac{l(l + 1)}{r}\,z_l(kr)\,Y_l^m(\theta, \varphi),$$

$$E_\theta = \frac{-km}{\sin\theta}\,z_l(kr)\,Y_l^m(\theta, \varphi), \qquad H_\theta = \sqrt{\frac{\epsilon_0}{\mu_0}}\,\frac{1}{r}\,\frac{d}{dr}\,[rz_l(kr)]\,\frac{\partial}{\partial\theta}\,[Y_l^m(\theta, \varphi)],$$

$$E_\varphi = -ikz_l(kr)\,\frac{\partial}{\partial\theta}\,[Y_l^m(\theta, \varphi)], \qquad H_\varphi = \sqrt{\frac{\epsilon_0}{\mu_0}}\,\frac{im}{r\sin\theta}\,\frac{d}{dr}\,[rz_l(kr)]\,Y_l^m(\theta, \varphi).$$

(13–61)

It is seen that these fields vanish identically for $l = 0$. According to their spatial symmetry they are classified as multipole fields of order 2^l; those for which the radial component of **E** does not vanish, Eqs. (13–58), (13–60), are electric multipoles in character, while Eqs. (13–59), (13–61) represent magnetic multipole fields. In Chapter 14 we shall examine these fields in relation to their sources.

The resonant fields of a spherical cavity can be written down immediately in terms of the particular choice of Bessel functions which satisfy the appropriate boundary conditions. For a cavity bounded by a perfectly conducting sphere, $z_l(kr)$ must be restricted to $j_l(kr)$ to ensure regularity at $r = 0$. The requirement that the tangential component of **E** vanish at the cavity wall puts different restrictions on k for the two modes. From Eqs. (13–60) it is seen that the equation to be satisfied by electric multipole fields is

$$\left[\frac{d}{dr}\left(rz_l(kr)\right)\right]_{r=r_0} = 0, \qquad (13\text{–}62)$$

where r_0 is the radius of the cavity. The eigenvalues of k for magnetic multipoles, described by Eqs. (13–61), are those for which

$$j_l(k_{l,n}r_0) = 0, \qquad (13\text{–}63)$$

where $k_{l,n}r_0$ is the nth zero of j_l.

13–9 Scattering by a sphere. The problem of scattering a plane electromagnetic wave by a sphere can be solved in terms of the two independent fields characterized by the subscripts E and M. To facilitate identification of the polar coordinate functions necessary to represent rectangular components of the incoming wave, it is convenient to specify the dependence on the azimuthal angle φ as $\sin m\varphi$ or $\cos m\varphi$. Since the fields themselves are identified by their radial components, they will be called even or odd according to whether $\cos m\varphi$ or $\sin m\varphi$ occurs in the radial component. Thus, for example, $\mathbf{E}_E$ and $\mathbf{H}_E$ are called even if*

$$E_r(l, m) = \frac{l(l + 1)}{r} z_l(kr)P_l^m (\cos \theta) \cos m\varphi, \qquad (13\text{–}64)$$

---

* It will be seen that the fields $\mathbf{E}(l, m;$ even or odd$)$, $\mathbf{H}(l, m;$ even or odd$)$ differ from those of Section 13–8 by a normalization factor,

$$\left[(2l + 1) \frac{(l - m)!}{(l + m)!}\right]^{1/2}.$$

This factor, very convenient in the computation of energy associated with each spherical mode, is somewhat awkward for the identification of a plane wave in terms of spherical waves. If inserted it would, of course, cancel out in the expression for the scattering cross section.

and $\mathbf{E}_M$, $\mathbf{H}_M$ are even if

$$H_r(l, m) = \sqrt{\frac{\epsilon_0}{\mu_0}} \frac{l(l+1)}{r} z_l(kr) P_l^m(\cos\theta) \cos m\varphi. \qquad (13\text{--}65)$$

Let us consider a plane-polarized plane wave advancing in the direction of the positive z-axis, with the electric vector confined to the xz-plane. We must express the incoming fields in terms of the fields given by Eqs. (13–58) and (13–59). The basic formula for the expansion of a plane wave propagated along the polar axis in spherical coordinates is the scalar series:

$$e^{ikr\cos\theta} = \sum_{l=1}^{\infty} i^l(2l+1)j_l(kr)P_l(\cos\theta). \qquad (13\text{--}66)$$

Now a vector in the x direction has spherical coordinate components given by

$$\mathbf{E}_x = E_0 e^{ikr\cos\theta}\hat{\mathbf{x}} = E_0 e^{ikr\cos\theta}(\sin\theta\cos\varphi\,\hat{\mathbf{r}} + \cos\theta\cos\varphi\,\hat{\boldsymbol{\theta}} - \sin\varphi\,\hat{\boldsymbol{\phi}}). \qquad (13\text{--}67)$$

But

$$\sin\theta\,e^{ikr\cos\theta} = -\frac{1}{ikr}\frac{\partial}{\partial\theta}(e^{ikr\cos\theta}), \qquad (13\text{--}68)$$

and

$$\frac{dP_l(\cos\theta)}{d\theta} = -P_l^1(\cos\theta). \qquad (13\text{--}69)$$

Equations (13–68) and (13–69) make it possible to identify at once the coefficient of $\mathbf{E}_E(l, m)$ in a series expansion of $\mathbf{E}_x$ by examination of the radial component, since $\mathbf{E}_M$ has no component along r. We note that the parameter m is unity throughout, and that the dependence on φ makes it necessary to choose $\mathbf{E}_E$ (even). To find the coefficient of $\mathbf{E}_M$ we may examine the expansion of the accompanying magnetic field:

$$H_y\hat{\mathbf{y}} = H_y(\sin\theta\sin\varphi\,\hat{\mathbf{r}} + \cos\theta\sin\varphi\,\hat{\boldsymbol{\theta}} + \cos\varphi\,\hat{\boldsymbol{\phi}}). \qquad (13\text{--}70)$$

Again by means of Eqs. (13–68) and (13–69), we can identify the radial component of Eq. (13–70) with a series in $\mathbf{H}_M(l, 1)$, noting that the choice of $\mathbf{H}_M$ (odd) is necessary. Since

$$\mathbf{E}_M = \frac{i}{k}\sqrt{\frac{\mu_0}{\epsilon_0}}\,\boldsymbol{\nabla}\times\mathbf{H}_M,$$

the identification is now complete, and the expression for a plane wave in terms of the (modified) "unit" fields of the previous section is

$$\mathbf{E}_x = E_0 \sum_{l=1}^{\infty} \frac{i^{l-1}}{k} \frac{2l+1}{l(l+1)} \left[\mathbf{E}_M(l, 1; \text{odd}) + \mathbf{E}_E(l, 1; \text{even}) \right], \qquad (13\text{-}71)$$

$$\mathbf{H}_y = -E_0 \sum_{l=1}^{\infty} \frac{i^{l+1}}{k} \frac{2l+1}{l(l+1)} \left[\mathbf{H}_M(l, 1; \text{odd}) + \mathbf{H}_E(l, 1; \text{even}) \right]. \qquad (13\text{-}72)$$

As already indicated in Eq. (13–66), the Bessel function in these expansions must be that which is regular at the origin, namely, $j_l(kr)$. The limitation of m to $m = 1$, evident in the limited dependence of the fields on the angle φ, means that the series are single sums over l.

If this plane wave falls on a sphere of radius a, whose center is at the origin of coordinates, there will be a similar series representing the scattered wave, with coefficients determined by the boundary conditions at the surface of the sphere. The angular dependence of $\mathbf{E}_E$, $\mathbf{E}_M$ in the outgoing wave must be the same for every value of l as in Eq. (13–71) if the boundary conditions are to be satisfied for all values of the angular variables, but $z_l(kr)$ must be chosen so as to behave as e^{ikr}/r for large r. This function is

$$h_l = j_l(kr) + in_l(kr) \simeq (-i)^{l+1} \frac{e^{ikr}}{kr}. \qquad (13\text{-}73)$$

Formally we may write the scattered electric field as

$$\mathbf{E}_s = E_0 \sum_{l=1}^{\infty} \frac{i^{l-1}}{k} \frac{2l+1}{l(l+1)} \left[a_l \mathbf{E}_M(l, 1; \text{odd}) + b_l \mathbf{E}_E(l, 1; \text{even}) \right], \qquad (13\text{-}74)$$

together with the corresponding expression for $\mathbf{H}_s$. For simplicity we shall consider only a *perfectly conducting* sphere, for which the sum of the tangential components of the incident and scattered electric fields must vanish at every point on the surface of the sphere for each term of the series. Comparison of Eqs. (13–74) and (13–71), together with the dependence of the "unit" fields on r, leads to evaluation of the a_l's and b_l's:

$$a_l = -\frac{j_l(ka)}{h_l(ka)},$$

$$b_l = -\left\{ \frac{(d/dr)[rj_l(kr)]}{(d/dr)[rh_l(kr)]} \right\}_{r=a}. \qquad (13\text{-}75)$$

Before examining these coefficients as functions of the size of the sphere and the wavelength of the radiation, let us look at the behavior of the outgoing fields. From the asymptotic formula for $h_l(kr)$, Eq. (13–73), and the components of $\mathbf{E}_E$, $\mathbf{H}_E$, $\mathbf{E}_M$, $\mathbf{H}_M$, it is evident that the radially directed fields fall off as $1/r^2$, whereas the transverse fields behave as $1/r$ for large r. The Poynting vector of the scattered wave is thus radial, as would be expected, and the total scattered radiation can be determined by integrating over a sphere of large radius. The evaluation of the time average of the scattered radiation,

$$\frac{dU_s}{dt} = \frac{1}{2}\operatorname{Re}\int |\mathbf{E}_s \times \mathbf{H}_s^*| r^2 \sin\theta\, d\theta\, d\varphi = \frac{1}{2}\operatorname{Re}\int (E_\theta H_\varphi^* - E_\varphi H_\theta^*) r^2\, d\Omega,$$

$$(13–76)$$

may be accomplished by means of the integral formulas for the associated Legendre functions:

$$\int_0^\pi \left(\frac{dP_l^1}{d\theta}\frac{dP_{l'}^1}{d\theta} + \frac{1}{\sin^2\theta}P_l^1 P_{l'}^1\right)\sin\theta\, d\theta = \frac{2[l(l+1)]^2}{2l+1}\delta_{l,l'},$$

$$\int_0^\pi \left(\frac{P_l^1}{\sin\theta}\frac{dP_{l'}^1}{d\theta} + \frac{P_{l'}^1}{\sin\theta}\frac{dP_l^1}{d\theta}\right)\sin\theta\, d\theta = 0.$$

The result of performing the integration indicated in Eq. (13–76) is

$$\frac{dU_s}{dt} = \frac{\pi E_0^2}{k_2}\sqrt{\frac{\epsilon_0}{\mu_0}}\sum_{l=1}^\infty (2l+1)(|a_l|^2 + |b_l|^2).\qquad(13–77)$$

Since the energy of the incident wave per unit area per unit time is

$$\frac{dU}{dt} = \frac{1}{2}E_0^2\sqrt{\frac{\epsilon_0}{\mu_0}},$$

the cross section for scattering can be written

$$\sigma = \frac{2\pi}{k^2}\sum(2l+1)(|a_l|^2 + |b_l|^2).\qquad(13–78)$$

While the coefficients a_l and b_l are tedious to compute, some idea of their general behavior may be gained by writing them in the form

$$a_l = ie^{i\gamma_l}\sin\gamma_l,\qquad b_l = ie^{i\gamma_l'}\sin\gamma_l',\qquad(13–79)$$

where

$$\tan\gamma_l = \frac{j_l(ka)}{n_l(ka)},\qquad \tan\gamma_l' = \left[\frac{d[rj_l(kr)]}{dr}\bigg/\frac{d[rn_l(kr)]}{dr}\right]_{r=a}.$$

In terms of these parameters, the scattering cross section is

$$\sigma = \frac{2\pi}{k^2} \sum_l (2l + 1)(\sin^2 \gamma_l + \sin^2 \gamma_l'), \qquad (13\text{-}80)$$

from which it is clear that the contributions of the two kinds of waves whose electric fields are represented by $\mathbf{E}_E$ and $\mathbf{E}_M$ oscillate as functions of ka. The total cross section reduces to a very simple expression only when $ka \ll 1$. In that case, the first term in the series expansions for the Bessel functions in Eq. (13-75) gives

$$|a| = \frac{(ka)^3}{3}, \qquad |b| = \frac{2(ka)^3}{3},$$

so that

$$\sigma = \frac{10\pi}{3} k^4 a^6. \qquad (13\text{-}81)$$

Thus, in the limit of low frequencies, the scattering cross section varies as the inverse fourth power of the wavelength, and is, as in the corresponding cylindrical scattering case, smaller than the geometrical cross section.

The scattering of a plane wave by a sphere of general electric and magnetic properties embedded in a dielectric follows the same pattern as the solution outlined here. The fields inside the sphere must also be considered, but the condition that the tangential components of both $\mathbf{E}$ and $\mathbf{H}$ be continuous across the boundary is sufficient to determine all the coefficients. Again in the low-frequency limit it is found that the cross section varies as the inverse fourth power of the wavelength.

Suggested References

J. A. Stratton, *Electromagnetic Theory*. The literature on waves in regions with metallic boundaries has grown very rapidly, but the principles and many of the methods involved in applications are treated in Chapter IX of this work. Chapters VI and VII deal with cylindrical and spherical waves.

P. M. Morse and H. Feshbach, *Methods of Theoretical Physics*. This is a comprehensive and elegant presentation of mathematical methods, including those for solving boundary value problems.

E. U. Condon, "Principles of Microwave Radio," *Rev. Modern Phys.* 14 (1942). An early review article, forming an excellent introduction to the physics of cavity and guide modes.

J. C. Slater, *Microwave Electronics*. An expanded and more recent (1950) version of a *Rev. Modern Phys.* (1946) article by the same author, stressing physical principles.

P. M. MORSE, *Vibration and Sound,* and H. LAMB, *Hydrodynamics* are interesting for comparing the cylindrical scattering of longitudinal waves with the results of Section 13–7.

M. BORN, *Optik.* Chapter VI contains a full account of spherical scattering, including the angular distribution for various ratios of wavelength to diameter.

A. SOMMERFELD, *Electrodynamics.* The boundary value problem emphasized here is that of waves on wires, and thus the work supplements both the present chapter and the references above.

E. L. INCE, *Ordinary Differential Equations.* A discussion of the eigenvalue problem will be found in Chapter 10 of this work.

E. JAHNKE AND F. EMDE, *Tables of Functions.* For numerical applications this is the best and most convenient short table of functions.

G. N. WATSON, *Theory of Bessel Functions.* The properties of Bessel functions will be found in many references, but for the proofs for series expansions and similar matters Watson is excellent.

EXERCISES

1. Show that at a conducting boundary the ratio of the normal component of the magnetic field to its tangential component is of the order of the skin depth to the wavelength of the oscillating fields.

2. A rectangular cavity of dimensions a, a, L, and walls of large conductivity σ is excited in the mode

$$E_z = E_0 \sin (\pi x/a) \sin (\pi y/a) e^{-i\omega t},$$

$$H_z = E_x = E_y = 0.$$

What is ω? Calculate the forces exerted on all the faces. What is Q for this mode?

3. Show that the absolute cutoff for TE modes in a wave guide of rectangular cross section is lower than for TM modes. Find the attenuation coefficient for the lowest TE mode.

4. Find the lowest frequency for a cylindrical cavity of circular cross section in terms of its radius and height. How would you estimate Q for this mode?

5. Find the fields and cross section for scattering from a conducting cylinder in the limit $ka \ll 1$ if the incoming wave is polarized so that E is along z, parallel to the axis of the cylinder.

6. A solution of the wave equation for a given orientation of the coordinate axes must be a solution for any other orientation of the axes. Consider the difference between $\psi(r, \theta, \varphi)$ and $\psi(r, \theta', \varphi')$ for an infinitesimal rotation to show that $\mathbf{r} \times \nabla \psi$ is a solution if ψ is a solution.

7. Calculate the total force exerted on the reflecting sphere of Section 13–9 under the conditions that $ka \ll 1$. Estimate the change in your result if the sphere were of copper instead of being a perfect conductor.

8. Consider a cubical cavity with perfectly conducting walls. Prove that for waves short in comparison with the dimension l of the cube the number of modes

within a frequency range $d\omega$ is given by

$$\frac{l^3 \omega^2 \, d\omega}{\pi^2 c^3},$$

or for $d\nu$ by

$$\frac{l^3 8\pi\nu^2 \, d\nu}{c^3}.$$

Does the number of modes per unit volume depend on the shape of the cavity? This formula is of great historical importance in the theory of "blackbody" radiation.

9. Prove that TEM waves $(E_z = 0 = H_z)$ may be propagated along a coaxial line with a speed equal to the speed of free electromagnetic waves in the insulating medium. What is the pattern of the transverse field?

CHAPTER 14

THE INHOMOGENEOUS WAVE EQUATION

14-1 The wave equation for the potentials. The relation of radiation fields to their sources is most readily found in terms of potential functions. In view of the law of induction, curl $\mathbf{E}$ is not zero in a changing magnetic field, so that it is no longer possible to derive the electric field solely from a scalar potential. The divergence of $\mathbf{B}$ vanishes under all conditions, however, and thus the magnetic field is still derivable from a vector potential:

$$\mathbf{B} = \boldsymbol{\nabla} \times \mathbf{A}. \tag{7-40}$$

If Eq. (7-40) is assumed valid, then the electric field can be written as the sum of the gradient of a scalar potential and a supplementary non-conservative contribution from the rate of change of the vector potential. That is, $\mathbf{E}$ may be derived from the scalar and vector potentials by

$$\mathbf{E} = -\boldsymbol{\nabla}\phi - \frac{\partial \mathbf{A}}{\partial t}, \tag{14-1}$$

which makes $\mathbf{E}$ conform to the relation of the third of Eqs. (9-7). As before, the divergence of $\mathbf{A}$ remains undefined, or at least remains undefined within an additive arbitrary function of position. Let us define the divergence of $\mathbf{A}$ by what is called the Lorentz condition,

$$\boldsymbol{\nabla} \cdot \mathbf{A} + \mu\epsilon \frac{\partial \phi}{\partial t} + \mu\sigma\phi = 0, \tag{14-2}$$

which in free space becomes

$$\boldsymbol{\nabla} \cdot \mathbf{A} + \frac{1}{c^2} \frac{\partial \phi}{\partial t} = 0. \tag{14-3}$$

The Lorentz condition is not quite the arbitrary subsidiary relation it appears at first sight. We shall see at once that it has the advantage of introducing complete symmetry between the scalar and vector potentials, i.e., it makes both potentials satisfy the same wave equation as that obeyed by the fields. Later we shall find that the Lorentz condition also assures a relativistic covariant relation between the scalar and vector potentials. To determine the equations satisfied by the potentials we need only introduce the defining equations (7-40) and (14-1) and the Lorentz condition into the first and fourth of Maxwell's equations, Eqs. (9-7).

We obtain the symmetrical set of equations:

$$\nabla^2 \mathbf{A} - \mu\epsilon \frac{\partial^2 \mathbf{A}}{\partial t^2} - \mu\sigma \frac{\partial \mathbf{A}}{\partial t} = -\mu\mathbf{j}', \qquad (14\text{-}4)$$

$$\nabla^2 \phi - \mu\epsilon \frac{\partial^2 \phi}{\partial t^2} - \mu\sigma \frac{\partial \phi}{\partial t} = -\rho/\epsilon. \qquad (14\text{-}5)$$

Here $\mathbf{j}'$ represents a current given by $\mathbf{j}' = \sigma\mathbf{E}'$, that part of the current density that is produced by the external electromotive forces. It does not contain any part of the current induced by the electric fields in conducting media and represented by the last term on the left side of Eq. (14-4).

Let us introduce the symbolic operator $\square$, known as the D'Alembertian, and defined by

$$\square = \nabla^2 - \mu\epsilon \frac{\partial^2}{\partial t^2}. \qquad (14\text{-}6)$$

In free space, Eqs. (14-4) and (14-5) then become simply

$$\square \mathbf{A} = -\mu_0 \mathbf{j}', \qquad (14\text{-}7)$$

$$\square \phi = -\rho/\epsilon_0, \qquad (14\text{-}8)$$

where $\mathbf{j}'$ and ρ are the sources of the field and are produced by external agents. These equations are known as the inhomogeneous wave equations. Their particular solutions are expressible in terms of integrals over the charge and current distributions, while their complementary solutions, namely, those of the homogeneous equations, are obviously just the wave solutions.

By inspection of Eqs. (7-40) and (14-1) it can be seen that the resultant electric and magnetic fields are unchanged by transformations of the type

$$\mathbf{A}' = \mathbf{A} - \nabla\psi, \qquad (14\text{-}9)$$

$$\phi' = \phi + \frac{\partial \psi}{\partial t}, \qquad (14\text{-}10)$$

where ψ is a function of the coordinates and the time. This means that if any physical law involving electromagnetic interaction is to be expressed in terms of the general electrodynamic potentials $\mathbf{A}$ and ϕ then such a physical law must be unaffected by a transformation of the type given by Eqs. (14-9) and (14-10). These transformations are usually known as gauge transformations, and a physical law that is invariant under such a transformation is said to be gauge invariant. The property of gauge invariance ensures that the physical law will not lead to consequences that cannot be expressed in the field formulation of the interaction of charges and currents.

14–2 Solution by Fourier analysis. Let us begin the investigation of the particular solutions of the inhomogeneous equations by reviewing the solution of the analogous static problem. In the static case, Eq. (14–8) reduces to Poisson's equation,

$$\nabla^2\phi = -\rho/\epsilon_0, \tag{1–35}$$

of which the particular solution is

$$\phi(x_\alpha) = \frac{1}{4\pi\epsilon_0} \int \frac{\rho(x'_\alpha)}{r(x_\alpha, x'_\alpha)} \, dv'.$$

What we seek now is the modification of the solution of Eq. (1–35) that is produced by the presence of the time-dependent term in Eq. (14–7).

The equations for both ϕ and $\mathbf{A}$ have the general form

$$\Box\psi(x_\alpha, t) = -g(x_\alpha, t). \tag{14–11}$$

Let us assume that the source function $g(x_\alpha, t)$ can be analyzed by the Fourier integral

$$g(x_\alpha, t) = \int_{-\infty}^{\infty} g_\omega(x_\alpha)e^{-i\omega t} \, d\omega, \tag{14–12}$$

which has the Fourier inversion

$$g_\omega(x_\alpha) = \frac{1}{2\pi} \int_{-\infty}^{\infty} g(x_\alpha, t)e^{i\omega t} \, dt. \tag{14–13}$$

Similarly, we may analyze the general potential $\psi(x_\alpha, t)$ into Fourier components,

$$\psi(x_\alpha, t) = \int_{-\infty}^{\infty} \psi_\omega(x_\alpha)e^{-i\omega t} \, d\omega, \tag{14–14}$$

with a corresponding inverse relation

$$\psi_\omega(x_\alpha) = \frac{1}{2\pi} \int_{-\infty}^{\infty} \psi(x_\alpha, t)e^{i\omega t} \, dt. \tag{14–15}$$

By substitution of Eqs. (14–12) and (14–14) into Eq. (14–11), we see that the Fourier component $\psi_\omega(x_\alpha)$ obeys the differential relation

$$\nabla^2\psi_\omega + \frac{\omega^2}{c^2}\psi_\omega = -g_\omega, \tag{14–16}$$

which is similar to Poisson's equation.

We may synthesize the solution of Eq. (14–16) by the superposition of unit point solutions corresponding to a source at the point x_α' given by $g_\omega(x_\alpha) = \delta(x_\alpha - x_\alpha')$, where $\delta(x_\alpha - x_\alpha')$ is the Dirac δ-function. Each unit source potential satisfies the equation

$$\nabla^2 G(x_\alpha, x_\alpha') + \frac{\omega^2}{c^2} G(x_\alpha, x_\alpha') = -\delta(x_\alpha - x_\alpha'), \qquad (14\text{--}17)$$

with G a function of both x_α and x_α'. The partial solution corresponding to the frequency ω of the total source is then given by the superposition

$$\psi_\omega(x_\alpha) = \int g_\omega(x_\alpha') G(x_\alpha, x_\alpha')\, dv'. \qquad (14\text{--}18)$$

To find G, we note that the solution of Eq. (14–17) will be spherically symmetric in r, the distance between points x_α and x_α', and hence at all points other than $r = 0$ it will be identical with the solution of the equation

$$\frac{1}{r} \frac{d^2(rG)}{dr^2} + k^2 G = 0,$$

where $k = \omega/c$ as usual. This equation can be integrated immediately:

$$G = \frac{A}{r} e^{\pm ikr}.$$

To evaluate the constant A, we consider the volume integral of Eq. (14–17) over the neighborhood of the singular point $r = 0$. Here G behaves as A/r, and the use of Eq. (1–11) in the integration leads to $-4\pi A = -1$. Therefore

$$G = \frac{1}{4\pi r} e^{\pm ikr} \qquad (14\text{--}19)$$

is the solution of Eq. (14–17), and substitution into Eq. (14–18) gives

$$\psi_\omega(x_\alpha) = \frac{1}{4\pi} \int \frac{g_\omega(x_\alpha')}{r(x_\alpha, x_\alpha')} e^{\pm ikr(x_\alpha, x_\alpha')}\, dv'. \qquad (14\text{--}20)$$

The effect of the second term in Eq. (14–16) is then simply to introduce a complex exponential factor in the Coulomb potential expression of Chapter 1.

In terms of Eq. (14–20) the time-dependent potential satisfying Eq. (14–11) is given by

$$\psi(x_\alpha, t) = \int \psi_\omega(x_\alpha) e^{-i\omega t}\, d\omega = \frac{1}{4\pi} \iint \frac{g_\omega(x_\alpha') e^{-i(\omega t \pm kr)}}{r(x_\alpha, x_\alpha')}\, d\omega\, dv'. \qquad (14\text{--}21)$$

We may introduce a new time

$$t'(x_\alpha, x'_\alpha) = t \pm r/c = t \pm kr/\omega, \qquad (14\text{--}22)$$

which corresponds to shifting the origin of time by an amount equal to the time it takes a light signal to be propagated from point x'_α to point x_α. If we now evaluate the Fourier transformation by using Eq. (14–12), we obtain

$$\psi(x_\alpha, t) = \frac{1}{4\pi} \int \frac{g(x'_\alpha, t \pm r/c)}{r(x_\alpha, x'_\alpha)} \, dv'. \qquad (14\text{--}23)$$

Note that the point of observation is contained explicitly in the denominator of the integrand and also, by means of Eq. (14–22), in the time at which the time-varying currents and charges are introduced into the integration. Mathematically, both plus and minus signs in Eq. (14–23) are valid, but only the minus sign appears to have physical significance. We are concerned with the effect at x_α of sources at x'_α, and the minus sign corresponds to the cause preceding the effect. Equation (14–23) with the minus sign is known as the retarded potential solution of the inhomogeneous wave equation. The solution with the plus sign is known as the advanced potential. Its use appears to violate elementary notions of causality: it implies that effect precedes cause in time. For this reason, we will ignore this solution at this time since it clearly violates experience; the field does not appear to precede the charges and currents which cause it. Nevertheless, the advanced potential solution is a mathematically valid solution of the field equations which are basically time-symmetric. We will reopen this question critically in Chapter 21, but will restrict our discussion for the time being to the use of the retarded solutions.

A retarded potential might be visualized as follows. Consider an observer located at the point x_α in space as seen in Fig. 14–1, and let a sphere whose center is at x_α contract with a radial velocity c such that it has just converged on the point at the time of observation t. The time at which this information-collecting sphere passes the source of the electric field at

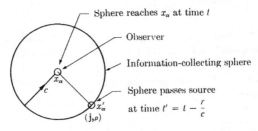

FIG. 14–1. Sphere collecting information for the retarded potential at x_α.

point x_α' is then the time at which the source produced the effect which is felt at x_α at time t.

If we denote by the rectangular bracket symbol [ ] that the variables contained within the bracket are to be evaluated at the retarded time t', then the integrals of the inhomogeneous wave equations (14–7) and (14–8) corresponding to a current distribution $\mathbf{j}(x_\alpha')$ and a charge distribution $\rho(x_\alpha')$ may be written

$$\mathbf{A}(x_\alpha, t) = \frac{\mu_0}{4\pi} \int \frac{[\mathbf{j}(x_\alpha')]}{r(x_\alpha, x_\alpha')} \, dv', \qquad (14\text{--}24)$$

$$\phi(x_\alpha, t) = \frac{1}{4\pi\epsilon_0} \int \frac{[\rho(x_\alpha')]}{r(x_\alpha, x_\alpha')} \, dv'. \qquad (14\text{--}25)$$

14–3 The radiation fields. The solutions of the inhomogeneous wave equation obtained in Section 14–2 were based on the existence of the Fourier transforms of the potentials and currents. They do not in principle apply to fields produced by sources radiating at a single frequency, since a strictly monochromatic source would imply radiation over an infinite amount of time. We may apply the results to monochromatic sources, however, if we recognize that these results should be formally derived by a limiting process, starting with a pulse of finite duration.

For a pulse of radiation finite in time we may write, according to Eq. (14–20),

$$\mathbf{A}_\omega(x_\alpha) = \frac{\mu_0}{4\pi} \int \mathbf{j}_\omega \frac{e^{ikr}}{r} \, dv', \qquad (14\text{--}26)$$

$$\phi_\omega(x_\alpha) = \frac{1}{4\pi\epsilon_0} \int \rho_\omega \frac{e^{ikr}}{r} \, dv', \qquad (14\text{--}27)$$

with

$$\mathbf{A}(x_\alpha, t) = \int_{-\infty}^{\infty} \mathbf{A}_\omega(x_\alpha) e^{-i\omega t} \, d\omega, \qquad (14\text{--}28)$$

$$\phi(x_\alpha, t) = \int_{-\infty}^{\infty} \phi_\omega(x_\alpha) e^{-i\omega t} \, d\omega, \qquad (14\text{--}29)$$

and corresponding relations for $\mathbf{j}_\omega$ and ρ_ω. The fields are real; thus the Fourier components $\mathbf{A}_\omega$, ϕ_ω obey the condition that

$$\mathbf{A}_\omega = \mathbf{A}_{-\omega}^*, \qquad \phi_\omega = \phi_{-\omega}^*, \qquad (14\text{--}30)$$

where * denotes the complex conjugate.

If the source is effectively monochromatic, the potentials and sources may be written

$$\mathbf{A}(t) = \mathbf{A}_0 e^{-i\omega t},$$

$$\phi(t) = \phi_0 e^{-i\omega t},$$

$$\mathbf{j}(t) = \mathbf{j}_0 e^{-i\omega t},$$

$$\rho(t) = \rho_0 e^{-i\omega t},$$

$$(14\text{–}31)$$

where the real part of each expression is implied. In accord with the remarks above we shall assume that the retarded potential solutions (14–26) and (14–27) are valid for the monochromatic amplitudes (14–31) as well as for the Fourier components. This equivalence is justified only for linear relations between the fields and the sources; for quadratic functions such as the radiated energy the two cases must be treated separately.

Let us now compute the fields. We note that our assumption of the Lorentz condition, Eq. (14–3), yields a relation between the potentials,

$$\mathbf{\nabla} \cdot \mathbf{A}_\omega - \frac{ik}{c} \phi_\omega = 0. \tag{14–32}$$

The magnetic field component $\mathbf{B}_\omega(x_\alpha)$ is given by

$$\mathbf{B}_\omega(x_\alpha) = \frac{\mu_0}{4\pi} \mathbf{\nabla} \times \int \mathbf{j}_\omega \frac{e^{ikr}}{r} \, dv'$$

$$= - \frac{\mu_0}{4\pi} \int \mathbf{j}_\omega(x_\alpha') \times \mathbf{\nabla} \left(\frac{e^{ikr}}{r} \right) dv'$$

$$= \frac{\mu_0}{4\pi} \left\{ \int \frac{\mathbf{j}_\omega \times \mathbf{r}}{r^3} e^{ikr} \, dv' - ik \int \frac{\mathbf{j}_\omega \times \mathbf{r}}{r^2} e^{ikr} \, dv' \right\}. \tag{14–33}$$

The first term of Eq. (14–33) is just the retarded form of the magnetostatic solution given by Eq. (7–41); this is called the "induction field." The second term, which depends on the distance as r^{-1}, is called the "radiation field." The Fourier components $\mathbf{B}_\omega$ can be resynthesized to give

$$\mathbf{B}(x_\alpha, t) = \int \mathbf{B}_\omega(x_\alpha) e^{-i\omega t} \, d\omega$$

$$= \frac{\mu_0}{4\pi} \int \left(\frac{[\mathbf{j}] \times \mathbf{r}}{r^3} + \frac{1}{c} \frac{[\dot{\mathbf{j}}] \times \mathbf{r}}{r^2} \right) dv', \tag{14–34}$$

where

$$[\dot{\mathbf{j}}(x_\alpha')] = \int (-i\omega) \mathbf{j}_\omega(x_\alpha') e^{i(kr - \omega t)} \, d\omega \tag{14–35}$$

is the time rate of change of the current at the time the "information collecting sphere" of Fig. 14–1 passed the source point x'_α.

We can now recognize physically why the retarded potentials yield field solutions which vary as $1/r$. The operator ∇ operating on the potentials compares the potentials at neighboring field points *at the same time*. The two information collecting spheres converging simultaneously on the two field points will not, however, have passed a given source point at the same time; hence the differential operator at the field point involves time differentiation at the source, as shown formally above.

The calculation of the electric field is slightly more complicated. We have from Eq. (14–1), with $\mu_0\epsilon_0 = 1/c^2$,

$$4\pi\epsilon_0 E_\omega = -\int \rho_\omega \nabla\left(\frac{e^{ikr}}{r}\right) dv' - \frac{1}{c^2}\int\left(-i\omega j_\omega \frac{e^{ikr}}{r}\right) dv'$$

$$= \int \frac{\rho_\omega \mathbf{r}}{r^3} e^{ikr}\, dv' - ik\int\left(\frac{\rho_\omega \mathbf{r}}{r} - \frac{\mathbf{j}_\omega}{c}\right)\frac{e^{ikr}}{r}\, dv'. \qquad (14\text{–}36)$$

The first term is simply the retarded Coulomb field. The second integral, which we shall call $\mathbf{I}_\omega$, contains the radiation term. To reduce this integral further we note that $\mathbf{j}_\omega$ and ρ_ω are not independent, but are connected by the continuity equation which expresses the conservation of charge:

$$\nabla' \cdot \mathbf{j}_\omega - i\omega\rho_\omega = 0. \qquad (14\text{–}37)$$

Therefore

$$\mathbf{I}_\omega = -\frac{1}{c}\left\{\int\left[\frac{(\nabla' \cdot \mathbf{j}_\omega)\mathbf{r}}{r} - ik\mathbf{j}_\omega\right]\frac{e^{ikr}}{r}\, dv'\right\}, \qquad (14\text{–}38)$$

of which the α-component is

$$I_{\omega\alpha} = -\frac{1}{c}\left\{\int\left[\frac{\partial j_{\omega\beta}}{\partial x'_\beta}\frac{r_\alpha}{r} - ikj_{\omega\alpha}\right]\frac{e^{ikr}}{r}\, dv'\right\},$$

where $r_\alpha = x_\alpha - x'_\alpha$. (Note that β is only a summation index.) This expression may be rewritten as

$$I_{\omega\alpha} = \frac{1}{c}\int\left[j_{\omega\beta}\frac{\partial}{\partial x'_\beta}\left(\frac{r_\alpha e^{ikr}}{r^2}\right) + ikj_{\omega\alpha}\frac{e^{ikr}}{r}\right] dv'$$

$$- \frac{1}{c}\int\frac{\partial}{\partial x'_\beta}\left(j_{\omega\beta}\frac{r_\alpha e^{ikr}}{r^2}\right) dv'. \qquad (14\text{–}39)$$

The last term of Eq. (14–39) is equivalent to a surface integral by Gausss' theorem; if $\mathbf{j}_\omega$ is bounded in space this integral will vanish. Using the usual relations

$$\frac{\partial r_\alpha}{\partial x'_\beta} = -\delta_{\alpha\beta}, \qquad \frac{\partial}{\partial x'_\alpha} [f(r)] = -f'(r) \frac{r_\alpha}{r},$$

we obtain from the remaining expression

$$\mathbf{I}_\omega = \frac{1}{c} \int \left[-\mathbf{j}_\omega \frac{e^{ikr}}{r^2} + \frac{2}{r^4} (\mathbf{j}_\omega \cdot \mathbf{r}) \mathbf{r} e^{ikr} \right] dv'$$

$$+ \frac{ik}{c} \int \left[- \frac{(\mathbf{j}_\omega \cdot \mathbf{r})\mathbf{r}}{r^3} e^{ikr} + \mathbf{j}_\omega \frac{e^{ikr}}{r} \right] dv', \qquad (14\text{--}40)$$

which may be written as

$$\mathbf{I}_\omega = \frac{1}{c} \int [(\mathbf{j}_\omega \cdot \mathbf{r})\mathbf{r} - \mathbf{r} \times (\mathbf{j}_\omega \times \mathbf{r})] \frac{e^{ikr}}{r^4} dv'$$

$$+ \frac{ik}{c} \int [\mathbf{r} \times (\mathbf{j}_\omega \times \mathbf{r})] \frac{e^{ikr}}{r^3} dv'. \qquad (14\text{--}41)$$

Note that only the last term varies with distance as $1/r$.

If we synthesize the field from its Fourier components we obtain from Eqs. (14–35) and (14–41)

$$4\pi\epsilon_0 \mathbf{E} = \int \frac{[\rho]\mathbf{r}}{r^3} dv' + \frac{1}{c} \int \frac{\{([\mathbf{j}] \cdot \mathbf{r})\mathbf{r} - \mathbf{r} \times ([\mathbf{j}] \times \mathbf{r})\}}{r^4} dv'$$

$$+ \frac{1}{c^2} \int \frac{([\dot{\mathbf{j}}] \times \mathbf{r}) \times \mathbf{r}}{r^3} dv'. \qquad (14\text{--}42)$$

The radiation fields alone are given by

$$\mathbf{B}_{\text{rad}} = \frac{1}{4\pi\epsilon_0 c^3} \int \frac{[\dot{\mathbf{j}}] \times \mathbf{r}}{r^2} dv',$$

$$\mathbf{E}_{\text{rad}} = \frac{1}{4\pi\epsilon_0 c^2} \int \frac{([\dot{\mathbf{j}}] \times \mathbf{r}) \times \mathbf{r}}{r^3} dv'. \qquad (14\text{--}43)$$

At large distances from the source these fields are transverse to the direction of propagation, and the ratio of electric field strength $\mathbf{E}$ to magnetic intensity $\mathbf{H}$ is $(\mu_0/\epsilon_0)^{1/2}$, just as in the case of plane waves.

14–4 Radiated energy. Let us now calculate the radiated energy corresponding to the fields of Eqs. (14–43). Here it is necessary to distinguish between a radiation pulse with Fourier field components $\mathbf{E}_\omega$ and $\mathbf{B}_\omega$ and the field components representing a monochromatic source, $\mathbf{E}_0$ and $\mathbf{B}_0$.

We consider first a pulse of radiation. The total radiated energy crossing a unit area at right angles to the direction of propagation is the time integral of the Poynting vector **N**. This time integral is

$$\int_{-\infty}^{\infty} \mathbf{N}(t)\, dt = \int_{-\infty}^{\infty} (\mathbf{E} \times \mathbf{H})\, dt$$

$$= \int_{-\infty}^{\infty} \int_{-\infty}^{\infty} \int_{-\infty}^{\infty} (\mathbf{E}_\omega \times \mathbf{H}_{\omega'}) e^{-i(\omega+\omega')t}\, d\omega\, d\omega'\, dt. \quad (14\text{–}44)$$

Now it is true in general that if $f(t)$ and $g(t)$ are two real functions with Fourier transforms f_ω and g_ω, the integral of their product over the time is given by

$$I = \int_{-\infty}^{\infty} f(t)g(t)\, dt = 2\pi \int_0^{\infty} (f_\omega g_\omega^* + f_\omega^* g_\omega)\, d\omega. \quad (14\text{–}45)$$

To show this we note that by definition

$$I = \int_{-\infty}^{\infty} \int_{-\infty}^{\infty} \int_{-\infty}^{\infty} e^{-i(\omega+\omega')t} f_\omega g_{\omega'}\, d\omega\, d\omega'\, dt. \quad (14\text{–}46)$$

The integral over the time, taken first, has the functional properties of $2\pi\, \delta(\omega + \omega')$, where $\delta(\omega + \omega')$ is the one-dimensional delta function. This can be seen by examining an integral of the form

$$J(a) = \int_{-\infty}^{\infty} e^{iat}\, dt, \quad (14\text{–}47)$$

which is obviously divergent for $a = 0$ and which for finite limits would oscillate rapidly as a function of the limits for $a \neq 0$. If we replace the limits by $\pm L$ we obtain

$$J(a) = \int_{-L}^{L} e^{iat}\, dt = (2 \sin aL)/a,$$

and

$$\int_{-\infty}^{\infty} J(a)\, da = 2 \int_{-\infty}^{\infty} \frac{\sin aL}{a}\, da = 2\pi \quad (14\text{–}48)$$

for all L. Hence $J(a)$ has the properties of $2\pi\, \delta(a)$, in the sense that both behave the same way as factors in an integrand. In this application, Eq. (14–46) becomes

$$I = 2\pi \int_{-\infty}^{\infty} f_\omega g_{-\omega}\, d\omega = 2\pi \left[\int_0^{\infty} f_\omega g_{-\omega}\, d\omega + \int_{-\infty}^{0} f_\omega g_{-\omega}\, d\omega \right]$$

$$= 2\pi \int_0^{\infty} (f_\omega g_\omega^* + f_\omega^* g_\omega)\, d\omega, \quad (14\text{–}49)$$

since $g_{-\omega} = g_\omega^*$ and $f_{-\omega} = f_\omega^*$ for real $g(t)$ and $f(t)$. The relation (14–45) is thus verified, and the results may be applied to compute the value of Eq. (14–44). If $f(t) = g(t)$, Eq. (14–49) becomes simply

$$\int_{-\infty}^{\infty} [f(t)]^2 \, dt = 4\pi \int_0^{\infty} |f_\omega|^2 \, d\omega. \tag{14–49'}$$

Now the Fourier components of the radiation fields, from Eqs. (14–34) and (14–41), are

$$\mathbf{B}_{\omega\text{rad}} = \frac{-ik\mu_0}{4\pi} \int \frac{\mathbf{j}_\omega \times \mathbf{r}}{r^2} e^{ikr} \, dv' = \frac{-i\mu_0}{4\pi} \int \frac{\mathbf{j}_\omega \times \mathbf{k}}{r} e^{ikr} \, dv',$$
$$\tag{14–43'}$$
$$\mathbf{E}_{\omega\text{rad}} = \frac{ik}{4\pi\epsilon_0 c} \int \frac{[\mathbf{r} \times (\mathbf{j}_\omega \times \mathbf{r})]}{r^3} e^{ikr} \, dv' = \frac{i}{4\pi\epsilon_0 c} \int \frac{[\mathbf{r} \times (\mathbf{j}_\omega \times \mathbf{k})]}{r^2} e^{ikr} \, dv',$$

where $\mathbf{k} = k\mathbf{r}/r$ is the propagation vector. With the substitution of these fields into Eq. (14–44), making use of Eq. (14–49), we obtain the total radiated energy U:

$$U = \int_S \int_{-\infty}^{\infty} \mathbf{N}(t) \cdot d\mathbf{S} \, dt$$
$$= \frac{1}{4\pi} \left(\frac{\mu_0}{\epsilon_0}\right)^{1/2} \int_S \int_\omega \left| \int \frac{\mathbf{j}_\omega \times \mathbf{k}}{r} e^{ikr} \, dv' \right|^2 \left(\frac{\mathbf{k} \cdot d\mathbf{S}}{k}\right) d\omega, \tag{14–50}$$

where S is a large sphere of radius R surrounding the origin and enclosing the source. Clearly the radiated energy should be independent of the particular surface S, although its angular distribution will depend on the source coordinates $\boldsymbol{\xi}(x'_\alpha)$. With reference to Fig. 14–2 we note that for R

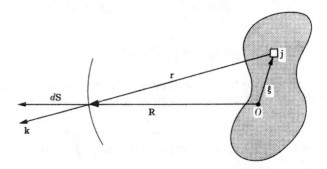

FIG. 14–2. Geometry for performing integration for radiated energy of a current system $\mathbf{j}$ at a distance $\boldsymbol{\xi}$ from an arbitrary origin 0.

large in comparison with the dimensions of the source,

$$kr \simeq kR - \mathbf{k} \cdot \boldsymbol{\xi}(x'_\alpha),$$

and

$$\frac{d\mathbf{S} \cdot \mathbf{k}}{r^2 k} = d\Omega,$$

where $d\Omega$ is the solid angle subtended by $d\mathbf{S}$ at the origin O. Hence, if we put

$$U = \int_0^\infty U_\omega \, d\omega, \qquad (14\text{–}51)$$

then

$$\frac{dU_\omega}{d\Omega} \, d\omega = \frac{1}{4\pi} \left(\frac{\mu_0}{\epsilon_0}\right)^{1/2} \left| \int (\mathbf{j}_\omega \times \mathbf{k}) e^{-i(\mathbf{k} \cdot \boldsymbol{\xi})} \, dv' \right|^2 d\omega \qquad (14\text{–}52)$$

is the energy radiated per unit solid angle in a frequency band $d\omega$. Note that the integral defining the strength of the radiating system involves the source coordinates only.

If the source is monochromatic, corresponding to a single frequency ω, we may define a rate P of radiated energy averaged over a cycle. By Eq. (11–22) the time average of the product of the real parts of two vectors. which vary as $e^{-i\omega t}$ is one-half the real part of the product of one vector and the complex conjugate of the other. Hence the radiation fields of Section 14–3 yield

$$\frac{dP}{d\Omega} = \frac{1}{32\pi^2} \left(\frac{\mu_0}{\epsilon_0}\right)^{1/2} \left| \int (\mathbf{j}_0 \times \mathbf{k}) e^{-i(\mathbf{k} \cdot \boldsymbol{\xi})} \, dv' \right|^2 \qquad (14\text{–}53)$$

for the radiated power per unit solid angle.

Let us apply this formula to the problem of the linear antenna, i.e., a straight wire carrying a time-varying current and which thus emits electromagnetic radiation. Such a problem is in general not soluble exactly, since the current distribution depends on the radiation field while the radiation field is given by the currents as specified above. An exact solution thus calls for consideration of the complete boundary value problem of radiator and field; this is soluble by analytical methods only for selected idealized geometries.*

If we are considering the ideal case of a straight thin wire fed by a power source at its center, then so long as the resistance and the radiation loss

---

* See, e.g., the biconical antenna as discussed in S. A. Schelkunoff, *Electromagnetic Waves*, D. Van Nostrand Co., New York, 1943, pp. 441 ff.

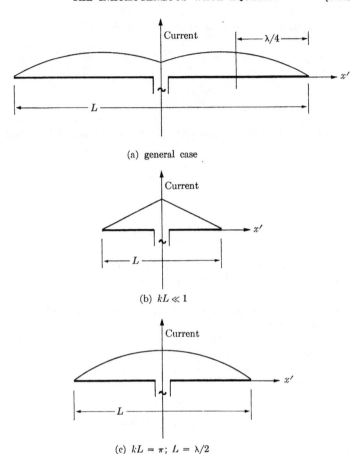

FIG. 14–3. Current distributions along ideal center-fed linear antennas of length L.

can be neglected, the current distribution along the wire is that of a lossless transmission line; such a line would support a standing wave of wavenumber k corresponding to the driving angular frequency ω such that $c = \omega/k$. If the wire, of total length L, is infinitely thin, the current at its ends must vanish (Fig. 14–3). Hence we can put

$$\mathbf{j}_0(x', y', z') = J_0\, \delta(y')\, \delta(z')\, \frac{\sin\,[k(L/2 - |x'|)]}{\sin\,(kL/2)}\, \hat{\mathbf{x}}', \qquad (14\text{–}54)$$

where J_0 is the crest current at the feed point $x' = 0$.

The volume integral in Eq. (14–53) thus becomes

$$\left| \int (\mathbf{j}_0 \times \mathbf{k}) e^{-i\mathbf{k}\cdot\boldsymbol{\xi}} \, dv' \right| = \frac{2J_0 k \sin \theta}{\sin (kL/2)} \int_0^{L/2} \sin \left[k \left(\frac{L}{2} - x' \right) \right] \cos (kx' \cos \theta) \, dx' \tag{14–55}$$

$$= \frac{2J_0}{\sin \theta \sin (kL/2)} \left\{ \cos \left[\frac{kL}{2} (\cos \theta) \right] - \cos \left(\frac{kL}{2} \right) \right\}.$$

Note that this quantity tends to

$$\tfrac{1}{2} J_0 (kL) \sin \theta \tag{14–55'}$$

as $kL \to 0$, i.e., as the antenna becomes short compared to the wavelength (Fig. 14–3b). According to (14–53), the total radiated power becomes, if we use $d\Omega = 2\pi \sin \theta \, d\theta$ and put $R_0 = (\mu_0/\epsilon_0)^{1/2}$,

$$P = \frac{R_0}{4\pi} J_0^2 \int_0^\pi \left\{ \frac{\cos [(kL/2) \cos \theta] - \cos (kL/2)}{\sin (kL/2) \sin \theta} \right\}^2 \sin \theta \, d\theta. \tag{14–56}$$

In the general case, this integral must be evaluated numerically or by expansion. In the limit $kL \to 0$ [see Eq. (14–56')], the power becomes

$$P = \frac{\pi}{12} R_0 J_0^2 \left(\frac{L}{\lambda} \right)^2. \tag{14–56'}$$

The quantity $P/(\tfrac{1}{2} J_0^2) = P/J_{\text{eff}}^2$ is usually called the radiation resistance R_{rad}; it is the resistance which would give the same power loss as that of the radiation if connected at the driving point. Thus, in the case $kL \to 0$,

$$R_{\text{rad}} = \frac{\pi}{6} R_0 \left(\frac{L}{\lambda} \right)^2 = 197(L/\lambda)^2 \text{ ohms.} \tag{14–57}^*$$

For the practically important case $kL = \pi$, $L = \lambda/2$ [Fig. 14–3(c)], Eq. (14–56) reduces to

$$P = \frac{R_0}{4\pi} J_0^2 \int_0^\pi \frac{\cos^2 [(\pi/2) \cos \theta]}{\sin^2 \theta} \sin \theta \, d\theta. \tag{14–58}$$

---

* This is less by a factor of four than the relation

$$R_{\text{rad}} = (2\pi/3) R_0 (L/\lambda)^2 = 789(L/\lambda)^2 \text{ ohms} \tag{14–57'}$$

given in many references. The difference is the factor $\frac{1}{2}$ in Eq. (14–55') which in turn originates from the triangular distribution in Fig. 14–3b. Equation (14–57') refers to the power radiated by an elementary antenna carrying a *uniform* current J.

This can be evaluated in terms of non-elementary functions; numerically,

$$R_{\text{rad}} = P/(J_0^2/2) = 73.1 \text{ ohms} \qquad (14\text{–}59)$$

for the "half-wave" antenna.

Equation (14–53) permits us to compute the radiated power and angular distribution of the radiation for any arbitrary *given* distribution of current in an antenna. For further examples, the reader is referred to the engineering literature.

14–5 The Hertz potential. The Lorentz condition of Eq. (14–3) which relates the scalar and vector potentials is consistent with the equation of continuity,

$$\boldsymbol{\nabla} \cdot \mathbf{j} + \frac{\partial \rho}{\partial t} = 0. \qquad (7\text{–}1)$$

In fact, it can be easily shown that if the retarded potentials of Eqs. (14–24) and (14–25) are assumed, then the Lorentz condition is a direct consequence of the equation of continuity. Since charges and currents cannot be specified independently, it is advantageous for the calculation of radiation fields to represent them by a single function chosen in such a way that the continuity equation is identically satisfied. The radiation field can similarly be represented by a single potential so chosen that the Lorentz condition is identically satisfied. The first requirement can be met by deriving the charge and current densities from a single vector function $\mathbf{p}(x_\alpha', t)$ of the source coordinates by the relations

$$\begin{aligned} \rho_{\text{true}} &= -\boldsymbol{\nabla} \cdot \mathbf{p}, \\ \mathbf{j}_{\text{true}} &= \frac{\partial \mathbf{p}}{\partial t}. \end{aligned} \qquad (14\text{–}60)$$

It is seen by inspection that the equation of continuity is satisfied by this choice of $\mathbf{p}$, but otherwise an arbitrary current distribution can be represented. The source function $\mathbf{p}$ is known as the polarization vector, and it is related to the true charges and true currents in the same way that the dielectric polarization $\mathbf{P}$ is related to the polarization charges and polarization currents. This is only a mathematical parallel, however, and it should be emphasized that $\mathbf{j}$ and ρ represent the true charges which constitute the external sources of the field, so that $\mathbf{p}$ is entirely different from the ordinary polarization $\mathbf{P}$.

A vector $\mathbf{Z}$ which combines in a similar way the potentials $\mathbf{A}$ and ϕ, and which at the same time implies the Lorentz condition, is defined by the equations

$$\mathbf{A} = \frac{1}{c^2} \frac{\partial \mathbf{Z}}{\partial t}, \qquad \phi = -\boldsymbol{\nabla} \cdot \mathbf{Z}. \qquad (14\text{–}61)$$

Thus $\mathbf{Z}$ is a "superpotential," and is commonly known as the polarization potential, or Hertz vector. By combining the definitions of $\mathbf{p}$ and $\mathbf{Z}$, we find that if the Hertz vector obeys the inhomogeneous wave equation with $\mathbf{p}$ as the source, then the ordinary potentials ϕ and $\mathbf{A}$ obey their respective wave equations with ρ and $\mathbf{j}$ as sources. That is, we have

$$\Box \mathbf{Z} = \nabla^2 \mathbf{Z} - \frac{1}{c^2} \frac{\partial^2 \mathbf{Z}}{\partial t^2} = -\frac{\mathbf{p}}{\epsilon_0}. \qquad (14\text{-}62)$$

Note that this is a three-component equation, whereas $\mathbf{A}$ and ϕ amount to a four-component potential.

The retarded potential solution of Eq. (14-62) is given by

$$\mathbf{Z}(x_\alpha) = \frac{1}{4\pi\epsilon_0} \int \frac{[\mathbf{p}(x_\alpha')]}{r(x_\alpha, x_\alpha')} \, dv'; \qquad (14\text{-}63)$$

the Fourier components are

$$\mathbf{Z}_\omega = \frac{1}{4\pi\epsilon_0} \int \frac{\mathbf{p}_\omega(x_\alpha')e^{ikr}}{r(x_\alpha, x_\alpha')} \, dv'. \qquad (14\text{-}63')$$

The fields can be derived from the Hertz vector by means of the defining equations, Eqs. (14-61). If we let

$$\mathbf{C} = \boldsymbol{\nabla} \times \mathbf{Z}, \qquad (14\text{-}64)$$

the magnetic field is given by

$$\mathbf{B} = \frac{1}{c^2} \frac{\partial \mathbf{C}}{\partial t} \qquad (14\text{-}65)$$

and the electric field by

$$\mathbf{E} = \boldsymbol{\nabla} \times \mathbf{C}. \qquad (14\text{-}66)$$

Equation (14-65) is quite general, but Eq. (14-66) is true only outside the source, where $\boldsymbol{\nabla} \cdot \mathbf{E} = 0$.

14-6 Computation of radiation fields by the Hertz method. We shall apply this method to the case where the fields are observed at distances large compared with the extent of the source distribution, and where in addition the extent of the source is reasonably small compared with the wavelength of the outgoing radiation. These restrictions, $\xi(x_\alpha') \ll r(x_\alpha, x_\alpha')$, $\xi(x_\alpha') \ll \lambda$, are equivalent to assuming that the phase shift of the outgoing wave over the current and charge distributions is small, and that the distance to the observer is large compared to the dimensions of the source.

Under these conditions, the function e^{ikr}/r is a slowly varying function relative to the variation of $\mathbf{p}$ itself in the integral of Eq. (14–63'). It is therefore natural to consider the expansion of this function as a power series in terms of the distance $\xi = |\boldsymbol{\xi}|$ of the source point from the origin of the charge distribution, and to study the asymptotic behavior of this expansion for large distances R of the observation point. The parameters of expansion will be taken as ratios of ξ and R to the wavelength $\lambdabar = 1/k$. Such an expansion can be derived by considering the expansion of e^{ikr}/r relative to a shifted origin,

$$\frac{e^{ikr}}{r} = ik \sum_{n=0}^{\infty} (2n + 1)P_n (\cos \Theta)j_n(k\xi)h_n(kR), \qquad (14\text{–}67)$$

which is valid if $R > \xi$. Here Θ is the angle between $\boldsymbol{\xi}$ and $\mathbf{R}$, and $P_n(\cos \Theta)$ is the Legendre polynomial of order n; $j_n(k\xi)$ and $h_n(kR)$ are the spherical Bessel and Hankel functions, defined in terms of ordinary cylinder functions by Eq. (13–54). By means of the addition theorem for spherical harmonics, $P_n(\cos \Theta)$ can be written in terms of the angular coordinates θ', φ' of the source and θ, φ of the field point:

$$P_n (\cos \Theta) = \sum_{m=-n}^{n} (-1)^m P_n^m (\cos \theta)P_n^{-m} (\cos \theta')e^{im(\varphi-\varphi')}. \qquad (14\text{–}68)$$

Therefore the Hertz potential can be written as

$$\mathbf{Z} = \frac{ike^{-i\omega t}}{4\pi\epsilon_0} \sum_n \sum_m (2n + 1)(-1)^m h_n(kR)P_n^m (\cos \theta)e^{im\varphi}$$
$$\times \int \mathbf{p}(\boldsymbol{\xi}) j_n(k\xi) P_n^{-m} (\cos \theta')e^{-im\varphi'} dv'. \qquad (14\text{–}69)$$

This series is exact. The ordinary multipole description of the source is obtained by replacing $j_n(k\xi)$ by the first term of its series expansion valid for $k\xi \ll 1$,

$$j_n(k\xi) \simeq \frac{2^n n!}{(2n + 1)!} (k\xi)^n. \qquad (14\text{–}70)$$

The spherical Hankel function may be written in terms of elementary functions, although this procedure is practical only for the first few values of n:

$$h_0(kR) = j_0(kR) + in_0(kR) = \frac{\sin kR}{kR} - i\frac{\cos kR}{kR} = -\frac{ie^{ikR}}{kR},$$

$$h_1(kR) = -\frac{1}{kR}\left(1 + \frac{i}{kR}\right) e^{ikR}. \qquad (14\text{–}71)$$

The radiation fields are correctly given by the asymptotic behavior of $h_n(kR)$ for $kR \gg 1$:

$$h_n(kR) \simeq (-i)^{n+1} \frac{e^{ikR}}{kR}. \qquad (14\text{–}72)$$

Ordinarily, the successive values of n become important only if lower order moments of the source distribution vanish; if two or more moments contribute sensibly to the radiation, further terms in the expansion for $j_n(k\xi)$ must be taken into account. The integrals over the source coordinates are then not moments in the elementary sense, although the nomenclature is retained.

If the wavelength is long compared with the dimensions of the source, so that the use of Eq. (14–70) is justified, the radiation fields corresponding to the nth term in the expansion may be computed from

$$\mathbf{Z}_\omega^{(n)} = (-i)^n \frac{e^{ikR}}{4\pi\epsilon_0 R} \frac{2^n n!}{(2n)!} \int (k\xi)^n \mathbf{p}_\omega P_n (\cos \Theta) \, dv' \qquad (14\text{–}73)$$

by means of Eqs. (14–64)–(14–66). If the source distribution is linear in space the source line may be taken as the polar axis, and $\Theta = \theta$; the Legendre function thus automatically defines the angular distribution of the corresponding fields. Otherwise Eq. (14–68) must be used to ascertain the dependence of the radiation field on its angular coordinates. Since the Legendre functions are orthogonal, there will be no interference between terms of the series when the field quantities are squared and integrated over solid angle, and we may neglect the phase shifts.

14–7 Electric dipole radiation. For $n = 0$, integration of $\mathbf{p}_\omega$ over the source coordinates gives $\int \mathbf{p}_\omega \, dv' = \mathbf{p}_1$, the electric dipole moment of the distribution. The corresponding Hertz vector, $\mathbf{Z}_\omega$, is simply

$$\mathbf{Z}_\omega (x_\alpha) = \frac{e^{ikR}}{4\pi\epsilon_0 R} \int \mathbf{p}_\omega(x_\alpha') \, dv' = \frac{e^{ikR}}{4\pi\epsilon_0 R} \mathbf{p}_1, \qquad (14\text{–}74)$$

and we may take the direction of $\mathbf{p}_1$ as parallel to the polar axis. To obtain the radiation field from the polarization potential of Eq. (14–74), we shall first find $\mathbf{C}$ as defined by Eq. (14–64). From the components of the polarization vector as shown in Fig. 14–4, it is clear that the spherical polar components of $\mathbf{Z}_{0\omega}$ are

$$Z_R = Z_{0\omega} \cos \theta = \frac{p_1 \cos \theta e^{ikR}}{4\pi\epsilon_0 R},$$

$$Z_\theta = - Z_{0\omega} \sin \theta = - \frac{p_1 \sin \theta e^{ikR}}{4\pi\epsilon_0 R}, \qquad (14\text{–}75)$$

$$Z_\varphi = 0.$$

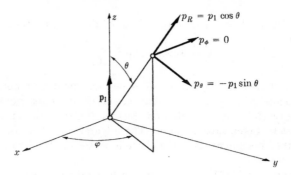

Fig. 14-4. Components of a polarization vector oriented along the polar axis.

It then follows from the expression for the curl in polar coordinates (Appendix III) that C_φ is the only component of $\mathbf{C}$:

$$C_\varphi = \frac{p_1 \sin \theta}{4\pi\epsilon_0 R}\left(\frac{1}{R} - ik\right)e^{ikR}. \tag{14-76}$$

Therefore the magnetic field is just

$$H_\phi = -\frac{i\omega}{4\pi}\, p_1 \sin \theta \left(\frac{1}{R^2} - \frac{ik}{R}\right)e^{ikR}, \tag{14-77}$$

and the components of the electric field, from Eq. (13-46), are

$$E_R = \frac{1}{R \sin \theta}\frac{\partial}{\partial\theta}(\sin \theta\, C_\varphi) = \frac{p_1 \cos \theta}{2\pi\epsilon_0 R^2}\left(\frac{1}{R} - ik\right)e^{ikR}, \tag{14-78}$$

$$E_\theta = -\frac{1}{R}\frac{\partial}{\partial R}(RC_\varphi) = \frac{p_1 \sin \theta}{4\pi\epsilon_0 R}\left(\frac{1}{R^2} - \frac{ik}{R} - k^2\right)e^{ikR}. \tag{14-79}$$

The two terms of Eq. (14-77) represent the induction field and the radiation field. The θ-component of the electric field has three terms: the first is the static dipole field, varying as the inverse cube of the distance from the dipole; the second term varies as the inverse square of the distance and is called the transition field. The transition field will not contribute to the radiated energy but it does contribute to the energy storage during the oscillation. The radiation fields alone are simply

$$H_\phi = -\frac{\omega k p_1 \sin \theta\, e^{ikR}}{4\pi R}, \tag{14-80}$$

$$E_\theta = -\frac{k^2 p_1 \sin \theta\, e^{ikR}}{4\pi\epsilon_0 R}. \tag{14-81}$$

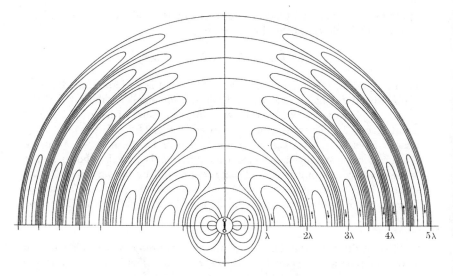

Fig. 14–5. Electric field lines produced by an oscillating dipole.

In vector form, the radiation fields become

$$\mathbf{E} = \frac{e^{ikR}}{4\pi\epsilon_0 R}[(\mathbf{p}_1 \times \mathbf{k}) \times \mathbf{k}], \qquad \mathbf{H} = -\frac{\omega e^{ikR}}{4\pi R}(\mathbf{p}_1 \times \mathbf{k}), \quad (14\text{–}82)$$

where $\mathbf{k}$ is, of course, along $\mathbf{R}$. The subscript ω is implicit throughout the calculation.

The equations for the electric field lines, or lines of force, can be expressed very simply in terms of $\mathbf{C}$. Since a line of force is defined by the condition that its line element $d\mathbf{s}$ be in the direction of $\mathbf{E}$, the equation of the field lines is just $d\mathbf{s} \times \mathbf{E} = d\mathbf{s} \times (\mathbf{\nabla} \times \mathbf{C}) = 0$. In this case, the equation is

$$\frac{\partial}{\partial R}(RC_\varphi \sin\theta)\,dR + \frac{\partial}{\partial\theta}(RC_\varphi \sin\theta)\,d\theta = 0, \qquad (14\text{–}83)$$

of which a solution is

$$RC_\varphi \sin\theta = \text{constant}.$$

On substituting the expression for C_φ from Eq. (14–76), we obtain

$$\left(\sqrt{R^{-2} + k^2}\right)\sin^2\theta\cos[kR - \omega t - \tan^{-1}(kR)] = \text{constant} \quad (14\text{–}84)$$

when the time-dependent term is included. This is an exact equation applying both in the radiation zone and in the induction zone. If we consider the radiation field alone, we find

$$\sin^2 \theta \cos (kR - \omega t) = \text{constant}$$

as the equation for the lines of force. This field is periodic radially, with radial spacing equal to the wavelength λ. For smaller values of kR the apparent phase velocity of the wave front, in accord with Eq. (14–84), is greater than c. The transition is indicated in Fig. 14–5, with the regularly spaced lines of force in the radiation field near the periphery. Accurate scale has been sacrificed in order that both fields may be shown in the same drawing; a famous series of diagrams exhibiting the way in which the electric lines of force are initiated in time is to be found in *Gesammelte Werke*, Band II, of H. Hertz, or in the translated volume entitled *Electric Waves*.

14–8 Multipole radiation. Let us now consider the significance of higher moments that contribute to the **Z** vector. The term $n = 1$ in the general expansion of Eq. (14–73),

$$Z_{1\omega} = \frac{e^{ikR}}{4\pi\epsilon_0 R} \int k\xi \mathbf{p} \cos \Theta \, dv', \qquad (14\text{–}85)$$

can be written as

$$Z_{1\omega} = \frac{ke^{ikR}}{4\pi\epsilon_0 R^2} \int \mathbf{p}(\xi \cdot \mathbf{R}) \, dv', \qquad (14\text{–}86)$$

where **p** is to be understood as $\mathbf{p}_\omega$. The components of ξ are x_α', and those of **R** are x_α. Then in tensor notation the integrand of Eq. (14–86) is $x_\alpha p_\beta x_\alpha'$, which is a function of two sets of parameters of the distribution, p_β and x_α'. It is convenient to break up this tensor into two parts, one symmetric and the other antisymmetric in the two parameters. That is, we put

$$x_\alpha p_\beta x_\alpha' = \frac{x_\alpha(p_\beta x_\alpha' + p_\alpha x_\beta') + x_\alpha(p_\beta x_\alpha' - p_\alpha x_\beta')}{2}. \qquad (14\text{–}87)$$

This process, much the same as that employed in Section 7–11, is analogous to the process often used in the mechanics of continua, namely, that of separating a general strain of an elastic solid into the sum of a pure strain represented by a symmetric tensor and a body rotation represented by an antisymmetric tensor. Putting the antisymmetric part of Eq. (14–87) back into vector notation, we may write the integral in Eq. (14–86) as

$$\tfrac{1}{2} \int \{[\mathbf{R} \times (\mathbf{p} \times \xi)]_\beta + x_\alpha(p_\beta x_\alpha' + p_\alpha x_\beta')\} \, dv'. \qquad (14\text{–}88)$$

Let us consider the antisymmetric term first. Since $\mathbf{R}$ is not a function of the primed variables of integration, it can be taken outside the integral sign, and the first term of expression (14–88) is just

$$\tfrac{1}{2}\mathbf{R} \times \int (\mathbf{p} \times \boldsymbol{\xi})\, dv'. \tag{14–89}$$

The significance of this integral can be easily recognized if we express the polarization vector in terms of the current. Since $\mathbf{p} = -\mathbf{j}/i\omega$ for sinusoidal time variation, (14–89) becomes

$$\frac{-\mathbf{R}}{2i\omega} \times \int \mathbf{j} \times \boldsymbol{\xi}\, dv'. \tag{14–90}$$

But the magnetic moment of a current distribution is

$$\mathbf{m} = \tfrac{1}{2} \int \boldsymbol{\xi} \times \mathbf{j}\, dv',$$

and hence (14–90) is given by

$$\frac{1}{i\omega} \mathbf{R} \times \mathbf{m}.$$

The polarization potential $\mathbf{Z}$ corresponding to the antisymmetric part of its original integrand is therefore

$$\mathbf{Z}_{\text{antisym.}} = \frac{k e^{ikR}(\mathbf{R} \times \mathbf{m})}{4\pi\epsilon_0 R^2 i\omega}. \tag{14–91}$$

The field can be computed readily from $\mathbf{Z}$ as given by Eq. (14–91). If we omit the induction terms, we obtain

$$\mathbf{B}_{\text{antisym.}} = \frac{1}{c^2}\frac{\partial \mathbf{C}}{\partial t} = \frac{i\mu_0 k^2 [\mathbf{R} \times (\mathbf{m} \times \mathbf{R})]e^{ikR}}{4\pi R^3}, \tag{14–92}$$

which, if we ignore the phase factor, is equivalent to

$$\mathbf{H}_{\text{antisym.}} = \frac{e^{ikR}}{4\pi R} [\mathbf{k} \times (\mathbf{m} \times \mathbf{k})]. \tag{14–93}$$

The magnetic field of Eq. (14–93) has exactly the same mathematical form as that of the electric field in Eq. (14–82) for the electric dipole. It is known as the magnetic dipole radiation field. Magnetic dipole radiation corresponds to a current distribution which, although it has no net oscillating electric dipole, does have a sinusoidally varying circulation of the charges.

Now let us consider the physical significance of the symmetrical term of Eq. (14–87). The corresponding Hertz vector is given by

$$Z_{\beta \text{ sym}} = \frac{ke^{ikR}}{4\pi\epsilon_0 R^2} x_\alpha \int \frac{1}{2} (p_\beta x'_\alpha + p_\alpha x'_\beta)\, dv'. \tag{14–94}$$

The symmetry of Eq. (14–94) will become more evident if we write it in terms of the charge density $\rho(x'_\alpha)$ and the coordinates of the charge density. Consider a quantity defined by [see Eq. (1–48)]

$$Q_{\alpha\beta} = \int \rho x'_\alpha x'_\beta\, dv' = -\int \frac{\partial p_\gamma}{\partial x'_\gamma} x'_\alpha x'_\beta\, dv', \tag{14–95}$$

which is known as the electric quadrupole moment of the charge distribution. If we integrate Eq. (14–95) by parts, and extend the volume of integration outside the charge distribution so that the integrated parts vanish, we find that in terms of the quadrupole moment Eq. (14–94) can be written

$$Z_\beta = \frac{ke^{ikR}}{8\pi\epsilon_0 R^2} x_\alpha Q_{\alpha\beta}. \tag{14–96}$$

Since the quadrupole moment is represented by the symmetric matrix of Eq. (14–95), it can also be represented by a family of quadrics derived from the quadratic form

$$x_\alpha x_\beta Q_{\alpha\beta} = C_2 = \text{constant.} \tag{14–97}$$

In terms of the parameter C_2 of this equation $\mathbf{Z}$ can therefore be written as

$$\mathbf{Z} = \frac{ke^{ikR}}{8\pi\epsilon_0 R^2} \nabla C_2, \tag{14–98}$$

indicating that the direction of $\mathbf{Z}$ is everywhere normal to the quadric surfaces of Eq. (14–97).

Let us calculate the components of the fields that correspond to a general quadrupole. We may choose a system of axes x, y, z, along the principal axes of the quadrupole quadric. The components of $\mathbf{Z}$ are then given by

$$\begin{Bmatrix} Z_x \\ Z_y \\ Z_z \end{Bmatrix} = \frac{ke^{ikR}}{8\pi\epsilon_0 R} \begin{Bmatrix} \sin\theta\cos\varphi\, Q_{xx} \\ \sin\theta\sin\varphi\, Q_{yy} \\ \cos\theta\, Q_{zz} \end{Bmatrix}. \tag{14–99}$$

The components of **C**, obtained by taking the curl of **Z**, are

$$C_\theta = \frac{ik^2 e^{ikR}}{16\pi\epsilon_0 R} \sin\theta \sin 2\varphi (Q_{xx} - Q_{yy}),$$

$$C_\varphi = \frac{ik^2 e^{ikR}}{32\pi\epsilon_0 R} \sin 2\theta \left[Q_{xx} + Q_{yy} - 2Q_{zz} - (Q_{yy} - Q_{xx})\cos 2\varphi \right].$$

(14–100)

And from **C** the electric field components are found to be

$$E_\theta = \frac{k^3 e^{ikR}}{32\pi\epsilon_0 R} \sin 2\theta \left[Q_{xx} + Q_{yy} - 2Q_{zz} - (Q_{yy} - Q_{xx})\cos 2\varphi \right],$$

$$E_\varphi = \frac{k^3 e^{ikR}}{16\pi\epsilon_0 R} \sin\theta \sin 2\varphi (Q_{yy} - Q_{xx}),$$

(14–101)

while the magnetic field components are

$$B_\theta = -\frac{\omega\mu_0 k^2 e^{ikR}}{16\pi R} \sin\theta \sin 2\varphi (Q_{yy} - Q_{xx}),$$

$$B_\varphi = \frac{\omega\mu_0 k^2 e^{ikR}}{32\pi R} \sin 2\theta \left[Q_{xx} + Q_{yy} - 2Q_{zz} - (Q_{yy} - Q_{xx})\cos 2\varphi \right].$$

(14–102)

Only the radiation fields are given by Eqs. (14–101) and (14–102), and there are no radial components.

Three features of the quadrupole radiation fields can be noted by inspection: (a) In general, the distributions will have two nodal cones where there is zero radiation field, compared with the single nodal line of a dipole distribution. (b) The fields depend only on the differences in the quadrupole moments $Q_{\alpha\beta}$; hence five, not six, second moments of the distribution specify the radiation field. If the moments are reduced to principal axes, then only two moments are necessary (in addition to the three angles necessary to define the orientation of the axes). We can demonstrate this by re-expressing Eq. (14–102) in terms of two of the components of the $D_{\alpha\beta}$ matrix defined in Eq. (1–51); we obtain

$$B_\theta = \frac{\omega\mu_0 k^2 e^{ikR}}{16\pi R} \sin\theta \sin 2\varphi \left(\frac{D_{xx} - D_{yy}}{3} \right);$$

$$B_\varphi = \frac{\omega\mu_0 k^2 e^{ikR}}{32\pi R} \sin 2\theta \left[(D_{xx} + D_{yy}) + \left(\frac{D_{xx} - D_{yy}}{3} \right)\cos 2\varphi \right].$$

(14–103)

(c) In case the quadrupole is a spheroidal distribution, only one moment is required to specify the radiation field in addition to the orientation of the axis of symmetry which we take as the z-axis. In this case, the only component of $\mathbf{C}$ and hence the magnetic field is

$$C_\varphi = \frac{ik^2 e^{ikR}}{16\pi\epsilon_0 R} \sin(2\theta)(Q_{xx} - Q_{zz}); \qquad (14\text{--}104)$$

and hence,

$$B_\varphi = \left(\frac{\omega k^2 \mu_0}{16\pi}\right) D \frac{e^{ikR}}{R} \sin(2\theta), \qquad (14\text{--}105)$$

where the quantity

$$D = 3Q_{xx} - (Q_{xx} + Q_{yy} + Q_{zz}) = Q_{xx} - Q_{zz}$$

as defined in Eq. (1–52) is often called *"the* quadrupole moment." Physically, the quadrupole moment arises from a pulsating charge distribution of such symmetry that the dipole moment remains zero during the pulsation. The simplest example of a quadrupole consists of two dipoles displaced a slight distance from each other and oscillating in opposition.

14–9 Derivation of multipole radiation from scalar superpotentials. As derived from the Hertz polarization potential defined by Eqs. (14–60), the magnetic multipole of a given order arises from a higher term in the expansion, Eq. (14–69) or Eq. (14–73), than the electric multipole of the same order. On the other hand, we could introduce a *magnetization* potential $\mathbf{Z}_m$, which satisfies the equation

$$\nabla^2 \mathbf{Z}_m - \frac{1}{c^2} \frac{\partial^2 \mathbf{Z}_m}{\partial t^2} = -\frac{\mathbf{m}}{\mu_0}, \qquad (14\text{--}106)$$

where $\mathbf{m}$ is the magnetization which represents the true source currents, with $\mathbf{j}_{\text{true}} = \nabla \times \mathbf{m}$. If the retarded potential solution of Eq. (14–106) is expanded in a way corresponding to that of the electric Hertz potential in Section 14–6, the magnetic dipole field is given by the term $n = 0$, in complete formal analogy to the electric dipole field in Section 14–7. The fields are of course the same as those derived for magnetic dipole radiation in Section 14–8. Instead of considering the polarization and magnetization "superpotentials" separately, it is also possible to introduce them together.[*]

It is sometimes desirable to utilize potentials which will exhibit clearly the formal geometrical symmetry between electric and magnetic multipoles of the same order, and in which quadrupole and higher order terms are not in principle more difficult to compute than dipole radiation. Such

---

[*] The full generalization of the vector superpotentials, and their relation to specialized scalar potentials, has been given by A. Nisbet, *Proc. Roy. Soc.* 231 **A**, pp. 250–263 (1955).

a set of potentials is equivalent to the scalar functions Π_E, Π_M which we have already used in Sections 13–8 and 13–9. A simple derivation of these potentials as related to source distributions has been given by Bouwkamp and Casimir.* We have noted in Section 13–8 that it is possible to express all fields in terms of their radial components in the absence of sources; the problem is to find the radial components of the fields in terms of the sources which produce them.

Let us consider the vacuum fields of the true sources $\mathbf{j}$ and ρ, with the assumption that the time dependence of all fields and sources is given by $e^{-i\omega t}$, not explicitly written. In order to make more convenient use of the spherical coordinate fields introduced in Chapter 13, we shall express the magnetic field as $\mathbf{H}$ instead of $\mathbf{B}$. The time dependent Maxwell equations are then

$$\boldsymbol{\nabla} \times \mathbf{E} = i\mu_0\omega\mathbf{H}$$
$$\boldsymbol{\nabla} \times \mathbf{H} = \mathbf{j} - i\omega\epsilon_0\mathbf{E}, \tag{14–107}$$

and the continuity equation becomes

$$\boldsymbol{\nabla} \cdot \mathbf{j} - i\omega\rho = 0. \tag{14–108}$$

Now according to Eq. (13–57), one of the two Debye potentials is simply proportional to $\mathbf{r} \cdot \mathbf{E}$ at a point whose radius vector from the origin is $\mathbf{r}$; similarly, the other Debye potential is proportional to $\mathbf{r} \cdot \mathbf{H}$. To find an equation for $\mathbf{r} \cdot \mathbf{H}$, say, consider the vector identity

$$\nabla^2(\mathbf{r} \cdot \mathbf{V}) = 2\boldsymbol{\nabla} \cdot \mathbf{V} - \mathbf{r} \cdot (\nabla^2\mathbf{V}), \tag{14–109}$$

where $\mathbf{V}$ is any vector field. If the term $k^2(\mathbf{r} \cdot \mathbf{V})$ is added to both sides of the equation, and $\mathbf{V}$ is taken as $\mathbf{H}$,

$$(\nabla^2 + k^2)(\mathbf{r} \cdot \mathbf{H}) = -\mathbf{r} \cdot (\boldsymbol{\nabla} \times \mathbf{j}), \tag{14–110}$$

since $\nabla^2\mathbf{H} + k^2\mathbf{H} = -\boldsymbol{\nabla} \times \mathbf{j}$ follows from combining Eqs. (14–107), and since $\boldsymbol{\nabla} \cdot \mathbf{H} = 0$. But, as an equation in the scalar function $(\mathbf{r} \cdot \mathbf{H})$, Eq. (14–110) is of exactly the same form as Eq. (14–16), and its solution is analogous to that for the ordinary retarded scalar potential. If we let the radius vector of the field point be $\mathbf{R}$, and that of the source distribution be $\boldsymbol{\xi}$, with $r = |\mathbf{R} - \boldsymbol{\xi}|$ the scalar distance from source point to field point,

$$(\mathbf{R} \cdot \mathbf{H}) = \frac{1}{4\pi} \int \frac{\boldsymbol{\xi} \cdot (\boldsymbol{\nabla}' \times \mathbf{j})e^{ikr}}{r} \, dv', \tag{14–111}$$

---

* C. J. Bouwkamp and H. B. G. Casimir, *Physica* **XX**, pp. 539–554 (1954).

and thus we have obtained an expression for the radial component of the magnetic field in terms of the source current.

Equation (14–109) may also be applied to the vector $\mathbf{E}$, but the result is complicated by the fact that the divergence of $\mathbf{E}$ does not necessarily vanish. A simpler expression for the radial component of the electric field results if $\mathbf{V}$ is taken to be $\mathbf{E} - \mathbf{j}/i\omega\epsilon_0$; substitution of the Maxwell relations then gives

$$(\nabla^2 + k^2)[\mathbf{r} \cdot (\mathbf{E} - \mathbf{j}/i\omega\epsilon_0)] = \frac{-i}{\omega\epsilon_0}\mathbf{r} \cdot [\nabla \times (\nabla \times \mathbf{j})]. \quad (14\text{–}112)$$

Again we have obtained an equation which is formally similar to Eq. (14–110) and thus to that for the scalar potential, so that we may write, in the notation of Eq. (14–111),

$$(\mathbf{R} \cdot \mathbf{E}) = \frac{\mathbf{R} \cdot \mathbf{j}}{i\omega\epsilon_0} + \frac{1}{4\pi i\omega\epsilon_0} \int \frac{\boldsymbol{\xi} \cdot [\nabla' \times (\nabla' \times \mathbf{j})]e^{ikr}}{r}\, dv'. \quad (14\text{–}113)$$

The first term on the right of Eq. (14–113) vanishes outside the source.

If Eq. (14–111) is expanded, by means of Eqs. (14–67) and (14–68), in terms of the field and source coordinates, assuming that $|\mathbf{R}| > |\boldsymbol{\xi}|$, we may write

$$(\mathbf{R} \cdot \mathbf{H}) = \frac{ik}{4\pi} \sum_l \sum_m (2l + 1)(-1)^m h_l(kR)P_l^m(\cos\theta)e^{im\varphi}$$

$$\times \int j_l(k\xi)P_l^{-m}(\cos\theta')e^{-im\varphi'}[\boldsymbol{\xi} \cdot (\nabla' \times \mathbf{j})]\, dv', \quad (14\text{–}114)$$

or, in terms of the functions Π_l^m of Eq. (13–52'),

$$(\mathbf{R} \cdot \mathbf{H}) = \frac{ik}{4\pi} \sum_l \sum_m (-1)^m \Pi_l^m(\mathbf{R}) \int [\boldsymbol{\xi} \cdot (\nabla' \times \mathbf{j})]\, \Pi_l^{-m}(\boldsymbol{\xi})\, dv'. \quad (14\text{–}115)$$

But

$$[\boldsymbol{\xi} \cdot (\nabla' \times \mathbf{j})]\, \Pi_l^{-m}(\boldsymbol{\xi}) = \mathbf{j} \cdot \nabla' \times [\boldsymbol{\xi}\, \Pi_l^{-m}(\boldsymbol{\xi})] + \nabla' \cdot [\mathbf{j} \times \boldsymbol{\xi}\, \Pi_l^{-m}(\boldsymbol{\xi})] \cdot$$

$$(14\text{–}116)$$

If Eq. (14–116) is substituted in the integrand of Eq. (14–115), the last term may be converted to an integral over a surface immediately *surrounding* the source; it therefore contributes nothing to the fields outside. Thus, outside the source,

$$(\mathbf{R} \cdot \mathbf{H}) = \frac{ik}{4\pi} \sum_l \sum_m (-1)^m \Pi_l^m(\mathbf{R}) \int \mathbf{j} \cdot \nabla' \times [\boldsymbol{\xi}\, \Pi_l^{-m}(\boldsymbol{\xi})]\, dv'. \quad (14\text{–}117)$$

Analogously it may be shown that (outside the source)

$$(\mathbf{R} \cdot \mathbf{E}) = \frac{1}{4\pi} \sqrt{\frac{\mu_0}{\epsilon_0}} \sum_l \sum_m (-1)^m \, \Pi_l^m(\mathbf{R}) \int \mathbf{j} \cdot \boldsymbol{\nabla}' \times \{\boldsymbol{\nabla}' \times [\boldsymbol{\xi} \, \Pi_l^{-m}(\boldsymbol{\xi})]\} \, dv'.$$

$$(14\text{–}118)$$

The identification of these fields with those of Section 13–8 is straightforward. There we noted that for each l, m fields of the type denoted by subscript M could be determined from H_r, and that the electric field so determined has no radial component. Similarly, fields of the type denoted by subscript E may be derived from E_r, and for these fields $H_r = 0$. Comparison of Eqs. (14–117) and (14–118) with the radial components of the "unit" fields given in Eqs. (13–60) and (13–61) enables us to determine the coefficients of the "unit" fields in a series expansion of the total field of any source distribution of current. Thus

$$\mathbf{E} = \sum_l \sum_m [a_l^m \, \mathbf{E}_E(l, m; \mathbf{R}) + b_l^m \, \mathbf{E}_M(l, m; \mathbf{R})], \quad (14\text{–}119)$$

where the coefficients are defined by

$$a_l^m = \frac{1}{4\pi} \sqrt{\frac{\mu_0}{\epsilon_0}} \int \mathbf{j} \cdot \boldsymbol{\nabla}' \times \{\boldsymbol{\nabla}' \times [\boldsymbol{\xi} \, \Pi_l^{-m}(\boldsymbol{\xi})]\} \, dv' \quad (14\text{–}120)$$

for electric multipoles, and, for magnetic multipoles,

$$b_l^m = \frac{ik}{4\pi} \sqrt{\frac{\mu_0}{\epsilon_0}} \int \mathbf{j} \cdot \boldsymbol{\nabla}' \times [\boldsymbol{\xi} \, \Pi_l^{-m}(\boldsymbol{\xi})] \, dv'. \quad (14\text{–}121)$$

The source strength of each multipole field is thus expressed as the integral over the source of the scalar product of a standard function and the current density. Note from Eq. (14–114) that the radial function implied in $\Pi_l^{-m}(\boldsymbol{\xi})$ is $j_l(k\xi)$, which yields the elementary multipole moments if substitution of the first term in its expansion for small $k\xi$ is justified. In this formulation the order of both electric and magnetic multipoles is 2^l. Since Eqs. (14–120) and (14–121) vanish for $l = 0$, it is evident, as in other formulations, that there are no radiating monopoles.

14–10 Energy and angular momentum radiated by multipoles. For the computation of radiated energy it suffices to use the asymptotic form of $h_l(kR)$ implied in the function $\Pi_l^m(\mathbf{R})$, and to ignore terms which fall off faster than $1/R$. The result is a generalization of the computation of scattered radiation intensity in Section 13–9. For electric multipoles, Eqs. (13–58) become

$$\mathbf{E}_E \simeq i^{-l} \frac{e^{ikR}}{R} \left[\frac{\partial Y_l^m(\theta, \varphi)}{\partial \theta} \hat{\theta} + \frac{im}{\sin \theta} Y_l^m(\theta, \varphi) \hat{\boldsymbol{\varphi}} \right]$$

$$= i^{-l} e^{ikR} \boldsymbol{\nabla} Y_l^m(\theta, \varphi),$$

$$\mathbf{H}_E \simeq i^{-l} \sqrt{\frac{\epsilon_0}{\mu_0}} \frac{e^{ikR}}{R} \left[\frac{\partial Y_l^m(\theta, \varphi)}{\partial \theta} \hat{\boldsymbol{\varphi}} - \frac{im}{\sin \theta} Y_l^m(\theta, \varphi) \hat{\boldsymbol{\theta}} \right]$$

$$= i^{-l} \sqrt{\frac{\epsilon_0}{\mu_0}} e^{ikR} [\hat{\mathbf{R}} \times \boldsymbol{\nabla} Y_l^m(\theta, \varphi)],$$

(14–122)

while for magnetic multipoles, Eqs. (13–59) become

$$\mathbf{E}_M \simeq -i^{-l} e^{ikR} [\hat{\mathbf{R}} \times \boldsymbol{\nabla} Y_l^m(\theta, \varphi)],$$

$$\mathbf{H}_M \simeq i^{-l} \sqrt{\frac{\epsilon_0}{\mu_0}} e^{ikR} \boldsymbol{\nabla} Y_l^m(\theta, \varphi).$$

(14–123)

For either type of multipole the time average of the Poynting vector corresponding to the "unit" field is given by

$$\frac{1}{2} \operatorname{Re}(\mathbf{E} \times \mathbf{H}^*) \simeq \frac{1}{2} \sqrt{\frac{\epsilon_0}{\mu_0}} |\mathbf{E}^2| \hat{\mathbf{R}} \simeq \frac{1}{2} \sqrt{\frac{\mu_0}{\epsilon_0}} |\mathbf{H}^2| \hat{\mathbf{R}}$$

$$\simeq \frac{1}{2} \sqrt{\frac{\epsilon_0}{\mu_0}} |\boldsymbol{\nabla} Y_l^m(\theta, \varphi)|^2 \hat{\mathbf{R}},$$

(14–124)

which is to be integrated over a sphere of radius R. But, in the limit of large R,

$$R^2 \int |\boldsymbol{\nabla} Y_l^m(\theta, \varphi)|^2 \, d\Omega = l(l+1) \int |Y_l^m(\theta, \varphi)|^2 \, d\Omega = 4\pi l(l+1). \quad (14\text{–}125)$$

Therefore the energy radiated per unit time by a "unit" multipole of order 2^l is

$$P = 2\pi \sqrt{\frac{\epsilon_0}{\mu_0}} l(l+1). \tag{14–126}$$

The standard fields denoted by subscripts E and M have been defined so that both yield the same power for a given l. It may be noted, however, that for a particular source distribution the magnetic multipole strength b_l^m is in general smaller than the corresponding electric multipole strength a_l^m by a factor ka, where a is of the order of magnitude of the dimensions of the source. With the harmonic time dependence assumed in this expansion, the factor $ka = \omega a/c$ is equivalent to v/c, where v is the velocity of the charge which gives rise to the current.

It can also be shown that the multipole fields transmit angular momentum. The angular momentum transported across a spherical surface of radius R in time dt is

$$d\mathbf{L} = \left\{ \int \frac{[\mathbf{R} \times (\mathbf{E} \times \mathbf{H})]}{c^2} R^2 \, d\Omega \right\} c \, dt, \qquad (14\text{–}127)$$

which may be readily rewritten as

$$d\mathbf{L} = \frac{R^2}{c} \int [\mathbf{E}(\mathbf{R} \cdot \mathbf{H}) - \mathbf{H}(\mathbf{R} \cdot \mathbf{E})] \, d\Omega \, dt. \qquad (14\text{–}128)$$

The integral is most simply evaluated by examination of the components of the "unit" fields, Eqs. (13–60) and (13–61). Let us first consider electric multipoles, for which

$$(\mathbf{R} \cdot \mathbf{E}) = l(l + 1)i^{-(l+1)} \frac{e^{ikR}}{kR} Y_l^m(\theta, \varphi) \qquad (14\text{–}129)$$

when the asymptotic form of the spherical Hankel function has been substituted for $z_l(kR)$. The φ component of the corresponding magnetic field will contribute nothing to $\mathbf{L}$, since

$$\frac{d}{d\theta} P_l^m (\cos \theta) = -P_l^{m+1} (\cos \theta),$$

and the associated spherical harmonics are orthogonal over the surface. But, in the absence of a radial component,

$$H_z = - \sin \theta \, H_\theta = -m \sqrt{\frac{\epsilon_0}{\mu_0}} i^{-(l+1)} \frac{e^{ikR}}{R} Y_l^m(\theta, \varphi), \quad (14\text{–}130)$$

and thus, for electric multipoles, the time average of dL_z is

$$\overline{dL_z} = \frac{1}{2} \frac{m}{kc} \sqrt{\frac{\epsilon_0}{\mu_0}} l(l + 1) \int |Y_l^m(\theta, \varphi)|^2 \, d\Omega \, dt$$

$$= 2\pi \frac{m}{kc} l(l + 1) \sqrt{\frac{\epsilon_0}{\mu_0}} \, dt. \qquad (14\text{–}131)$$

An identical expression is obtained for magnetic multipoles. It is of interest to compare Eq. (14–131) with Eq. (14–126). We note that the ratio of the time average radiated momentum to radiated power is

$$\frac{\overline{dL_z/dt}}{P} = \frac{m}{kc} = \frac{m}{\omega}. \qquad (14\text{–}132)$$

Thus the z component of angular momentum bears a fixed ratio to the energy radiated by a multipole of given m, and $m \neq 0$ corresponds to circularly polarized light. This is a purely classical result, but if the radiated energy is assumed to consist of photons each having energy $\hbar\omega$, the z component of the angular momentum per photon is $m\hbar$. Equation (14–132) does not, however, distinguish between spin and orbital angular momentum, and its application to individual photons is not fully justified.

Suggested References

J. A. Stratton, *Electromagnetic Theory*. Chapter VIII on radiation includes the retarded potential solution and multipole expansion of the Hertz solution.

P. M. Morse and H. Feshbach, *Methods of Theoretical Physics* includes much highly useful material on the solution of the inhomogeneous wave equation.

G. A. Schott, *Electromagnetic Radiation* contains many examples.

Somewhat different but equivalent versions of the symmetric multipole expansion are given in appendices by J. M. Blatt and V. F. Weisskopf, *Theoretical Nuclear Physics*, and W. Heitler, *The Quantum Theory of Radiation*, 3rd ed.

Exercises

1. Show that a sphere charged in spherical symmetry and oscillating purely radially will not radiate.

2. Show that if we take the solution of the wave equation as

$$\phi = (\phi_{\text{retarded}} + \phi_{\text{advanced}})/2,$$
$$\mathbf{A} = (\mathbf{A}_{\text{retarded}} + \mathbf{A}_{\text{advanced}})/2,$$

we obtain zero for the total energy radiated by an oscillating system.

3. The current in a circular loop of radius a is given by

$$I = I_0 \sin(n\varphi)e^{-i\omega t}.$$

Show that, except for a phase factor, the vector potential in the radiation zone is

$$A_\varphi = \frac{\mu_0 I_0}{2} \frac{a}{r} J_n'(ka \sin\theta) \sin n\varphi\, e^{i(kr-\omega t)}.$$

4. An electric dipole lies in the xy-plane at the origin and rotates about the z-axis with constant angular velocity. Find the cylindrical coordinate components of the magnetic field intensity for large distances from the dipole.

5. Consider two dipoles oriented at right angles to each other and oscillating 90° out of phase but at the same frequency. (a) Calculate the energy U emitted in a time dt. (b) Consider a spherical absorber at a large distance R. The angular momentum imparted to this screen in time dt is

$$\mathbf{L} = \left[\int \frac{\mathbf{r} \times (\mathbf{E} \times \mathbf{H})}{c^2}\, dS \right] c\, dt,$$

where the integral is to be taken over the surface of the sphere. Calculate **L** and compare with the result of (a). Use the complete dipole fields.

6. A linear quadrupole oscillator consists of charges $-e$, $+2e$, $-e$, with the positive charge stationary at the origin and the negative charges at z_1 and z_2 given by

$$z_1 = -z_2 = a \cos \tfrac{1}{2}\omega t.$$

Compute the fields at large distances and find the average rate at which energy is radiated.

7. Instead of oscillating, the two negative charges of problem 6 rotate with constant angular velocity about the positive charge at the origin, maintaining a fixed distance a from $+2e$ on the line of the three charges. Find the components of the quadrupole moment and the average rate of radiation.

8. Consider a center-fed antenna of total length $L = \lambda/4$ with a current distribution

$$J = J_0 \cos \frac{2\pi\xi}{\lambda}, \qquad -\lambda/8 < \xi < \lambda/8.$$

Calculate the radiated power corresponding to the linear electric dipole and octupole moments and add. Compare with the exact result.

9. Two equal positive charges q are describing circles on the opposite ends of a diameter $2a$, with an angular velocity ω. The charge density is thus

$$\rho(\rho, z, \varphi) = q \, \delta(z) \, \delta(\rho - a)[\delta(\varphi - \omega t) + \delta(\varphi + \pi - \omega t)].$$

Calculate (a) the Fourier components j_ω of current; (b) the quadrupole moment Fourier component; (c) the radiation pattern if $\omega a \ll c$.

10. Prove the vector identity, Eq. (14–109).

CHAPTER 15

THE EXPERIMENTAL BASIS FOR THE THEORY OF SPECIAL RELATIVITY

15–1 Galilean relativity and electrodynamics. We have found that electromagnetic fields are propagated *in vacuo* with a velocity $c = 1/\sqrt{\mu_0\epsilon_0}$ which is *per se* a characteristic constant of the theory. This is a feature of Maxwell's equations that is at variance with the laws of classical mechanics. Classical mechanics contains no characteristic constants and can be scaled with respect to all physical quantities; Maxwell's equations can be scaled in relation to length and time individually but not as to velocity. This means that Maxwell's theory did not share a noteworthy property of Newtonian mechanics, namely, that uniform motion is indistinguishable from no motion at all—the laws of motion hold equally in all rigid coordinate frames moving with uniform velocity with respect to each other. Mathematically this can be stated by noting that if a set of primed coordinates is moving with a velocity v along the x-axis of the plane coordinates of Fig. 15–1, substitution of the coordinate relations

$$x' = x - vt, \qquad y' = y, \qquad z' = z, \qquad t' = t, \qquad (15\text{–}1)$$

leaves the equations of motion unchanged. This can be seen by direct substitution in the basic equation of motion for n mass points whose interaction is describable by means of potential functions depending on their separation. For the ith particle, this equation is

$$m_i\ddot{\mathbf{r}}_i = -\sum_{j\neq i}^{n} \boldsymbol{\nabla}_i[V(|\mathbf{r}_i - \mathbf{r}_j|)], \qquad (15\text{–}2)$$

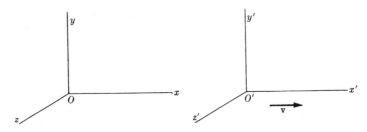

Fig. 15–1. Primed frame moving with respect to unprimed coordinate frame.

and it is obvious that Eq. (15–2) would be exactly the same in the primed coordinates. This result depends on the existence of a "universal time scale"; i.e., the last member of Eq. (15–1) is held self-evident. Equations (15–1) are known as the Galilean transformation, and it can thus be said that the laws of classical mechanics are invariant under a Galilean transformation. A coordinate frame in which a body on which no forces are acting is unaccelerated is called an "inertial" frame; we can therefore formulate the invariance of the laws of mechanics under the transformation of Eqs. (15–1) by saying that in classical mechanics all inertial frames are equivalent. This statement is sometimes known as the principle of Galilean relativity.

Since the absence of forces can be detected only by the absence of acceleration unless *all* sources of force are known, an inertial frame is not strictly definable. But there is a more immediate difficulty with Maxwell's equations: the form of the wave equation is not preserved by the substitution of Eqs. (15–1); electromagnetic effects will therefore presumably not be the same if observed from different frames moving with a constant velocity relative to one another. Specifically, the velocity of propagation of a plane wave *in vacuo* would not retain its value $c = 1/\sqrt{\mu_0 \epsilon_0}$. If we accept the basic correctness of Maxwell's equations and classical kinematic laws, it follows that there exists a unique privileged frame of reference, the classical "ether frame," in which Maxwell's equations are valid and in which light is propagated with the velocity c.

Since the principle of Galilean relativity does apply to the laws of mechanics but does not apply to electrodynamics, we are forced to choose among the following alternatives:

(a) A principle of relativity exists for mechanics, but not for electrodynamics. A preferred inertial frame (the ether frame) exists in electrodynamics.

(b) A principle of relativity exists for both mechanics and electrodynamics, but electrodynamics is not correct in the Maxwell formulation.

(c) A principle of relativity exists for both mechanics and electrodynamics, but the laws of mechanics in the Newtonian form need modification.

The choice between these possibilities can be made only on the basis of experimental results. We shall see by analysis of the relevant experiments that alternative (c), in the form of the special theory of relativity, is essentially correct. A reference list of the basic experiments is given at the end of the chapter, and further references may be found in the more comprehensive treatments of relativity also listed. The fundamental experiments fall into three main classes: (a) attempts to locate a preferred inertial frame for the laws of electrodynamics; (b) attempts to obtain deviations from the laws of classical electrodynamics; and (c) attempts to observe deviations from classical mechanics.

15-2 The search for an absolute ether frame. It can be seen in an elementary way that two charges, q and $-q$, fixed at the ends of a rigid rod and set in motion with velocity $\mathbf{v}$, would interact magnetically in the same way as two current elements of magnitude $J\,d\mathbf{l} = q\mathbf{v}$. The forces on the two current elements would be directed oppositely, but in general they would not be collinear. Under these conditions the rod, if free to turn, would tend to set itself at right angles to the velocity. It is of interest to compute the size of this effect. By Ampère's law (with reference to Fig. (15-2),

$$\mathbf{F} = \frac{\mu_0}{4\pi}\, q^2\, \frac{\mathbf{v}\times(\mathbf{v}\times\mathbf{L})}{L^3}, \tag{15-3}$$

where $\mathbf{L}$ represents both the length and the direction of the rod. The magnitude of the force given by Eq. (15-3) may be written

$$F = \frac{1}{4\pi\epsilon_0}\, \frac{q^2}{L^2}\, \frac{v^2}{c^2}\sin\theta, \tag{15-4}$$

and its direction is perpendicular to $\mathbf{v}$ in the plane of $\mathbf{L}$ and $\mathbf{v}$. Equation (15-4) indicates that the effect is of order $(v/c)^2$ in comparison with the electrostatic interaction of the two charges. The corresponding value of the torque, i.e., the net value of the couple shown in Fig. 15-2, is

$$FL\cos\theta = \frac{1}{4\pi\epsilon_0}\, \frac{q^2 v^2 \sin\theta\cos\theta}{c^2 L}$$

$$= \frac{1}{8\pi\epsilon_0}\, \frac{q^2}{L}\, \frac{v^2}{c^2}\sin 2\theta. \tag{15-5}$$

Now if there is a preferred frame of coordinates it seems very unlikely that it is the frame with respect to which the earth is at rest, and the translatory motion of the earth should produce such a torque. An experiment

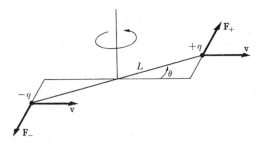

FIG. 15-2. Equal and opposite charges moving with velocity $\mathbf{v}$. $\mathbf{F}_+$ and $\mathbf{F}_-$ are not collinear.

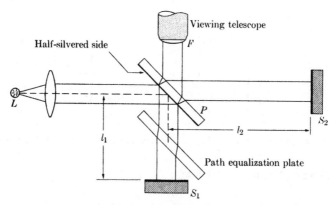

FIG. 15–3. The Michelson-Morley experiment.

involving a delicately suspended parallel plate capacitor was performed by Trouton and Noble, and repeated later with even greater accuracy, but it failed to give any indication of a torque, even though its magnitude with v equal to the known velocity of the earth in its orbit should be easily measurable.

The most famous attempt to localize the ether frame was the Michelson-Morley experiment. Light from a source L in Fig. 15–3 is split into two beams by a half-silvered mirror at P. The beams are reflected at mirrors S_1 and S_2 respectively and return through the half-silvered mirror to the telescope at F, where interference fringes are observed. Let us assume that relative to a stationary ether the instrument is moving with a velocity v parallel to S_1P. By the arguments of classical physics, the time required for light to traverse the path PS_1P is

$$t_1 = l_1 \left(\frac{1}{c - v} + \frac{1}{c + v} \right) = \frac{2l_1}{c(1 - \beta^2)} , \qquad (15\text{--}6)$$

where $\beta = v/c$. In computing the time required for the path PS_2P, we must take account of the fact that P will move through the distance δ (see Fig. 15–4) while the light travels from P to S_2. δ is given by

$$\frac{\delta}{\sqrt{\delta^2 + l_2^2}} = \frac{v}{c} ; \qquad \delta = \frac{\beta l_2}{\sqrt{1 - \beta^2}} . \qquad (15\text{--}7)$$

Hence, for path PS_2P,

$$t_2 = \frac{2}{c} \sqrt{l_2^2 + \delta^2} = \frac{2l_2}{c\sqrt{1 - \beta^2}} . \qquad (15\text{--}8)$$

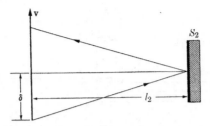

FIG. 15–4. Detail of the Michelson-Morley experiment in the calculation of δ.

The difference Δ in optical path is

$$\Delta = c(t_1 - t_2) = \frac{2}{\sqrt{1 - \beta^2}} \left(\frac{l_1}{\sqrt{1 - \beta^2}} - l_2 \right). \qquad (15\text{–}9)$$

If the instrument is rotated through 90°, l_1 and l_2 interchange roles, and

$$t_1' = \frac{2l_1/c}{\sqrt{1 - \beta^2}}, \qquad t_2' = \frac{2l_2/c}{1 - \beta^2}.$$

The new difference in optical path is

$$\Delta' = c(t_1' - t_2') = \frac{2}{\sqrt{1 - \beta^2}} \left(l_1 - \frac{l_2}{\sqrt{1 - \beta^2}} \right).$$

Thus, on rotating the apparatus through 90°, we should expect the interference pattern to shift by n fringes, where n is given by

$$n = \frac{\Delta' - \Delta}{\lambda} = \frac{2(l_1 + l_2)}{\lambda\sqrt{1 - \beta^2}} \left(1 - \frac{1}{\sqrt{1 - \beta^2}} \right)$$

$$\simeq -\frac{(l_1 + l_2)}{\lambda} \beta^2, \qquad (15\text{–}10)$$

since v is small compared with c.

No such shift was observed by Michelson and Morley. The estimated accuracy of their result was 10 km/sec, i.e., the velocity of the earth relative to any "ether frame" must be less than 10 km/sec, although the velocity of the earth in its orbit is roughly 30 km/sec. Much discussion resulted from the experiments of Miller, which gave positive indication of a velocity of about 10 km/sec apparently directed toward a certain point in space, but improved technique has confirmed the absence of a fringe shift, and a new analysis of Miller's data has shown that they are not inconsistent with those of other observers. A summary of the various

TABLE 15-1

TRIALS OF THE MICHELSON-MORLEY EXPERIMENT

Observer	Year	Place	l	$(2l/\lambda)(v/c)^2$	A	Ratio
Michelson[a]	1881	Potsdam	120 cm	0.04 fringe	0.01 fringe	2
Michelson & Morley[b]	1887	Cleveland	1100	0.40	0.005	40
Morley & Miller[c]	1902-04	Cleveland	3220	1.13	0.0073	80
Miller[d]	1921	Mt. Wilson	3200	1.12	0.04	15
Miller[e]	1923-24	Cleveland	3200	1.12	0.015	40
Miller[f] (sunlight)	1924	Cleveland	3200	1.12	0.007	80
Tomaschek[g] (starlight)	1924	Heidelberg	860	0.3	0.01	15
Miller[h]	1925-26	Mt. Wilson	3200	1.12	0.044	13
Kennedy[i]	1926	Pasadena & Mt. Wilson	200	0.07	0.001	35
Illingworth[j]	1927	Pasadena	200	0.07	0.0002	175
Piccard & Stahel[k]	1927	Mt. Rigi	280	0.13	0.003	20
Michelson et al.[l]	1929	Mt. Wilson	2590	0.9	0.005	90
Joos[m]	1930	Jena	2100	0.75	0.001	375

[a] A. A. Michelson, *Am. J. Sci.* **22**, 120 (1881); *Phil. Mag.* **13**, 236 (1882).

[b] A. A. Michelson and E. W. Morley, *Am. J. Sci.* **34**, 333 (1887); *Phil. Mag.* **24**, 449 (1887).

[c] E. W. Morley and D. C. Miller, *Phil. Mag.* **9**, 680 (1905); *Proc. Am. Acad. Arts Sci.* **41**, 321 (1905).

[d] D. C. Miller, *Data sheets of observations*, Dec. 9-11, 1921 (unpublished).

[e] D. C. Miller, *Observations*, Aug. 23-Sept. 4, 1923; June 27-July 26, 1924 (unpublished).

[f] D. C. Miller, Observations with Sunlight on July 8-9, 1924, *Proc. Nat. Acad. Sci.* **11**, 311 (1925).

[g] R. Tomaschek, *Ann. Physik* **73**, 105 (1924).

[h] D. C. Miller, *Rev. Mod. Phys.* **5**, 203 (1933).

[i] R. J. Kennedy, *Proc. Nat. Acad. Sci.* **12**, 621 (1926); *Astrophys. J.* **68**, 367 (1928).

[j] K. K. Illingworth, *Phys. Rev.* **30**, 692 (1927).

[k] A. Piccard and E. Stahel, *Compt. rend.* **183**, 420 (1926); **184**, 152, 451 (1927).

[l] A. A. Michelson, F. G. Pease, and F. Pearson, *Nature* **123**, 88 (1929); *J. Opt. Soc. Am.* **18**, 181 (1929).

[m] G. Joos, *Ann. Physik* **7**, 385 (1930); *Naturwiss.* **38**, 784 (1931).

trials of the Michelson-Morley experiment is given in Table 15-1, to which is appended a list of the appropriate journal references.* In these experi-

---

* "New Analysis of the Interferometer Observations of Dayton C. Miller," by R. S. Shankland, S. W. McCuskey, F. C. Leone, and G. Kuerti, *Rev. Modern Phys.* **27**, 167 (1955). Table 15-1 is reproduced here through the courtesy of these authors.

ments the interferometer arms were equal, and $(2l/\lambda)(v/c)^2$, by the ether theory, is the shift expected due to the earth's motion. The amplitude of the second harmonic of the fringe shift actually found by each observer is given in the column headed A. To provide a comparison of accuracy of the various trials, the last column gives the ratio of the shift expected pre-relativistically to $2A$. It is clear that a null result can be accepted with confidence.

15-3 The Lorentz-Fitzgerald contraction hypothesis.

We have seen that the search for a preferred frame for electrodynamics without further modification of either electrodynamics or mechanics was unsuccessful. An attempt to preserve the concept of the preferred ether frame despite the negative result of the Michelson-Morley experiment led to the contraction hypothesis. Lorentz and Fitzgerald postulated that as a result of motion relative to the stationary ether all bodies are contracted by the factor $\sqrt{1 - \beta^2}$ in the direction of motion. On this hypothesis, l_1 in Fig. 15-3 is actually equal to $l_1^0 \sqrt{1 - \beta^2}$, where l_1^0 is the length of l_1 when at rest with respect to the ether, while $l_2 = l_2^0$. Hence, Eq. (15-9) becomes

$$\Delta = \frac{2}{\sqrt{1 - \beta^2}} (l_1^0 - l_2^0) \qquad (15\text{-}11)$$

and no fringe shift would be obtained by rotation through 90°. Furthermore, if $l_1^0 \simeq l_2^0$, as was the case in the original experiments, no fringe shift would occur as a result of changing the velocity. If $l_1^0 \neq l_2^0$, however, even with the Lorentz contraction a fringe shift given approximately by

$$n = \frac{l_1^0 - l_2^0}{\lambda} (\beta'^2 - \beta^2) \qquad (15\text{-}12)$$

would be expected from a velocity change, as indicated by the factor $(\beta'^2 - \beta^2)$.

An interferometer using a path difference essentially as long as coherence of the source permitted was constructed by Kennedy. The square of the velocity of the instrument is presumably given by

$$v^2 = c^2 \beta^2 = (\mathbf{v}_E + \mathbf{v}_R + \mathbf{v}_S)^2,$$

where $\mathbf{v}_E$ is the velocity of the earth with respect to the sun, $\mathbf{v}_R$ is the surface velocity of the earth due to its rotation, and $\mathbf{v}_S$ is the velocity of the sun. Every twelve hours this quantity should change by

$$\Delta v^2 = 4(\mathbf{v}_S + \mathbf{v}_E) \cdot \mathbf{v}_R,$$

and every six months it should change by

$$\Delta v^2 = 4(\mathbf{v}_S + \mathbf{v}_R) \cdot \mathbf{v}_E.$$

Neither effect was observed, in contradiction to the Lorentz contraction hypothesis.

15–4 "Ether drag." A further alternative in which the concept of the ether could be reconciled with the Michelson-Morley result was to consider the ether frame attached to ponderable bodies. This would automatically give a null result for terrestrial interferometer experiments or the Trouton-Noble experiment. The assumption of a local ether, however, is in direct contradiction to two well-established phenomena.

The first is the aberration of "fixed" stars. Due to the motion of the earth about the sun, distant stars appear to move in orbits approximately 41″ in angular diameter. Consider a star at the zenith of the ecliptic. If this star is to be observed through a telescope the telescope tube must be tilted toward the direction of the earth's motion by an angle α, as shown in Fig. 15–5. It is seen from the figure that, classically,

$$\tan \alpha = v/c, \tag{15–13}$$

and with 30 km/sec for the velocity of the earth in its orbit, $\alpha = 10^{-4} = 20.5″$, in agreement with observation. If the ether were dragged by the earth in its motion we should expect no aberration to occur.

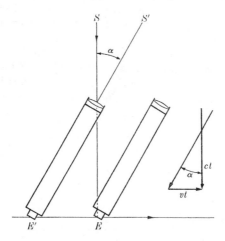

FIG. 15–5. Aberration of starlight. SE is the actual path of the light from the star at zenith and $S'E'$ is the apparent path.

The second contradiction arises in connection with the propagation of electromagnetic waves in a moving medium. We have seen in Section 11–4 that only the part of the velocity of light which depends on the polarization current $\partial \mathbf{P}/\partial t$ is affected by the motion of the medium, while the $\epsilon_0(\partial \mathbf{E}/\partial t)$ term remains unchanged. If the ether moved with the medium, the propagation velocity would be simply $c \pm v$, instead of being given by the Fresnel-Fizeau coefficient. This conclusion is contrary to observation. The "ether drag" hypothesis thus leads to discrepancies of the first order in v/c.

These considerations make the idea of a preferred frame appear unacceptable, even when it is only locally stationary. We are therefore led to the alternative that a principle of relativity must be valid in electrodynamics. This is equivalent to demand for modification of either electrodynamics or classical mechanics. We shall first consider the attempts to modify electrodynamics in such a way as to escape the paradoxes associated with the idea of an ether frame.

15–5 Emission theories. In the so-called emission theories electrodynamics is modified by supposing that the velocity of a light wave remains associated with the source rather than with a local or universal frame. In all emission theories it is postulated that the velocity of light is c/n relative to the original source, and independent of the state of the transmitting medium of refractive index n. The theories differ among themselves in predicting what happens to the velocity of light upon reflection from a moving mirror. After a reflection three alternatives are possible:

(1) The velocity remains c/n relative to the original source.

(2) The velocity becomes c/n relative to the mirror (the so-called ballistic theory).

(3) The velocity becomes c/n relative to the mirror image.

The first alternative was proposed by Ritz, and is the only one of the three theories that does not lead to coherence difficulties with reflected light. Ritz retained two of Maxwell's equations:

$$\nabla \cdot \mathbf{B} = 0$$

and

$$\nabla \times \mathbf{E} = -\frac{\partial \mathbf{B}}{\partial t},$$

with $\mathbf{E}$ and $\mathbf{B}$ derived from

$$\mathbf{B} = \nabla \times \mathbf{A}, \qquad \mathbf{E} = -\nabla \phi - \frac{\partial \mathbf{A}}{\partial t},$$

as before. But he replaced the two equations involving sources by

$$\phi = \frac{1}{4\pi\epsilon_0} \int \frac{\rho[x'_\alpha, t - r/(c + v_r)]}{r} \, dv', \qquad (15-14)$$

$$\mathbf{A} = \frac{\mu_0}{4\pi} \int \frac{\mathbf{j}[x'_\alpha, t - r/(c + v_r)]}{r} \, dv', \qquad (15-15)$$

where v_r is the component of the velocity of the source in the direction of the vector joining the source and the point of observation. Equations (15–14) and (15–15) replace the ordinary retarded potential solutions given by Eqs. (14–24) and (14–25). In this way, fields due to a moving source are definable.

In comparison with ordinary electromagnetic theory all three emission theories predict differences of the first order in v/c for experiments on the interference between light beams reflected from moving mirrors. Thomson, Majorana, and Stewart performed such experiments and obtained results in disagreement with the second and third of the emission theories but agreeing, within experimental error, with the predictions of Ritz's theory. This apparent agreement is due to the fact that as applied to any closed system of interfering beams the Ritz theory yields results differing only by terms of the second order in v/c from those to be expected if the velocity of light is constant. According to Ritz, if the source is moving toward the mirror with velocity v the time required to traverse a given distance l on the forward trip is $l/(c + v)$, while on the return trip, after reflection from the mirror, it is $l/(c - v)$. For the total elapsed time this gives

$$\Delta t = \frac{2l}{c(1 - \beta^2)},$$

which differs only by terms in the second order from the expression for constant velocity of light, $t = 2l/c$. Thus terrestrial moving source and mirror experiments fail to give a first-order contradiction to the Ritz emission theory.

There are, however, two extraterrestrial phenomena which contradict any form of emission theory. The dynamics of eclipsing binary stars has been carefully analyzed by de Sitter. If the velocity of light depends additively on the velocity of the source, it is evident that the time for light to reach the earth from the approaching star should be smaller than that from the receding member of the doublet. De Sitter showed that this would have the effect of introducing a spurious eccentricity into the orbit as calculated by the laws of mechanics. Actually, no such effect is observed; in fact, de Sitter concluded that if $v_{\text{light}} = c + kv_{\text{star}}$, then $k < 0.002$. The second piece of extraterrestrial evidence is the experience

TABLE 15–2

Theory		Light propagation experiments							Experiments from other fields					
		Aberration	Fizeau convection coefficient	Michelson-Morley	Kennedy-Thorndike	Moving sources and mirrors	De Sitter spectroscopic binaries	Michelson-Morley, using sunlight	Variation of mass with velocity	General mass-energy equivalence	Radiation from moving charges	Meson decay at high velocity	Trouton-Noble	Unipolar induction, using permanent magnet
Ether theories	Stationary ether, no contraction	A	A	D	D	A	A	D	D	N	A	N	D	D
	Stationary ether, Lorentz contraction	A	A	A	D	A	A	A	A	N	A	N	A	D
	Ether attached to ponderable bodies	D	D	A	A	A	A	A	D	N	N	N	A	N
Emission theories	Original source	A	A	A	A	A	D	D	N	N	D	N	N	N
	Ballistic	A	N	A	A	D	D	D	N	N	D	N	N	N
	New source	A	N	A	A	D	D	A	N	N	D	N	N	N
Special theory of relativity		A	A	A	A	A	A	A	A	A	A	A	A	A

Legend: A, the theory agrees with experimental results.
D, the theory disagrees with experimental results.
N, the theory is not applicable to the experiment.

TABLE 15–3

	Emission theory	Classical ether theory	Special theory of relativity
Reference system	No special reference system	Stationary ether is special reference system	No special reference system
Velocity dependence	The velocity of light depends on the motion of the source	The velocity of light is independent of the motion of the source	The velocity of light is independent of the motion
Space-time connection	Space and time are independent	Space and time are independent	Space and time are interdependent
Transformation equations	Inertial frames in relative motion are connected by a Galilean transformation	Inertial frames in relative motion are connected by a Galilean transformation	Inertial frames in relative motion are connected by a Lorentz transformation

of Miller that the Michelson-Morley experiment is not affected when light from the sun is used. The Ritz theory would predict complications of the interference pattern due, for example, to the sun's rotation.

15–6 Summary. In this chapter we have examined various kinds of experimental evidence for the incompatibility of electrodynamics and Newtonian mechanics. A summary of the relevant experiments, including some whose bearing on the subject will not become evident until we have discussed the theory of relativity, is given in Table 15–2. Their relation to attempts to reconcile them with theory is indicated, and Table 15–3 lists the basic assumptions of the three general alternatives given in Section 15–1. It is clear from Table 15–2 that the experimental basis for the theory of relativity, which modifies Newtonian mechanics, consists essentially of observations in contradiction to all reasonable alternatives. There is no single experiment that *proves* relativity theory. The experiments do present evidence that:

(1) The existence of an ether, either stationary or carried convectively, is undemonstrable.

(2) Modifications of electrodynamics, such as the emission theories, are untenable.

It is then plausible to conclude that the basic laws of mechanics need modification.

In 1905, compatible with the experimental facts known at that time, Einstein proposed the following postulates as a solution:

1. The laws of electrodynamics (including, of course, the propagation of light with the velocity c in free space), as well as the laws of mechanics, are the same in all inertial frames.

2. It is impossible to devise an experiment defining a state of absolute motion, or to determine for any physical phenomena a preferred inertial frame having special properties.

If the laws of physics conform to these two postulates all the experiments listed in Table 15–2 are explicable. In the next chapter we shall begin to examine the implications of these postulates, known as the postulates of special relativity.

The postulates of special relativity leave the question of the meaning of an "inertial frame" in the same unsatisfactory state as in the principle of Galilean relativity. Specifically, the characterization of an inertial frame in terms of absence of acceleration in the absence of forces remains essentially a "circular" argument, since it leaves open the definition of a force. An inertial frame in this context is a frame in which bodies remain in uniform motion if the influence of known (!) forces is negligible. This clearly unsatisfactory definition cannot be avoided unless the broader framework of general relativity is introduced.

ADDITIONAL PAPERS ON THE EXPERIMENTAL BASIS OF SPECIAL RELATIVITY

1. TROUTON AND NOBLE, *Phil. Trans.* **A202**, 165 (1903); *Proc. Roy. Soc.* (London) **72**, 132 (1903). Reporting attempts to find torque on a charged capacitor.

2. LORENTZ, "Versuch einer Theorie der elektrischen und optischen Erscheinungen in bewegten Körpern," Leiden, 1895. The sections on the contraction hypothesis are reprinted in *The Principle of Relativity*, listed below.

3. KENNEDY AND THORNDIKE, *Phys. Rev.* **42**, 400 (1932). Interferometer with unequal arms. Null result to ± 10 km/sec.

4. RITZ, *Ann. Chim. et Phys.* **13**, 145 (1908). Original source emission theory.

5. TOLMAN, *Phys. Rev.* **31**, 26 (1910); *Phys. Rev*, **35**, 136 (1912). THOMSON, J. J., *Phil. Mag.* **19**, 301 (1910). STEWART, *Phys. Rev.* **32**, 418 (1911). Discussions of various emission theories.

6. COMSTOCK, *Phys. Rev.* **30**, 267 (1910); DE SITTER, *Proc. Amsterdam Acad.* **15**, 1297 (1913), and **16**, 395 (1913). Spectroscopic work on binary stars.

7. MAJORANA, *Phil. Mag.* **35**, 163 (1918), and **37**, 145 (1919). Moving source and mirror experiments.

8. KENNEDY, *Phys. Rev.* **47**, 965 (1935). Critical discussion of geometrical effects of high order in the Michelson-Morley experiment.

9. IVES, *J. Opt. Soc. Am.* **28**, 215 (1938). Relativistic Doppler shift.

Further references may be found in the books on relativity theory listed below.

SUGGESTED REFERENCES

R. BECKER, *Theorie der Elektrizität*, Band II. The experimental basis of the special theory of relativity is discussed with great clarity, pp. 255–269 particularly.

P. G. BERGMANN, *An Introduction to the Theory of Relativity*. The first three chapters of this excellent book form a very readable account of the necessity for special relativity.

A. EINSTEIN, H. A. LORENTZ, H. MINKOWSKI, AND H. WEYL, *The Principle of Relativity*, "A Collection of Original Memoirs of the Special and General Theory of Relativity, together with Notes by A. Sommerfeld." Especially appropriate here are the section on "Michelson's Interference Experiment" by Lorentz and parts of his paper on "Electromagnetic Phenomena in a System Moving with any Velocity Less than that of Light."

R. TOLMAN, *Relativity, Thermodynamics, and Cosmology*. Although the experiments are not discussed in detail, a summary is included in Chapter II, together with references to original papers.

C. MØLLER, *The Theory of Relativity*.

EXERCISES

1. Find what the wave equation becomes under a Galilean transformation.

2. The torque given by Eq. (15–5) may be written as $\frac{1}{2}U_{el}(v/c)^2 \sin 2\theta$, where U_{el} is the electrostatic energy of the system of opposite charges. Show that for a parallel plate capacitor, actually used by Trouton and Noble, the torque is given by $U_{el}(v/c)^2 \sin 2\theta$, differing from Eq. (15–5) by a factor of 2.

3. Find the torque to be expected on a parallel plate capacitor suspended at an angle of 45° to the direction of the earth's orbital velocity, 30 km/sec. Take a plate area of 100 cm², separation 1 cm, and a charging voltage of 10 kilovolts.

CHAPTER 16

RELATIVISTIC KINEMATICS AND THE
LORENTZ TRANSFORMATION

In addition to Einstein's formal postulates, there are two auxiliary principles implicit in the theory of relativity, the recognition of which is important for the necessary revision of physical concepts. The first is "invariance of the sense of time." The asymmetry of experience in time has been the subject of much discussion. From the standpoint of formal physics there is only one concept which is asymmetric in the time, namely, entropy. But this makes it reasonable to assume that the second law of thermodynamics can be used to ascertain the sense of time independently in any frame of reference; i.e., we shall take the positive direction of time to be that of statistically increasing disorder, or increasing entropy and degradation of heat. The second assumption might be called "the invariance of proper quantities," or the "law of reproducibility of proper quantities." By this we mean that, whenever a measuring experiment is performed, the length of a standard (such as the wavelength of a given spectral line or a crystal lattice spacing) and the rate of a fundamental clock (such as a radioactive decay period) shall be the same as seen by an observer at rest relative to the standard. These additional postulates make it possible for us to retain exact and reproducible meaning for length and time in the course of analyzing the consequences of the postulates of special relativity.

16-1 The velocity of light and simultaneity. We have seen in Chapter 15 that there is much experimental evidence for the conclusion that a principle of relativity exists for all fields of physics, including electrodynamics. This implies, among other things, that the velocity of propagation, c, of plane electromagnetic waves in free space must be independent of the observer's inertial frame. Plausible as it may seem, this statement runs grossly contrary to intuition. Consider, for example, a light pulse starting from a point O, and consider the event as recorded by observers stationed in two frames, one frame containing O at the origin, while the other moves relative to O with a velocity v. Let the origins of the two frames coincide at the emission of the pulse. According to the principle of relativity, *both* observers must see the light wave propagating as a spherical wave centered at their *respective* origins! If we hold that the position of the wave front is an event permitting description independently in space and in time, then we cannot accept this statement as true. The independence of the velocity of light of the particular frame therefore requires a revision of the customary practice of specifying the position coordinates of an event with reference to the particular frame while specifying the time of the event on a "universal" time scale. The paradox of the two wave fronts hinges on the assumption that simultaneity is independent of the frame of the observer. If a disagreement as to the simultaneity of passage through a set of points were permitted to exist, then presumably a kinematics could be constructed in which a spherical light wave would be seen in both frames of reference.

Consequently, we are led to re-examine the concept of simultaneity. If we must abandon the existence of a universal time as not corresponding to reality, then we must establish a mechanism whereby simultaneity can be established in a given frame. The mechanism must be such that a measurement of the velocity of light in the particular frame using its time and distance scale must give c. This means that the only way in which simultaneity can be defined is by means of the velocity of light itself. This conclusion gives c a much more fundamental significance than just the velocity of propagation of electromagnetic waves: it introduces c into all the relations of physics. For example, the utilization of c as the defining element of simultaneity precludes the existence of the "ideal rigid body" of mechanics; if there were such a body its ends would move simultaneously as observed from any frame, and it could therefore be used to establish a "universal time," in violation of our former conclusion.

We therefore consider two instants of time t_1 and t_2 observed at two points x_1 and x_2 in a particular frame as simultaneous if a light wave emitted at the geometrically measured mid-point between x_1 and x_2 arrives at x_1 at the time t_1 and at x_2 at the time t_2.

This definition will automatically ensure that a light pulse emitted at the origin will reach all equidistant points simultaneously and that the wave surface is therefore a sphere in a particular reference frame. Simultaneity of two events at two spatially separated points thus has no significance independent of the frame. The relation of the time intervals observed by two different frames will depend on the spatial interval between the events, and the Galilean transformation, Eqs. (15–1), which transformed temporal intervals as observed by two frames independently of spatial coordinates cannot be in agreement with the simultaneity definition in terms of c. We must therefore attempt to derive the corresponding transformations from an $(x, y, z; t)$ frame to an $(x', y', z'; t')$ frame which will supersede the Galilean transformation. Such a transformation must remain linear, to assure mathematical equivalence of all points in space and time, but the spatial and temporal coordinates need not transform independently.

16–2 Kinematic relations in special relativity. The transformation we seek will give the relations between the space-time coordinates of an arbitrary event $(x, y, z; t)$ as observed in a coordinate frame we may designate as Σ, and the space-time coordinates $(x', y', z'; t')$ in a frame Σ' moving with uniform velocity relative to Σ. The transformation must obey the postulates of special relativity, given in Section 15–6, for an event of any type. To see how the basic postulates of relativity necessitate the nature of the transformation, we can construct a set of "thought experiments" (*Gedanken Experimente*) so as to introduce one feature of the transformation at a time.

Experiment I. *Comparison of parallel measuring sticks oriented perpendicular to their direction of relative motion* (see Fig. 16–1). It is explicitly assumed that the properties of a given body of specified structure are independent of its past history when observed in a frame where the body is at rest. (This is called the "proper frame" of the body, and the length of a rod when measured in a frame in which the rod is at rest is called its "proper length.") It is therefore possible to demand that the two measuring sticks, OP and $O'P'$, of Fig. 16–1 be of equal proper length, whether they can be brought to rest with respect to each other or not; for instance, it could be specified that the length of each rod should be a given number of wavelengths of a particular spectral line measured in each frame. Both rods are at right angles to $\mathbf{v}$ (assumed along the x-axis), the velocity of the system Σ' as measured in frame Σ. (To an observer in Σ', Σ will have a velocity $-\mathbf{v}$.)

Let the two systems approach each other so that the mid-points of the rods, M and M', will coincide at passing. Considering Σ' to be moving and Σ to be stationary, let light signals be sent from O' and P' when O' and P' coincide with the y-axis. Since $O'M$ remains equal to $P'M$ during the motion, O' and P' will appear to cross the y-axis simultaneously in both systems. Similarly, O and P will cross the y'-axis simultaneously in both systems. Since the time of observation for both ends is defined iden-

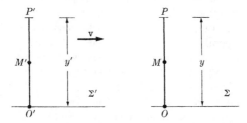

Fig. 16–1. Comparison of parallel measuring sticks oriented perpendicular to their direction of relative motion. $\mathbf{v}$ is the velocity of the Σ' frame measured in Σ.

tically in both frames, both observers can compare the positions of the end markers at the time of crossover and arrive at the same result. Hence *both* observers would conclude either that $OP \leq O'P'$ or $OP \geq O'P'$. Both systems are equivalent in every way, so that an asymmetric relation would provide a means of determining absolute motion, a possibility ruled out by the relativity postulate. We therefore conclude that lengths at right angles to the motion are the same in both frames, and the transformation for such lengths is

$$y' = y, \qquad z' = z, \qquad (16\text{--}1)$$

just as in the Galilean transformation.

EXPERIMENT II. *Comparison of clock rates.* In comparing the rates of clocks attached to coordinate frames in relative motion, we must recognize that it is impossible to compare *one* clock in Σ with *one* clock in Σ', since the clocks will not stay in coincidence. We must compare *two* clocks in Σ with *one* clock in Σ', and have the two clocks in Σ synchronized by means of light signals. Let a source of light be at the same position as the clock in Σ', as shown in Fig. 16–2. A light signal emitted normal to **v** is reflected by a mirror which is normal to the z'-axis and at a distance z_0' from the source, and returns to the clock. For an observer at rest in Σ' the time interval between sending and receiving the pulse is

$$\Delta t' = \frac{2z_0'}{c}. \qquad (16\text{--}2)$$

Observers at rest in Σ can record the time interval Δt between the same two events with two clocks spaced a distance $v\,\Delta t$ apart along x, the direction of the motion. Since c is independent of the frame, this interval will be given by

$$c\,\Delta t = 2\sqrt{(z_0)^2 + \left(\frac{v\,\Delta t}{2}\right)^2}, \qquad \Delta t = \frac{2z_0}{c}\frac{1}{\sqrt{1-\beta^2}}. \qquad (16\text{--}3)$$

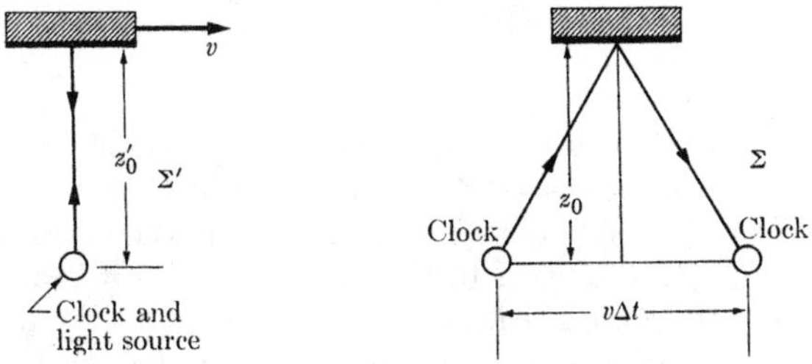

FIG. 16–2. Comparison of clock rates.

Since $z_0 = z_0'$ it follows that

$$\Delta t = \frac{\Delta t'}{\sqrt{1 - \beta^2}}, \tag{16-4}$$

where $\Delta t'$ is the proper time inteval, i.e., the time interval between two events occurring at the same place in the Σ' frame. Δt is not a proper time interval, for it is measured by two different clocks at different places in Σ. As in the case of a proper length, discussed above, a proper time interval is a definite function of the physical nature of the clock; e.g., a particular radioactive decay constant, or the natural frequency of a crystal of specified proper dimensions, is a constant in the frame where such time intervals are observable at a single point, i.e., in a frame where such a "clock" is at rest.

Note that the observer in Σ will find that his time interval Δt is longer than the proper time interval. This phenomenon is known as *time dilation*. For example, the lifetime of a high velocity meson disintegrating in flight appears lengthened, to a ground observer, by an amount depending on the meson velocity. On the other hand, the lifetime in the proper frame, i.e., the frame at rest with respect to the meson, is an invariant.

EXPERIMENT III. *Comparison of lengths parallel to the direction of motion.* Let us consider a rod of length x_0' in the frame in which it is at rest, i.e., a rod of proper length x_0'. In the Σ frame its length x_0 would be the distance between the ends of the rod observed "simultaneously" in the sense of the simultaneity definition in terms of c. To disentangle the length comparison from the simultaneity calculation, let us consider the following event: a light source S' at one end of the rod in frame Σ' sends a light pulse to a mirror M' at the other end, where it is reflected back to the source. The time interval between the emission and return of the signal is $\Delta t'$. Note that $\Delta t'$ is a proper time interval, being observable with a single clock at one point. Evidently,

$$\Delta t' = \frac{2x_0'}{c}. \tag{16-5}$$

Seen from Σ, these same events appear more complicated. With reference to Fig. 16–3, the source S' was at S_0 at the time of emission, and the mirror M' was at M_0. At the time of reflection, the mirror M' has moved to M, and the pulse returns to the source S' when it is at S_2. The time interval Δt is therefore measured between the points S_0 and S_2 with two clocks, as in Experiment II. Δt is not a proper time, and Eq. (16–4) will apply.

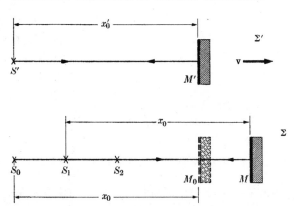

FIG. 16-3. Comparison of lengths parallel to the direction of motion.

Now, by definition, we mean by x_0 the distance $S_0 M_0 = S_1 M$. Since M_0 has moved to M with velocity v while the light moved from S_0 to M with velocity c, we have

$$S_0 M = x_0 + \frac{v}{c} S_0 M, \qquad S_0 M = \frac{x_0}{1 - \beta}.$$

Similarly, since the source has moved from S_1 to S_2 with velocity v while the light has traveled from M to S_2 with velocity c,

$$M S_2 = x_0 - \frac{v}{c} M S_2, \qquad M S_2 = \frac{x_0}{1 + \beta}.$$

Hence,

$$\Delta t = \frac{S_0 M + M S_2}{c}$$

$$= \frac{2 x_0}{c(1 - \beta^2)}, \tag{16-6}$$

and with the use of Eqs. (16-5) and (16-4), we obtain

$$x_0 = x_0' \sqrt{1 - \beta^2}. \tag{16-7}$$

This relation, called the Lorentz contraction, is asymmetrical in x_0 and x_0', since it gives the relation between measurement of a proper length x_0' (at rest) in Σ' and an improper length x_0 (not at rest) in Σ. The length x_0 in Σ has been definable only by the assumption of the constancy of the velocity of light. Equation (16-7) is identical with the hypothesis of Section 15-3, but here it is accompanied by a time dilation quite foreign to the postulate of Lorentz and Fitzgerald.

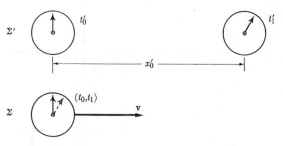

FIG. 16–4. Synchronization of clocks.

EXPERIMENT IV. *The synchronization of clocks.* By appropriate design of the conceptual experiments II and III, we have been able to derive from the postulates of relativity the transformation of temporal and spatial intervals from proper to nonproper frames. The question of how two clocks, synchronized in frame Σ' but separated by a distance x_0', would appear to an observer in Σ requires further consideration.

Consider two clocks synchronized in Σ' and located a distance x_0' apart, as indicated in Fig. 16–4. Let there be a single clock in Σ which will record the times t_0 and t_1 when it passes the ends of x_0'. The corresponding times as recorded on the two clocks in Σ' are denoted by t_0' and t_1'. Since $t_1 - t_0$ is a proper time interval in Σ, we can apply Eq. (16–4) in reverse, i.e.,

$$t_1' - t_0' = \frac{t_1 - t_0}{\sqrt{1 - \beta^2}}. \tag{16–8}$$

To observers in Σ, the time intervals on both of the Σ' clocks individually would appear too large by the same ratio, but they will not be in time with each other to a single observer. He would observe an error in synchronization of amount δ, that is,

$$t_1 - t_0 = \frac{t_1' - t_0' + \delta}{\sqrt{1 - \beta^2}}. \tag{16–9}$$

To determine δ, we need only note that in both frames the relative velocity must be the same, since the frames are equivalent. Hence

$$t_1' - t_0' = x_0'/v,$$
$$t_1 - t_0 = x_0/v. \tag{16–10}$$

But x_0' is a proper length in Σ', so that

$$x_0 = x_0'\sqrt{1 - \beta^2}. \tag{16–7}$$

By combining Eqs. (16–9), (16–10), and (16–7), we obtain

$$\delta = - \frac{x_0'\beta^2}{v}. \qquad (16\text{–}11)$$

The meaning of the negative sign is that to an observer in Σ the second clock in Σ' indicates a time later than the first. A whole series of clocks, equally spaced and synchronized in Σ', would appear successively ahead by an amount δ to an observer in Σ as he went by with velocity v.

By means of these four conceptual experiments we have demonstrated that the postulates of special relativity lead to four kinematic relations:

I. Spatial intervals transverse to the direction of relative motion are invariant.

II. A time interval Δt between two events measured in a frame moving with velocity $\pm v$ relative to a frame in which the corresponding time interval $\Delta\tau$ is proper (i.e., the two events occur at one place) is given by

$$\Delta t = \frac{\Delta\tau}{\sqrt{1 - \beta^2}}. \qquad (16\text{–}12)$$

III. The length Δx of a rod measured in a frame moving with velocity $v_x = \pm v$ relative to a frame in which the rod is at rest and has the proper length ΔL is given by

$$\Delta x = \Delta L\sqrt{1 - \beta^2}. \qquad (16\text{–}13)$$

IV. Two clocks, synchronous and separated by a distance ΔL in a given frame, appear out of synchronism as observed from a frame moving with a relative velocity v to the clock frame by an amount

$$\delta = - \frac{\Delta L v}{c^2}. \qquad (16\text{–}14)$$

16–3 The Lorentz transformation. The separate kinematic effects derived in the previous section from the fundamental postulates of relativity can be combined to give the general relations between the time and space coordinates of a particular event as observed from inertial frames in relative motion.

A point event at P moving with the Σ' frame, as indicated in Fig. 16–5, occurs at time t' and with coordinates x', y', z' in Σ'. Let us consider this same event as observed from a frame Σ in motion relative to Σ'. For simplicity, we may choose the x- and x'-axis as the direction of relative motion, and let the origins and the zero point of time be so chosen that at $t = t' = 0$ the two origins coincide. The time t or t' therefore means the

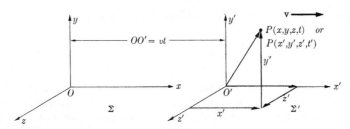

Fig. 16–5. Derivation of the Lorentz transformation equations for Σ and Σ'.

time elapsed since the coincidence of origins as measured on clocks in Σ or Σ', respectively. We shall say that v is positive if the origin of Σ' moves along the positive x-direction in Σ.

To observers in Σ, $OO' = vt$, but x', a proper length in Σ', is shortened by the Lorentz contraction. Hence

$$x = vt + x'\sqrt{1 - \beta^2} \qquad (16\text{–}15)$$

or

$$x' = \frac{x - vt}{\sqrt{1 - \beta^2}}. \qquad (16\text{–}16)$$

Again from the point of view of Σ, clocks located at P and O' (and synchronized in Σ') are out of synchronism by an amount

$$\Delta t' = \frac{x'v}{c^2}. \qquad (16\text{–}17)$$

The clocks at O and O' were synchronous at $t = t' = 0$. Since that time, according to Σ, the Σ' clocks have been running slow, i.e., at a rate which must be dilated by $1/\sqrt{1 - \beta^2}$ to make it equal to the rate of the clock on Σ. Combining this effect with that of Eq. (16–17), we have

$$t = \frac{t' + x'v/c^2}{\sqrt{1 - \beta^2}}. \qquad (16\text{–}18)$$

By the use of Eq. (16–16) this can be reduced to

$$t' = \frac{t - xv/c^2}{\sqrt{1 - \beta^2}}. \qquad (16\text{–}19)$$

Equations (16–18) and (16–19) indicate that except for the sign of v, Σ and Σ' are equivalent, in agreement with the second relativity postulate.

It can also be shown from Eqs. (16–18) and (16–16) that

$$x = \frac{x' + vt'}{\sqrt{1 - \beta^2}},\qquad (16\text{–}20)$$

in agreement with Eq. (16–16) except for the sign of v.

Equations (16–16), (16–20), (16–18), and (16–19), together with the consequence of Experiment I of Section 16–2 that lengths perpendicular to the motion are unaffected, constitute the general transformations we have sought, subject to the restrictions as to choice of origin and orientation of axes given above. For convenience, we may summarize what is known as the Lorentz transformation:

$$x' = \frac{x - vt}{\sqrt{1 - \beta^2}},$$

$$y' = y,$$

$$z' = z,\qquad (16\text{–}21)$$

$$t' = \frac{t - xv/c^2}{\sqrt{1 - \beta^2}}.$$

It is easy to show algebraically that if the Lorentz transformation is valid,

$$x^2 + y^2 + z^2 - c^2t^2 = x'^2 + y'^2 + z'^2 - c^2t'^2. \qquad (16\text{–}22)$$

This means that if a light signal is propagated in all directions with velocity c from O at $t = 0$, as observed in frame Σ, then a light signal is propagated from O' in all directions with velocity c at $t' = 0$, as observed in Σ'. The transformations are therefore in agreement with the first postulate, and resolve the apparent paradox mentioned at the beginning of the chapter.

The Galilean transformation equations, (15–1), do not satisfy Eq. (16–22). For two events, $(x_1, y_1, z_1; t_1)$ and $(x_2, y_2, z_2; t_2)$, Eqs. (15–1) yield

$$(x'_1 - x'_2)^2 + (y'_1 - y'_2)^2 + (z'_1 - z'_2)^2$$
$$= (x_1 - x_2)^2 + (y_1 - y_2)^2 + (z_1 - z_2)^2,$$
$$(t'_1 - t'_2) = (t_1 - t_2),$$

showing that in prerelativistic physics the spatial interval and the temporal interval between two events are *independently* invariant. In special relativity it is the *combined* space-time interval

$$(x_1 - x_2)^2 + (y_1 - y_2)^2 + (z_1 - z_2)^2 - c^2(t_1 - t_2)^2$$

which is invariant. In terms of the differential interval between two events the quantity

$$ds^2 = c^2 \, dt^2 - dx^2 - dy^2 - dz^2 \qquad (16\text{--}23)$$

is invariant under a Lorentz transformation.

The differential interval ds defined by Eq. (16–23) involves both space and time, but not symmetrically. If in any frame

$$dx^2 + dy^2 + dz^2 < c^2 \, dt^2, \qquad (16\text{--}24)$$

then an inertial frame can be found in which the spatial part of ds is zero, i.e., in which the two events occur at the same place. In that frame, $ds = c \, dt$ is just c times a proper time interval, as defined earlier. If the space-time interval satisfies the inequality (16–24), we say that the interval is "time-like," and ds/c represents the proper time interval between the events. Conversely, if in any frame

$$dx^2 + dy^2 + dz^2 > c^2 \, dt^2, \qquad (16\text{--}25)$$

then a frame can be found in which $dt = 0$; in that frame the two events are simultaneous and $i \, ds$ is their mutual distance. Hence, if ds satisfies the inequality (16–25), the interval is space-like, and $i \, ds$ represents the proper length of the increment. No Lorentz transformation with real β can invalidate either of the inequalities (16–24) or (16–25), and the time-like or space-like nature of an interval ds is an invariant.

It can be shown algebraically that two successive Lorentz transformations with velocity parameters β_1 and β_2 are equivalent to a single Lorentz transformation of parameter

$$\beta = \frac{\beta_1 + \beta_2}{1 + \beta_1 \beta_2}. \qquad (16\text{--}26)$$

The Lorentz transformations thus form a mathematical "group" with "commutative" properties.

It is possible to obtain the Lorentz transformation equations in several ways by simply using the demand that the interval ds of Eq. (16–23) be invariant and that the transformations be linear; the second condition arises from the fact that all points in space and time should have identical transformation character if only inertial frames are considered. We must also demand, in accord with the relativity principle, that if $x = f(x', t', v)$, then $x' = f(x, t, -v)$. As an example of such a derivation, let us begin by assuming

$$kx = x' - vt', \qquad kx' = x + vt, \tag{16–27}$$

where k is to be an even function of the velocity. By simple algebra,

$$x'^2 - x^2 = v^2(t'^2 - t^2)/(1 - k^2). \tag{16–28}$$

To make Eq. (16–28) agree with the invariant interval, Eq. (16–22), we must set

$$c^2(1 - k^2) = v^2$$

or

$$k = \sqrt{1 - \beta^2}, \tag{16–29}$$

giving the Lorentz transformation. A more rigorous formal derivation, based on the invariance of ds^2, is given in the following section.

16–4 Geometric interpretations of the Lorentz transformation. The Lorentz transformation treats x and t as equivalent variables. It was suggested by Minkowski that ct be introduced simply as a fourth coordinate. Let us put

$$x^1 = x, \qquad x^2 = y, \qquad x^3 = z, \qquad x^4 = ct, \tag{16–30}$$

as a set of variables in four-dimensional space. (Superscripts rather than subscripts are used here for a reason that will become clear in the following chapters.) The space-time interval ds is therefore given by

$$ds^2 = -(dx^1)^2 - (dx^2)^2 - (dx^3)^2 + (dx^4)^2. \tag{16–31}$$

The Lorentz transformation is thus, in a general sense, the set of linear transformations in four-dimensional space which leaves ds^2 invariant.

The interval ds can be written more symmetrically, although the result is physically less obvious, if we introduce

$$X^4 = ix^4, \qquad dS = i\, ds. \tag{16–32}$$

Equation (16–23) becomes

$$dS^2 = (dx^1)^2 + (dx^2)^2 + (dx^3)^2 + (dX^4)^2. \tag{16–33}$$

Since our choice of coordinates in the Lorentz transformation leaves dx^2 and dx^3 unaffected, it will suffice to consider the invariance of the two-dimensional interval

$$dS^2 = (dx^1)^2 + (dX^4)^2. \tag{16–34}$$

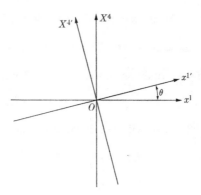

FIG. 16–6. Rotation of the complex coordinate axes through the angle θ.

This interval is invariant to translations of the origin, and also to rotations of the coordinate axes in the x^1–X^4 plane. Let the coordinate axes be rotated through an angle θ, as shown in Fig. 16–6. It is clear that

$$x'^1 = x^1 \cos \theta + X^4 \sin \theta,$$
$$X'^4 = -x^1 \sin \theta + X^4 \cos \theta. \tag{16–35}$$

Putting $\theta = i\varphi$, and transforming back to ordinary space and time variables by Eqs. (16–30) and (16–32), we obtain

$$x' = x \cosh \varphi - ct \sinh \varphi,$$
$$ct' = -x \sinh \varphi + ct \cosh \varphi. \tag{16–36}$$

Equations (16–36) are identical with the Lorentz transformation equations if we put

$$\sinh \varphi = \frac{\beta}{\sqrt{1 - \beta^2}} \; ; \qquad \cosh \varphi = \frac{1}{\sqrt{1 - \beta^2}} \; ; \qquad \tanh \varphi = \beta. \tag{16–37}$$

Thus the Lorentz transformation is simply a rotation in the four-dimensional space x^1, x^2, x^3, X^4. An "event" is therefore conveniently described by the four coordinates in such a space-time system, where temporal and spatial coordinates are entirely equivalent. Equation (16–26), which gives the β equivalent to two successive Lorentz transformations, simply corresponds to the addition formula for $\tanh \varphi$:

$$\tanh (\varphi_1 + \varphi_2) = \frac{\tanh \varphi_1 + \tanh \varphi_2}{1 + \tanh \varphi_1 \tanh \varphi_2} . \tag{16–38}$$

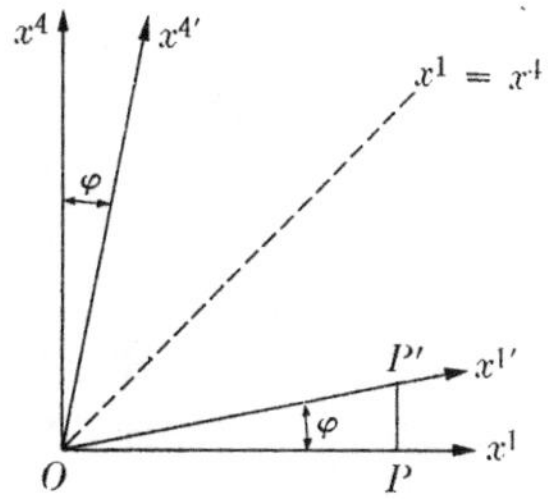

FIG. 16-7. Lorentz transformed axes in a Minkowski diagram, with PP' illustrating the relativity of simultaneity.

The representation of the Lorentz transformation as a rotation in the four-dimensional space x, y, z, ict is a useful concept, but it is artificial from a physical viewpoint. Let us investigate the geometrical representation of the Lorentz transformation in the real four-dimensional space of x, y, z, ct. We shall plot only $x^1 = x$ and $x^4 = ct$, so as to permit representation in a plane. The trajectory of an event so plotted as a function of space and time is called a world line, and the diagram itself is called the Minkowski diagram. The world line of a ray of light *in vacuo* is the line $x^1 = x^4$, at 45° to each of the original axes in Fig. 16-7. Under a Lorentz transformation, these axes will transform into x'^1 and x'^4 by Eqs. (16-36), but the world line of the light ray is unchanged. Figure 16-7 shows that simultaneity is a relative concept; all events located on the x'^1 axis are simultaneous in Σ', but not in Σ. Thus event P' is simultaneous with event O to an observer at rest in Σ', but occurs later, at time $t = x^4/c = PP'/c$, to an observer at rest in Σ.

Considerable care is necessary in the general interpretation of the Minkowski diagram. In contrast to the diagram of Fig. 16-6, where we artificially produced a "Euclidian" geometry by the imaginary transformation Eqs. (16-32), the intervals cannot be measured by the sum of the squares of the coordinate intervals. A substitute for distance measurement in this real space can be obtained by noting that the family of hyperbolas

$$(x^1)^2 - (x^4)^2 = (x'^1)^2 - (x'^4)^2 = \text{constant} \qquad (16\text{-}39)$$

lays out a convenient net which permits comparison of the various quantities involved. Such a net is indicated in Fig. 16-8.

Let us see how the various phenomena of relativistic kinematics are interpreted on this diagram. The Lorentz contraction, Eq. (16-7), gives the transformation of a proper length x'^1 in Σ' to the Σ frame. Consider a rod OP' at rest in Σ'. The world line of the end point P' is moving parallel to the $x'^4(ct')$-axis, since it is at rest (proper) in Σ'. Similarly,

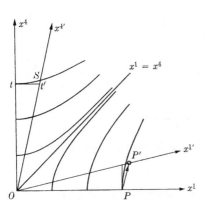

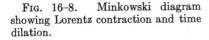

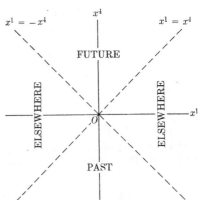

FIG. 16–8.   Minkowski diagram showing Lorentz contraction and time dilation.

FIG. 16–9.   Minkowski diagram showing the division of space-time by the light cone into past, future and elsewhere for an observer at the origin.

the point O is moving along the x'^4 axis. The rod is measured in Σ when the origins coincide and when the end points are simultaneous in Σ, i.e., along the x^1-axis. The length of the rod in Σ is thus the length OP. In comparing OP and OP', we must be careful to refer the measurements to the hyperbolic grid. It is easily seen that the hyperbola

$$(x^1)^2 - (x^4)^2 = OP^2 = (x'^1)^2 - (x'^4)^2$$

crosses $Ox^{1'}$ between O and P', and hence $OP < OP'$, in accordance with the Lorentz contraction of proper lengths observed from a moving frame,

In order to consider the time dilation, let us take a single clock at rest in Σ' at $x' = t' = x = t = 0$. As time progresses, the time interval relative to $t' = t = 0$ will be represented by a world line moving along the x'^4-axis and, in Σ', is thus just Ot'. O and t' are not at the same spatial point in Σ, and the point considered simultaneous with t' will be at t; there tt' is parallel to the x^1-axis. But in the hyperbolic "metric," $Ot = OS$, and $Ot > Ot'$. Hence an observer at rest in Σ will observe a longer elapsed time than the proper time interval measured in Σ', in agreement with the time dilation of Eq. (16–4).

The Minkowski diagram shows the symmetry between the Σ and Σ' frames despite the apparent asymmetry of the time dilation and Lorentz contraction. In our examples, Σ' was taken as the proper frame for both spatial and temporal intervals. If we had chosen Σ as the proper frame, PP' would have been parallel to Ot, and tt' parallel to OP', which would have reversed the contraction and dilation relations.

A Lorentz transformation with $\beta > 1$ is complex and thus physically impossible. (We shall re-examine the significance of this statement later.) On the Minkowski diagram, this means that neither of the primed axes can pass the line $x^1 = -x^4$, which in four-dimensional space is a cone, called the light cone. Hence a time-like interval with $ds^2 = (ds^4)^2 - (dx^1)^2 > 0$ cannot become space-like in any frame, but can become purely temporal if referred to the proper frame. For the same reason, a space-like interval $ds^2 < 0$ cannot become time-like in any frame, but can become purely spatial if referred to a proper frame.

The light cone $x^1 = \pm x^4$ thus divides the Minkowski space into regions, as indicated in Fig. 16–9, that have invariant significance for an observer at the origin, i.e., at a given time and place. The temporal region labeled "past" represents events whose temporal interval relative to the origin is negative from any inertial frame. Similarly, positive temporal intervals are confined to the region labeled "future," regardless of the frame of the observer at O. Events in the region called "elsewhere" are spatially separated from the observer, but can be transformed into one another by spatial rotations and translations.

16–5 Transformation equations for velocity. The transformation equations for the velocity of a moving point can be found by taking the derivatives of the Lorentz transformation equations with respect to t and t'. We shall continue to use the symbols v and $\beta = v/c$ to denote the velocity of frame Σ' relative to frame Σ, evaluated in Σ, and introduce $u_x = dx/dt$, $u_x' = dx'/dt'$, etc., to denote velocities in a given frame. The Lorentz transformation from Σ to Σ' has been found to be

$$x' = \frac{x - vt}{\sqrt{1 - \beta^2}},$$

$$y' = y, \qquad z' = z, \tag{16–40}$$

$$t' = \frac{t - \beta x/c}{\sqrt{1 - \beta^2}}.$$

Differentiating Eqs. (16–40) with respect to t', we obtain

$$\frac{dx'}{dt'} = u_x' = \frac{u_x - v}{\sqrt{1 - \beta^2}} \frac{dt}{dt'}, \tag{16–41}$$

$$\frac{dy'}{dt'} = \frac{dy}{dt} \frac{dt}{dt'}, \qquad \frac{dz'}{dt'} = \frac{dz}{dt} \frac{dt}{dt'}. \tag{16–42}$$

Also,

$$\frac{dt'}{dt} = \frac{1 - \beta u_x/c}{\sqrt{1 - \beta^2}}.$$ (16-43)

Equation (16-43) enables us to write Eqs. (16-41) and (16-42) in terms of the unprimed variables:

$$u_x' = \frac{u_x - v}{1 - u_x v/c^2},$$ (16-44)

$$u_y' = \frac{\sqrt{1 - \beta^2}}{1 - u_x v/c^2} u_y, \qquad u_z' = \frac{\sqrt{1 - \beta^2}}{1 - u_x v/c^2} u_z.$$ (16-45)

Equation (16-44), the "longitudinal velocity addition formula," is in agreement with Eq. (16-26) for successive Lorentz transformations, since u_x may represent the motion of the origin of another Lorentz frame relative to the Σ' frame.

Equation (16-43), the relation between time intervals dt and dt', has some interesting consequences. If it were possible to make $(dt'/dt) < 0$ by a suitable choice of u, then the temporal sequence of two events would be reversed between the two frames under consideration. This would be a logical contradiction if (a) the two events represent cause and effect, and (b) the *sense* of time has an invariant significance. With the assumption of (b), made at the beginning of this chapter, we are forced to conclude that, in order for the sequence between cause and effect to be preserved in all frames, in *any* particular frame

$$u \leq c,$$ (16-46)

where u represents the velocity of propagation of any event which can connect cause and effect. Obviously, phase velocities, or velocities of geometrical significance only, are not affected by this restriction. The restriction does apply to the relative velocities v of possible inertial frames, so that it is not necessary to consider the kinematical significance of the Lorentz transformation when $\sqrt{1 - \beta^2}$ is complex.

The velocity transformation equations apply very simply to the velocity of light in a moving medium and to stellar aberration.

Let a homogeneous isotropic medium of index of refraction n move with velocity v along the positive x-axis with respect to an observer at rest in Σ. The velocity of light with respect to the frame in which the medium is at rest is $u' = c/n$. With respect to an observer at rest in Σ, the velocity is u, where u is obtained from Eq. (16-44) by reversing the sign of v,

$$u = \frac{c/n + v}{1 + v/nc} \simeq \frac{c}{n} + v(1 - 1/n^2)$$ (16-47)

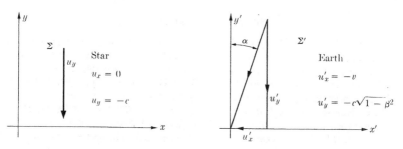

Fig. 16–10. Relativistic explanation of aberration of starlight.

to the first order in v/c. This is in agreement with the experimental facts and the classical electrodynamic result derived in Section 11–4. In the classical derivation of this equation, however, a relatively complicated mechanism was involved: the effect was attributed to reradiation from the moving secondary radiators in the fluid. Here Eq. (16–47) has been derived without any detailed information concerning the mechanism. We shall frequently meet situations in which an end result is demanded by relativistic considerations, but where the physical mechanism of attaining the result is far from obvious.

The aberration of distant stars can also be derived from the velocity transformation equations. Let a ray of starlight approach the earth in a direction perpendicular to the earth's velocity, as indicated in Fig. 16–10. In the frame of the star the process is simply the emission of light with velocity components $u_y = -c$, $u_x = 0$. In the earth's frame, the velocity components as given by Eqs. (16–44) and (16–45) are

$$u_y' = -c\sqrt{1 - v^2/c^2}, \qquad u_x' = -v.$$

Therefore the angle of incidence to the normal is given by

$$\tan \theta' = \frac{u_x'}{u_y'} = \frac{\beta}{\sqrt{1 - \beta^2}},$$

$$\sin \theta' = \beta.$$

(16–48)

We have seen in Chapter 15 that a mechanical emission picture or the assumption of a stationary ether gives $\tan \theta' = \beta$, which is, in practice, indistinguishable from Eq. (16–48).

SUGGESTED REFERENCES

There are many excellent introductions to the concepts of special relativity and the Lorentz transformation. The list below includes treatments most familiar to the present authors and those likely to be most easily accessible.

R. BECKER, *Theorie der Elektrizitat*, Band II, especially, pp. 266–282. The Minkowski geometry is particularly well presented.

P. G. BERGMANN, *An Introduction to the Theory of Relativity*. Chapter IV introduces the Lorentz transformation in a very lucid way.

V. FOCK, *The Theory of Space, Time and Gravitation*. The first chapter provides a thorough and elegant introduction to relativity theory.

R. C. TOLMAN, *Relativity, Thermodynamics, and Cosmology*, through p. 34, or, if it is available, *The Theory of the Relativity of Motion*, which provides an excellent introduction to the subject.

A. EINSTEIN AND H. MINKOWSKI in *The Principle of Relativity*. Original papers, carefully translated.

A. EINSTEIN, *The Meaning of Relativity*. This is not a textbook nor one designed to be read hurriedly. It will repay careful study.

C. MØLLER, *The Theory of Relativity*, Chapters I and II. This is a recent book to which we shall refer in later chapters.

Of the many more general textbooks which contain clear and competent introductions to special relativity, we may mention

G. JOOS, *Theoretical Physics*, Chapter X.

R. B. LINDSAY AND H. MARGENAU, *Foundations of Physics*, Chapter 7.

EXERCISES

1. Does the choice of relative velocity along the x-axis and coincidence of origins at $t = t' = 0$ put any essential limitations on the Lorentz transformation as given by Eqs. (16–21)? How, in detail, would you handle an arbitrarily directed relative velocity and an arbitrary pair of origins in the coordinate frames in making a Lorentz transformation?

2. With relation to the Minkowski diagram of Fig. 16–8, interpret a proper length and a proper time in Σ as observed in the primed system Σ'.

3. Show that the transformation of $\dot{u}^2$ from a proper frame to a general frame is

$$\dot{u}^2 \rightarrow [\dot{u}^2 - (\dot{\mathbf{u}} \times \boldsymbol{\beta})^2]\gamma^6,$$

where $\gamma = 1/\sqrt{1 - \beta^2}$.

CHAPTER 17

COVARIANCE AND RELATIVISTIC MECHANICS

In Chapter 16 we have investigated the bearing of the principles of special relativity on the laws of kinematics. If the principles are valid they must apply to all fields of physics; by no experiment of any kind should it be possible to detect a preferred inertial frame. The bearing of the principles of relativity on other fields could be introduced by designing appropriate *Gedanken Experimente*, as we did in kinematics. Or we could attempt to obtain transformation relations for physical quantities by applying the Lorentz transformation to the time and space coordinates of the pertinent prerelativistic equations and then trying to deduce transformation relations for the remaining quantities. Both of these approaches are useful. The thought experiment approach remains closest to physical concepts, and we shall make use of it in discussing collisions. The direct transformation process is frequently tedious, although it was the method used by Einstein in his original work to deduce the transformation equations for the electromagnetic fields and to show that Maxwell's equations are in agreement with relativistic principles.

There is a third method which is by far the most powerful and efficient in extending relativity. This approach is to rewrite the equations in a *form* which explicitly makes evident how the quantities would behave under a change to a different inertial frame. If an equation has a form which is invariant to a change in inertial frame, then an experiment based on this equation obviously could not give a result depending on the particular frame of reference. The equation then describes a phenomenon which would be in agreement with the principles of special relativity. An equation written in such a way that its form is independent of the choice of inertial frame is said to be *Lorentz covariant*.

We shall investigate this method sufficiently to be able to deduce the basic relativistic relations in mechanics and electrodynamics.

17–1 The Lorentz transformation of a four-vector. The Lorentz transformation can be written as a linear transformation of the interval components from the origin to point x^j, $(x^1, x^2, x^3, x^4 = ct)$, in a four-dimensional space. Making use of the summation convention for repeated indices, we may write

$$x'^i = Q_j^i \, x^j, \tag{17–1}$$

where Q^i_j is given by the matrix

$$Q^i_j = \begin{pmatrix} \gamma & 0 & 0 & -\beta\gamma \\ 0 & 1 & 0 & 0 \\ 0 & 0 & 1 & 0 \\ -\beta\gamma & 0 & 0 & \gamma \end{pmatrix}. \tag{17-2}$$

Here $\beta = v/c$ as usual, and for convenience in writing,

$$\gamma = \frac{1}{\sqrt{1 - \beta^2}}. \tag{17-3}$$

We have already noted that if the components of the vector x^j transform according to Eq. (17-1), then an experiment involving x^j cannot yield a preferred frame. If, therefore, any physical relation is written as a vector equation in four-space where the vector components transform in accordance with Eq. (17-1), then such an equation is said to be written in Lorentz covariant form.

If we solve Eq. (17-1) for x^j, we obtain

$$x^j = (Q^i_j)^{-1} x'^i, \tag{17-4}$$

where

$$(Q^i_j)^{-1} = \begin{pmatrix} \gamma & 0 & 0 & +\beta\gamma \\ 0 & 1 & 0 & 0 \\ 0 & 0 & 1 & 0 \\ +\beta\gamma & 0 & 0 & \gamma \end{pmatrix} \tag{17-5}$$

is the inverse matrix of Eq. (17-2), i.e., the matrix of the transformation corresponding to relative motion of the frames with opposite velocity. If a quantity with four components A_j transforms as the reverse transformation of the x^j, i.e., as

$$A'_i = (Q^i_j)^{-1} A_j, \tag{17-6}$$

then a relation equating components of the type A_j is also Lorentz covariant. To restate these two cases: any quantities A_j or B^j are Lorentz covariant if under change of inertial frame they transform as Eq. (17-6) or as

$$B'^i = Q^i_j B^j, \tag{17-7}$$

respectively. A_j is called a covariant four-vector and B^j is called a contravariant four-vector. (It is unfortunate that the word "covariant" is used in two different ways, but this usage is a matter of accepted convention.) In the language of matrices Q^i_j is related to the inverse transformation by the equation

$$(Q^k_j)^{-1} Q^i_j = \delta^i_k, \tag{17-8}$$

where

$$\delta^i_k = \begin{pmatrix} 1 & 0 & 0 & 0 \\ 0 & 1 & 0 & 0 \\ 0 & 0 & 1 & 0 \\ 0 & 0 & 0 & 1 \end{pmatrix} \tag{17-9}$$

is a matrix representing identity.

We are not here concerned with general tensor analysis *per se*, nor with the general transformation of coordinates. In the Lorentz transformation of special relativity the transformation coefficients Q^i_j are constants and thus independent of the coordinates. In most of the references this transformation with constant coefficients is derived as a special case of the general transformation of coordinates in which the coefficients themselves are functions of the coordinates. In the general analog of Eq. (17–1) a differential coordinate element dx^j transforms like

$$dx'^i = \left(\frac{\partial x'^i}{\partial x^j}\right) dx^j. \tag{17-10}$$

Covariant and contravariant vector components transform as

$$A'_i = \left(\frac{\partial x^j}{\partial x'^i}\right) A_j, \qquad B'^i = \left(\frac{\partial x'^i}{\partial x^j}\right) B^j, \tag{17-11}$$

respectively. In the general theory of relativity it is necessary to use Eqs. (17–10) and (17–11), since the transformation coefficients may be functions of the coordinates. The distinction between covariant and contravariant entities takes on added significance in that case. In special relativity it is clear from Eq. (17–5) that the covariant vector A_j simply transforms like $-x$, $-y$, $-z$, ct. Even when the transformation coefficients are constants, however, the algorism of general tensor analysis is extremely useful, and we shall follow the standard forms.

17–2 Some tensor relations useful in special relativity. For our purpose, we may define as tensors all quantities which themselves maintain definite transformation properties when the coordinates undergo a Lorentz transformation. The simplest is an invariant, or scalar, which can be specified by a single number. We have seen, for example, that $ds^2 = c^2\,dt^2 - dx^2 - dy^2 - dz^2$ is such an invariant. A quantity of this kind is called a tensor of rank zero.

The contravariant and covariant four-vectors of Section 17–1 are tensors of rank one. A contravariant tensor of rank two is a collection of sixteen quantities which transforms like

$$T'^{kl} = Q_i^k Q_j^l T^{ij}, \tag{17–12}$$

i.e., like the product of two contravariant vectors. Similarly, a covariant tensor of second rank is a set of sixteen components which transform like

$$T'_{kl} = (Q_i^k)^{-1}(Q_j^l)^{-1} T_{ij}, \tag{17–13}$$

i.e., like the product of two covariant vectors. A mixed tensor of second rank transforms like the product of a contravariant and a covariant vector,

$$T'^l_k = Q_j^l (Q_i^k)^{-1} T_i^j. \tag{17–14}$$

Thus the pattern becomes clear: any quantity of the type $T^{l_1 l_2 \cdots l_n}_{k_1 k_2 k_3 \cdots k_m}$ which transforms like

$$T'^{l_1 l_2 \cdots l_n}_{k_1 k_2 k_3 \cdots k_m} = Q_{i_1}^{l_1} Q_{i_2}^{l_2} \cdots Q_{i_n}^{l_n} (Q_{j_1}^{k_1})^{-1}(Q_{j_2}^{k_2})^{-1} \cdots (Q_{j_m}^{k_m})^{-1} \, T^{i_1 i_2 \cdots i_n}_{j_1 j_2 \cdots j_m}, \tag{17–15}$$

i.e., simply like products of covariant and contravariant four-vector components, is called a tensor of rank $(m + n)$. It follows that the product of a tensor of rank m and one of rank n is a tensor of rank $m + n$. We shall rarely be concerned with tensors of higher than second rank in our physical applications.

Since the transformations involved are linear, the sum or difference of any two tensors of equal rank is a tensor of the same rank, and equations can be made only between quantities of the same kind. Such tensor equations will be Lorentz covariant. In order to extend the principles of relativity by rewriting other laws of physics in the form of tensor equations, we shall need some of the formal rules of tensor manipulation.

1. When a mixed tensor has a contravariant index which is the same as a covariant index the implied summation reduces the rank of the tensor by two. This process is called *contraction*. The contraction of a tensor T_j^i, namely, T_i^i, is an invariant. We shall see that if we contract the vector gradient (a tensor of rank 2) we obtain the divergence (a tensor of rank 0).

2. We have seen that the "line element" in special relativity, $ds^2 = -dx^2 - dy^2 - dz^2 + c^2\,dt^2$, is a scalar invariant. ds^2 can be written in the form

$$ds^2 = g_{ij}\,dx^i\,dx^j, \tag{17–16}$$

where

$$g_{ij} = \begin{pmatrix} -1 & 0 & 0 & 0 \\ 0 & -1 & 0 & 0 \\ 0 & 0 & -1 & 0 \\ 0 & 0 & 0 & +1 \end{pmatrix} \tag{17–17}$$

is called the "metric tensor" corresponding to the line element. (In general tensor analysis g_{ij} is a function of the coordinates.) It can easily be proved by transformation, using Eq. (17–5), that g_{ij} is actually a tensor.

3. By means of the relation

$$B_i = g_{ij}B^j \tag{17–18}$$

each member of a set of covariant components B_i can be associated with a contravariant tensor component. For the special form of g_{ij} given by Eq. (17–17) this process will simply reverse the sign of the first three components. The covariant components dx_i corresponding to the basic contravariant interval dx^i are such that

$$dx^i\, dx_i = ds^2. \tag{17–19}$$

4. With reference to Eqs. (17–11), it can be seen by writing $\partial/\partial x'^i = (\partial x^j/\partial x'^i)\, \partial/\partial x^j$ that the derivative $\partial/\partial x^i$ operating on a tensor transforms like an additional covariant tensor factor. (This is not true if the g_{ij} are functions of the coordinates.) The operation of the "gradient" thus increases the rank of the tensor by one (covariant index): if T^0 is a scalar, its gradient $\partial T^0/\partial x^i$ is a covariant vector, and the increment of the scalar can be written as a tensor relation by contraction,

$$dT^0 = \frac{\partial T^0}{\partial x^i}\, dx^i. \tag{17–20}$$

Similarly, if T^k is a contravariant vector, $\partial T^k/\partial x^i$ is a mixed tensor of the second rank, while $\partial T^i/\partial x^i$, the four-divergence of T^i, is a scalar invariant.

5. If a four-vector obeys the relation

$$\frac{\partial T^i}{\partial x^i} = 0, \tag{17–21}$$

and if the components of T^i are different from zero only in a finite spatial region, then the integral over three-dimensional space,

$$I = \int T^4\, dv, \tag{17–22}$$

is an invariant. Let us prove this theorem. The four-dimensional analog of Gauss's theorem states that

$$\int \frac{\partial T^i}{\partial x^i}\, d^4x = \int T^i\, dS_i, \tag{17–23}$$

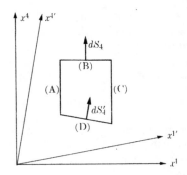

Fig. 17-1. Showing region of integration for proving the theorem of Eq. (17-22).

where $d^4x = dx^1\, dx^2\, dx^3\, dx^4 = dv\, dx^4$, and dS_i is an element of three-dimensional "surface" normal to T^i in four-space. The region over which the integration of Eq. (17-23) is to be performed is indicated in Fig. 17-1, where the surfaces (A) and (C) are chosen so that the spatial components of T^i vanish on (A) and (C). This can be done, for it was assumed that the region where the spatial components of T^i do not vanish is finite. (B) is chosen normal to the x^4-axis, while (D) is chosen normal to the x'^4-axis. It then follows from Eqs. (17-21) and (17-23) that

$$\int T^4\, dS_4 = \int T'^4\, dS_4'.$$

Since $dS_4 = dv$ and $dS_4 = dv'$, the ordinary three-dimensional volume element, this means that

$$I = \int T^4\, dv \qquad (17\text{-}22)$$

is an invariant under a Lorentz transformation. Similarly, it can be proved that if a tensor of rank two satisfies the relation

$$\frac{\partial T^{ij}}{\partial x^i} = 0, \qquad (17\text{-}24)$$

then

$$B^j = \int T^{4j}\, dv \qquad (17\text{-}25)$$

is a four-vector. Equations (17-21) and (17-24) will later be recognized as "conservation laws."

6. A tensor possesses symmetry with respect to two contravariant indices, i and j, or two covariant indices i and j, if the interchange of i and j leaves all the components of the tensor unchanged. A tensor is said to

be antisymmetric, or *skew-symmetric*, if such an interchange of i and j simply changes the sign of all components. All symmetry properties are invariant under a transformation of coordinates.

These relations will suffice for our covariant description of particle mechanics, although we have refrained from anticipating all the problems involved in the covariant description of electromagnetic fields. Before attempting the extension of relativistic principles to mechanics, let us note that a covariant relation in physics can be generated in three ways:

(a) The relation is known in a special inertial frame, such as the proper frame where the system under consideration is at rest. If it is possible to express this law in the form of a tensor equation which reduces to the correct relation in the special frame, then this tensor equation has general significance.

(b) A known tensor relation is converted into a new tensor equation by a covariant tensor operation. The simplest example is multiplication by an invariant, but we have also noted tensor multiplication, contraction, and covariant differentiation as covariant tensor operations.

(c) An equation is obtained from a relation valid in a special frame by transforming those quantities whose transformation properties are known, and deducing the correct transformations for the remaining quantities. These quantities are then expressible in tensor form. This process has been referred to at the beginning of the chapter as that of direct transformation and it is usually very tedious.

17–3 The conservation of momentum. Let us use the above considerations to formulate the law of conservation of momentum for mass points. The formal relations so derived will then be compared with the result of a "thought experiment" involving an inelastic collision as observed from various frames.

In prerelativistic physics, a particle of mass m which is moving with velocity **u** possesses linear momentum **p**, where

$$\mathbf{p} = m\mathbf{u}. \tag{17–26}$$

The vector **u** does not constitute the first three components of a four-vector, for $\mathbf{u} = d\mathbf{r}/dt$, and while $d\mathbf{r}$ represents three components of the four-vector $dx^i = (d\mathbf{r}, c\,dt)$, dt is not an invariant. [If the spatial components of a four-vector conform to a standard three-dimensional vector, we shall, in enumerating the components of the four-vector, use regular vector notation, as in $x^i = (\mathbf{r}, ct)$.] However, if we divide the contravariant vector dx^i by the invariant line element ds, we do obtain a four-vector

$$u^i = \frac{dx^i}{ds}, \tag{17–27}$$

known as the four-velocity. The quantity ds is the proper time interval multiplied by c, or

$$ds = c \, dt \sqrt{1 - \frac{u^2}{c^2}}, \qquad (17\text{-}28)$$

and therefore the components of u^i are given by

$$u^i = \left(\frac{\mathbf{u}}{c\sqrt{1 - u^2/c^2}}, \quad \frac{1}{\sqrt{1 - u^2/c^2}} \right). \qquad (17\text{-}29)$$

We note that u^i so defined is dimensionless, instead of having the dimensions of ordinary velocity.*

A covariant expression corresponding to linear momentum can be generated by multiplying Eq. (17-27) by an invariant quantity $m_0 c^2$ which is assumed to be characteristic of the particle. Here m_0 is the rest mass, i.e., the mass measured in the proper frame of the particle. The four-momentum thus defined is

$$p^i = m_0 \frac{dx^i}{ds} c^2. \qquad (17\text{-}30)$$

Defined in this way, p^i is Lorentz covariant; we note that its dimensions are those of energy. If we want to assure that the law of conservation of momentum shall be preserved in the framework of relativity for two particles interacting at a point, then in order that this law be independent of the choice of inertial frame it must take the form

$$p_1^i + p_2^i = \text{constant} \qquad (17\text{-}31)$$

before and after the collision. Equation (17-31) replaces the classical law

$$m_1 u_1 + m_2 u_2 = \text{constant}, \qquad (17\text{-}32)$$

in which it is assumed that m_1 and m_2 are constant.

This discussion is restricted to direct interaction between particles, rather than including "interaction at a distance," for two reasons. The total momentum of two separated particles at a "given" time has no mean-

---

* In constructing four-vectors or tensors we must make sure that the various members of the matrix representing the tensor have the same dimensions; this is done by using appropriate powers of c. Just how this is done is a matter of convention, and several texts differ from our convention by powers of c. For example, u^i may be consistently defined as $u^i = (\mathbf{u}/\sqrt{1 - u^2/c^2}, c/\sqrt{1 - u^2/c^2})$, which does give u^i the dimension of velocity. The convention of choosing units of length and time to make $c = 1$ is also frequently used, with resulting simplification.

ing in relativity, and furthermore all interactions are propagated with finite velocity. Therefore exchange of momentum between separated particles has meaning only if each particle conserves momentum with a field acting on it, or if the interaction is carried by a particle interacting in succession with the two mass points. Strictly speaking, each of the mass points considered here must have zero extension in order that the discussion be rigorous.

The components of the four-momentum defined by Eq. (17–30) are

$$p^i = \left(\frac{m_0 \mathbf{u} c}{\sqrt{1 - u^2/c^2}}, \quad \frac{m_0 c^2}{\sqrt{1 - u^2/c^2}}\right). \tag{17-33}$$

where $\mathbf{u}$ is the ordinary velocity in a given frame. The first three components have the form

$$c\mathbf{p} = cm\mathbf{u}, \tag{17-34}$$

where

$$m = \frac{m_0}{\sqrt{1 - u^2/c^2}}. \tag{17-35}$$

Hence, if we require that the law of conservation of momentum be maintained in its classical form, Eq. (17–32), and that it be Lorentz covariant, the mass of a particle is no longer an invariant but will depend on its velocity in the particular reference frame. The "variation of mass with velocity" is an immediate consequence of formulating the law of conservation of momentum in a covariant manner.

17–4 Relation of energy to momentum and to mass. The relativistic form of the law of conservation of momentum implies not only the conservation of the three spatial components of p^i but also the conservation of the fourth component,

$$p^4 = \frac{m_0 c^2}{\sqrt{1 - u^2/c^2}} = mc^2. \tag{17-36}$$

Let us investigate the physical significance of this quantity. The time rate of change of p^4 in a given frame is

$$\frac{dp^4}{dt} = \frac{d}{dt}\left(\frac{m_0 c^2}{\sqrt{1 - u^2/c^2}}\right) = \mathbf{u} \cdot \frac{d}{dt}\left(\frac{m_0 \mathbf{u}}{\sqrt{1 - u^2/c^2}}\right), \tag{17-37}$$

or

$$\frac{dp^4}{dt} = \mathbf{u} \cdot \frac{d\mathbf{p}}{dt}. \tag{17-38}$$

If we continue to measure the (three-dimensional) force by the time rate of change of momentum

$$\mathbf{F} = \frac{d\mathbf{p}}{dt},$$ (17–39)

then

$$\mathbf{F} \cdot \mathbf{u} = \mathbf{u} \cdot \frac{d\mathbf{p}}{dt}$$ (17–40)

represents the rate at which work is being done in a particular system. Hence if the law of conservation of energy is to hold in a particular frame, and if E denotes the energy in that frame,

$$\frac{dp^4}{dt} = \frac{dE}{dt} ; \quad p^4 = E + \text{constant.}$$ (17–41)

Since energy manifests itself only when energy changes occur, we lose no physical significance if we put

$$E = mc^2 = \frac{m_0c^2}{\sqrt{1 - u^2/c^2}} .$$ (17–42)

We are thus led to conclude that energy as measured by "work content" and mass as measured by the momentum for a given velocity are interchangeable concepts; when one exists so does the other. Neither mass nor energy is an invariant; the magnitude of both depends on the frame of the observer by the relation of Eq. (17–42). We have shown that the change in mc^2 corresponds to work done by mechanical forces; that it corresponds generally to change in energy under any mechanism that might be involved implies an additional assumption whose justification rests with experiment. Experience in other fields of physics, particularly in nuclear physics where the fractional mass changes become very large, certainly proves beyond any reasonable doubt that Eq. (17–42) is valid in this more general interpretation.

For small velocities, E reduces to the classical kinetic energy plus the "rest energy" m_0c^2. By expansion of Eq. (17–42) for small values of u/c, we obtain

$$E = m_0c^2\left[1 + \frac{1}{2}\left(\frac{u}{c}\right)^2 + \frac{3}{8}\left(\frac{u}{c}\right)^4 + \cdots\right]$$

$$= m_0c^2 + \frac{1}{2}m_0u^2 + \frac{3}{8}m_0\frac{u^4}{c^2} + \cdots .$$

In general, relativistic mechanics reduces to Newtonian mechanics in the limit of small velocities, as indeed it must.

Relativistically, the conservation of energy and the conservation of momentum are not independent principles; one demands the other for a covariant formulation. The invariant related to the energy-momentum four-vector,

$$p^i = (c\mathbf{p}, E), \qquad (17\text{–}43)$$

in the same way that ds is related to dx^i, may be easily ascertained:

$$p^i p_i = E^2 - c^2 p^2 = (m_0 c^2)^2. \qquad (17\text{–}44)$$

This relation between energy and momentum is valid in any frame, and in a proper frame with $\mathbf{p} = 0$ we have simply $E = m_0 c^2$.

The energy-momentum vector p^i transforms like any other contravariant vector. In accordance with Eq. (17–2), we have

$$c p'_x = \frac{c p_x - \beta E}{\sqrt{1 - \beta^2}}, \qquad (17\text{–}45)$$

$$p'_y = p_y; \quad p'_z = p_z, \qquad (17\text{–}46)$$

$$E' = \frac{E - \beta c p_x}{\sqrt{1 - \beta^2}}, \qquad (17\text{–}47)$$

where $E = p^4$. From these transformation relations, it can be seen that any transfer of energy implies transfer of mass and therefore momentum. If E is the energy of a system in its proper frame (zero momentum), then from the primed system we shall observe a momentum

$$p'_x = \frac{-\beta E}{c \sqrt{1 - \beta^2}} = \frac{-v}{c^2} E'. \qquad (17\text{–}48)$$

According to Eq. (17–48), we must associate momentum with any agency that transmits energy, whatever the form of the energy. Consider a motor M driving a load L at a distance x, as shown in Fig. 17–2. If the motor transfers energy at the rate dE/dt to the load, the mass of L increases correspondingly. Mass is being transmitted over a distance x at a rate of

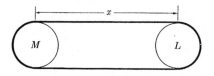

Fig. 17–2. Motor M driving load L at distance x.

$(1/c^2) \, dE/dt$, and the system thus has a momentum

$$p = \frac{x}{c^2} \frac{dE}{dt}.$$

In general, if energy is being absorbed by a body at a given rate the momentum of the body increases, and to conserve over-all momentum we must associate a momentum density per unit volume, $\mathbf{g}$, with any agent that transmits energy at the rate $\mathbf{N}$ per unit area in a given direction, where

$$\mathbf{g} = \frac{\mathbf{N}}{c^2}. \tag{17-49}$$

This relation is in agreement with our discussion of radiation pressure in Chapter 11, and with our considerations of momentum balance in Section 10–6. When we considered electromagnetic radiation, which represents an energy flow $\mathbf{N}$, we were forced to attribute a momentum density as given by Eq. (17–49) to the electromagnetic field. This was really because we refused to accept the ether as capable of sustaining a volume force. The failure to detect an ether by the experiments outlined in Chapter 15 strengthens the conviction that the system of radiation and absorption is a *closed* system. The result is that Eq. (17–49) is required by the conservation of momentum. Conversely, the assumptions of conservation of momentum and the absence of an ether can be used to derive the mass-energy equivalence. In this way, the relation $E = mc^2$ can be obtained without introducing the entire relativistic kinematics.

17–5 The Minkowski force. The force $\mathbf{F} = d\mathbf{p}/dt$ in a given frame is not the spatial part of a four-vector. On the other hand, a quantity known as the *Minkowski force*,

$$F^i = \frac{dp^i}{ds} = \frac{d}{ds} (c\mathbf{p}, \, mc^2)$$

$$= \left[\frac{1}{\sqrt{1 - u^2/c^2}} \frac{d\mathbf{p}}{dt}, \, \frac{1}{\sqrt{1 - u^2/c^2}} \frac{d(mc)}{dt} \right], \tag{17-50}$$

is a contravariant four-vector. The components of F^i can be written as

$$F^i = \left(\frac{\mathbf{F}}{\sqrt{1 - u^2/c^2}}, \, \frac{\mathbf{F} \cdot \mathbf{u}/c}{\sqrt{1 - u^2/c^2}} \right), \tag{17-51}$$

where $\mathbf{F}$ and $\mathbf{u}$ are the ordinary three-dimensional force and velocity of

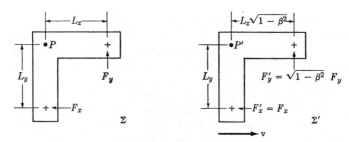

FIG. 17–3. Right-angled lever in a moving coordinate system.

Eqs. (17–39) and (17–40). The transformation laws for force can be derived from the four-vector character of F^i. We shall restrict ourselves to the case where **F** is proper in what we have called the Σ frame, i.e., $\mathbf{u} = 0$, $F^i = (\mathbf{F}, 0)$. In the Σ' frame $u'_x = v$, since $u_x = 0$, and we obtain

$$F'_x = F_x, \tag{17–52}$$

$$F'_y = F_y\sqrt{1 - \beta^2}, \tag{17–53}$$

$$F'_z = F_z\sqrt{1 - \beta^2}. \tag{17–54}$$

An interesting application of the transformation equations for **F** is afforded by the equilibrium of the right-angled lever shown in Fig. 17–3. To an observer at rest in Σ, the lever is in static equilibrium under the action of F_x and F_y as shown, so that

$$F_xL_y = F_yL_x. \tag{17–55}$$

To an observer at rest in Σ', we should expect the lever to remain in rotational and translational equilibrium, for otherwise the two inertial frames would be distinguishable. Using the transformation equations for lengths and forces, we find that to an observer in Σ' a net torque, T', is acting on the lever in a counterclockwise direction, where

$$T' = F_xL_y - (\sqrt{1 - \beta^2}\, F_y)(\sqrt{1 - \beta^2}\, L_x) = \frac{F_xL_yv^2}{c^2}. \tag{17–56}$$

This torque results in no rotation, however, for F'_x is doing work on the lever at the rate $F'_xv = F_xv$, and the angular momentum of the lever is increasing at the rate

$$\frac{dM}{dt} = \frac{(F_xv)vL_y}{c^2} = \frac{F_xL_yv^2}{c^2}. \tag{17–57}$$

Hence to an observer in Σ', even though neither the torque nor the change in angular momentum is zero, the existing torque exactly balances the gain in angular momentum and the equilibrium condition is preserved as an invariant property. (In this discussion we have omitted all mention of the mechanism by which forces are transmitted through the lever. The laws of elasticity are also profoundly modified by relativity: the lever cannot be treated as a rigid body, since the velocity of propagation of an impulse is limited. A more detailed discussion, taking account of these modifications, does not alter the above conclusions, however.)

Before leaving the subject of forces, let us consider briefly the motion of a particle under the influence of external forces. From

$$\mathbf{F} = \frac{d\mathbf{p}}{dt} = \frac{d}{dt}\left(\frac{m_0\mathbf{u}}{\sqrt{1 - u^2/c^2}}\right), \tag{17-39}$$

we obtain

$$\mathbf{F} = \frac{m_0}{\sqrt{1 - u^2/c^2}}\frac{d\mathbf{u}}{dt} + \frac{m_0 u\mathbf{u}}{c^2(1 - u^2/c^2)^{3/2}}\frac{du}{dt}. \tag{17-58}$$

This shows that the acceleration of a particle requires a component of force parallel to the velocity as well as that parallel to the acceleration. For low velocities the second member of the right side of Eq. (17–58) is much smaller than the first term. When u/c is very small in comparison with unity, Eq. (17–58) can be approximated by $\mathbf{F} = m_0\,d\mathbf{u}/dt$, the classical form of Newton's second law of motion.

17–6 The collision of two similar particles. Without reference to tensor methods the dependence of mass on velocity can be deduced by use of the Lorentz transformation and the requirement that in a two-particle collision momentum be conserved. This is an example of the *Gedanken Experimente* process of extending the principles of relativity.

Consider a head-on collision between two point particles of equal properties. We may choose Σ' as the center of mass frame in which the collision is symmetrical, i.e., they appear to approach with equal velocities, as in Fig. 17–4. After an infinitesimal period of coalescence they will move apart in opposite directions with velocities of equal magnitude in this frame.

We now postulate that momentum and mass are to be conserved in any frame. Hence, in Σ,

$$m_1 + m_2 = M, \tag{17-59}$$

$$m_1 u_1 + m_2 u_2 = Mv, \tag{17-60}$$

where M is the combined mass during collision. From the longitudinal

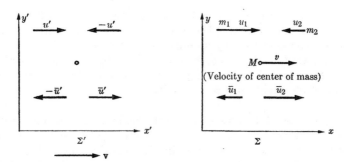

FIG. 17–4. Collision of two similar particles. The bars designate quantities after the collision; v is the velocity of Σ' in Σ.

velocity addition relations, we have

$$u_1 = \frac{u' + v}{1 + u'v/c^2}, \qquad u_2 = \frac{-u' + v}{1 - u'v/c^2}. \tag{17-61}$$

By eliminating M from Eqs. (17–59) and (17–60), and making use of Eq. (17–61), we obtain

$$\frac{m_1}{m_2} = \frac{1 + u'v/c^2}{1 - u'v/c^2} = \frac{\sqrt{1 - u_2^2/c^2}}{\sqrt{1 - u_1^2/c^2}}. \tag{17-62}$$

Hence, in order to preserve the postulated conservation laws in all frames, we must have

$$m = \frac{m_0}{\sqrt{1 - u^2/c^2}}, \tag{17-63}$$

where m_0 is the mass in a frame proper to the particle. This is in agreement with Eq. (17–35), and the further deductions follow as above.

This description is of particular interest, since it enables us to obtain from Eq. (17–61) the mass M during the collision:

$$M = m_0 \left(\frac{1}{\sqrt{1 - u_1^2/c^2}} + \frac{1}{\sqrt{1 - u_2^2/c^2}} \right)$$

$$= \frac{2m_0}{\sqrt{1 - v^2/c^2} \sqrt{1 - u'^2/c^2}}. \tag{17-64}$$

This is larger than $2m_0/\sqrt{1 - v^2/c^2}$, which would be the mass of the two particles of rest mass m_0 moving with velocity v. This increased mass represents the increase in energy of the two particles during collision owing to the stored elastic energy or to the energy increase where it is not all released again in kinetic form. The distinction between inelastic and elastic collisions therefore essentially disappears so far as the first part of a collision is concerned. If the two particles separate after the collision, a knowledge of the energy changes during the impact is necessary for the determination of the final velocities in any frame.

17-7 The use of four-vectors in calculating kinematic relations for collisions. Our general considerations of Section 17–3 have indicated that in a collision between two particles, the four components of the quantity p^i as given by Eq. (17–33) are conserved; hence analysis of collision kinematics can be carried out by the appropriate algebraic operations using these components. It is frequently more expedient, and also physically more instructive, to handle the problem with the four-vector operations themselves. Let us consider two examples of physical interest.

EXAMPLE I. *Collision of two particles of rest mass m_{10} and m_{20}; "center-of-mass" energy.* Let the four-momenta of the two particles be p_1^i and p_2^i; at the time of collision the total four-momentum is

$$p^i = p_1^i + p_2^i. \tag{17-65}$$

The invariant

$$p^i p_i = (E_1 + E_2)^2 - c^2(\mathbf{p}_1 + \mathbf{p}_2)^2 \tag{17-66}$$

must be the same in whatever frame the components of $p^i = (c\mathbf{p}, E)$ are evaluated. Of particular interest is the frame in which $\mathbf{p}_1 + \mathbf{p}_2 = 0$, usually known as the "center-of-mass frame," although "zero-momentum frame" would be more accurate. In that frame the quantity

$$E = (E_1 + E_2)_{\text{center of mass}} - (m_{10} + m_{20})c^2 \tag{17-67}$$

is the energy excess available for "inelastic" processes. Hence, in general,

$$E + (m_{10} + m_{20})c^2 = [(E_1 + E_2)^2 - c^2(\mathbf{p}_1 + \mathbf{p}_2)^2]^{1/2}, \tag{17-68}$$

where E_1, E_2, $\mathbf{p}_1$ and $\mathbf{p}_2$ may be evaluated in any frame. Specifically, if, in the frame of interest, the particle of rest mass m_{20} is initially at rest, then $\mathbf{p}_2 = 0$ and $E_2 = m_{20} c^2$, so that

$$E = \{[(m_{10} + m_{20})c^2]^2 + 2T m_{20} c^2\}^{1/2} - (m_{10} + m_{20})c^2, \tag{17-69}$$

where $T = E_1 - m_{10} c^2$ is the kinetic energy of the moving particle in

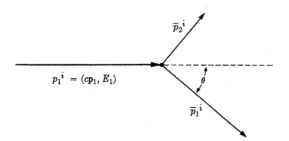

FIG. 17–5. Kinematics of elastic collision.

this frame. This shows that a laboratory kinetic energy given by

$$T = \frac{E^2 + 2(m_{10} + m_{20})c^2 E}{2m_{20}\, c^2} \tag{17-70}$$

is required to have an energy E available for inelastic processes in a collision reaction with a stationary target.

The reader will find it instructive to compare the foregoing analysis with an attempt to derive Eq. (17–70) from Eq. (17–64) by noncovariant methods.

EXAMPLE II. *Kinematics of the elastic collision between two particles of rest mass m_{10} and m_{20}.* Let us again, as in Section 17–6, use bars to denote quantities after collision. Conservation of four-momentum gives

$$p_1^i + p_2^i = \overline{p}_1^i + \overline{p}_2^i. \tag{17-71}$$

Consider the frame of reference in which the particle of rest mass m_{20} is initially at rest, i.e., $p_2^i = (0, m_{20}\, c^2)$, and let the particle of rest mass m_{10} be "scattered" at an angle θ with its original momentum (see Fig. 17–5). The invariant $\overline{p}_2^i\, \overline{p}_{2_i}$ (see Eq. (17–44)), is, with the use of Eq. (17–71),

$$\overline{p}_2^i\, \overline{p}_{2_i} = (m_{20}\, c^2)^2 = (p_1^i + p_2^i - \overline{p}_1^i)(p_{1_i} + p_{2_i} - \overline{p}_{1_i}) \tag{17-72}$$

$$= 2(m_{10}\, c^2)^2 + (m_{20}\, c^2)^2 - 2c^2\, \mathbf{p}_1 \cdot \overline{\mathbf{p}}_1 + 2E_1 m_{20}\, c^2$$

$$-2E_1\overline{E}_1 - 2\overline{E}_1 m_{20}\, c^2. \tag{17-73}$$

Solving for $\cos \theta = (\mathbf{p}_1 \cdot \overline{\mathbf{p}}_1)/p_1\overline{p}_1$, we obtain

$$\cos \theta = \frac{m_{20}\, c^2(T_1 - \overline{T}_1) - T_1\overline{T}_1 - m_{10}\, c^2(T_1 + \overline{T}_1)}{c^2 p_1\overline{p}_1} \tag{17-74}$$

as the relation between $\cos \theta$ and the kinetic energies T_1 and $\overline{T}_1$. We may note that

$$cp_1 = [T_1(T_1 + 2m_{10} \, c^2)]^{1/2} \qquad (17\text{-}75)$$

and the similar relation between $\overline{p}_1$ and $\overline{T}_1$; Eq. (17-74) can in general be simplified for special values of the rest masses, using Eq. (17-75).

SUGGESTED REFERENCES

P. G. BERGMANN, *An Introduction to the Theory of Relativity*, Chapters 5 and 6. The tensor calculus is presented in such a way that it will be applicable to the general theory of relativity.

R. C. TOLMAN, *Relativity, Thermodynamics, and Cosmology*, Chapters III and IV. Tolman and his collaborators have been responsible for such "thought experiments" as the collision of two similar particles.

L. LANDAU AND E. LIFSHITZ, *The Classical Theory of Fields*, Chapters 1 and 2. We shall have further occasion to refer to this excellent book on relativistic electrodynamics. The authors stress rather formally the internal logic of the subject, and begin with the principle of relativity and the variation principle.

V. FOCK, *The Theory of Space, Time and Gravitation*. The first four chapters are devoted to the theory of special relativity, and include some features not found elsewhere.

C. MøLLER, *The Theory of Relativity*, Chapter IV. Chapter VI includes the mechanics of elastic continua, but its study is better postponed until after the relativistic formulation of electrodynamics, Chapter V.

H. GOLDSTEIN, *Classical Mechanics*, Chapter 6. Many textbooks on mechanics include a section on relativity, and this is certainly one of the best.

Of the many excellent mathematical books on tensor calculus we recommend:

C. E. WEATHERBURN, *An Introduction to Riemannian Geometry and the Tensor Calculus*.

J. L. SYNGE AND A. SCHILD, *Tensor Calculus*.

EXERCISES

1. Let a particle of initial energy E and rest energy E_0 hit a like particle at rest. Show that if $E \gg E_0$, the maximum energy available in the zero momentum frame is $\sqrt{2EE_0}$.

2. A π^0 meson of rest energy E_0 moving with velocity u in the laboratory, disintegrates into two γ-rays. Calculate (a) the energy distribution of γ-rays from π^0 mesons in the laboratory, (b) the distribution in angle of one of the γ-rays against the other in the laboratory.

3. If a particle of initial kinetic energy T_0 and rest energy E_0 strikes a like particle at rest, show that the kinetic energy T of the particle scattered at an angle θ is

$$T = \frac{T_0 \cos^2 \theta}{1 + (T_0 \sin^2 \theta)/2E_0}.$$

4. To produce a "proton pair" an energy of $2M_0c^2 \doteq 1862$ Mev is required in the center-of-mass system. What is the minimum energy (a) of a proton, and (b) of an electron, needed to produce such an event by striking a proton at rest?

5. In experiments on proton-neutron scattering the scattered and recoil particles are at 90° to each other for nonrelativistic velocities. Calculate the first order correction (assumed small) to this angle for velocities large enough to make such a correction necessary.

6. Prove Eq. (17–38).

7. Derive the equation for the wavelength shift in the Compton effect. Use Eq. (17–74) and compare with the three-vector analysis.

8. Let $F(\tau)$ be the (three-dimensional) proper force acting on a particle in its own rest frame, given as a function of the proper time τ measured at the particle. Assume one-dimensional motion only. Show that in any given frame the velocity u is given by

$$\frac{u}{c} = \sinh\left[\sinh^{-1}\left(\frac{u_0}{c}\right) + \frac{1}{m_0c}\int_0^\tau F(\tau')\,d\tau'\right],$$

where u_0 is the initial velocity.

9. Consider a traveler in a spaceship accelerated such that the force $F(\tau)$ as given in Exercise 8 has the constant value $F(\tau) = m_0 g$, where g has the usual value of the acceleration due to gravity. Calculate the distance traversed from rest by the traveler during his lifetime (assumed to be 60 years).

CHAPTER 18

COVARIANT FORMULATION OF ELECTRODYNAMICS

We recall that the Lorentz transformation was introduced by consideration of the propagation of an electromagnetic wave. Actually, the homogeneous equation governing electromagnetic wave propagation is already in covariant form, since the D'Alembertian operator

$$\Box = -\frac{\partial}{\partial x_i}\frac{\partial}{\partial x^i} \tag{18-1}$$

is an invariant. In general, Maxwell's equations and their consequences lend themselves very simply to covariant description. This follows from the fact that no modifications at all are necessary in the laws of electrodynamics to make them agree with the requirements of relativity. The covariant formulation of the space-time coordinates in the equations automatically puts the rest of the equation into covariant form.

18-1 The four-vector potential. Relativistically, it is clear that charge density and current are simply different aspects of the same thing. If we have a "proper" charge density ρ_0 in a frame where such charges are at rest, then the contravariant vector

$$j^i = \rho_0 \frac{dx^i}{ds} \tag{18-2}$$

has the components

$$j^i = \left(\rho \frac{\mathbf{u}}{c}, \rho\right), \tag{18-3}$$

where

$$\rho = \frac{\rho_0}{\sqrt{1 - u^2/c^2}}. \tag{18-4}$$

Hence the transformation equations for charge and current densities follow automatically. And since charge and current densities are components of a single four-vector j^i, we are led to combine the inhomogeneous wave equations, Eqs. (14-7) and (14-8), in an analogous way. Expressed as

$$\Box \phi = -\rho/\epsilon_0, \tag{18-5}$$

$$\Box c\mathbf{A} = -\rho\mathbf{u}/c\epsilon_0, \tag{18-6}$$

these equations are equivalent to

$$\Box \phi^i = -\frac{j^i}{\epsilon_0}, \tag{18-7}$$

where

$$\phi^i = (c\mathbf{A}, \phi). \tag{18-8}$$

The scalar and vector potentials are therefore no longer quantities permitting independent description; they are different aspects of the same thing. It follows that the same can be said of electric and magnetic fields: in some sense they, too, are different aspects of the same thing.

The equation of continuity of charge and current takes the simple covariant form

$$\frac{\partial j^i}{\partial x^i} = 0, \tag{18-9}$$

and the Lorentz condition, Eq. (14-3), becomes its counterpart:

$$\frac{\partial \phi^i}{\partial x^i} = 0. \tag{18-10}$$

In terms of the covariant components of ϕ^i, the gauge transformations, Eqs. (14-9) and (14-10), combine as the single equation

$$\phi_i' = \phi_i + \frac{\partial}{\partial x^i}(c\psi), \tag{18-11}$$

giving both the correct transformation character and the correct sign. The derivation of the field from the ϕ^i and laws of physical consequence must not depend on the choice of the scalar function ψ.

Equation (18-4), which gives the transformation from a charge density at rest to a charge density in a nonproper frame, is such as to ensure the invariance of total charge. A spatial volume element dv is related to a proper spatial volume dv_0 by

$$dv = dv_0\sqrt{1 - u^2/c^2}, \tag{18-12}$$

since only one dimension suffers a Lorentz contraction. Hence,

$$\rho\,dv = \rho_0\,dv_0 \tag{18-13}$$

and the charge within a given boundary remains invariant. The electronic charge e thus remains a universal constant in the theory of relativity. Since no charges have been found in nature which are not integral multiples of e, total charge could be measured by a counting operation which is

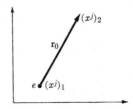

FIG. 18–1. Source point and field point in a proper frame.

presumably also an invariant. These facts are all in agreement with experiment. The invariance of total charge is also a direct consequence of the theorem stated in Eq. (17–22) combined with Eq. (18–9).

Let us obtain the integral of the inhomogeneous wave equation, Eq. (18–7), corresponding to an (invariant) point charge e at a point $(x^j)_1$. We know that in a proper frame, as in Fig. 18–1, the integral of Eq. (18–7) is simply the Coulomb potential,

$$\phi_0^i = \left(0, \frac{e}{4\pi\epsilon_0} \frac{1}{r_0}\right), \tag{18–14}$$

where r_0 is the proper vector distance from the source to the field point. The potential signal at $(x^j)_2$ is to be measured at the time corresponding to the retardation condition

$$R^j R_j = 0 = -r^2 + c^2 t^2, \tag{18–15}$$

where

$$R^j = (x^j)_2 - (x^j)_1 = (\mathbf{r}, ct) = (\mathbf{r}, r). \tag{18–16}$$

To make Eq. (18–14) valid in any frame, we seek to write the potential solution in tensor form such that it reduces to the Coulomb potential if $\mathbf{u} = \mathbf{0}$. In a proper frame, the four-velocity, $u^i = dx^i/ds$, has the components

$$u_0^i = (\mathbf{0}, 1). \tag{18–17}$$

Now the invariant, $u^i R_i$, can be evaluated in the proper frame where $R_0^i = (\mathbf{r}_0, r_0)$, and is just

$$u^i R_i = r_0. \tag{18–18}$$

Equation (18–14) can then be written in tensor form,

$$\phi^i = \frac{e}{4\pi\epsilon_0} \frac{u^i}{u^j R_j}, \tag{18–19}$$

subject to the condition $R_j R^j = 0$. This equation is now valid in any frame, whether proper or not.

In order to write Eq. (18–19) in terms of three-dimensional vectors, we recall that

$$u^j = \left(\frac{\mathbf{u}}{c\sqrt{1 - u^2/c^2}} , \frac{1}{\sqrt{1 - u^2/c^2}} \right).$$ (17–29)

Hence,

$$u^j R_j = \frac{-\mathbf{r} \cdot \mathbf{u}}{c\sqrt{1 - u^2/c^2}} + \frac{r}{\sqrt{1 - u^2/c^2}},$$ (18–20)

and Eq. (18–19) then has the components

$$\phi^i = \frac{e}{4\pi\epsilon_0} \left(\frac{\mathbf{u}}{cs} , \frac{1}{s} \right),$$ (18–21)

where

$$s = r - \frac{\mathbf{r} \cdot \mathbf{u}}{c}.$$ (18–22)

The potentials of Eq. (18–21) are called the Liénard-Wiechert potentials, and later we shall derive them from the retarded potentials of Chapter 14. In Eq. (18–21) $\mathbf{u}$ enters as the velocity of the observer relative to the frame in which the charge was at rest at the time of "emission" of the signal.

18–2 The electromagnetic field tensor. In writing down the covariant derivation of the fields from the potentials a problem arises which has its origin in ordinary three-dimensional vector analysis: $\mathbf{E}$ and $\mathbf{B}$ are not vectors of the same kind. The prototype for an ordinary vector is the displacement of a point in space. Such vectors are called *polar* vectors, and $\mathbf{u}, \mathbf{F}, \mathbf{E}, \mathbf{D}$, etc., are of this type (since charge is assumed to be a scalar, and not a "pseudoscalar" as defined in Section 1–1). The scalar products of these vectors are unchanged by all orthogonal transformations of the coordinates—translations, rotations, and reflections—although if any or all of the coordinate axes are changed in sign the corresponding vector components change sign. The same cannot be said of the "vector product" of two polar vectors: the very definition of vector product depends on whether $\mathbf{A} \times \mathbf{B}$ is to be taken in the right-handed or left-handed sense. (The usual prototype for a vector product represents the area of the parallelogram defined by $\mathbf{A}$ and $\mathbf{B}$, but the "sense" of such a surface is arbitrary.) It is clear that if all three coordinate components of $\mathbf{A}$ and $\mathbf{B}$ are changed in sign no change at all occurs in $\mathbf{A} \times \mathbf{B}$. Quantities that transform like vector products are called *axial* vectors, and examples are torque, angular momentum, $\mathbf{B}, \mathbf{H}$, and $\mathbf{M}$. They transform like ordinary vectors under translations and proper rotations of the coordinate system, but under any changes involving reflections of the coordinates the relative sign between axial and polar vectors is reversed. The scalar product of an axial vector and a polar vector is not a true scalar, but a pseudoscalar: it changes sign under an inversion of the axes. Physical vector relations can, of course, equate only vectors of the same kind.

Axial vectors cannot form the spatial components of a four-vector. On the other hand, the components of $\mathbf{C} = \mathbf{A} \times \mathbf{B}$ can be expressed by two indices, those of the components of $\mathbf{A}$ and $\mathbf{B}$:

$$C_{\alpha\beta} = A_\alpha B_\beta - A_\beta B_\alpha = -C_{\beta\alpha}, \tag{18-23}$$

where

$$C_{12} = C_z, \qquad C_{23} = C_x, \qquad C_{31} = C_y, \tag{18-24}$$

in a right-handed Cartesian coordinate system. For example, the equation $\nabla \times \mathbf{E} = -\dot{\mathbf{B}}$ is written in this notation as

$$\frac{\partial E_\beta}{\partial x^\alpha} - \frac{\partial E_\alpha}{\partial x^\beta} = -\dot{B}_{\alpha\beta}, \tag{18-25}$$

and $\nabla \times \mathbf{A} = \mathbf{B}$ is written as

$$\frac{\partial A_\beta}{\partial x^\alpha} - \frac{\partial A_\alpha}{\partial x^\beta} = B_{\alpha\beta}, \tag{18-26}$$

where α and β are restricted to 1, 2, 3, and ordinary three-dimensional components are involved on the left side of the equations. On the other hand, $\mathbf{E}$ is polar; it can be derived from the potentials by

$$E_\alpha = -\frac{\partial \phi}{\partial x^\alpha} - \frac{\partial A_\alpha}{\partial t}, \qquad (\alpha = 1, 2, 3). \tag{18-27}$$

Examination of Eqs. (18–26) and (18–27) in the light of tensor requirements leads us to introduce a four-dimensional antisymmetric field tensor F_{ij} which, as a function of the covariant four-vector potential

$$\phi_i = (-c\mathbf{A}, \phi), \tag{18-28}$$

is

$$F_{ij} = \frac{\partial \phi_j}{\partial x^i} - \frac{\partial \phi_i}{\partial x^j}. \tag{18-29}$$

The contravariant tensor would be given by

$$F^{ij} = \frac{\partial \phi^j}{\partial x_i} - \frac{\partial \phi^i}{\partial x_j}. \tag{18-30}$$

In conformity with Eqs. (18–26) and (18–27), the components of F_{ij} are

$$
F_{ij} = \begin{array}{c} \\ i \\ \downarrow \end{array} \overset{j \rightarrow}{\begin{pmatrix} 0 & -cB_z & +cB_y & -E_x \\ +cB_z & 0 & -cB_x & -E_y \\ -cB_y & +cB_x & 0 & -E_z \\ +E_x & +E_y & +E_z & 0 \end{pmatrix}} . \qquad (18\text{--}31)
$$

The components of the corresponding contravariant tensor F^{ij} are

$$
F^{ij} = \begin{array}{c} \\ i \\ \downarrow \end{array} \overset{j \rightarrow}{\begin{pmatrix} 0 & -cB_z & +cB_y & +E_x \\ +cB_z & 0 & -cB_x & +E_y \\ -cB_y & +cB_x & 0 & +E_z \\ -E_x & -E_y & -E_z & 0 \end{pmatrix}} = g^{in} g^{jm} F_{nm}. \quad (18\text{--}32)
$$

Since it is possible to write the fields covariantly in terms of the potential, it should be possible to write Maxwell's equations themselves in covariant form. It is easily verified that the source equations,

$$
\mathbf{\nabla} \cdot \mathbf{E} = \rho/\epsilon_0,
$$
$$
\mathbf{\nabla} \times \mathbf{B} = \mu_0 \left(\rho\mathbf{u} + \epsilon_0 \frac{\partial \mathbf{E}}{\partial t} \right), \qquad (18\text{--}33)
$$

correspond to

$$
\frac{\partial F^{ji}}{\partial x^j} = \frac{j^i}{\epsilon_0}, \qquad (18\text{--}34)
$$

which is just the covariant divergence of the field tensor. Agreement of Eq. (18–34) with the equation of continuity, Eq. (18–9), is obvious from the antisymmetric character of the field tensor. The other field equations are a little more awkward:

$$
\mathbf{\nabla} \cdot \mathbf{B} = 0 \quad \text{and} \quad \mathbf{\nabla} \times \mathbf{E} = -\frac{\partial \mathbf{B}}{\partial t} \qquad (18\text{--}35)
$$

correspond to

$$
\frac{\partial F_{ij}}{\partial x^k} + \frac{\partial F_{jk}}{\partial x^i} + \frac{\partial F_{ki}}{\partial x^j} = 0. \qquad (18\text{--}36)
$$

The left side of Eq. (18–36) will vanish identically unless $i \neq j \neq k$. Moreover, permutation of indices will not change the content, and hence only four of the sixty-four equations formally represented by Eq. (18–36) are nontrivial and distinct.

Equation (18–36) can be written in a form resembling Eq. (18–34) (which is equivalent to a four-vector) by taking account of the fact that it is a completely antisymmetric tensor of third rank. The completely antisymmetric tensor of fourth rank, P^{ijkl}, is defined so that its components are zero unless $i \neq j \neq k \neq l$, and equal to ± 1 according to whether $ijkl$ is an even or odd permutation of 1234. P^{ijkl} transforms like a tensor under translations and proper rotations of the coordinates, but not under reflections or inversions (improper rotations): changing the sign of an odd number of coordinate axes does not change the components of P^{ijkl} under this definition, contrary to what is expected of a tensor. It is thus not strictly a tensor, but a *pseudotensor*. The pseudotensor formed by multiplying the antisymmetric field tensor by P^{ijkl} is called its *dual*:

$$G^{ij} = P^{ijkl}F_{kl}. \tag{18–37}$$

It can be seen that G^{ij} is constructed like F^{ij} except that **E** and c**B** are interchanged. In terms of G^{ij}, Eq. (18–36) is equivalent to

$$\frac{\partial G^{ij}}{\partial x^i} = 0. \tag{18–38}$$

This equation is written so as to permit the introduction of magnetic poles and pole currents if they exist. Since G^{ij} is a pseudotensor, any hypothetical pole four-vector to be added to Eq. (18–38) is a pseudovector.

The tensor expression for the fields immediately permits a derivation of the Lorentz transformation of the fields. From Eq. (17–12),

$$F'^{ij} = Q_k^i\, Q_l^j\, F^{kl}, \tag{18–39}$$

and thence we can easily derive the relations

$$E'_{||} = E_{||}, \tag{18–40}$$

$$B'_{||} = B_{||}, \tag{18–41}$$

$$\mathbf{E}'_\perp = \gamma(\mathbf{E}_\perp + \mathbf{v} \times \mathbf{B}_\perp), \tag{18–42}$$

$$\mathbf{B}'_\perp = \gamma(\mathbf{B}_\perp - \mathbf{v} \times \mathbf{E}_\perp/c^2), \tag{18–43}$$

where $\gamma = 1/\sqrt{1 - \beta^2}$, as in Chapter 17, and $E_{||}$, $B_{||}$ and $\mathbf{E}_\perp$, $\mathbf{B}_\perp$ are the components of **E** and **B** parallel and normal to **v**, respectively.

Equations (18–42) and (18–43) can be interpreted physically in a fairly simple way. The terms linear in **v** (except for the factor γ) should be essentially classical, i.e., describable by Maxwell's equations without explicit use of relativistic arguments. Equation (18–42) corresponds to the fact that to order β^2 a particle moving relative to a magnetic field experi-

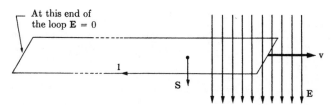

FIG. 18–2. Rectangular path of integration moving through an inhomogeneous electric field.

ences an electric field $\mathbf{E} + \mathbf{v} \times \mathbf{B}$, as has been discussed in detail in Chapter 9. To interpret Eq. (18–43), we may consider the electric flux through a rectangle moving through a region containing an electric field, as shown in Fig. 18–2. The rectangle extends outside the region where $\mathbf{E}$ is different from zero. In exact analogy to the magnetic case discussed in Section 9–3, the line integral of the magnetic field $\mathbf{B}'$ seen by an observer moving with the rectangle would be governed by the rate of change of electric flux through this rectangle as well as the real current J linking the rectangle. The rate of change of electric flux is given by

$$\oint \mathbf{E} \cdot (\mathbf{v} \times d\mathbf{l}) = - \oint (\mathbf{v} \times \mathbf{E}) \cdot d\mathbf{l}. \tag{18–44}$$

Therefore the circulation of $\mathbf{B}'$ is given by

$$\oint (\mathbf{B}' \cdot d\mathbf{l}) = \frac{1}{c^2} \frac{d}{dt} \int \mathbf{E} \cdot d\mathbf{S} + \mu_0 J = - \frac{1}{c^2} \oint (\mathbf{v} \times \mathbf{E}) \cdot d\mathbf{l} + \mu_0 J. \tag{18–45}$$

In terms of the rest coordinates, this is equivalent to

$$\mathbf{B} = \mathbf{B}' + \frac{1}{c^2} (\mathbf{v} \times \mathbf{E}), \tag{18–46}$$

in agreement with Eq. (18–43) for small β.

The transformation equations for the fields are of considerable value in the solution of practical problems involving the motion of electrons and ions in electromagnetic fields. It is frequently possible to transform away either the electric or the magnetic field by choosing a suitable Lorentz frame, whereupon the solution may be much simplified.

18–3 The Lorentz force in vacuum. The Lorentz force per unit volume, $\mathbf{f} = \rho(\mathbf{E} + \mathbf{u} \times \mathbf{B})$, is the space part of a four-vector,

$$f^i = F^{ik} j_k = \left(\mathbf{f}, \frac{\mathbf{u} \cdot \mathbf{f}}{c} \right). \tag{18–47}$$

The fourth component of f^i is $1/c$ times the power expended by the electric field per unit volume. But the total force acting on a charge occupying a volume δv, so that $\delta q = \rho \, \delta v$, is not the space part of a four-vector, since it is given by

$$\mathbf{F} = \mathbf{f} \, \delta v = \delta q (\mathbf{E} + \mathbf{u} \times \mathbf{B}). \tag{18-48}$$

In a proper frame, since δq is an invariant,

$$\mathbf{F}_0 = \delta q \mathbf{E}_0 \tag{18-49}$$

and hence, in general, from Eqs. (18-40) and (18-42),

$$F_{||} = F_{0||}, \tag{18-50}$$

$$\mathbf{F}_\perp = \gamma^{-1} \mathbf{F}_{0\perp}. \tag{18-51}$$

These equations are in agreement with the mechanical force transformations, Eqs. (17-52), (17-53), (17-54). It follows that equilibrium between mechanical forces and electrical forces is invariant to the choice of frame— the nature of the force does not affect its transformation properties. Moreover, we are justified in defining force by the relation $\mathbf{F} = d\mathbf{p}/dt$ if for $\mathbf{F}$ we use the Lorentz force, Eq. (18-48), and thus Eq. (17-50) is the relativistically correct expression for the equation of motion of a charged particle in an electromagnetic field.

The equation of motion of a particle of charge q in an external electromagnetic field is, in any Lorentz frame,

$$\frac{d}{dt}\left[\frac{m_0\mathbf{u}}{(1 - u^2/c^2)^{1/2}}\right] = q(\mathbf{E} + \mathbf{u} \times \mathbf{B}), \tag{18-52}$$

which is the space component of the covariant equation

$$m_0 c^2 \frac{du^i}{ds} = qF^{ik}u_k. \tag{18-53}$$

18-4 Covariant description of sources in material media. From the electron theory viewpoint the ordinary Maxwell equations are the result of averaging electrical quantities over regions large enough to permit macroscopic observation. We have seen that vacuum electrodynamics with charges and currents may be described in a simple covariant manner. It should then be possible to write the equations of electrodynamics in material media in covariant form. The new element in macroscopic electrodynamics of material media is that the four-vector current j^i may have four nonvanishing components even in a frame in which the medium is at rest. In such a frame,

$$(j^i) = \left(\frac{\mathbf{j}^0}{c}, \rho^0\right), \tag{18–54}$$

where $\mathbf{j}^0$ is the current density in the proper frame. Let us now consider the form of the components of j^i in a nonproper frame. To correspond to the vacuum case, j^i must retain the components

$$(j^i) = \left(\frac{\mathbf{j}}{c}, \rho\right) \tag{18–55}$$

in any frame, and hence, in general,

$$j_x = \gamma(j_x^0 - \rho^0 v), \tag{18–56}$$

$$\rho = \gamma\left(\rho^0 - \frac{j_x^0 v}{c^2}\right), \tag{18–57}$$

where v is the velocity (taken along x) of the arbitrary frame measured in the proper frame.

Equation (18–56) is physically clear: it contains the convective current due to transport of charge and the contraction factor which assures the invariance of the charge. Equation (18–57) is less obvious physically: it says that a substance which carries a current but is electrically neutral ($\rho^0 = 0$) in a proper frame does not remain neutral when observed from another inertial frame. This effect can be understood in terms of the kinematics of the moving charges. Let us consider equal densities of positive and negative charges in the proper frame Σ^0. For simplicity, let the positive charges be at rest (although this is not necessary to the argument) and the negative charges be in motion. The world lines of the plus charges (shown as dashes) and of the minus charges (shown solid) as observed in Σ^0 and in an arbitrary frame Σ are indicated in the Minkowski diagram, Fig. 18–3. The charge density in Σ^0 is measured by counting

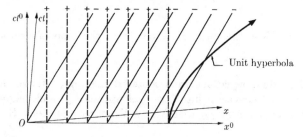

Fig. 18–3. Minkowski diagram showing the possibility of unequal charge densities in two Lorentz frames.

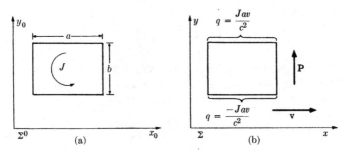

FIG. 18-4. Electric moment acquired by a current loop when viewed from a moving frame.

$+$ and $-$ charges simultaneously in Σ^0, i.e., on the x^0-axis; the charge density in Σ is measured by counting $+$ and $-$ charges along the x-axis, i.e., averaging them simultaneously in Σ. It is seen that the density of $-$ charges along the x-axis is decidedly decreased relative to that along the x^0-axis, while the density of $+$ charges has changed very little. (Note, however, that density of charge is measured by counting charges per unit length defined by the intercept of the unit hyperbola with the respective axes, as shown in the figure.) Therefore a net positive charge is found in Σ, corresponding to the negative current in the neutral proper frame, in agreement with Eq. (18–57).

One consequence of this effect, resulting directly from the difference in simultaneity between Σ^0 and Σ, is the fact that a neutral stationary current loop in Σ^0 acquires an electric moment when observed in Σ. Consider the rectangular current loop of Fig. 18–4, carrying a current J. From the point of view of Σ the legs parallel to x^0 will carry charges of $\pm Jav/c^2$ respectively, if we ignore the terms quadratic in v. Thus the system has an electric moment

$$|\mathbf{P}| = \frac{vabJ}{c^2} = \frac{|\mathbf{v} \times \mathbf{m}|}{c^2}, \qquad (18\text{–}58)$$

where $\mathbf{m} = J\mathbf{S}$ is the magnetic moment of the loop of vector area $\mathbf{S}$, $|\mathbf{S}| = ab$.

18–5 The field equations in a material medium. If we divide the charge-current four-vector into "true" components j^i and magnetization-polarization components j^i_M, then Maxwell's source equations, Eq. (18–34), become

$$\frac{\partial F^{il}}{\partial x^i} = \frac{j^l + j^l_M}{\epsilon_0}, \qquad (18\text{–}59)$$

while Eq. (18–36) remains unchanged. As before, the F^{ij} are given by the matrix of Eq. (18–32) in terms of **B** and **E**. It is desirable, as in the three-dimensional formulation, to write Eq. (18–59) in terms of the true charges and currents as external sources only, and to incorporate the "induced" charges and currents into the fields. This can be done by introducing the moment tensor M^{ij} by the equation

$$j_M^l = \frac{\partial M^{il}}{\partial x^i}. \tag{18–60}$$

In order to correspond to the inaccessible sources of Chapter 7, the components of j_M^i must be

$$j_M^l = \left(\frac{\mathbf{j}_M + \mathbf{j}_P}{c}, \rho_P\right) = \left(\frac{\boldsymbol{\nabla} \times \mathbf{M}}{c} + \frac{1}{c}\frac{\partial \mathbf{P}}{\partial t}, -\boldsymbol{\nabla} \cdot \mathbf{P}\right). \tag{18–61}$$

The components of M^{ij} are then given by

$$
(M^{ij}) = \begin{array}{c} \\ i \\ \downarrow \end{array}
\overset{\displaystyle j \rightarrow}{
\begin{pmatrix}
0 & -M_z/c & M_y/c & -P_x \\
M_z/c & 0 & -M_x/c & -P_y \\
-M_y/c & M_x/c & 0 & -P_z \\
P_x & P_y & P_z & 0
\end{pmatrix}}. \tag{18–62}
$$

Corresponding to the three-dimensional relations

$$\mathbf{H}/c\epsilon_0 = c\mathbf{B} - \mathbf{M}/c\epsilon_0, \tag{18–63}$$

$$\mathbf{D}/\epsilon_0 = \mathbf{E} + \mathbf{P}/\epsilon_0, \tag{18–64}$$

we may introduce a new field H^{ij} defined by

$$H^{ij}/\epsilon_0 = F^{ij} - M^{ij}/\epsilon_0. \tag{18–65}$$

The source equations, Eq. (18–59), then become simply

$$\frac{\partial H^{ij}}{\partial x^i} = j^j, \tag{18–66}$$

where

$$
(H^{ij}) = \begin{array}{c} \\ i \\ \downarrow \end{array}
\overset{\displaystyle j \rightarrow}{
\begin{pmatrix}
0 & -H_z/c & H_y/c & D_x \\
H_z/c & 0 & -H_x/c & D_y \\
-H_y/c & H_x/c & 0 & D_z \\
-D_x & -D_y & -D_z & 0
\end{pmatrix}}. \tag{18–67}
$$

The signs in the defining equations of the auxiliary fields correspond to the way in which the equivalent currents and charges are derived from the moments.

Equations (18–36) and (18–66) are the covariant Maxwell equations related by Eq. (18–65), which shows the connection between the fields H^{ij} and F^{ij} and the moments M^{ij}. If the moments M^{ij} and the currents j^i, rather than being given functions, are expressed in terms of the fields by "constitutive equations" (e.g., permeability, dielectric constant and conductivity), the covariant equations are not very useful, since such constants are defined in the frame in which the medium is at rest. A covariant definition of these constants is possible, but involves considerable complexity.

18–6 Transformation properties of the partial fields. The transformation properties of the moments follow directly from the covariant formulation, as in Eqs. (18–40) through (18–43). We obtain for the transformation from the plain to the primed system moving with relative velocity v,

$$P'_{||} = P_{||}, \tag{18–68}$$

$$M'_{||} = M_{||}, \tag{18–69}$$

$$\mathbf{P}'_{\perp} = \gamma \left(\mathbf{P}_{\perp} - \frac{\mathbf{v} \times \mathbf{M}_{\perp}}{c^2} \right), \tag{18–70}$$

$$\mathbf{M}'_{\perp} = \gamma (\mathbf{M}_{\perp} + \mathbf{v} \times \mathbf{P}_{\perp}). \tag{18–71}$$

Equation (18–68) is to be expected, since $P_{||}$ is the product of an (invariant) charge and a distance divided by a volume, both contracted in the same ratio. A similar argument applies to Eq. (18–69). The term $\mathbf{v} \times \mathbf{P}$ in Eq. (18–71) is, apart from the factor γ, nonrelativistic, arising from the fact that convection of a polarized medium corresponds to a net circulation of charge. We have met this effect before (Chapter 9) in the discussion of Maxwell's equations in moving media from a nonrelativistic point of view. Consider an infinite polarized slab, shown in Fig. 18–5. From Σ, a moving frame, this slab possesses opposing surface currents corresponding to a uniformly magnetized medium.

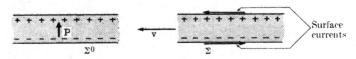

FIG. 18–5. A polarized slab in motion corresponds to a magnetized medium.

Equation (18–70) has no nonrelativistic counterpart. It represents, as already shown, the effect of the net charges when a current in a neutral stationary conductor is viewed from a nonproper frame. The extra electric moment is that predicted by Eq. (18–58), and arises from the relativistic definition of simultaneity. This equivalent electric moment resolves the apparent paradox of the unipolar induction generator. We concluded in Section 9–5 that a current would flow when a conductor, in contact with take-off brushes, moves transverse to a magnetic field. No difficulties arise so long as the source of **B** is external; the effect should, however, persist if **B** is due to a permanent magnetization of the bar itself. Since the permanent magnetization is to be describable in terms of equivalent Ampèrian currents alone, the question arose as to how such currents could produce an electrostatic effect when viewed from a moving frame. Now we see that this description leads to no difficulties, since we can interpret the electrostatic field as due to the equivalent moment $\mathbf{v} \times \mathbf{M}/c^2$. Inasmuch as this equivalent moment is a consequence of the relativistic definition of simultaneity, unipolar induction is fundamentally a relativistic effect.

Practical devices for producing unipolar induction ordinarily involve rotation, and consequently some remarks should be added here concerning rotary motion or accelerated motion in general. A full treatment of accelerated frames is beyond the scope of this book, but some consideration is necessary to clarify the applicability of statements made here and in Section 9–5 to actual experiments.

Contrary to what is true of frames differing in motion only by uniform relative velocity (Lorentz frames), experiments *can* be performed that distinguish rotating (accelerated) frames. Here we have been dealing with the special (inertial) frames in which the laws of special relativity are valid. Which frame is an inertial frame is presumably defined by gravitational forces, and therefore "a preferred frame as to rotation" is defined by the location of the bulk of the masses in the universe. Strictly speaking, there is no frame in which special relativity is exactly valid, since no frame can be found in which gravitational accelerations or "equivalent" (i.e., indistinguishable) inertial accelerations vanish. For all electrical phenomena, however, the surface of the earth is a satisfactory Lorentz frame to a very good approximation; note that the tests of the special theory are concerned with the effects of the linear, not the curvilinear, motion of the earth's surface.

But there are large scale electromagnetic effects which distinguish noninertial frames produced by rotation of matter in the earth's frame. Many paradoxes result if one assumes that such phenomena should be reciprocal in the rotating frame and that of the earth. Let us consider the device analogous in rotation to the sliding-bar experiment of Section 9–5, the

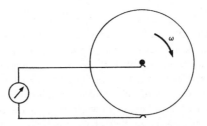

FIG. 18–6. Faraday disk illustrating unipolar induction in rotation.

"Faraday disk" illustrated in Fig. 18–6. Its behavior conforms in some details to the conclusions drawn in Section 9–5. All the results set forth in Table 9–1, which gives the electromotive force corresponding to various possibilities of motion of the disk, the source of magnetic field, and the observing apparatus, are preserved. In particular, the important conclusion is retained that motion (rotation in this case) of the source of magnetic field does not affect any physical process, so long as such motion does not produce a time-varying field. Again the electromotive force can be calculated directly from Faraday's law, Eq. (9–2), if the change in flux is computed through an orbit which moves with the actual carrier of the current; here this path is split between the two frames. Also, consideration of the "effective force on an electron" gives the right answer. Note, however, that the effective electric field $\mathbf{E}_r$ in the rotating frame is given by $\mathbf{E}_r = \omega B \mathbf{r}$ for low velocities; this expression has a nonvanishing divergence, and thus a volume charge is developed. This is at variance with our transformation laws for linear motion, and is an indication that the "absolute" rotational motion of the disk (i.e., the motion relative to an inertial frame) can in principle be determined.

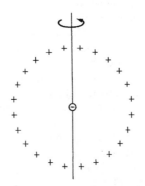

FIG. 18–7. Rotating charge distribution of zero net charge.

Another experiment which more obviously distinguishes absolute rotational frames is that shown in Fig. 18-7. A negative point charge is surrounded by a positive spherical surface distribution of equal total charge. If this system is not rotating, $E = B = 0$ outside the sphere, in accord with Gauss's theorem. But if the charge is rotating the positive charges constitute ring currents, and thus the entire assembly develops a magnetic moment. Hence B is different from zero, both inside and outside the sphere. Again our transformation laws fail to apply to nonrectilinear motion. It is not possible to discuss the problem consistently in both the stationary and the rotating frames without recourse to general relativity.*

It should be emphasized, however, that within the framework of special relativity our earlier considerations do permit us to describe all fields in all Lorentz frames, whether the *sources* are accelerated or not.

SUGGESTED REFERENCES

Excellent accounts will be found in the previously listed books by Becker, Bergmann, Fock, Møller, and Tolman. Landau and Lifshitz approach the subject from the consideration of the behavior of a particle in a field. In addition we should list the following works:

A. SOMMERFELD, *Electrodynamics*, Part III, Theory of Relativity and Electron Theory. Electrodynamics is the third volume of Professor Sommerfeld's justly famous Lectures on Theoretical Physics.

A. EINSTEIN, a translation of the original paper "On the Electrodynamics of Moving Bodies" in *The Principle of Relativity*.

W. PAULI, *Relativitätstheorie*, Encyclopädie der Mathematischen Naturwissenschaften, Vol. V, or in special reprint. A comprehensive classical reference on the subject since 1921.

EXERCISES

Show that:

1. If E and B are perpendicular in one Lorentz frame, they are perpendicular in all Lorentz frames.

2. If $|E| > |cB|$ in any one Lorentz frame, then $|E| > |cB|$ in any other Lorentz frame, and vice versa.

3. If the angle between E and B is acute (or obtuse) in one Lorentz frame, it is acute (or obtuse) in any other Lorentz frame.

4. If E is perpendicular to B but $|E| \neq |cB|$, then there is a frame in which the field is either purely electric or purely magnetic.

---

* See L. I. Schiff, *Proc. Nat. Acad. Sci.* **25,** 391 (1939).

5. If a particle of charge q moves in a plane perpendicular to a magnetic field **B** not varying in time, show that the radius of curvature ρ is given by $eB\rho = p$, where p is the magnitude of the (relativistic) momentum of the particle. If B is also constant in space, show that plane motion is a circle described with angular frequency

$$\omega = \frac{eB}{m} = \frac{eB}{m_0\gamma},$$ (18-72)

where m_0 is the rest mass of the particle, and $\gamma = m/m_0$.

CHAPTER 19

THE LIÉNARD-WIECHERT POTENTIALS AND THE FIELD OF A UNIFORMLY MOVING ELECTRON

The mathematical "machinery" of covariance is a powerful tool for deriving the consequences of electromagnetic theory, as we have seen, but the resulting expressions must be translated into the ordinary space and time variables of the observer in order to be compared with experiment. Most of the results of classical radiation theory were derived before the advent of relativity theory, it being assumed that the frame of the observer was one in which Maxwell's equations are valid. It is instructive, in view of the physical interpretation of the formulas, to go through some of these derivations, especially now that we have seen from relativistic considerations that their validity is much more general than could have been originally supposed.

19–1 The Liénard-Wiechert potentials. Let us consider the application of the retarded potentials, Eqs. (14–24) and (14–25), to compute the radiation from an electron. Now in classical electrodynamics the only thing known about the electron is that it has a certain total charge, and any calculation of its radiation field cannot involve details of how this charge may be distributed geometrically within the electron. On the other hand, it is impossible to assume that the charge has zero physical extent without introducing various mathematical divergences. But certain features of the radiation field are actually independent of the radius of the electron, provided only that it is small compared with the other dimensions of the radiation field. In our discussion of the electron and its behavior we shall assume that it has a finite radius, but we shall ascribe physical significance only to those properties which are independent of the magnitude of the radius. A finite radius of an electron has no meaning in the sense of the radius of a rigid structure; as noted previously, a "rigid" dimension for an object is relativistically self-contradictory. On the other hand, two possibilities are not excluded: (a) a covariantly defined scalar distance δ exists such that if signals are communicated over a four-vector distance $|r_i r^i| < \delta$, the laws of electrodynamics (or, more generally, quantum electrodynamics) have to be modified, presumably still within the framework of a covariant field theory (or its quantum-mechanical equivalent); (b) relativity does not remain valid over distances less than some critical

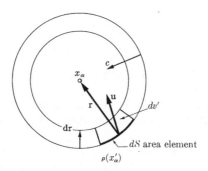

FIG. 19-1. Element dS of sphere collecting information from charge in motion with velocity **u**.

quantity. There is no reason to believe that the "classical electron radius" $r_0 = e^2/4\pi\epsilon_0 mc^2$ introduced in the following chapter has any relation to such a possible limit of validity of electrodynamics.

Prerelativistically, considerable care had to be exercised in applying the concept of retarded potential to a system whose total charge is known. If $[\rho]$ is the retarded charge density to be substituted in Eq. (14–25) to give the correct potential $\phi(x_\alpha, t)$, then $\int[\rho] \, dv'$ does *not* in general represent the total charge of the system. The reason is that the various contributions to the integrand $[\rho] \, dv'$ are evaluated at different times, and during the time the information-collecting sphere of Fig. 19–1 sweeps over the charge distribution the charges may move so as to appear more or less dense than they should to give a correct value for the total charge. This can be illustrated by a detailed consideration of the process. Consider the sphere of Fig. 19–1 converging onto the point of observation x_α with a velocity c, and let it gather information as to the charge density within a certain charge system as it sweeps across the system. If the charge system has an average velocity component in the same direction as the motion of the converging sphere the volume integral of the retarded charge density will be in excess of the total charge. If the charge distribution has an average velocity component in opposition to the velocity of the contracting sphere the integral will give a result less than the total charge of the system.

The situation is analogous to the problem of taking a census of the population of a country. Let us assume that a group of census takers converges upon the "information center" with a certain speed, measuring the population density by counting the people every day as they travel. The correct population will differ from the total of the census takers' information depending on whether the population had a net migration trend with the census takers (in which case the true population is less than the sum of the reported densities) or against the census takers. Similarly, the retarded potential of an approaching charge will be larger than that of a receding charge at the same distance from the observer, since the approaching charge stays longer within the information-collecting sphere.

Let us consider the radiation field of an electron whose velocity is comparable to c. We shall assume that our electron is a system about which we know (a) that the total charge is e, and (b) that within an unspecified but small volume all parts of the electron's charge are moving systematically with a velocity $\mathbf{u}$. Let $\mathbf{r}$ be the radius vector from a point where there is charge to the point of observation on which the sphere of Fig. 19–1 is converging; $\mathbf{r}$ is the "retarded" value of the radius, measured at the time of passage of the sphere. If the electron is at rest, the amount of charge that the sphere will cross during the time dt as it shrinks by dr is given by $[\rho]\, dS\, dr$. On the other hand, if $\mathbf{u}$ is different from zero a quantity of charge which is less than $[\rho]\, dS\, dr$ by the amount $[\rho]\, dS(\mathbf{u}\cdot\mathbf{r}/r)\, dt$ will be crossed by the sphere. Thus the amount of charge observed by this means of collecting information is

$$de = [\rho]\, dv' - [\rho]\,\frac{\mathbf{u}\cdot\mathbf{r}}{r}\, dS\, dt. \tag{19–1}$$

But $dt = dr/c$, and $dS\, dr = dv'$, so that

$$de = [\rho]\, dv' - [\rho]\,\frac{\mathbf{u}\cdot\mathbf{r}}{rc}\, dv'. \tag{19–2}$$

Solving for the retarded charge density, we obtain

$$[\rho]\, dv' = \frac{de}{1 - \mathbf{u}\cdot\mathbf{r}/cr}$$

or

$$\frac{[\rho]\, dv'}{r} = \frac{de}{r - \mathbf{u}\cdot\mathbf{r}/c}. \tag{19–3}$$

The retarded potentials of Eqs. (19–25) and (19–24) thus become

$$\phi = \frac{1}{4\pi\epsilon_0}\int\frac{de}{r - \mathbf{r}\cdot\mathbf{u}/c}, \tag{19–4}$$

$$\mathbf{A} = \frac{\mu_0}{4\pi}\int\frac{\mathbf{u}\, de}{r - \mathbf{r}\cdot\mathbf{u}/c}. \tag{19–5}$$

At the limit of a point charge the distance-dependent terms are slowly varying, and since $\int de = e$, the known electronic charge, the potentials of a point charge are

$$\phi = \frac{1}{4\pi\epsilon_0}\left[\frac{e}{r - \mathbf{r} \cdot \mathbf{u}/c}\right], \tag{19-6}$$

$$\mathbf{A} = \frac{\mu_0}{4\pi}\left[\frac{e\mathbf{u}}{r - \mathbf{r} \cdot \mathbf{u}/c}\right]. \tag{19-7}$$

Equations (19-6) and (19-7) are known as the Liénard-Wiechert potentials of a single electron. Note that these expressions are independent of the extent of the charge, and therefore independent of any detailed electronic model. They are, in fact, just Eqs. (18-21), but the significance of the velocity is quite different. Here the velocity is that of the charge in its retarded position relative to the special coordinate frame in which it had to be assumed, prior to relativity theory, that Maxwell's equations were valid. In Chapter 18 we saw that the velocity was that of the observer relative to the frame in which the charge was at rest at the time the signal was emitted. All the detailed calculations of fields from moving charges made on the assumption that there is one frame in which the wave equation is correct are equivalent to those resulting from a covariant formulation of electrodynamics if we thus re-interpret the velocity.

19-2 The fields of a charge in uniform motion. Since the relation of the "retarded" position to the "present" position of an electron is not, in general, known, the Liénard-Wiechert potentials ordinarily permit only the evaluation of the fields in terms of the retarded positions and velocities of the charges. If the motion is uniform, however, it is possible to express the potentials and the fields in terms of the "present" position of the charge, i.e., the position of the charge at time t that the information-collecting sphere of Fig. 19-1 converges at the point of observation.

Consider an electron, as in Fig. 19-2, moving with a uniform velocity in the x-direction. Let us evaluate the Liénard-Wiechert denominator

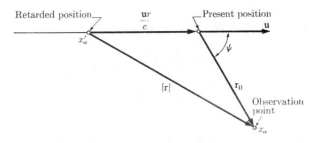

FIG. 19-2. Potential and field coordinates of an electron in uniform motion.

$s = r - (\mathbf{r} \cdot \mathbf{u})/c$ in terms of the present position of the electron. Since by the geometry of the figure, $\mathbf{r}_0 \times \mathbf{u} = \mathbf{r} \times \mathbf{u}$, it can be easily verified that

$$s^2 = r_0^2 - \left(\frac{\mathbf{r}_0 \times \mathbf{u}}{c}\right)^2. \tag{19–8}$$

The denominator s can then be expressed explicitly in terms of the present position coordinates x_0, y_0, z_0, or in terms of r_0 and the angle ψ between u and r_0:

$$s = \sqrt{x_0^2 + y_0^2 + z_0^2 - \frac{u^2}{c^2}(y_0^2 + z_0^2)}$$

$$= \sqrt{x_0^2 + \left(1 - \frac{u^2}{c^2}\right)(y_0^2 + z_0^2)}$$

$$= r_0\sqrt{1 - \frac{u^2}{c^2}\sin^2 \psi}. \tag{19–9}$$

With this substitution for the denominator of Eqs. (19–6) and (19–7) the Liénard-Wiechert potentials are given in terms of the present coordinates of the electron.

The fields are easy to compute explicitly in this case. The electric field is

$$\mathbf{E} = -\boldsymbol{\nabla}\phi - \frac{\partial \mathbf{A}}{\partial t}. \tag{14–1}$$

The time derivative can be evaluated in terms of the spatial derivative by noting that the field must be carried by the uniformly moving charge. A stationary observer will see a field at time $t + dt$ which existed at a position displaced from him by $-\mathbf{u}\,dt$ at time t. Hence the time derivative can be replaced by

$$\frac{\partial}{\partial t} = -u\frac{\partial}{\partial x}. \tag{19–10}$$

With the use of this relation in Eq. (19–1), the components of the electric field become simply:

$$E_x = \frac{ex_0}{4\pi\epsilon_0 s^3}\left(1 - \frac{u^2}{c^2}\right), \tag{19–11}$$

$$E_y = \frac{ey_0}{4\pi\epsilon_0 s^3}\left(1 - \frac{u^2}{c^2}\right), \tag{19–12}$$

$$E_z = \frac{ez_0}{4\pi\epsilon_0 s^3}\left(1 - \frac{u^2}{c^2}\right). \tag{19–13}$$

Although **A** has only an x-component, the electric field is symmetrical (except for the denominator) in its three components. Note that the field is directed toward the "present" position of the electron (for a negative electron), and not toward the "retarded" position. Vectorially the electric field is given by

$$\mathbf{E} = \frac{e}{4\pi\epsilon_0 s^3}\left[\mathbf{r} - \frac{\mathbf{u}r}{c}\right]_{\text{ret}}\left(1 - \frac{u^2}{c^2}\right) = \frac{e\mathbf{r}_0(1 - u^2/c^2)}{4\pi\epsilon_0 s^3}$$

$$= \frac{e\mathbf{r}_0}{4\pi\epsilon_0 r_0^3}\frac{(1 - u^2/c^2)}{[1 - (u^2/c^2)\sin^2\psi]^{3/2}}, \qquad (19\text{--}14)$$

which shows explicitly the direction of the field.

The magnetic field of the uniformly moving electron, as computed from $\mathbf{B} = \nabla \times \mathbf{A}$, turns out to be

$$\mathbf{B} = \frac{\mu_0 e}{4\pi}\frac{\mathbf{u} \times \mathbf{r}_0}{s^3}\left(1 - \frac{u^2}{c^2}\right) = \frac{1}{c^2}\mathbf{u} \times \mathbf{E}. \qquad (19\text{--}15)$$

For low velocities, $u \to 0$, $s \to r$, Eqs. (19–14) and (19–15) reduce to the Coulomb and Biot-Savart fields. For high velocities, $u \to c$, both field magnitudes depend on the angle between the direction of motion of the electron and the radius vector $\mathbf{r}_0$. From Eq. (19–14) it is seen that the electric field is increased in a direction at right angles to the direction of motion in the ratio of $1/\sqrt{1 - u^2/c^2}$, while in the direction of motion the field is decreased in the ratio $(1 - u^2/c^2)$. At very high velocities the field thus resembles more and more the field in a plane wave. For a short time, as a high-velocity electron passes an observer, he sees a purely transverse electric and magnetic field. Note, however, that a uniformly moving charge is nonradiating in the sense that its field does not represent an energy loss. This can be shown by direct evaluation of the Poynting vector corresponding to the fields given by Eqs. (19–14) and (19–15).

These formulas are identical with those obtained by applying a Lorentz transformation to the fields of a static charge, with no restriction on the velocity. (The proof of this statement is left to a problem.) Prior to relativity theory, however, considerable care had to be exercised in the application of these results, and the assumption that the observer is at rest with respect to the particular frame of reference in which Maxwell's equations were assumed valid was very troublesome. The theory of relativity has eliminated these difficulties by justifying the interpretation of **u** as the relative velocity between the electron and an observer.

19–3 Direct solution of the wave equation. It is instructive to see how our conclusions regarding the field of a uniformly moving charge can be derived from the inhomogeneous wave equations. Equations (14–7) and (14–8) are to be solved subject to the subsidiary Lorentz condition of Eq. (14–3). The field of an electron moving with uniform velocity must be carried convectively along with the electron, which implies that the time and space derivatives are not independent. This fact can be expressed mathematically by a generalization of Eq. (19–10):

$$\frac{\partial}{\partial t} = -\mathbf{u} \cdot \boldsymbol{\nabla}. \tag{19–16}$$

This means that any field parameter at a given point changes by the same amount in a time dt as at a fixed time it would differ from the same field parameter evaluated at a distance $-u\,dt$ along the direction of motion of the electron.

Let us consider the inhomogeneous equation

$$\Box\psi(x_\alpha, t) = -g(x_\alpha) \tag{14–11}$$

under the condition that the source charge moves with constant velocity in the x-direction and thus does not depend explicitly on the time. Then, by Eq. (19–16), the wave equation becomes

$$\left(1 - \frac{u^2}{c^2}\right)\frac{\partial^2\psi}{\partial x^2} + \frac{\partial^2\psi}{\partial y^2} + \frac{\partial^2\psi}{\partial z^2} = -g(x, y, z). \tag{19–17}$$

This equation can be transformed by a change of variables

$$x_1 = \frac{x}{\sqrt{1 - u^2/c^2}}; \quad y_1 = y; \quad z_1 = z; \tag{19–18}$$

to a simple electrostatic Poisson equation,

$$\nabla_1^2\psi = -g(\sqrt{1 - u^2/c^2}\, x_1, y_1, z_1), \tag{19–19}$$

of which the solution is the ordinary Coulomb potential,

$$\psi(x_{1_\alpha}) = \frac{1}{4\pi}\int \frac{g(x'_{1_\alpha})\,dv'_1}{r(x'_{1_\alpha}, x_{1_\alpha})}. \tag{19–20}$$

If we transform back to the original variables, this solution becomes

$$\psi(x_\alpha) = \frac{1}{4\pi}\int \frac{g(x'_\alpha)}{s}\,dv', \tag{19–21}$$

where

$$s = \sqrt{(x - x')^2 + (1 - u^2/c^2)[(y - y')^2 + (z - z')^2]}. \quad (19\text{--}22)*$$

Explicitly, then, for a charge e moving with a velocity $\mathbf{u}$,

$$\phi = \frac{e}{4\pi\epsilon_0 s}, \quad (19\text{--}23)$$

$$\mathbf{A} = \frac{e\mathbf{u}}{4\pi\epsilon_0 c^2 s}. \quad (19\text{--}24)$$

Equations (19–23) and (19–24) represent the same potentials as those obtained from the Liénard-Wiechert expressions. In this derivation, however, the question regarding the propagation velocity of the corresponding wave and the relation between present and retarded potentials does not enter, since by a suitable transformation we have reduced the equation to be solved, Eq. (19–11), to the static equation, Eq. (19–19). It is obvious that the purely mathematical process of Eq. (19–18) is in reality a Lorentz transformation, in which we transform the observer's position to a frame that is at rest relative to the electron whose field is to be computed.

19–4 The "convection potential." The force which would be exerted by these fields on another electron moving with a velocity u parallel to that of the original electron producing the field is presumably given by the Lorentz expression:

$$\mathbf{F} = e(\mathbf{E} + \mathbf{u} \times \mathbf{B}).$$

If the fields are computed from the potentials of Eqs. (19–23) and (19–24), this force is given by

$$\mathbf{F} = \frac{e^2}{4\pi\epsilon_0}\left[-\boldsymbol{\nabla}\left(\frac{1}{s}\right) + (\mathbf{u}\cdot\boldsymbol{\nabla})\frac{\mathbf{u}}{c^2 s} + \frac{\mathbf{u}}{c^2}\times\left(\boldsymbol{\nabla}\times\frac{\mathbf{u}}{s}\right)\right], \quad (19\text{--}25)$$

which, by expansion of the vector product, becomes

$$\mathbf{F} = -\frac{e^2}{4\pi\epsilon_0}\boldsymbol{\nabla}\left(\frac{1 - u^2/c^2}{s}\right). \quad (19\text{--}26)$$

Equation (19–26) can be written in the form

$$\mathbf{F} = -\boldsymbol{\nabla}\psi, \quad (19\text{--}27)$$

where

$$\psi = \frac{e^2(1 - u^2/c^2)}{4\pi\epsilon_0 s} \quad (19\text{--}28)$$

---

* Note that $x - x'$ in Eq. (19–22) corresponds to the "present" coordinate x_0 in Eq. (19–9), and likewise for the other coordinates.

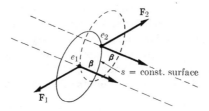

Fig. 19–3. Showing forces between two point charges as derived from the convection potential.

is called the convection potential. The force of one electron on the other is thus derivable from a scalar potential, ψ, but this scalar potential does not have spherical symmetry about the position of the electron producing the field. In particular, since the direction of the force must be perpendicular to the surface of equal convection potential, we would conclude that the force $\mathbf{F}_2$ exerted by the electron e_1 at (x_1, y_1, z_1) on the electron e_2 at (x, y_2, z_2) is perpendicular to the ellipsoid

$$s = \sqrt{(x_1 - x_2)^2 + (1 - u^2/c^2)[(y_1 - y_2)^2 + (z_1 - z_2)^2]}$$

$$= \text{constant}, \tag{19–29}$$

as shown in Fig. 19–3.

On the other hand, the reaction force $\mathbf{F}_1$ on the electron e_1 is perpendicular to the corresponding ellipsoid (shown by the dashed line in the figure) referred to the co-moving electron e_2. Hence, except when the line between the electrons is parallel or perpendicular to the direction of motion, the forces of action and reaction do not appear to be collinear. Therefore if the two electrons were connected by a rigid bar there would be a couple acting about an axis perpendicular to the plane of the line joining the electrons and the direction of motion. This will be recognized as the torque also predicted by Ampère's law when current elements are substituted for the moving charges, and which Trouton and Noble attempted to measure. The paradox produced by the observation of a null effect indicates the difficulties in interpreting the velocity of moving charges in pre-relativistic electrodynamics.

The torque predicted here is real enough to an observer moving with a velocity $\mathbf{u}$ relative to the two charges, and should in that case be measurable if there were no mechanical considerations involved. We have already noted that the assumption of a "rigid" bar is not consistent with the theory of relativity. What relativity means is that mechanical quantities obey the same transformation laws, whether their origin is mechanical or electrical. Hence forces derived from elastic stresses would depend on the velocity in the same way as those corresponding to the Lorentz force. The problem as a whole is similar to that in which the torque is balanced by the gain in angular momentum: in every case equilibrium is a property invariant under a Lorentz transformation. It is clear that to an observer moving along with the charges they do not actually constitute current elements, and the observed interaction will be just the static Coulomb force.

19–5 The virtual photon concept. We have shown that the electromagnetic field of a rapidly moving charge approaches the field of a plane wave as the velocity approaches c. It is of considerable interest to express this correspondence in a quantitative way.*

The electromagnetic field due to a passing charge represents a pulse in time which corresponds to a distribution in frequency of the energy contained in the field. Clearly, the integral of this frequency spectrum will simply represent the total energy of the field of the charge. This integral will, of course, be divergent unless the field is "cut off" at small distances.

Let us first make a Fourier analysis of the electric field component transverse to the velocity **u**. We have, from Eq. (19–13),

$$\frac{4\pi\epsilon_0}{e} E_\perp(t) = \frac{\gamma^{-2} b}{[(ut)^2 + \gamma^{-2}b^2]^{3/2}}, \tag{19-30}$$

where $\gamma = (1 - u^2/c^2)^{-1/2}$ and b is the perpendicular distance between the particle trajectory and the observer (see Fig. 19–4). The Fourier components of this field are given by

$$E_{\omega\perp} = \frac{1}{2\pi} \int_{-\infty}^{\infty} E_\perp(t)e^{i\omega t}\, dt$$

$$= \frac{eb\gamma^{-2}}{8\pi^2\epsilon_0} \int_{-\infty}^{\infty} \frac{e^{i\omega t}}{[(ut)^2 + \gamma^{-2}b^2]^{3/2}}\, dt$$

$$= \frac{e}{8\pi^2\epsilon_0 bu} \int_{-\infty}^{\infty} \frac{e^{i(\omega b/\gamma u)\xi}}{(\xi^2 + 1)^{3/2}}\, d\xi, \tag{19-31}$$

where $\xi = \gamma ut/b$. This can be written as

$$E_{\omega\perp} = \frac{e}{4\pi^2\epsilon_0 bu}\left[\left(\frac{\omega b}{\gamma u}\right) K_1\left(\frac{\omega b}{\gamma u}\right)\right], \tag{19-32}$$

---

* The close relation between interactions produced by moving charged particles and those due to incident electromagnetic waves was first pointed out by E. Fermi, *Z. Physik* **29**, 315 (1924).

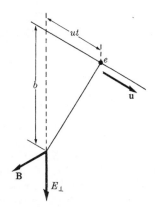

FIG. 19–4. For computing the transverse field of a moving electron.

where K_1 is the Bessel function of the second kind with imaginary argument. The function $K_1(x)$ behaves like $1/x$ near $x = 0$ and goes to zero asymptotically as e^{-x}. Hence we can approximate Eq. (19–31) by

$$E_{\omega\perp} = \frac{e}{4\pi^2\epsilon_0 bu}, \quad \omega < \gamma u/b,$$
$$E_{\omega\perp} = 0, \qquad \omega > \gamma u/b. \tag{19–33}$$

This relation is quite reasonable, since due to the transverse contraction of the field the effective "time of passage" is of order $b/\gamma u$.

In an electromagnetic wave half the energy is carried by the electric and half by the magnetic field. Hence the total energy is

$$U = \epsilon_0 \int E_\perp^2 \, dv = \epsilon_0 \int \left(\int E_\perp^2 u \, dt \right) 2\pi b \, db. \tag{19–34}$$

By using the theorem proved in Eq. (14–45) we may write this energy in terms of Fourier components as

$$U = \int_0^\infty U_\omega \, d\omega = 4\pi\epsilon_0 u \int \left(\int_0^\infty |E_{\omega\perp}|^2 \, d\omega \right) 2\pi b \, db$$
$$= \frac{2}{\pi} \frac{e^2}{4\pi\epsilon_0 u} \int_0^\infty \int_{b_{\min}}^{\gamma u/\omega} \frac{db}{b} \, d\omega, \tag{19–35}$$

where $b_{\min}$ is an arbitrary lower cutoff for the Coulomb field. Hence,

$$U_\omega = \frac{2}{\pi} \frac{e^2}{4\pi\epsilon_0 u} \ln\left(\frac{\gamma u}{\omega b_{\min}} \right) \tag{19–36}$$

gives the energy distribution of the equivalent plane wave field in frequency. Clearly, b_{min} remains undefined within the framework of classical theory.

It is attractive to translate Eq. (19–36) into the number of "equivalent photons" N_ω by the relation $U_\omega \, d\omega = \hbar\omega N_\omega \, d\omega$. Let us take $u \simeq c$. Then

$$N_\omega \, d\omega = (2\alpha/\pi) \ln \left(\gamma c/\omega b_{min}\right) (d\omega/\omega), \qquad (19\text{–}37)$$

where $\alpha = e^2/4\pi\epsilon_0\hbar c \sim 1/137$ is the fine structure constant.

Equation (19–37) is very useful in relating the probability of processes induced by electrons, or by other particles which act essentially only through their electromagnetic field, to the probability of processes induced by electromagnetic radiation. Consider an electron of momentum $\mathbf{p}_1$ reacting such that after the process its momentum is $\mathbf{p}_2$. To give validity to this approximate treatment, we must assume that the momentum and energy changes are small relative to the initial energy. Quantum mechanically, b_{min} is then of order of the distance below which the position of the electron cannot be defined according to the uncertainty principle, i.e., $b_{min} \sim \hbar/|\mathbf{p}_1 - \mathbf{p}_2|$. Hence, the number of equivalent photons N_ω absorbed in the process in question is

$$N_\omega \, d\omega = \frac{2\alpha}{\pi} \ln \left(\frac{A\gamma c}{\omega\hbar} \, |\mathbf{p}_1 - \mathbf{p}_2|\right) \frac{d\omega}{\omega}, \qquad (19\text{–}38)$$

where A is a constant of order unity. Since $\hbar\omega = E_1 - E_2$, where E_1 and E_2 are the energies corresponding to the momenta $\mathbf{p}_1$ and $\mathbf{p}_2$, and since $\gamma = E_1/m_0c^2$, we have

$$N_\omega \, d\omega = \frac{2\alpha}{\pi} \ln \left[A \left(\frac{E_1}{m_0c^2}\right) \frac{|c\mathbf{p}_1 - c\mathbf{p}_2|}{E_1 - E_2} \right] \frac{d\omega}{\omega}. \qquad (19\text{–}39)$$

This formula gives a numerical basis for relating photon- and electron-induced processes in terms of the actual experimental parameters; the numerical factor $2\alpha/\pi = 0.00464$ shows that the number of equivalent photons per electron is small.

Suggested References

R. Becker, *Theorie der Elektrizität*, Band II. Our treatment parallels that found in Section 11 of this work.

L. Landau and E. Lifshitz, *The Classical Theory of Fields*. The Liénard-Wiechert potentials are derived from the four-dimensional tensor formulation.

A. Sommerfeld, *Electrodynamics*. The differential equation for the four-potential is solved by integration in the complex plane.

J. A. Stratton, *Electromagnetic Theory*. Another presentation of the complex integration of the wave equation. Sommerfeld has pointed out that the method was devised (by G. Herglotz) in 1904, before the theory of relativity, on the basis of mathematical elegance.

EXERCISES

1. Obtain the general expression for the field of a uniformly moving charge by making a Lorentz transformation of the static Coulomb field. How do you reconcile the statement that the field is decreased in the line of motion with Eq. (18–40), according to which $E'_{||} = E_{||}$?

2. Find the torque on the charges of Fig. 19–3 as derived from the convection potential and compare the answer with that found in Chapter 15.

CHAPTER 20

RADIATION FROM AN ACCELERATED CHARGE

We shall now consider the fields of a charge e in the general case for which the retarded position as a function of the retarded time is assumed given; in other words, $x'_\alpha(t')$ is known, where t' is the time at which a signal propagated with velocity c is emitted at x'_α so as to arrive at x_α at time t. Note that this statement implies the existence of a definite means for establishing the relation between the time of emission, t', and the time t of reception at the field point. Prior to relativity theory, it was simply assumed that the problem is solved in the frame for which the wave equation involving c is valid, and hence t and t' are connected by c. Relativity extends the validity of this formulation to all frames. The additional difference between the classical and relativistic formulation is conceptual. In classical physics it is assumed that there exists a universal time scale, so that the connection between t' and t could be verified independently; hence the validity of the assumption that a frame is, in fact, the one for which the wave equation is correct could be independently verified. Relativity denies this hypothesis. Moreover, as in Chapter 19, we shall have to consider the meaning of the velocity $\mathbf{u}$ in the two cases. In classical theory $\mathbf{u}$ is the velocity of the source point in the frame in which the wave equation is valid; in special relativity $\mathbf{u}$ is the velocity in any inertial frame. We are carrying through the calculations in three-dimensional language so that the results may be more readily interpreted. A covariant expression for the fields is given in Chapter 21 [Eq. (21–98)].

20–1 Fields of an accelerated charge. Let us compute the complete electric and magnetic fields of a charge e for which $x'_\alpha(t')$ is given. The retarded values of the velocity and acceleration of the charge,

$$u_\alpha = dx'_\alpha/dt', \qquad \dot{u}_\alpha = d^2x'_\alpha/dt'^2, \tag{20–1}$$

are thus also known. Vectorially,

$$\frac{d\mathbf{r}}{dt'} = -\mathbf{u}, \qquad \frac{d^2\mathbf{r}}{dt'^2} = \dot{\mathbf{u}}. \tag{20–2}$$

The Liénard-Wiechert potentials are given by the usual formulas,

$$\phi(x_\alpha, t) = \frac{e}{4\pi\epsilon_0}\, \frac{1}{s}\,, \tag{19–6}$$

$$\mathbf{A}(x_\alpha, t) = \frac{e}{4\pi\epsilon_0 c^2}\, \frac{\mathbf{u}}{s}\,, \tag{19–7}$$

where $s = r - (\mathbf{u} \cdot \mathbf{r})/c$ is a function of both the field point and the retarded source point coordinates. The field and source point variables are connected by the retardation condition:

$$r(x_\alpha, x'_\alpha) = [\textstyle\sum(x_\alpha - x'_\alpha)^2]^{1/2} = c(t - t'). \tag{20–3}$$

When we derive the fields $\mathbf{E}$ and $\mathbf{B}$ from the potentials of Eqs. (19–6) and (19–7) in the usual way,

$$\mathbf{B} = \boldsymbol{\nabla} \times \mathbf{A},$$

$$\mathbf{E} = -\boldsymbol{\nabla}\phi - \frac{\partial \mathbf{A}}{\partial t}\,,$$

we must notice that the components of the vector operator $\boldsymbol{\nabla}$ are partial derivatives at constant time t, and therefore *not* at constant time t'. Partial differentiation with respect to x_α compares the potentials at neighboring points at the same time, but these potential signals originated from the charge at different times. Similarly, the partial derivative with respect to t implies constant x_α, and hence refers to the comparison of potentials at a given field point over an interval of time during which the coordinates of the source will have changed. Since only the time variation with respect to t' is given (in the original description of the problem) we must transform $\partial/\partial t|_{x_\alpha}$ and $\boldsymbol{\nabla}|_t$ to expressions in terms of $\partial/\partial t'|_{x_\alpha}$ in order to compute the fields. This procedure is necessary because it is, in general, impossible in the case of an accelerated source charge to express the potentials in terms of the "present position" alone, as we did in the case of uniform motion.

To obtain the required transformation for the derivatives, we note that since x'_α is assumed given as a function of t', we have from Eq. (20–3),

$$r[x_\alpha, x'_\alpha(t')] = f(x_\alpha, t') = c(t - t'), \tag{20–4}$$

which is a functional relation between x_α, t, and t'. In order to relate $\partial/\partial t$ to $\partial/\partial t'$, we need only hold x_α fixed in this relation and note that

$$\left(\frac{\partial r}{\partial t'}\right)_{x_\alpha} = -\,\frac{\mathbf{r} \cdot \mathbf{u}}{r}\,. \tag{20–5}$$

Then

$$\frac{\partial r}{\partial t} = c\left(1 - \frac{\partial t'}{\partial t}\right) = \frac{\partial r}{\partial t'}\frac{\partial t'}{\partial t} = -\frac{\mathbf{r}\cdot\mathbf{u}}{r}\frac{\partial t'}{\partial t}, \qquad (20\text{-}6)$$

or

$$\frac{\partial t'}{\partial t} = \frac{1}{1 - \mathbf{r}\cdot\mathbf{u}/rc} = \frac{r}{s}. \qquad (20\text{-}7)$$

Therefore

$$\frac{\partial}{\partial t} = \frac{r}{s}\frac{\partial}{\partial t'} \qquad (20\text{-}8)$$

is the desired transformation for the time derivatives. Similarly, for the vector operator $\boldsymbol{\nabla}$,

$$\boldsymbol{\nabla}r = -c\boldsymbol{\nabla}t' = \boldsymbol{\nabla}_1 r + \frac{\partial r}{\partial t'}\boldsymbol{\nabla}t' = \frac{\mathbf{r}}{r} - \frac{\mathbf{r}\cdot\mathbf{u}}{r}\boldsymbol{\nabla}t', \qquad (20\text{-}9)$$

where by $\boldsymbol{\nabla}_1$ we mean differentiation with respect to the first argument of the function f in Eq. (20-4), that is, differentiation at constant retarded time t'. Therefore, from Eq. (20-9),

$$\boldsymbol{\nabla}t' = -\frac{\mathbf{r}}{sc}, \qquad (20\text{-}10)$$

and, in general,

$$\boldsymbol{\nabla} = \boldsymbol{\nabla}_1 - \frac{\mathbf{r}}{sc}\frac{\partial}{\partial t'}. \qquad (20\text{-}11)$$

Equations (20-8) and (20-11) constitute the required transformation of the differential operators from the coordinates of the field point to those of the radiator.

The computation of the electric field from the Liénard-Wiechert potentials thus becomes

$$\frac{4\pi\epsilon_0}{e}\mathbf{E} = \frac{1}{s^2}\boldsymbol{\nabla}s - \frac{\partial}{\partial t}\frac{\mathbf{u}}{sc^2}$$

$$= \frac{1}{s^2}\boldsymbol{\nabla}_1 s - \frac{\mathbf{r}}{cs^3}\frac{\partial s}{\partial t'} - \frac{r}{s^2c^2}\dot{\mathbf{u}} + \frac{r\mathbf{u}}{c^2s^3}\frac{\partial s}{\partial t'}. \qquad (20\text{-}12)$$

Using $\boldsymbol{\nabla}_1 s = \mathbf{r}/r - \mathbf{u}/c$ and Eq. (20-5), and then collecting terms, we find that the two terms containing $\dot{\mathbf{u}}$ combine to a vector triple product, so that

$$\frac{4\pi\epsilon_0}{e}\mathbf{E} = \frac{1}{s^3}\left(\mathbf{r} - \frac{r\mathbf{u}}{c}\right)\left(1 - \frac{u^2}{c^2}\right) + \frac{1}{c^2s^3}\left\{\mathbf{r}\times\left[\left(\mathbf{r} - \frac{r\mathbf{u}}{c}\right)\times\dot{\mathbf{u}}\right]\right\}. \qquad (20\text{-}13)$$

Similarly,

$$\frac{4\pi\epsilon_0 c^2}{e}\mathbf{B} = \boldsymbol{\nabla}\times\frac{\mathbf{u}}{s} = -\frac{\mathbf{r}\times\dot{\mathbf{u}}}{cs^2} + \frac{\mathbf{u}\times\mathbf{r}}{s^2}\left[\frac{1}{r} + \frac{1}{s}\left(\frac{\mathbf{r}\cdot\mathbf{u}}{rc} + \frac{\mathbf{r}\cdot\dot{\mathbf{u}}}{c^2} - \frac{u^2}{c^2}\right)\right]$$

$$= -\frac{\mathbf{r}\times\dot{\mathbf{u}}}{cs^2} + \frac{\mathbf{u}\times\mathbf{r}}{s^3}\left(1 - \frac{u^2}{c^2} + \frac{\mathbf{r}\cdot\dot{\mathbf{u}}}{c^2}\right)$$

$$= \frac{\mathbf{u}\times\mathbf{r}}{s^3}\left(1 - \frac{u^2}{c^2}\right) + \frac{1}{cs^3}\frac{\mathbf{r}}{r}\times\left\{\mathbf{r}\times\left[\left(\mathbf{r} - \frac{r\mathbf{u}}{c}\right)\times\dot{\mathbf{u}}\right]\right\}. \quad (20\text{–}14)$$

On comparison with Eq. (20–13), we see that

$$\mathbf{B} = \frac{\mathbf{r}\times\mathbf{E}}{rc}. \quad (20\text{–}15)$$

Thus the magnetic field is always perpendicular to $\mathbf{E}$ and to the *retarded* radius vector $\mathbf{r}$.

The electric field is composed of two separate parts. The first term in Eq. (20–13) varies as $1/r^2$ for large distances, and is formally identical with the "convective" field of a uniformly moving charge. We might call $\mathbf{r}_u = \mathbf{r} - r\mathbf{u}/c$ the "virtual present radius vector," i.e., the position the charge would occupy "at present" if it had continued with uniform velocity from the point x_α' (see Fig. 20–1). In terms of $\mathbf{r}_u$ the $1/r^2$ field is simply

$$\mathbf{E}_{\text{induction}} = \frac{e\mathbf{r}_u}{4\pi\epsilon_0 s^3}\left(1 - \frac{u^2}{c^2}\right), \quad (20\text{–}16)$$

which is identically Eq. (19–14) for a charge in uniform motion. It is analogous to the quasi-static or induction field which we discussed in connection with the radiation from variable current and charge systems in Chapter 14. Equation (20–16) represents a nonradiating term in the sense

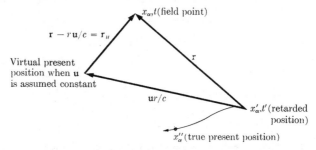

FIG. 20–1. Position parameters for the field of a charge in arbitrary motion.

that it does not contribute to the energy flow over an infinitely distant surface. If the charge is accelerated, the field is neither static nor convective, and there is a net change in field energy which must be supplied. This energy loss will cause a reaction on whatever outside force is responsible for the electron's motion. We shall return to a detailed consideration of this reaction force in the next chapter.

The second term in Eq. (20–13),

$$\mathbf{E}_{\text{radiation}} = \frac{e}{4\pi\epsilon_0 s^3 c^2} \, \mathbf{r} \times (\mathbf{r}_u \times \dot{\mathbf{u}}), \qquad (20\text{–}17)$$

is of order $1/r$ and therefore does represent a radiation field in the sense of contributing to the energy flux over a large sphere. Similar considerations hold for the two terms of Eq. (20–14). Let us look at some important special cases of this radiation field.

20–2 Radiation at low velocity. In case the velocity is so small that u/c is negligible in comparison with unity, $\mathbf{r}_u \simeq \mathbf{r}$ and $s \simeq r$, so that

$$\mathbf{E}_{\text{rad}} = \frac{e}{4\pi\epsilon_0 c^2 r^3} \, \mathbf{r} \times (\mathbf{r} \times \dot{\mathbf{u}}), \qquad (20\text{–}18a)$$

$$\mathbf{B}_{\text{rad}} = \frac{e}{4\pi\epsilon_0 c^3 r^2} \, \dot{\mathbf{u}} \times \mathbf{r}. \qquad (20\text{–}18b)$$

Equations (20–18a, b) represent a field which is formally identical to that [Eq. (14–82)] of a radiating electric dipole of moment equal to $\dot{\mathbf{u}}e/\omega^2$. The angular distribution of the radiated energy is therefore simply the $\sin^2\theta$ distribution discussed in Section 14–7 (see Fig. 20–2). To obtain the rate

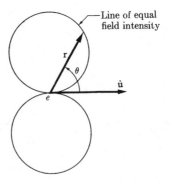

Fig. 20–2. Angular distribution of radiation field of a low velocity electron. The pattern is symmetrical about the axis of θ.

of energy loss from the accelerated charge, we integrate over a sphere the Poynting vector corresponding to Eqs. (20–18) and find

$$-\frac{dU}{dt'} = \frac{e^2\dot{u}^2}{6\pi\epsilon_0 c^3}. \tag{20–19}*$$

20–3 The case of $\dot{\mathbf{u}}$ parallel to $\mathbf{u}$. If $\dot{\mathbf{u}}$ and $\mathbf{u}$ are along the same direction, regardless of whether u is small, the radiation fields are simply

$$\mathbf{E} = \frac{e}{4\pi\epsilon_0 c^2 s^3}\,\mathbf{r} \times (\mathbf{r} \times \dot{\mathbf{u}}), \tag{20–20a}$$

$$\mathbf{B} = \frac{er}{4\pi\epsilon_0 c^3 s^3}\,\dot{\mathbf{u}} \times \mathbf{r}. \tag{20–20b}$$

Equations (20–20) differ only by the factor $r^3/s^3 = [1 - (u/c)\cos\theta]^{-3}$ from the slow electron (dipole) case of Eqs. (20–18). The qualitative effect of this factor is to increase the radiated energy in the forward direction, as shown in Fig. 20–3.

To calculate quantitatively the angular distribution of the radiated energy, we must look carefully at what is meant by "the rate of radiation" of the charge. This is the amount of energy lost by the electron in a time interval dt' during the emission of the signal, i.e., the rate of energy loss $-dU/dt'$ of the electron itself. At a given field point the Poynting vector $\mathbf{N}$ represents the energy flow per unit time measured as t. Therefore the

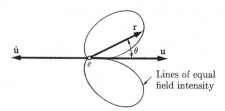

FIG. 20–3. Sketch of the radiation field of an electron decelerated along its line of motion, $\mathbf{u}$ comparable to c.

---

* In Gaussian units (see Appendix I),

$$\frac{-dU}{dt'} = \frac{2e^2\dot{u}^2}{3c^3}.$$

All equations for radiation rates in this and the succeeding chapters are written so that they can be reduced to Gaussian units by putting $\left(\dfrac{e^2}{4\pi\epsilon_0}\right)_{\text{mks}} = (e^2)_{\text{Gaussian}}.$

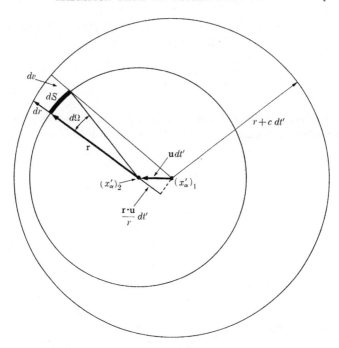

FIG. 20-4. Location of energy radiated by an electron as it moves from $(x'_\alpha)_1$ to $(x'_\alpha)_2$.

rate of energy loss of the electron into a given solid angle $d\Omega$ is given by

$$- \frac{dU(\theta)}{dt'} \, d\Omega = |\mathbf{N}| \frac{dt}{dt'} r^2 \, d\Omega = |\mathbf{E} \times \mathbf{H}| \frac{dt}{dt'} r^2 \, d\Omega$$

$$= \epsilon_0 c E^2 r^2 \frac{dt}{dt'} \, d\Omega = \epsilon_0 c E^2 \frac{s}{r} r^2 \, d\Omega, \qquad (20\text{-}21)$$

when we have used Eq. (20-7). Hence the directional rate of energy loss for the accelerated electron is

$$- \frac{dU(\theta)}{dt'} = \frac{\dot{u}^2}{c^3} \left(\frac{e^2}{16\pi^2 \epsilon_0} \right) \frac{\sin^2 \theta}{[1 - (u/c) \cos \theta]^5} . \qquad (20\text{-}22)$$

The correction s/r can be physically interpreted as follows. The energy emitted by the electron in a time dt' is located in the volume between two spheres, one of radius r about a point $(x'_\alpha)_2$ and the other of radius $r + c \, dt'$ about a point $(x'_\alpha)_1$ which is a distance $-u \, dt'$ from the second position of the source, as indicated in Fig. 20-4. Consider the element

dv of this volume subtending a solid angle $d\Omega = dS/r^2$ at $(x'_\alpha)_2$. Since $dr = c\, dt' - [(\mathbf{r} \cdot \mathbf{u})/r]\, dt'$,

$$dv = dS\, dr = dS \left(c - \frac{\mathbf{u} \cdot \mathbf{r}}{r}\right) dt' = \frac{cs}{r}\, dS\, dt'. \qquad (20\text{–}23)$$

Therefore the energy $dU\, d\Omega$ contained in this volume within the solid angle $d\Omega$ is

$$dU\, d\Omega = \frac{\epsilon_0 E^2 + \mu_0 H^2}{2} \frac{cs}{r}\, dS\, dt' = \epsilon_0 E^2 \frac{cs}{r}\, dS\, dt' \qquad (20\text{–}24)$$

in agreement with Eq. (20–21).

The principal application of Eqs. (20–20) and (20–21) is to the calculation of radiation from an electron which is decelerated, it being assumed in this approximation that the direction of motion does not change. Deceleration radiation is known as *bremsstrahlung*. For an exact classical calculation it would be necessary to put into the equations the precise variation of acceleration with time, using the stopping power of the material upon which the electron impinges. For a simplified discussion, let us assume that $\dot{u}$ is constant while the velocity decreases from u_0 to 0. This gives a resultant pulse of radiated energy:

$$dU = \frac{-e^2 \sin^2 \theta\, \dot{u}}{(4\pi)^2 \epsilon_0 c^3} \int_{u_0}^{0} \frac{\dot{u}\, dt}{(1 - u/c \cos \theta)^5}\, d\Omega,$$

$$-dU = \frac{e^2 \sin^2 \theta\, \dot{u}}{64\pi^2 \epsilon_0 c^2 \cos \theta} \left\{\frac{1}{[1 - (u_0/c) \cos \theta]^4} - 1\right\} d\Omega. \qquad (20\text{–}25)$$

Thus the angular distribution of the pulse as a whole is also tipped forward in the direction of the motion. The radiation is polarized with the electric vector lying in the plane of the radius vector and the direction of deceleration. Equation (20–25) can be used to estimate the total efficiency of a low-voltage x-ray tube. In practice, however, both the angular distribution and the polarization of the outgoing radiation are greatly modified by scattering of the electrons in the target material.

The frequency spectrum of the outgoing radiation can be obtained by Fourier analysis of the radiation field. Let us assume for simplicity that a change Δu in the velocity takes place in a very short time $\Delta t'$, and that $u \ll c$. If the radiation takes place at time t_0, the wave field during the short time interval may be written

$$E(t) = \frac{e \sin \theta}{4\pi \epsilon_0 c^2 r} \frac{\Delta u}{\Delta t'} = \frac{e \sin \theta}{4\pi \epsilon_0 c^2 r} \Delta u\, \delta(t - t_0), \qquad (20\text{–}26)$$

where we have expressed $\dot{u}$ as a δ-function,

$$\dot{u} = \delta(t - t_0) \, \Delta u, \quad \int_{-\infty}^{\infty} \dot{u} \, dt = \Delta u.$$

If we put

$$E(t) = \int_{-\infty}^{\infty} E_\omega e^{-i\omega t} \, d\omega,$$

then

$$E_\omega = \frac{1}{2\pi} \int E(t) e^{i\omega t} \, dt = \frac{e \sin \theta \, \Delta u}{8\pi^2 \epsilon_0 c^2 r} e^{i\omega t_0}, \qquad (20\text{--}27)$$

which, except for phase, is independent of ω. Now the total energy loss is obtained by integrating the Poynting vector over the surface of a sphere and over the time during which the change in velocity takes place. For $u \ll c$,

$$-U = \int -\frac{dU}{dt'} \, dt' = \epsilon_0 c \iint_{-\infty}^{\infty} E^2 \, dt \, dS. \qquad (20\text{--}28)$$

The frequency of the energy loss, corresponding to a field whose Fourier components are E_ω, is then obtained as in Section 14–4.

$$\int_{-\infty}^{\infty} E^2 \, dt = 4\pi \int_0^{\infty} |E_\omega|^2 \, d\omega \qquad (20\text{--}29)$$

follows from Eq. (14–45). Hence the energy loss in a frequency band $d\omega$ is

$$-U_\omega \, d\omega = 4\pi \epsilon_0 c \int_S |E_\omega|^2 \, dS \, d\omega. \qquad (20\text{--}30)$$

For our specific spectrum, Eq. (20–27),

$$-U_\omega \, d\omega = \frac{e^2}{4\pi\epsilon_0 c} \left(\frac{\Delta u}{c}\right)^2 \frac{d\omega}{(2\pi)^2} \int_0^{2\pi} d\varphi \int_0^{\pi} \sin^2 \theta \sin \theta \, d\theta$$

$$= \frac{e^2}{3\pi\epsilon_0 c} \left(\frac{\Delta u}{c}\right)^2 \frac{d\omega}{2\pi}. \qquad (20\text{--}31)$$

Equation (20–31) indicates that the spectrum of radiation is constant on a frequency scale. Even classically the extent of the spectrum to high ω is due only to the simplifying assumption of zero collision time: a Fourier analysis of a finite collision time process will automatically remove the very high frequency components. Actually the spectrum will be cut off at

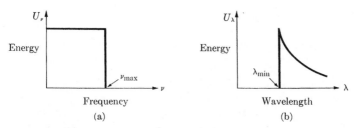

FIG. 20–5. Ideal spectrum of bremsstrahlung as a function of frequency and as a function of wavelength.

the point where the kinetic energy of the electron is equivalent to a single quantum of radiation:

$$\text{K.E.} = \tfrac{1}{2}mu^2 = h\nu_{\max}. \tag{20–32}$$

With such a maximum the ideal spectrum is shown on a frequency scale and again on a wavelength scale in Fig. 20–5. By means of Planck's hypothesis, Eq. (20–31) can be expressed in terms of the number of quanta, $dN = -U_\omega \, d\omega/\hbar\omega$, that are "shaken off" during the velocity change. This gives

$$dN_\omega = \frac{e^2}{4\pi\epsilon_0\hbar c}\frac{2}{3\pi}\left(\frac{\Delta u}{c}\right)^2\frac{d\omega}{\omega} = \frac{1}{137}\frac{2}{3\pi}\left(\frac{\Delta u}{c}\right)^2\frac{d\omega}{\omega}. \tag{20–33}$$

Thus an infinite number of zero energy quanta would be emitted, although the total energy is finite.

20–4 Radiation when the acceleration is perpendicular to the velocity (radiation from circular orbits). There are important practical applications of the case of a charge moving in a circle of radius a with a constant angular velocity ω_0, as indicated in Fig. 20–6. Here we have shown the circular orbit to be in the xy-plane with z perpendicular to the orbit. Let the observation point O be chosen such that the radius vector $\mathbf{r}$ from the source point $P(x'_a)$ to the point $O(x_a)$ is parallel to the zy-plane. Let θ be the angle between $\mathbf{u}$ and $\mathbf{r}$ as before; let $\varphi = \omega_0 t'$ be the angle between the x-axis and the radius vector which is parallel to $\dot{\mathbf{u}}$. We have

$$u = a\omega_0, \qquad \mathbf{r}_u = \mathbf{r} - \frac{\mathbf{u}r}{c},$$

$$\dot{u} = a\omega_0^2, \qquad \mathbf{u}\cdot\mathbf{r} = ur\cos\theta,$$

$$\dot{\mathbf{u}}\cdot\mathbf{r} = \dot{u}r\cos\theta\,\frac{\sin\varphi}{\cos\varphi}.$$

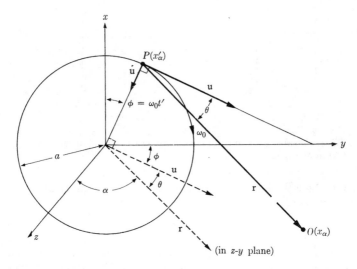

FIG. 20–6. Charge at $P(x'_\alpha)$ with acceleration at right angles to its velocity (motion in a circle of radius a). The vector $\mathbf{u}$ is in the xy-plane; the vector $\mathbf{r}$ is parallel to the zy-plane (see dotted line).

Vector algebra shows

$$[\mathbf{r} \times (\mathbf{r}_u \times \dot{\mathbf{u}})]^2 = -(\dot{\mathbf{u}} \cdot \mathbf{r})^2 r^2 \left(1 - \frac{u^2}{c^2}\right) + \dot{u}^2 r^4 \left(1 - \frac{\mathbf{r} \cdot \mathbf{u}}{rc}\right)^2$$

$$= \dot{u}^2 r^4 \left[\left(1 - \frac{u}{c} \cos\theta\right)^2 - \left(1 - \frac{u^2}{c^2}\right) \tan^2\varphi \cos^2\theta\right],$$

(20–34)

and thus

$$-\frac{dU(\theta, \varphi)}{dt'} d\Omega$$

$$= \frac{e^2 \dot{u}^2}{16\pi^2\epsilon_0 c^3} \frac{[1 - (u/c)\cos\theta]^2 - [1 - (u^2/c)^2]\tan^2\varphi \cos^2\theta}{[1 - (u/c)\cos\theta]^5} d\Omega. \quad (20\text{–}35)$$

Note that both φ and θ vary in time as the particle rotates, and that θ refers to a rotating coordinate system. We can reduce Eq. (20–35) in terms of a single time-varying quantity $\varphi = \omega_0 t'$ if we put

$$\cos\theta = \sin\alpha \cos\varphi. \quad (20\text{–}36)$$

Here, according to Fig. 20–6, α is the (fixed) angle between the axis of

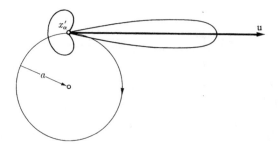

Fig. 20–7. Sketch of the radiation field of Fig. 20–6 in the plane of motion for $|\mathbf{u}|$ comparable with c.

rotation (z-axis) of the electron and the direction of observation. In terms of this variable,

$$- \frac{dU}{dt'} \, d\Omega$$

$$= \frac{e^2 \dot{u}^2}{16\pi^2 \epsilon_0 c^3} \frac{[1 - (u/c)\sin\alpha\cos\varphi]^2 - [1 - (u^2/c^2)]\sin^2\varphi\sin^2\alpha}{[1 - (u/c)\sin\alpha\cos\varphi]^5} \, d\Omega$$

$$= \frac{e^2 \dot{u}^2}{16\pi^2 \epsilon_0 c^3} \frac{[1 - (u^2/c^2)]\cos^2\alpha + [(u/c) - \sin\alpha\cos\varphi]^2}{[1 - (u/c)\sin\alpha\cos\varphi]^5} \, d\Omega, \qquad (20\text{–}37)$$

after simplification.

The resultant pattern has zeros in the plane of the circle at

$$\theta = \cos^{-1}(u/c).$$

For large velocities the radiation is very much more intense in the forward direction than in any other, as indicated in Fig. 20–7. As u approaches c the radiation becomes a sharp forward ray.

Integration of Eq. (20–37) gives the total rate of radiation of a charge e moving in a circle of radius a with constant angular velocity ω_0:

$$-\frac{dU}{dt'} = P = \frac{e^2 \dot{u}^2}{6\pi\epsilon_0 c^3} \frac{1}{[1 - (u^2/c^2)]^2}$$

$$= \frac{2}{3} \frac{r_0}{c} \frac{a^2 \omega_0^4}{c^2} \gamma^4 (m_0 c^2) = \frac{2}{3} \left(\frac{\omega_0^2 r_0}{c} \right) \beta^2 \gamma^4, \qquad (20\text{–}38)$$

where $r_0 = e^2/4\pi\epsilon_0 m_0 c^2$ is called the "classical electron radius" (for reasons which are considered in Chapter 21), $\gamma = m/m_0$, and $\beta = a\omega_0/c = u/c$.

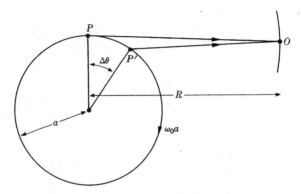

FIG. 20-8. Signals arriving at an observation point O from two successive points P and P' along the orbit of an electron traveling in a circle of radius a with velocity $\omega_0 a$. The time interval of arrival at O is shorter by a factor $1 - (u/c)$ than the time interval $\Delta\theta/\omega_0$ of emission.

If the circular motion takes place in an orbit of radius a perpendicular to a magnetic field **B**, then ω_0 is given by $\omega_0 = eB/m_0\gamma$ [see Eq. (18-72) from Exercise 5, Chapter 18], and thus the radiated power is

$$\frac{P}{m_0c^2} = \frac{2}{3}\left(\frac{eB}{m_0}\right)^2\left(\frac{r_0}{c}\right)(\beta\gamma)^2, \qquad (20\text{-}39)$$

and the radiated energy ΔU per revolution in the field is

$$\frac{\Delta U}{m_0c^2} = \frac{4\pi}{3}\left(\frac{r_0}{a}\right)\beta^3\gamma^4. \qquad (20\text{-}40)$$

Let us consider the frequency distribution of the emitted radiation. Since the exact computation is difficult, we shall first discuss the qualitative features of the frequency spectrum for the extremely relativistic case ($\gamma \gg 1$). Consider an observer O receiving the radiation pulse as the electron moves through an angle $\Delta\theta$ (Fig. 20-8). The pulse is emitted during a time $\Delta t' = \Delta\theta/\omega_0$ but is received in a time interval

$$\Delta t = \frac{\Delta\theta}{\omega_0}\left(1 - \frac{u}{c}\right) \approx \frac{\Delta\theta}{2\omega_0\gamma^2} \qquad (20\text{-}41)$$

if $\gamma \gg 1$. The angular interval $\Delta\theta$ over which emission reaches the observer is itself of order $\Delta\theta = 1/\gamma$, however; this can be seen by noting that the denominator of Eq. (20-35) contains a power of the factor $[1 - (u/c)\cos\theta]$

(the Liénard-Wiechert denominator).   If $\gamma \gg 1$, this factor can be approximated by

$$1 - \frac{u}{c} \cos \theta \sim \frac{1}{2}\left(1 - \frac{u}{c}\right)\left(1 + \frac{u}{c}\right) + \frac{\theta^2}{2} = \frac{1}{2}\left(\frac{1}{\gamma^2} + \theta^2\right). \quad (20\text{–}42)$$

Hence, the angular width of the angular distribution is of order γ^{-1}, and the time pulse reaching the observer will be approximately of length (with omission of numerical factors)

$$\Delta t \approx 1/\omega_0 \gamma^3. \qquad (20\text{–}43)$$

Since the Fourier analysis of width $\Delta \omega$ of the pulse obeys the relation

$$\Delta \omega \, \Delta t \approx 1, \qquad (20\text{–}44)$$

we find

$$\Delta \omega = \gamma^3 \omega_0. \qquad (20\text{–}45)$$

Thus, the emitted frequency spectrum will contain harmonics of frequency up to γ^3 times that of the fundamental.

Let us now determine the frequency spectrum of the radiation.   The vector potential is given according to Eq. (19–7) by

$$\frac{4\pi\epsilon_0 c^2}{e} \mathbf{A}(x_\alpha, t) = \frac{\mathbf{u}}{s}, \qquad (20\text{–}46)$$

where

$$s = r[1 - (u/c) \sin \alpha \cos (\omega_0 t')]; \qquad (20\text{–}47)$$

note that s and $\mathbf{u}$ are given as functions of t'.   The two times, t and t', are connected by the retardation condition (20–8):

$$\frac{\partial}{\partial t} = \frac{r}{s}\frac{\partial}{\partial t'} = \left[1 - \frac{u}{c}\sin \alpha \cos (\omega_0 t')\right]^{-1}\frac{\partial}{\partial t'}; \qquad (20\text{–}48)$$

thus, with the integration constant chosen in accordance with Fig. 20–8,

$$t = t' - \left(\frac{u}{c\omega_0}\right) \sin \alpha \sin (\omega_0 t') + \left(\frac{R}{c}\right). \qquad (20\text{–}49)$$

We note that because of the periodic character of the motion, the frequency spectrum will be a line spectrum of frequencies $\omega = l\omega_0$, where l is the order of the harmonic.   Thus the Fourier amplitude $\mathbf{A}_l(1 \leq l < \infty)$

of each harmonic is given by*

$$\frac{4\pi\epsilon_0 c^2}{e} \mathbf{A}_l(x_\alpha) = \frac{\omega_0}{\pi} \int_0^{2\pi/\omega_0} \frac{\mathbf{u}(t')}{s} e^{+i\omega t} \, dt \tag{20-50}$$

$$= e^{ikR} \frac{\omega_0}{\pi} \int_0^{2\pi/\omega_0} \frac{\mathbf{u}(t')}{r} \exp\left\{i\omega\left[t' - \frac{u}{c\omega_0}\sin\alpha\sin(\omega t')\right]\right\} dt',$$

where $k = \omega/c$. According to Fig. 20-6, $\mathbf{u}$ has two components,

$$u_x = -u\sin(\omega_0 t') \qquad \text{and} \qquad u_y = u\cos(\omega_0 t');$$

thus at large distances, where r can be considered constant and equal to R over the range of integration, we have, choosing the angle φ for a variable,

$$A_{xl} = -\frac{eue^{ikR}}{4\pi^2\epsilon_0 c^2 R} \int_0^{2\pi} \exp\left[il\left(\varphi - \frac{u}{c}\sin\alpha\sin\varphi\right)\right] \sin(l\varphi)\, d\varphi,$$

$$A_{yl} = \frac{eue^{ikR}}{4\pi^2\epsilon_0 c^2 R} \int_0^{2\pi} \exp\left[il\left(\varphi - \frac{u}{c}\sin\alpha\sin\varphi\right)\right] \cos(l\varphi)\, d\varphi. \tag{20-51}$$

These integrals are in standard Bessel function form and can be written as

$$A_{xl} = -\frac{ieue^{ikR}}{2\pi\epsilon_0 c^2 R} J_l'\left(\frac{lu}{c}\sin\alpha\right),$$

$$A_{yl} = l\frac{eue^{ikR}}{2\pi\epsilon_0 c^2 R}\left\{\left[J_l\left(\frac{lu}{c}\sin\alpha\right)\right]\bigg/\left(\frac{lu}{c}\sin\alpha\right)\right\}. \tag{20-52}$$

The time average of the Poynting flux corresponding to the lth harmonic is

$$\overline{\mathbf{N}_l} = \frac{\epsilon_0 c^3}{2}|\boldsymbol{\nabla}\times\mathbf{A}|^2 \tag{20-53}$$

$$= \left(\frac{e^2}{4\pi\epsilon_0}\right)\left(\frac{ck^2}{2\pi}\right)\frac{1}{R^2}\left[\cot^2\alpha\, J_l^2\left(\frac{lu}{c}\sin\alpha\right) + \frac{u^2}{c^2}J_l'^2\left(\frac{lu}{c}\sin\alpha\right)\right].$$

---

* The Fourier component is normalized to conform to the conventions of Section 14-2, i.e., such that

$$\mathbf{A}(x_\alpha, t) = \text{Re}\left[\sum_{l=0}^{\infty}\mathbf{A}_l\exp(-il\omega_0 t)\right].$$

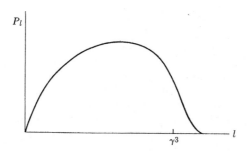

Fig. 20–9. Typical spectrum for the radiation from a relativistic electron moving in a circle; P_l is the power emitted at an harmonic order l.

Hence the power radiated into a solid angle interval $d\Omega = 2\pi \sin \alpha \, d\alpha$ corresponding to the lth harmonic is

$$\frac{dP_l}{d\Omega} d\Omega$$

$$= \left(\frac{e^2}{4\pi\epsilon_0 c}\right)\left(l^2\omega_0^2\right)\left[\cot^2 \alpha \, J_l^2\left(\frac{lu}{c}\sin \alpha\right) + \frac{u^2}{c^2}J_l'^2\left(\frac{lu}{c}\sin \alpha\right)\right]\sin \alpha \, d\alpha. \tag{20–54}$$

This gives the general expression for the spectral and angular distributions of the emitted radiation. Using various approximations or numerical methods, one can integrate Eq. (20–54) over solid angle. It can be shown* that in the relativistic limit ($\gamma \gg 1$), the integrated spectrum takes the approximate form

$$P_l = 0.52 \left(\frac{e^2}{4\pi\epsilon_0 c}\right)\omega_0^2 l^{1/3}, \qquad 1 \ll l \ll \gamma^3,$$

$$P_l = \frac{1}{2\sqrt{\pi}}\left(\frac{e^2}{4\pi\epsilon_0 c}\right)\omega_0^2\left(\frac{l}{\gamma}\right)^{1/2}e^{(-2/3)(l/\gamma^3)}, \qquad l \gg \gamma^3. \tag{20–55}$$

The spectrum is shown schematically in Fig. 20–9. Note that the spectrum cuts off approximately at the harmonic order γ^3, as predicted by the qualitative discussion earlier in this section.

The calculations of this section apply to the radiation from a single point charge (the electron, in practical applications) moving in a circle. If more than a single charge is present, then the amplitude of the radiation

---

* See, e.g., L. Landau and E. Lifshitz, *The Classical Theory of Fields*, Addison-Wesley, 1951, pp. 216 ff.

[e.g., the vector potential Eq. (20–52)] must be superposed. For those harmonics where the phase differences between the N electrons in a "bunch" is negligibly small, the radiation will be coherent, i.e., the total radiated intensity will be N^2 times that emitted by a single electron; the energy loss per electron will thus be increased N-fold. This effect can become quite significant in high-energy circular electron accelerators. It is interesting to note that if the charges occupy the entire circle *continuously*, then inspection of Eq. (20–50) indicates that the total radiation field will vanish and only terms in R^{-2} will remain in the vector potential; this implies that a continuous ring current will not radiate; a metallic wire loop carrying a steady current falls into this category, since the electrostatic lattice forces keep the distribution of electrons uniform. On the other hand, a ring current of *free* electrons (such as in a betatron) distributed in a circle will radiate because of density fluctuations along the circle. The intensity of such fluctuations is proportional to the square root of the coherent intensity, i.e., to $\sqrt{N^2} = N$, and the phases are uncorrelated. Thus a ring current of free electrons radiates as the sum of the radiation intensities of the individual electrons; the mean radiation loss per electron is thus equal to that of a single electron.

20–5 Radiation with no restrictions on the acceleration or velocity. The general radiation field, Eq. (20–17), together with the correction to the rate of radiation that leads to Eq. (20–21), gives the general expression for the directional rate of radiation:

$$- \frac{dU}{dt'} \, d\Omega = \frac{e^2 r}{16\pi^2 \epsilon_0 s^5 c^3} \left[\mathbf{r} \times (\mathbf{r}_u \times \dot{\mathbf{u}}) \right]^2 d\Omega. \qquad (20\text{–}56)$$

Since the radiation fields vanish when $\mathbf{r}_u = \mathbf{r} - \mathbf{u}r/c$ is parallel to $\dot{\mathbf{u}}$, there are, in general, two nodal lines such as we have noted in the special cases. The position of the nodes can be constructed graphically, as shown in Fig. 20–10. Construct a circle of radius r about the charge at O, and lay off $OQ = \mathbf{u}r/c$. A line through Q parallel to $\dot{\mathbf{u}}$ with end points A and B on the circle will then represent both values of $\mathbf{r}_u$ for which the fields vanish. OA and OB are thus the two nodal lines, which always lie in the plane of $\mathbf{u}$ and $\dot{\mathbf{u}}$.

It is an elementary but somewhat complicated matter to integrate Eq. (20–56) over the total solid angle. Probably the simplest method is to choose the direction of $\mathbf{u}$ as the polar axis and first perform the azimuthal integration. The final result of the integration is

$$- \frac{dU}{dt'} = \frac{e^2}{6\pi\epsilon_0 c^3} \frac{\dot{u}^2 - (\mathbf{u} \times \dot{\mathbf{u}})^2/c^2}{(1 - u^2/c^2)^3} \qquad (20\text{–}57)$$

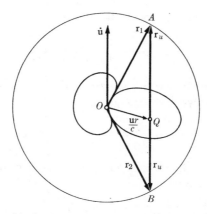

FIG. 20–10. Graphical construction of nodal lines in the radiation field.

for the total rate of radiation. It is easy to check that the radiation rates given in the previous sections are special cases of Eq. (20–57). That Eq. (20–57) can be derived by Lorentz transformation from Eq. (20–19) is shown in the problems.

20–6 Classical cross section for bremsstrahlung in a Coulomb field. At low electron energies Eq. (20–25) gives the approximate yield of "soft" x-rays if empirical values of $\dot{u}$ are used. In that case, collisions of the primary electrons with nuclei are rare, and the principal source of energy loss and momentum transfer consists of electron-electron collisions. But single collisions between particles of like e/m do not produce either electric dipole or magnetic dipole radiation, since displacement and circulation of the center of charge are inconsistent with the conservation of linear and angular momentum. (As an example, see problem 4 in the exercises at the end of this chapter.) Hence, in the low-energy range, where electron-electron collisions predominate, the radiation is computed by the collective effects of the electrons in retarding the incident particle. At higher energies, and particularly for heavy targets, the x-ray yield must be considered as being due to radiation during the deflection of the electron in the Coulomb field of a nucleus. For electron collisions with nuclei, dipole radiation is possible, and in these circumstances the effect of single collisions predominates. We can make a simple nonrelativistic calculation, involving only the transverse acceleration, which contains the essential physical features of a more detailed treatment.

Consider an electron of velocity u passing a nucleus of charge Ze at a distance b. If the time of closest approach is taken as $t = 0$, the magnitude of the force as a function of the time is

$$F = m\dot{u} = \frac{Ze^2}{4\pi\epsilon_0(b^2 + u^2t^2)} \,, \tag{20-58}$$

so that the (small) transverse component of acceleration suffered by the electron is approximately

$$\dot{u} = \frac{Ze^2}{4\pi\epsilon_0 m} \frac{b}{(b^2 + u^2t^2)^{3/2}} \,. \tag{20-59}$$

To calculate the radiation loss and the spectral distribution, we must make a Fourier analysis of $\dot{u}$:

$$\dot{u}_\omega = \frac{1}{2\pi}\int_{-\infty}^{\infty} \dot{u}(t)e^{+i\omega t}\,dt = \frac{1}{2\pi}\frac{2Ze^2b}{4\pi\epsilon_0 m}\int_0^\infty \frac{\cos\omega t\,dt}{(b^2 + u^2t^2)^{3/2}} \,. \tag{20-60}$$

Now the Bessel function $K_1(\omega b/v)$ is represented by the integral

$$K_1\left(\frac{\omega b}{u}\right) = \frac{b}{u\omega}\int_0^\infty \frac{\cos\omega t\,dt}{[(b/u)^2 + t^2]^{3/2}} \,, \tag{20-61}$$

and thus

$$\dot{u}_\omega = \frac{1}{2\pi}\frac{Ze^2}{4\pi\epsilon_0 m}\frac{2\omega}{u^2}K_1\left(\frac{\omega b}{u}\right). \tag{20-62}$$

The function $K_1(\omega b/u)$ has the property that $K_1(\omega b/u) \simeq u/\omega b$ for $\omega b/u < 1$, and for large values of the argument it decreases exponentially. We may therefore take

$$\dot{u}_\omega = \frac{1}{\pi}\frac{Ze^2}{4\pi\epsilon_0 m}\frac{1}{ub} \,, \qquad \omega < u/b,$$
$$\dot{u}_\omega = 0, \qquad \omega > u/b. \tag{20-63}$$

The energy radiated into an angle $d\Omega$ is then, from Eq. (20–30) with the substitution of the radiation of the radiation field appropriate to $\dot{u}_\omega$,

$$-U_\omega\,d\omega\,d\Omega = \frac{Z^2e^2r_0^2}{4\pi\epsilon_0}\frac{1}{\pi^2}\frac{c}{u^2b^2}\sin^2\theta\,d\omega\,d\Omega, \tag{20-64}$$

where $r_0 = e^2/4\pi\epsilon_0 mc^2$ is the classical electron radius. The total energy of the radiation per frequency interval $d\omega$ is

$$-U_\omega\,d\omega = \frac{2Z^2e^2r_0^2}{3\pi^2\epsilon_0}\frac{c}{u^2b^2}\,d\omega. \tag{20-65}$$

Equation (20–65) can be expressed as a cross section for the emission of quanta of energy $\hbar\omega$, if we take account of all possible distances b:

$$d\sigma = \int_{b_{\min}}^{b_{\max}} \frac{U_\omega \, d\omega}{\hbar\omega} 2\pi b \, db = \frac{4Z^2 r_0^2}{3\pi\epsilon_0} \frac{e^2}{\hbar c} \frac{c^2}{u^2} \ln\left(\frac{b_{\max}}{b_{\min}}\right) \frac{d\omega}{\omega}$$

$$= \frac{16\alpha}{3} \frac{Z^2 r_0^2}{u^2/c^2} \ln\left(\frac{b_{\max}}{b_{\min}}\right) \frac{d\omega}{\omega}, \tag{20–66}$$

in which it must be remembered that $\omega < u/b_{\min}$. The limit $b_{\max}$ is defined by the effective limit of the Coulomb field due to "screening" by atomic electrons surrounding the nucleus. The limit $b_{\min}$ is equivalent to cutting off the spectrum for $\hbar\omega$ greater than the kinetic energy of the incoming electron. Equation (20–66) is only approximate, but it illustrates the dependence on Z, r_0, and $\alpha = e^2/4\pi\epsilon_0\hbar c$ which is retained in the quantum-mechanical calculation.

20–7 Čerenkov radiation. We have seen in Chapter 19 and again in Section 20–1 that a charge moving uniformly *in vacuo* does not radiate. The Poynting vector corresponding to the induction fields, Eq. (20–16), and the related magnetic field given by Eq. (20–15), falls off too rapidly with increasing r to contribute to a surface integral at large distances. The situation may be different, however, in a dielectric medium for which $\epsilon > \epsilon_0, \kappa > 1$. Let us consider the field of a moving charge in such a non-conducting medium, where for simplicity we let $\mu = \mu_0$. With the more general definition of the D'Alembertian operator given by Eq. (14–6), the inhomogeneous equation for the vector potential is just

$$\Box\mathbf{A} = \nabla^2\mathbf{A} - \frac{n^2}{c^2} \frac{\partial^2\mathbf{A}}{\partial t^2} = -\mu_0\mathbf{j}. \tag{14–7}$$

The solutions are the same as the vacuum solutions except that c is replaced by c/n, where the index of refraction $n = \sqrt{\epsilon/\epsilon_0} = \sqrt{\kappa}$, and thus the retarded time is $t' = t - rn/c$. The relativistic limitation on the velocity of a particle is $u < c$, but if an electron with velocity nearly equal to c enters a dielectric medium its velocity may be greater than that of light in the medium. In these circumstances the retardation denominator may vanish, and in the direction for which $1 - un/c \cos\theta = 0$ the field intensities are infinite. In a cone defined by the angle $\theta = \cos^{-1}(c/nu)$ with the direction of motion, we might therefore expect to find radiation. Such radiation was first observed by Čerenkov in 1934. Note that if $u > c/n$ the transformation of Eq. (14–7) to an equivalent static solution, as discussed in Section 19–2, is impossible.

For a quantitative estimate of Čerenkov radiation it is convenient to use the Fourier solutions of Chapter 14. The current density $\mathbf{j}$ corresponding to a charge e moving with velocity $\mathbf{u}$ can be represented by

$$\mathbf{j} = e\mathbf{u} \, \delta(x' - ut) \, \delta(y') \, \delta(z'). \tag{20-67}$$

This expression evidently localizes the charge correctly at the source point (x', y', z'), and makes $\int \mathbf{j} \, dy' \, dz' \, dt = e\hat{\mathbf{x}}$, where $\hat{\mathbf{x}}$ is a unit vector in the direction of motion. The Fourier transform of $\mathbf{j}$ required for substitution into Eq. (14-52) is given by

$$\mathbf{j}_\omega = \frac{1}{2\pi} \int \mathbf{j} e^{i\omega t} \, dt = \frac{e}{2\pi} \, \delta(y') \, \delta(z') e^{i\omega x'/u}\hat{\mathbf{x}}. \tag{20-68}$$

Since in the wave zone

$$|\mathbf{E}| = \sqrt{\frac{\mu_0}{\epsilon}} \, |\mathbf{H}| = \frac{1}{n} \sqrt{\frac{\mu_0}{\epsilon_0}} \, |\mathbf{H}|, \tag{20-69}$$

Eq. (14-52) must be divided by n to give the radiated energy in a dielectric. If we let x', y', z' be the components of the source vector $\boldsymbol{\xi}$, with $k = \omega n/c$ as defined in Eq. (11-59) and θ the angle between $\mathbf{k}$ and $\mathbf{u}$, the energy loss U_ω in a frequency band $d\omega$ is given by

$$U_\omega \, d\omega = \left[\int \frac{dU_\omega}{d\Omega} \, d\Omega \right] d\omega$$

$$= \frac{e^2 \omega^2 n}{16\pi^3 \epsilon_0 c^3} \left[\int \left| \int_{-\infty}^{\infty} e^{i[(\omega x'/u) - kx' \cos\theta]} \, dx' \right|^2 \sin^2\theta \, d\Omega \right] d\omega. \tag{20-70}$$

The integral over x' is essentially a δ-function, $\delta(1 - nu/c \cos\theta)$, again indicating infinite fields in the direction $\theta = \cos^{-1} (c/nu)$. But this form is easily manageable: an infinite path is not available to the electron, and the integral can be taken from $-X$ to X:

$$\int_{-X}^{X} e^{i\omega[1 - (nu/c) \cos\theta]x'/u} \, dx' = \frac{2 \sin \{\omega[1 - (nu/c) \cos\theta]X/u\}}{\omega[1 - (nu/c) \cos\theta]/u}. \tag{20-71}$$

Therefore

$$U_\omega = \frac{1}{2\pi^2} \frac{e^2 n\omega^2}{\epsilon_0 c^3} \int_{-1}^{+1} \sin^2\theta \, \frac{\sin^2 \{[1 - (nu/c) \cos\theta]\omega X/u\}}{\omega^2[1 - (nu/c) \cos\theta]^2/u^2} \, d (\cos\theta). \tag{20-72}$$

The angular integral still has some of the characteristics of a δ-function: it has a sharp maximum at $\cos\theta = c/nu$. If we use this value of θ in the slowly varying function $\sin^2 \theta$, and remember that the limits do not matter so long as they include the one maximum, we may write the integral in Eq. (20–72) as

$$\left(1 - \frac{c^2}{n^2u^2}\right) \int_{-\infty}^{\infty} \frac{\sin^2 \{[1 - (nu/c) \cos \theta]\omega X/u\}}{\omega^2[1 - (nu/c) \cos \theta]^2/u^2} \, d(\cos \theta)$$

$$= \left(1 - \frac{c^2}{n^2u^2}\right) \frac{cX}{\omega n} \int_{-\infty}^{\infty} \frac{\sin^2 x}{x^2} \, dx = \frac{cX\pi}{\omega n}\left(1 - \frac{c^2}{n^2u^2}\right). \quad (20\text{–}73)$$

The radiation loss in the frequency interval $d\omega$ is then

$$U_\omega \, d\omega = \frac{e^2 X}{2\pi\epsilon_0 c^2}\left(1 - \frac{c^2}{n^2u^2}\right) \omega \, d\omega. \quad (20\text{–}74)$$

The corresponding energy loss per unit length of path becomes

$$\frac{\Delta U_\omega \, d\omega}{\Delta L} = \frac{e^2}{4\pi\epsilon_0 c^2}\left(1 - \frac{c^2}{n^2u^2}\right) \omega \, d\omega, \quad (20\text{–}75)$$

since the total path considered is $2X$. In terms of the number of quanta of energy $\hbar\omega$, Eq. (20–75) becomes

$$\frac{\Delta N}{\Delta L} \, d\omega = \frac{e^2}{4\pi\epsilon_0 \hbar c}\left(1 - \frac{c^2}{n^2u^2}\right) \frac{d\omega}{c} = \alpha\left(1 - \frac{c^2}{n^2u^2}\right) \frac{d\omega}{c}, \quad (20\text{–}76)$$

which is independent of the frequency except insofar as the index of refraction depends on the frequency. Since the velocity of propagation for high frequencies in a dielectric approaches its vacuum value, $n(\omega)$ will become smaller than c/u for sufficiently large values of ω, and the total number of quanta is finite.

Suggested References

Radiation from accelerated charges is treated in practically all books on electrodynamics. Three of the best are:

R. Becker, *Theorie der Elektrizität*, Band II.

L. Landau and E. Lifshitz, *The Classical Theory of Fields*.

A. Sommerfeld, *Electrodynamics*.

Briefer accounts will be found in:

J. A. Stratton, *Electromagnetic Theory*. The emphasis here is on antennas and antenna arrays, but the accelerated charge fields are also derived.

W. R. Smythe, *Static and Dynamic Electricity*. The radiation fields of a charge are included in Chapter XIV, together with special relativity.

For a treatment of the Čerenkov effect, see L. I. Schiff, *Quantum Mechanics* and Landau and Lifshitz, *Electrodynamics of Continuous Media*, Section 86.

EXERCISES

1. Equation (20–19) is valid in a proper frame. Show that a Lorentz transformation of *both* sides of the equation leads to Eq. (20–57). Show that Eq. (20–38) in turn is a special case of Eq. (20–57).

Hint: Note that $dU/dt = dU^0/dt^0$, i.e., that dU/dt is an invariant for a radiating point charge.

2. An electron synchrotron operates in a magnetic field of $1.5 W/m^2$. At what electron energy is the mean energy lost per turn equal to its own energy?

3. An electron moves in a uniform field **B** in a helical orbit with axis along **B**. Show that the radiated power is reduced relative to that radiated by an electron in a circular orbit by a factor $\cos^2 \delta$ where δ is the "pitch angle" of the helix (defined such that $\tan \delta$ is the ratio of the axial to azimuthal velocity components).

4. Show that the angular distribution of the total radiation emitted when one particle, e_1, m_1, passes another, e_2, m_2, at relative velocity $u \ll c$, is given by

$$-\frac{dU}{d\Omega} = \frac{\text{const}}{b^3} \left(\frac{e_1}{m_1} - \frac{e_2}{m_2} \right)^2 (4 - n_x^2 - 3n_y^2),$$

where n_x and n_y are the components of a unit vector **n** directed along the direction of the radiation and b is the impact parameter, assumed constant during the motion. The velocity u is in the x-direction, and the two charges lie in the xy-plane. Assume that the motion remains rectilinear during the collision.

5. Consider a charge e passing a heavy charge Ze with impact parameter b. The velocity is not necessarily small. Assuming that the angular change in path is small, calculate the transverse momentum imparted.

6. From the result of problem 5 calculate the mean square scattering angle of a charge passing through a thickness t of material containing N_v charges of magnitude Ze per unit volume. Assume that the impact parameters available range from b_1 to b_2.

7. Calculate the classical cross section (per unit frequency interval of the emitted radiation) for radiation from an electron colliding with a perfectly hard sphere of radius a with $u \ll c$. Assume the sphere to be transparent to radiation.

8. Estimate the output of Čerenkov radiation for a 350-Mev neutron traveling in glass of refractive index 1.9. (The current associated with the neutron is due to its magnetic dipole moment.) Compare this answer with the Čerenkov radiation from a proton of the same energy under the same conditions.

Hint: Put $\mathbf{j} = \nabla \times \mathbf{M}$ in Eq. (20–68) and integrate by parts.

9. Show that the rate of energy loss by radiation is negligible relative to the energy gain of a particle accelerated by an electric field **E** if the velocity **u** remains parallel to **E**. Assume that the field **E** is limited by practical considerations to 10^8 volts/meter.

CHAPTER 21

RADIATION REACTION AND COVARIANT FORMULATION OF THE CONSERVATION LAWS OF ELECTRODYNAMICS

It was shown in Chapter 10 that the law of conservation of momentum could be reconciled with an electrical interaction of the form

$$\mathbf{F} = e(\mathbf{E} + \mathbf{u} \times \mathbf{B}), \tag{21-1}$$

only if a momentum of density

$$\mathbf{g} = \frac{\mathbf{N}}{c^2} = \frac{\mathbf{E} \times \mathbf{H}}{c^2} \tag{21-2}$$

is assumed for the electromagnetic field. Similarly, the electromagnetic field carries an energy density

$$U_v = \frac{\mathbf{E} \cdot \mathbf{D} + \mathbf{H} \cdot \mathbf{B}}{2}, \tag{21-3}$$

in order that energy be conserved in a closed system containing both matter and radiation. Let us put these considerations into covariant form.

21–1 Covariant formulation of the conservation laws of vacuum electrodynamics. The conservation laws of Chapter 10 may be written in three-dimensional tensor language as

$$\frac{\partial N_\alpha}{\partial x_\alpha} + \rho E_\alpha u_\alpha + \epsilon_0 \frac{\partial}{\partial t} \left[\frac{(E^2 + c^2 B^2)}{2} \right] = 0, \tag{21-4}$$

$$\frac{\partial T_{\alpha\beta}^M}{\partial x_\beta} - \epsilon_0 \frac{\partial}{\partial t} (\mathbf{E} \times c\mathbf{B})_\alpha = \rho[E_\alpha + (\mathbf{u} \times \mathbf{B})_\alpha]. \tag{21-5}$$

Here $\mathbf{N} = \mathbf{E} \times \mathbf{H}$ is the Poynting vector, and $T_{\alpha\beta}^M = E_\alpha D_\beta + H_\alpha B_\beta - \frac{1}{2} \delta_{\alpha\beta}(E_\gamma D_\gamma + H_\gamma B_\gamma)$ is the Maxwell stress tensor. Equation (21–4) balances energy of radiation loss with the rate of mechanical and thermal work and the time rate of change of the electromagnetic field energy. Equation (21–5) balances the volume force with the rate of increase of the momentum of matter and the electromagnetic field. These two equations can be combined into a single relation by introducing the energy-momentum tensor of the electromagnetic field defined formally by the symmetric matrix

$$T^{ij} = \begin{pmatrix} -T_{11}^M & -T_{12}^M & -T_{13} & cG_x \\ -T_{12}^M & -T_{22}^M & -T_{23}^M & cG_y \\ -T_{13}^M & -T_{23}^M & -T_{33}^M & cG_z \\ cG_x & cG_y & cG_z & W \end{pmatrix}, \tag{21-6}$$

where $\mathbf{G} = \mathbf{N}/c^2$ is the momentum density* of the field and

$$W = \tfrac{1}{2}(\epsilon_0 E^2 + \mu_0 H^2)$$

is the energy density. If T^{ij} is a tensor with the components as given, the conservation laws, Eqs. (21–4) and (21–5), are equivalent to the simple covariant relation

$$\partial T^{ij}/\partial x^j = -f^i, \tag{21-7}$$

where

$$(f^i) = \rho[(\mathbf{E} + \mathbf{u} \times \mathbf{B}), \quad (\mathbf{E} \cdot \mathbf{u}/c)] \tag{21-8}$$

is the four-vector representing the Lorentz force per unit volume and the rate of work per unit volume of the electromagnetic field on the charges and currents.

To show that $T^{ij} = g^{ik}T_k^j$ is actually a tensor, we note that T_j^i can be generated from F^{ij} of Eq. (17–32) by the tensor operation

$$T_j^i = \epsilon_0[F^{ik}F_{kj} - \tfrac{1}{4}\delta_j^i(F^{kl}F_{kl})]. \tag{21-9}$$

The algebraic correctness of Eq. (21–9) can be easily checked. It is seen that the second term on the right side of Eq. (21–9) is simply the invariant "trace" of F^{kl}, namely

$$F^{kl}F_{kl} = 2(c^2B^2 - E^2). \tag{21-10}$$

The invariance of $(c^2B^2 - E^2)$ again shows that the ratio between the electric and magnetic fields in a plane electromagnetic wave in space is a constant.

---

* The relation $\mathbf{G} = \mathbf{N}/c^2$ can be taken as a consequence of the symmetry tensor $T^{ij} = T^{ji}$.

21-2 Transformation properties of the "free" radiation field. The conservation laws, like the field equations, are in agreement with relativity theory without modification. Moreover, a Lorentz transformation of Eq. (21-6) again shows the equivalence of energy transport and flow of momentum. A number of other interesting and important conclusions can be drawn from the covariant conservation laws and the form of the energy momentum tensor.

1. If "state of equilibrium" is to be an invariant property of a system, we must conclude not only that electromagnetic forces, energy, momenta, etc., must be describable by a tensor relation of the form $f^j = -\partial T^{ij}/\partial x^i$, but also that the totality of such mechanical quantities must obey an equation of this form. Hence, for example, Eq. (21-7) will be a valid equation, in a medium under elastic stress, with W as the mass density (including elastic energy density), G the mechanical momentum density, and $T^M_{\alpha\beta}$ the elastic stresses. This formulation yields the transformation equations for all quantities entering into the mechanics of continua. Note particularly that the mass density of a continuous medium cannot be treated as a scalar, or even as a component of a four-vector like the electrical charge density, but is the $(4, 4)$ component of the mechanical energy momentum tensor. The quantity $m = \int W \, dv$ has the same transformation character as a point mass. These facts are consistent with the existence of a fundamental unit of charge and the apparent lack of existence of a fundamental unit of mass. The tensor component character of mass density is of importance in the formulation of the gravitational action of matter in the general theory of relativity.

2. Let us consider a volume v containing *totally* a quantity of free electromagnetic radiation but *no* charges or currents. The energy tensor of Eq. (21-6) then obeys the conservation law

$$\partial T^{ij}/\partial x^i = 0, \tag{17-24}$$

and hence, according to the theorem of Eq. (17-25),

$$G^i = \left(\int c\mathbf{G} \, dv, \int W \, dv \right) = \int T^{4i} \, dv \tag{21-11}$$

is a contravariant four-vector. Thus the momentum and energy of a radiation pulse totally contained within a *finite* volume has the same transformation properties as a material point particle. For the *total* field of a charge this is not true, since Eq. (17-24) is not satisfied; we shall return to this point shortly. For a plane electromagnetic wave the invariant

$$G^i G_i = W^2 - c^2 G^2 \tag{21-12}$$

is zero, and hence the equivalent particle properties of such a wave correspond to zero rest mass. Thus radiation propagated with velocity c can obey the particle transformation laws and still yield finite momenta and energy, which would be impossible if the rest mass were other than zero. All these facts are consistent with the "light quantum" concept.

3. The "phase" of an electromagnetic wave is defined by the relation

$$E = E_0 e^{i\varphi} = E_0 e^{i(\mathbf{k}\cdot\mathbf{r}-\omega t)}, \tag{21-13}$$

where $\mathbf{k}$ is the wave propagation vector and ω is the angular frequency. The zero point of a field must be an invariant physical fact, and hence one would expect φ to be an invariant. Therefore we can write

$$\varphi = k_i x^i \tag{21-14}$$

with

$$k_i = \left(\mathbf{k}, -\frac{\omega}{c}\right), \qquad k^i = -\left(\mathbf{k}, \frac{\omega}{c}\right). \tag{21-15}$$

Since $k_i k^i = 0$, k^i transforms exactly like an energy momentum vector of a particle of zero rest mass. This suggests making the momentum and energy of a light quantum proportional to $\mathbf{k}$ and ω respectively, as has been done in quantum theory.

Equation (21–15), in defining the transformation character of $\mathbf{k}$ and ω, provides a simple method of obtaining the relativistically correct expressions for the Doppler shift and for the aberration of starlight. Consider a source at rest in a frame Σ^0 radiating in the x-direction such that $k_x^0 = \omega^0/c$. In a Σ frame we obtain from the Lorentz transformation applied to k^4:

$$\omega = c\gamma\left(\frac{\omega^0}{c} - \beta k_x^0\right) = \frac{\omega^0(1-\beta)}{(1-\beta^2)^{1/2}} = \omega^0\left(\frac{1-\beta}{1+\beta}\right)^{1/2}. \tag{21-16}$$

This is the relativistic formula for the Doppler shift. The expression for aberration can be obtained similarly from the spatial components of k^i. The relation

$$\sin\theta = \beta \tag{21-17}$$

is obtained for the aberration angle θ, in agreement with Eq. (16–48).

21–3 The electromagnetic energy momentum tensor in material media.
We have seen that the covariant formulation of the conservation laws for vacuum electrodynamics is both straightforward and fruitful. The separate formulation for the macroscopic fields in material media, however, involves some apparent ambiguities. We shall consider this problem only briefly.

The conservation laws of vacuum electrodynamics, Eq. (21–7), may be written

$$-f^j = \frac{\partial T^{ij}}{\partial x^i} = j_l F^{jl}, \tag{21–18}$$

where j^l is the four-vector current density. If we assume that the volume force f^j in material media is similarly given by $j_l F^{jl}$, where j^l is interpreted as the true (macroscopic) current and charge density satisfying Eq. (18–66), we must introduce a new energy momentum tensor. The equation corresponding to Eq. (21–18) becomes

$$-f^j = \frac{\partial S^{ij}}{\partial x^i}, \tag{21–19}$$

where

$$S^{ij} = g^{jk}(H^{il}F_{kl} - \tfrac{1}{4}\,\delta_k^i H^{ml}F_{ml}), \tag{21–20}$$

with H^{il} as defined in Section 18–5. Equation (21–19) is covariant, since S^{ij} as given by Eq. (21–20) is a tensor, and algebraically the conservation laws are correctly given. The only argument against the correctness of this energy momentum tensor is based on the fact that it is not symmetric: $S^{4i} \neq S^{i4}$ for dielectrics and permeable materials. It would seem that if Maxwell's macroscopic equations are indeed just the result of averaging the microscopic vacuum quantities, no difference in symmetry properties could arise; and for many years a "corrected" (i.e., symmetrized) energy momentum tensor was preferred by most physicists. More recently, however, it has been pointed out that there is no reason for demanding symmetry in the case of a partial field which also leaves out of consideration all mechanical properties, and that the only valid demand on S^{ij} is that it correspond to the correct Lorentz volume force in all instances. The situation is fully discussed by Møller (see references), who shows that the evidence tends to favor the formulation of Eq. (21–20) due originally to Minkowski. The total energy momentum tensor, including mechanical as well as electrical quantities, is of course symmetrical, corresponding to a closed system. This follows, since the relation $\mathbf{G} = \mathbf{N}/c^2$ should be obeyed by the total mechanical properties of the system.

21–4 Electromagnetic mass. In Section 21–2 we considered the energy momentum tensor of the *"free"* electromagnetic field and concluded that its integrals over a finite spatial region defined an energy momentum vector corresponding to a particle of zero rest mass. A necessary condition for this conclusion was that a finite boundary could be found such that the fields and thus the components of the energy momentum tensor would vanish on the boundary. This condition can be satisfied for the "free radiation field," but not for the total field of an electric charge. Partly for historical reasons we shall examine this problem in two parts to separate the different questions involved: in this section we restrict ourselves to the nonrelativistic case ($u \ll c$) and, moreover, consider only electromagnetic forces. The more general case is considered in Sections 21–9 through 21–12. Our discussion will lead to the complete relativistic equation of motion of a point charge in Section 21–13.

Consider a charge, assumed for simplicity to be spherically symmetrical in its rest frame, and let it move with a small velocity **u** along the x-direction in a general frame. In the rest frame Σ^0 the energy momentum tensor has the components

$$T^{ij} = \begin{pmatrix} -(T^M_{\alpha\beta})^0 & 0 \\ 0 & W_0 \end{pmatrix}, \tag{21–21}$$

where W_0 is the electrostatic field energy density. If we carry only terms to first order in $\beta = u/c$, we obtain from the appropriate transformation (giving Σ^0 a positive velocity in the general frame Σ):

$$T^{14} = \beta[W_0 - (T^M_{11})^0], \tag{21–22}$$

$$T^{44} = W_0. \tag{21–23}$$

If we carry only terms linear in u/c, the spherical symmetry of the fields in Σ^0 will be retained in Σ, and to this order integration over space will not extend over different values in Σ and Σ^0. Hence the momentum and energy of the field in frame Σ are given by

$$cG_x = \int T^{14} \, dv = \beta \int [W_0 - (T^M_{11})^0] \, dv, \tag{21–24}$$

$$U = \int T^{44} \, dv = \int W_0 \, dv. \tag{21–25}$$

But if U_0 is the total electrostatic field energy in Σ^0, then

$$\int W_0 \, dv = U_0, \tag{21–26}$$

and

$$\int (T^M_{11})^0 \, dv = \epsilon_0 \int (E^2_{x0} - \tfrac{1}{2}E^2_0) \, dv$$

$$= -\frac{\epsilon_0}{6} \int E^2_0 \, dv = -\frac{U_0}{3}, \tag{21–27}$$

where we have used the fact that $E_x^2 = E^2/3$ if averaged over a spherically symmetric field. Clearly, $G_y = G_z = 0$. Hence

$$\mathbf{G} = \left(\frac{4}{3}\,\frac{U_0}{c^2}\right)\mathbf{u}, \tag{21–28}$$

and

$$U = U_0. \tag{21–29}$$

Equation (21–28) corresponds to a "mass" equal to $4U_0/3c^2$, while the normal relativistic relation (17–42) would prohibit the factor 4/3. The total electromagnetic field of a charge, therefore, cannot be considered simply as a mechanical object, in contrast to the electromagnetic field of a free electromagnetic wave. Note that this conclusion applies to the energy of the entire field, including both Coulomb fields and radiation fields. In the following sections we shall examine this problem in a less formal manner, still assuming that $u \ll c$.

21–5 Electromagnetic mass—qualitative considerations. If we consider the energy and momentum balance of the fields due to an electron, we are immediately led to the following qualitative conclusions:

1. If an electron is in uniform motion its field will contribute to the momentum, since for a small virtual change in velocity the momentum of both particle and field would change simultaneously.

2. If an electron radiates by virtue of an acceleration produced by an external force, the external force must supply both the energy and the momentum required by the change in fields. This can be done only by means of a reaction force, produced by action of the radiation field on the electron itself.

To arrive at somewhat more quantitative conclusions, let us first consider the electron in uniform motion with $u \ll c$. If the velocity is changed by a small amount $\delta\mathbf{u}$, the vector potential changes by

$$\delta\mathbf{A} = \frac{\mu_0 e}{4\pi r}\,\delta\mathbf{u}. \tag{21–30}$$

Consider the magnetic flux passing through an area in a narrow strip normal to $\delta\mathbf{u}$ and extending from the electron to infinity, as indicated in Fig. 21–1. The flux change through the area of the strip indicated in the figure is given by

$$\delta\Phi = \oint \delta\mathbf{A} \cdot d\mathbf{l}. \tag{21–31}$$

This flux change will produce an electric field at the position of the electron in a direction such as to oppose the change in velocity. Specifically, a

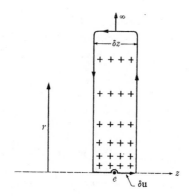

Fig. 21-1. Path of integration for qualitative computation of the reaction of an electron to a small change in velocity.

reaction force **F** will be produced which is given by

$$\mathbf{F} = e\mathbf{E} = -e\frac{\partial \mathbf{A}}{\partial t} = -\frac{\mu_0 e^2}{4\pi r_0}\frac{d\mathbf{u}}{dt} = -\left(\frac{e^2}{4\pi\epsilon_0}\right)\frac{1}{r_0 c^2}\frac{d\mathbf{u}}{dt}, \quad (21\text{--}32)$$

where r_0 is some minimum value of r at which the vector potential of Eq. (21–30) is assumed to break down owing to the structure of the electron. (The reaction would be infinite if a point electron were assumed.) Hence the action, in exact analogy to the "back electromotive force" of an electrical circuit, produces an "inertial reaction,"

$$\mathbf{F} = -m\frac{d\mathbf{u}}{dt}, \quad (21\text{--}33)$$

where m is an effective mass to be identified as

$$m \simeq \left(\frac{e^2}{4\pi\epsilon_0}\right)\frac{1}{r_0 c^2}, \quad (21\text{--}34)$$

or, in Gaussian units, $m \simeq e/r_0 c^2$.

Equation (21–34) for the "inertial mass" of electromagnetic origin was obtained by considering the reaction of the magnetic induction field on the electron. A similar estimate is obtained by noting that the electrostatic energy

$$U_0 = \tfrac{1}{2}\int \epsilon_0 E^2 \, dv \quad (21\text{--}35)$$

of a spherical charge distribution of charge e and radius r_0 is of order

$$U_0 \simeq \frac{e^2}{4\pi\epsilon_0} \frac{1}{r_0}, \qquad (21\text{–}36)$$

as has been verified in Chapter 1. Hence the relativistic mass-energy relation, Eq. (17–42), yields again the mass estimate given by Eq. (21–34).

Note that these considerations have yielded only mass estimates, since r_0 indicates only the dimensions of a possible model. The relation between U_0 and m can, however, be established in a manner not dependent on a model, if we use the general consideration of Chapter 10 for the momentum density $\mathbf{g} = (\mathbf{E} \times \mathbf{H})/c^2$. The total momentum of the field should then be given by

$$\mathbf{G} = \int \frac{\mathbf{E} \times \mathbf{H}}{c^2} \, dv. \qquad (21\text{–}37)$$

Again, if we take the field to be that of a point charge, the integration will diverge, so that a lower limit corresponding to a finite radius r_0 of the electron must be introduced. For low velocities this integration can be carried out very easily, using the quasi-static fields for which

$$\mathbf{H} = \epsilon_0 \mathbf{u} \times \mathbf{E}. \qquad (21\text{–}38)$$

For a symmetrical charge distribution the momentum,

$$\mathbf{G} = \int \frac{\epsilon_0 \mathbf{E} \times (\mathbf{u} \times \mathbf{E})}{c^2} \, dv = \epsilon_0 \int \frac{\mathbf{u}E^2 - (\mathbf{u} \cdot \mathbf{E})\mathbf{E}}{c^2} \, dv, \quad (21\text{–}39)$$

is axially symmetric about the direction of the velocity, and it is clear that $\mathbf{G}$ itself is along $\mathbf{u}$, which we may take in the x-direction. The second term in the integrand is therefore just $\mathbf{u}E_x^2$. For low velocities, $\mathbf{E}$ is spherically symmetric, and

$$\mathbf{G} = \int_{r_0}^{\infty} \frac{\epsilon_0(E_y^2 + E_z^2)\mathbf{u} \, dv}{c^2} = \frac{2\mathbf{u}}{3} \int_{r_0}^{\infty} \frac{\epsilon_0 E^2 \, dv}{c^2} = \frac{4}{3} \frac{U_0}{c^2}, \qquad (21\text{–}40)$$

if we make use of Eq. (21–35). Thus Eq. (21–40), which is the same as Eq. (21–28), is in qualitative agreement with an electromagnetic mass as given by Eq. (21–34), but violates the exact relativistic mass-energy relationship. An additional mass, $-U_0/3c^2$, whose origin is not electromagnetic, is needed to account for the observed mass, which obeys the relativistically correct equations. This extra mass (or energy) presumably represents the nonelectromagnetic binding which must be present to make the charge system of the electron stable; its need was pointed out by Poincaré.

21–6 The reaction necessary to conserve radiated energy. We have seen that the mass associated with the momentum of the field is due to a reaction on whatever tends to change the velocity of an electron in steady motion. Let us now obtain the reaction force which must be present if the electron is accelerated, giving rise to a rate of energy loss

$$-\frac{dU}{dt} = \frac{e^2 \dot{u}^2}{6\pi\epsilon_0 c^3} \qquad (20\text{--}19)$$

for nonrelativistic velocities. For energy to be conserved, the reaction force $\mathbf{F}$ should satisfy the condition

$$\mathbf{F} \cdot \mathbf{u} + \frac{e^2}{6\pi\epsilon_0} \frac{\dot{u}^2}{c^3} = 0. \qquad (21\text{--}41)$$

This equation obviously has no solution for $\mathbf{F}$ which can be instantaneously correct for all times, since $\mathbf{u}$ and $\dot{\mathbf{u}}$ are basically uncorrelated. We must be content with a solution representing an average over a sufficiently long period of time. This means that we have only an average energy balance between the force and the radiation field, and that an extra fluctuation will be available which will be stored in the induction field. On this basis, Eq. (21–41) becomes

$$\int_{t_1}^{t_2} (\mathbf{F} \cdot \mathbf{u}) \, dt + \int_{t_1}^{t_2} \frac{e^2}{6\pi\epsilon_0 c^3} \dot{u}^2 \, dt = 0. \qquad (21\text{--}42)$$

Integrating by parts, we obtain

$$\int_{t_1}^{t_2} \left(\mathbf{F} - \frac{e^2 \ddot{\mathbf{u}}}{6\pi\epsilon_0 c^3} \right) \cdot \mathbf{u} \, dt + \left[\frac{e^2 \mathbf{u} \cdot \dot{\mathbf{u}}}{6\pi\epsilon_0 c^3} \right]_{t_1}^{t_2} = 0. \qquad (21\text{--}43)$$

The integrated term represents the "fluctuation" referred to above; for periodic motion, or for an acceleration occurring over a limited time, it will not affect the integrated energy balance. On the average, energy will be conserved if we put

$$\mathbf{F}_{\text{react}} = \frac{e^2 \ddot{\mathbf{u}}}{6\pi\epsilon_0 c^3} \qquad (21\text{--}44)$$

as the radiation reaction force.

This reaction is in addition to that demanded by the conservation of momentum. The total electromagnetic reaction on the electron itself would then be

$$\mathbf{F} = \frac{e^2 \ddot{\mathbf{u}}}{6\pi\epsilon_0 c^3} - m_{\text{el. mag.}} \dot{\mathbf{u}}, \qquad (21\text{--}45)$$

where $m_{\text{el. mag.}}$ is the electromagnetic mass given by Eq. (21–40),

$$m_{\text{el. mag.}} = \frac{4}{3}\frac{U_0}{c^2}.\tag{21-46}$$

21-7 Direct computation of the radiation reaction from the retarded fields. It is instructive to compute Eq. (21–45) by direct integration of the interaction of the radiation field of one part of the electron with the other parts. This calculation, originally due to Lorentz, shows very clearly the limitations of the theory.

For this calculation we shall make the following assumptions:

(1) We shall choose a frame such that any element de of the electron, on which another element de' acts, is at rest: $\mathbf{u}(t)_{de} = 0$.

(2) None of the quantities $\mathbf{u}$, $\dot{\mathbf{u}}$, $\ddot{\mathbf{u}}$, etc., changes very much during the time it takes for an electromagnetic signal to cross the electron. This is equivalent to the conditions that $u \ll c$, $\dot{u} \ll c^2/r_0$, $\ddot{u} \ll \dot{u}c/r_0$, etc. The solution is effectively a power series in $\tau_0 = r_0/c = e^2/4\pi\epsilon_0 mc^3$ times the operator $\partial/\partial t$.

(3) The fields will be derived from the retarded Liénard-Wiechert potentials (cf. Section 21–12).

(4) Only terms not containing r_0 explicitly will be considered as having physical significance.

(5) The electron charge distribution has spherical symmetry.

Consider an element of charge de at x_α affected by the element de' at x'_α, as shown in Fig. 21–2. From Eq. (20–13) we have the electric field at x due to de':

$$4\pi\epsilon_0\, d\mathbf{E}(t) = \frac{de'}{s^3}\left\{\frac{1}{c^2}\mathbf{r}\times\left[\left(\mathbf{r} - \frac{\mathbf{u}(t')r}{c}\right)\times\dot{\mathbf{u}}(t')\right]\right.$$
$$\left. + \left[1 - \left(\frac{u(t')}{c}\right)^2\right]\left(\mathbf{r} - \frac{\mathbf{u}(t')r}{c}\right)\right\}.\tag{21-47}$$

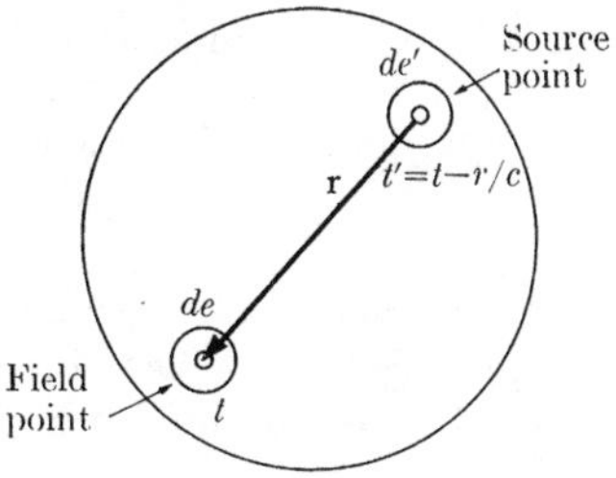

FIG. 21–2. For computing the interaction between one part of the electron and the other parts.

Since the field is expressed in terms of the electron's condition at t', the problem would be insoluble without knowledge of the electron's entire past if we allow arbitrary motions. Treatment of the problem is made possible only by the restrictions of item 2 above. With these limitations on the motion, it is possible to refer all velocities and accelerations generating signals as time t' to the time t at which the signals arrive at de. To find the total self-force the integration must be carried out at constant time t, and thus at time t' which is variable over the structure. The change from t' to t is accomplished by expanding the functions of $t' = t - r/c$ in powers of r/c, remembering that $\mathbf{u}(t) = 0$, in accord with assumption 1.

$$\dot{\mathbf{u}}(t') = \dot{\mathbf{u}}(t) - \frac{r}{c}\ddot{\mathbf{u}}(t) + \cdots, \tag{21-48}$$

$$\mathbf{u}(t') = -\frac{\dot{\mathbf{u}}(t)r}{c} + \frac{\ddot{\mathbf{u}}(t)}{2}\left(\frac{r}{c}\right)^2 - \cdots, \tag{21-49}$$

$$s^{-3} = r^{-3}\left[1 - \frac{3\dot{\mathbf{u}}(t)\cdot\mathbf{r}}{c^2} + \frac{3r}{2}\left(\frac{\ddot{\mathbf{u}}(t)\cdot\mathbf{r}}{c^3}\right)\cdots\right]. \tag{21-50}$$

Let us carry terms only to order $(r/c)^3$; thus $[u(t')/c]^2$ can be neglected, and the resulting expression is considerably simplified. In terms of t, Eq. (21–47) then becomes

$$4\pi\epsilon_0\,d\mathbf{E} = de'\left[\frac{\mathbf{r}(\mathbf{r}\cdot\dot{\mathbf{u}})}{r^3c^2} - \frac{\dot{\mathbf{u}}}{rc^2} - \frac{\mathbf{r}(\mathbf{r}\cdot\ddot{\mathbf{u}})}{r^2c^3} + \frac{\ddot{\mathbf{u}}}{c^3} + \frac{\mathbf{r}}{r^3}\right.$$

$$\left. - \frac{3\mathbf{r}(\dot{\mathbf{u}}\cdot\mathbf{r})}{r^3c^2} + \frac{3}{2}\frac{\mathbf{r}}{r^2c^3}(\ddot{\mathbf{u}}\cdot\mathbf{r}) + \frac{\dot{\mathbf{u}}}{rc^2} - \frac{\ddot{\mathbf{u}}}{2c^3}\right]$$

$$= de'\left[-\frac{2\mathbf{r}(\dot{\mathbf{u}}\cdot\mathbf{r})}{r^3c^2} + \frac{1}{2}\frac{\mathbf{r}(\ddot{\mathbf{u}}\cdot\mathbf{r})}{r^2c^3} + \frac{\mathbf{r}}{r^3} + \frac{\ddot{\mathbf{u}}}{2c^3}\right]. \tag{21-51}$$

It will be easier to take advantage of our assumption of spherical symmetry if we write Eq. (21–51) in tensor notation:

$$dE_\alpha = \frac{de'}{4\pi\epsilon_0}\left[-\frac{2r_\alpha(\dot{u}_\beta r_\beta)}{r^3c^2} + \frac{1}{2}\frac{r_\alpha(\ddot{u}_\beta r_\beta)}{r^2c^3} + \frac{r_\alpha}{r^3} + \frac{\ddot{u}_\alpha}{2c^3}\right]. \tag{21-52}$$

Now the average of r_α over a spherically symmetric region is zero, and the average of $r_\alpha r_\beta$ is simply $\frac{1}{3}r^2\,\delta_{\alpha\beta}$. Therefore the third term of Eq. (21–52) will contribute nothing to the final integral, and we may write

$$\overline{dE_\alpha} = \frac{de'}{4\pi\epsilon_0}\left(-\frac{2}{3}\frac{\dot{u}_\alpha}{c^2r} + \frac{2}{3}\frac{\ddot{u}_\alpha}{c^3}\right). \tag{21-53}$$

Integrating over de' and de, we obtain for the reaction force

$$\mathbf{F} = \iint de\, d\mathbf{E} = \frac{e^2 \ddot{\mathbf{u}}}{6\pi\epsilon_0 c^3} - m_{\text{el. mag.}}\, \dot{\mathbf{u}}, \qquad (21\text{--}54)$$

where

$$m_{\text{el. mag.}} = \frac{4}{3c^2} \iint \frac{de\, de'}{8\pi\epsilon_0 r} = \frac{4}{3c^2}\, U_0, \qquad (21\text{--}55)$$

with U_0 again representing the electrostatic energy of the electron in its own field. This is in complete agreement with Eq. (21–46), the result required by the conservation laws.

In Eq. (21–54) the term proportional to $\ddot{\mathbf{u}}$ is independent of the extent of the electron, and therefore presumably independent of the detailed structure of the electron. The mass term is indeed structure-dependent, but its relation to the electrostatic energy is not. If we had carried the calculation to higher order in r/c the additional terms would form an ascending series in r_0 which, according to our assumption 4, could not be assumed to have physical significance.

21–8 Properties of the equation of motion. Let us consider a charged particle subject to an external force $\mathbf{F}_{\text{ext}}$ in addition to the self-force $\mathbf{F}$ given by Eq. (21–45). The equation of motion is then

$$\mathbf{F}_{\text{ext}} - \frac{e^2 \ddot{\mathbf{u}}}{6\pi\epsilon_0 c^3} - m_{\text{el. mag.}}\dot{\mathbf{u}} = m_{\text{non-el. mag.}}\,\dot{\mathbf{u}}, \qquad (21\text{--}56)$$

where $m_{\text{non-el. mag.}}$ is the mass of the particle not associated with its electromagnetic field. Separation of the mass into electromagnetic and nonelectromagnetic parts is presumably not meaningful, and hence we write Eq. (21–56) as

$$m_0 \dot{\mathbf{u}} - \frac{e^2 \ddot{\mathbf{u}}}{6\pi\epsilon_0 c^3} = \mathbf{F}_{\text{ext}}, \qquad (21\text{--}57)$$

where m_0 is the *empirical* rest mass. We shall first study the consequences of Eq. (21–57), continuing to restrict our attention to nonrelativistic velocities, $u \ll c$.

The complementary solutions of Eq. (21–57) contain terms whose time dependence is given by

$$e^{+3t/2\tau_0}, \qquad (21\text{--}58)$$

where $\tau_0 = r_0/c = e^2/4\pi\epsilon_0 m_0 c^3 = 10^{-23}$ second. Such solutions (corresponding to a "run-away" electron) are clearly inadmissible physically. Hence we must either question the validity of Eq. (21–57) or supplement it arbitrarily with conditions which exclude the run-away solutions. Insofar as classical theory applies it is hardly reasonable to question the validity of Eq. (21–57), since the second term is independent of any particular model, and since the use of the experimentally determined mass bypasses our ignorance of its theoretical basis.

Let us therefore examine the conditions under which the run-away solutions of Eq. (21–57) can be avoided. If we multiply by the integrating factor exp $(-3t/2\tau_0)$ and integrate, we obtain

$$e^{-3t/2\tau_0}\dot{\mathbf{u}}(t) = e^{-3t_0/2\tau_0}\dot{\mathbf{u}}(t_0) - \frac{3}{2\tau_0} \int_{t_0}^{t} \frac{\mathbf{F}(t')}{m_0} e^{-3t'/2\tau_0} dt', \quad (21\text{–}59)$$

or

$$\dot{\mathbf{u}}(t) = e^{(3/2\tau_0)(t-t_0)}\dot{\mathbf{u}}(t_0) - \frac{3}{2\tau_0} \int_{t_0}^{t} \frac{\mathbf{F}(t')}{m_0} e^{-3(t'-t)/2\tau_0} dt', \quad (21\text{–}60)$$

where t_0 is an arbitrary constant. It is seen by inspection of Eq. (21–60) that the run-away solutions will be avoided only if we choose $t_0 = \infty$, for then $[\dot{\mathbf{u}}(t) \exp (-3t/2\tau_0)] \rightarrow 0$ as $t \rightarrow \infty$. Under this assumption,

$$\dot{\mathbf{u}}(t) = \frac{3}{2\tau_0} \int_{t}^{\infty} \frac{\mathbf{F}(t')}{m_0} e^{-3(t'-t)/2\tau_0} dt'. \quad (21\text{–}61)$$

Thus the acceleration of the particle at a time t is determined by the time average of the driving force $\mathbf{F}$ over a time interval of order $\tau_0 = 10^{-23}$ sec *after* the time t. Whether this conclusion, which apparently violates the causal relationship over a time of 10^{-23} sec, would rule out solutions such as Eq. (21–61) is a question which must be left open, since classical considerations would in general not apply over such short time intervals. The suggestion has been made* to consider Eq. (21–61) as a more fundamental equation than (21–57).

21–9 Covariant description of the mechanical properties of the electromagnetic field of a charge.

In the previous sections we have examined the low-velocity theory of the electromagnetic field of a charge, and in so doing we have noted several difficulties. These difficulties are of two kinds: (a) the divergence of the energy integrals for a point charge, and (b) the relativistically incorrect transformation properties of the energy and momentum of the electromagnetic field. The first difficulty is fundamentally the result of ignorance of an actual structure of the charge, and

---

* F. Rohrlich, "The Classical Electron," in Brittin and Downs, eds., *Lectures Delivered at the Summer Institute for Theoretical Physics, University of Colorado, Boulder, 1959,* Vol. 2 of *Lectures in Theoretical Physics,* Interscience Publishers, Inc., New York. See also *Am. J. Phys.* **28,** 639 (1960).

can be bypassed by effectively substituting the empirical mass values for the divergent integrals in combination with nonelectromagnetic sources of mass. The second difficulty arises from the fact that the expression (21–11),

$$G^i = \int T^{4i} \, dv, \tag{21–62}$$

is basically not covariant unless the field is "free" in the sense of Section 21–4. Identification of this quantity with the energy-momentum vector of the field is based on the discussion of Sections 10–5 and 10–6, in which energy and momentum were balanced between "field" and "matter" for an arbitrary volume in three-dimensional space. Although such an expression gives the correct macroscopic conservation laws, we see now that the distinction between matter and field cannot be operationally defined. It is thus appropriate to question whether we can substitute a fully covariant expression for Eq. (21–62) which will give the correct relativistic transformation properties for the mechanical aspects of the electromagnetic field by itself.

The covariant generalization of a three-dimensional volume dv is a hypersurface in four-space which must be expressible as a four-vector, dJ^i. This vector must be timelike and reduce to the value

$$dJ^i = (\mathbf{0}, \, dv) \tag{21–63}$$

in a frame for which the volume to be described is at rest. Hence we put

$$dJ^i = u^i \, dv^0, \tag{21–64}$$

where u^i is the usual four-velocity, so that dv^0 is now the "proper" volume element and thus an invariant.

A covariant generalization of Eq. (21–62) which substitutes the covariant hypersurface element for the three-dimensional element is

$$G^i_c = \int T^{ji} \, dJ_j = \int u_j T^{ji} \, dv^0. \tag{21–65}$$

This integral is readily evaluated for a charge which is spherically symmetrical in the rest frame. We obtain, in the notation of Section 21–4,

$$
\begin{aligned}
G^4_c &= \frac{1}{(1-\beta^2)^{1/2}} \int T^{44} \, dv^0 + \frac{\beta}{(1-\beta^2)^{1/2}} \int T^{41} \, dv^0 \\
&= \frac{1}{(1-\beta^2)^{1/2}} \int \left\{ \frac{W_0 - \beta^2 (T^M_{11})^0}{1-\beta^2} - \frac{\beta^2 [W_0 - (T^M_{11})^0]}{1-\beta^2} \right\} dv^0 \\
&= (1-\beta^2)^{-1/2} \int W_0 \, dv^0 = \gamma U_0,
\end{aligned}
\tag{21–66}
$$

and

$$G_c^1 = -\frac{\beta}{(1-\beta^2)^{1/2}} \int T^{11} \, dv^0 + \frac{1}{(1-\beta^2)^{1/2}} \int T^{14} \, dv^0$$

$$= \frac{1}{(1-\beta^2)^{1/2}} \int \left\{ -\beta \frac{[-(T_{11}^M)^0 + \beta^2 W_0]}{1-\beta^2} + \frac{\beta[W_0 - (T_{11}^M)^0]}{1-\beta^2} \right\} dv^0$$

$$= \frac{\beta}{(1-\beta^2)^{1/2}} \int W_0 \, dv^0 = \beta\gamma U_0. \tag{21-67}$$

Thus, we find, as expected,

$$G_c^i = (\beta\gamma U_0, \, \gamma U_0), \tag{21-68}$$

in agreement with the relativistic transformation laws, since in the rest frame, by definition,

$$G_c^{i0} = (0, \, U_0). \tag{21-69}$$

Hence, if we define the momentum and energy of the field by Eq. (21–65), a fully relativistic theory results.

Note that this formulation does not alter in any direct way the problem of the stability of a charge under electrical forces. In the rest frame the volume integral of the Maxwell stresses remains different from zero, i.e.,

$$\int (T_{11}^M)^0 \, dv = -U_0/3, \tag{21-70}$$

and there is still need for the "Poincaré binding force," as was pointed out at the end of Section 21–4, to stabilize an elementary charge. The covariant formulation of the energy-momentum vector as given by (21–65) has, however, made it possible to distinguish between electromagnetic and nonelectromagnetic mechanical properties in a covariant manner.

21–10 The relativistic equations of motion. Let us write a covariant equation which reduces to Eq. (21–57) in a proper frame. The general form of such an equation must be

$$m_0 c^2 \frac{du^i}{ds} - F_{\text{react}}^i = F_{\text{ext}}^i, \tag{21-71}$$

where F_{ext}^i is an external Minkowski four-force (see Section 17–5), and F_{react}^i is the covariant generalization of

$$\frac{e^2 \ddot{\mathbf{u}}}{6\pi\epsilon_0 c^3}. \tag{21-72}$$

Such a generalization need *not* be simply $(e^2/6\pi\epsilon_0)(d^2u^i/ds^2)$, but could be

$$F^i_{\text{react}} = \frac{e^2}{6\pi\epsilon_0}\left(\frac{d^2u^i}{ds^2} + Su^i\right), \qquad (21\text{--}73)$$

where S is some scalar. (Note that, as before, u^i is dimensionless, and ds is the proper time element multiplied by c.) The force F^i_{react} must have the property that the reaction and energy loss is conserved in all frames; this corresponds to the condition that

$$F^i_{\text{react}} \, u_i = 0. \qquad (21\text{--}74)$$

Hence

$$S = -u_i \frac{d^2u^i}{ds^2}$$

$$= -\frac{d}{ds}\left(u_i\frac{du^i}{ds}\right) + \frac{du_i}{ds}\frac{du^i}{ds}. \qquad (21\text{--}75)$$

The first term on the right-hand side of Eq. (21–75) vanishes:

$$u_i \, (du^i/ds) = d(u^i u_i)/ds = 0,$$

since $u^i u_i = 1$. Hence, the complete covariant equation of motion is

$$m_0c^2\frac{du^i}{ds} - \frac{e^2}{6\pi\epsilon_0}\left(\frac{d^2u^i}{ds^2} + u^i\cdot\frac{du_j}{ds}\frac{du^j}{ds}\right) = F^i_{\text{ext}}. \qquad (21\text{--}76)$$

This is usually called the Dirac* radiation reaction equation. Note that the last term of the left-hand side is simply

$$u^iP/c, \qquad (21\text{--}77)$$

where

$$P = -\frac{e^2c}{6\pi\epsilon_0}\left(\frac{du^i}{ds}\frac{du_i}{ds}\right) = \frac{e^2}{6\pi\epsilon_0c^3}\,(\dot u^0)^2 \qquad (21\text{--}78)$$

is the (invariant) rate of radiation of energy of the charge [as given by Eq. (20–19)]. Equation (21–76) can be written in the symmetrical form, using (21–75),

$$m_0c^2\frac{du^i}{ds} - \frac{e^2}{6\pi\epsilon_0}\,u^j\left(u_j\frac{d^2u^i}{ds^2} - u^i\frac{d^2u_j}{ds^2}\right) = F^i_{\text{ext}}. \qquad (21\text{--}79)$$

In Section 21–12 we shall try to understand this equation in a more fundamental way.

---

* P. A. M. Dirac, *Proc. Roy. Soc.* (London) **A167,** 148 (1938).

21–11 The integration of the relativistic equation of motion. The covariant equation (21–76) retains the difficulty associated with the non-relativistic equation (21–57) by requiring special boundary conditions on the integrals of the motion to circumvent the "run-away" solutions; again, these boundary conditions result in an integral which apparently violates causality for a time of order τ_0. In addition, the integral loses its simplicity relative to the nonrelativistic solution (21–61) (which reduces to a simple quadrature). The analogous solution becomes an integral equation

$$m_0 c^2 \frac{du^i}{ds} = \frac{3}{2\tau_0 c} \int_s^\infty \left[F^i_{\text{ext}}(s') - \left(\frac{u^i P}{c} \right) \right] e^{-(s'-s)/(2\tau_0 c/3)} \, ds', \quad (21\text{–}80)$$

where P is the radiation rate (21–78). It can be verified by direct differentiation that the integral equation (21–80) satisfies the differential equation (21–76); in addition, the run-away solutions are avoided.

Note that in the integral equation a reaction four-force $u^i P$ is subtracted from the external force; otherwise the equation corresponds to (21–61). This type of reaction four-force acts only during the time in which energy loss by radiation occurs. Equation (21–80) thus gives a more logical form of the radiation reaction than the usual expression which involves the second derivative of the velocity.

21–12 Modification of the theory of radiation to eliminate divergent mass integrals. Advanced potentials. In the previous sections we have bypassed the question of the meaning of the electromagnetic mass (which could not be computed in a satisfactory manner) by combining the electromagnetic mass with an (equally uncomputable) nonelectromagnetic mass; the sum was introduced into the equations as the *empirical* mass. This procedure led to equations of motion that are satisfactory once the choice of initial conditions is restricted as discussed above.

Let us now consider an alternative approach to the problem (leading to the same ultimate equation of motion) which gives finite results even for a *point* electron, but still remains within the framework of Maxwell's equations. This paradoxical program can be resolved by noting that in Chapter 14 [Eq. (14–23)] we found both "advanced" and "retarded" potential solutions which satisfied the field equations; we dropped the advanced solutions on "logical" grounds as violating experience. Let us now examine this matter critically. We make the following observations:*

(1) If we consider the field of an isolated radiator A, there is no question but that it is described correctly by the retarded solution, which we will

---

* In the following discussion we use F symbolically to stand for the field tensor F^{ij}.

call F_{ret}^A. Only this field agrees with experience at reasonable distances from the radiator and agrees with our elementary notion of causality, *once the sense of time is established by observation of the behavior of more complex systems.*

(2) If we compute the self-force by the Lorentz calculation of Section 21–7 by replacing the retarded fields F_{ret} by

$$F_{\text{react}} = \tfrac{1}{2}(F_{\text{ret}} - F_{\text{adv}}), \qquad (21\text{--}81)$$

all terms odd in r in Eqs. (21–48), (21–49), and (21–50) disappear, and thus the mass term in Eq. (21–54) would vanish. Only the radiation reaction term

$$e^2\ddot{\mathbf{u}}/6\pi\epsilon_0 c^3 \qquad (21\text{--}82)$$

remains in Eq. (21–54). Hence, under this prescription, the reaction force remains finite even for a point electron. The linear combination (21–81) of fields remains finite if computed for a point charge even at the position of the point charge. We can understand the reason for the finite result by noting that the elementary potential corresponding to a wave number k is given by

$$G_{\text{ret}} = \frac{1}{4\pi r}\, e^{+ikr},$$

$$G_{\text{adv}} = \frac{1}{4\pi r}\, e^{-ikr}, \qquad (21\text{--}83)$$

according to Eq. (14–19); hence,

$$\frac{G_{\text{ret}} - G_{\text{adv}}}{2} = \frac{1}{4\pi r}\, \frac{e^{ikr} - e^{-ikr}}{2} = \frac{1}{4\pi}\, \frac{\sin(kr)}{r}, \qquad (21\text{--}84)$$

which tends to the finite limit $k/4\pi$ as r tends to zero.

(3) The basic differential equations of classical mechanics are symmetric under reversal of the sense of time. Specifically, observation of the motion of a single particle could not give an indication as to the sense of time. Only the statistical behavior of systems with a large number of degrees of freedom single out a direction of time. Similarly, in electrodynamics the field equations are basically symmetric in time, and asymmetry is introduced only by our arbitrary rejection of the advanced solution. This rejection has no logical basis without establishment of the direction of the sense of time, using presumably the direction of statistically increasing disorder of a complex system. A time-symmetric solution would have the form

$$F_{\text{sym}} = \tfrac{1}{2}(F_{\text{ret}} + F_{\text{adv}}). \qquad (21\text{--}85)$$

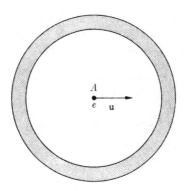

FIG. 21–3. Single charge surrounded by absorber.

It can be shown that a completely covariant "action at a distance" theory of electrodynamics can be formulated in terms of the symmetric combination of fields. It would be logically preferable if the actual field solutions were of the symmetric form (21–85), but the apparently time-asymmetric but certainly correct field F_{ret}^A of an isolated radiator can be constructed from the symmetric field (21–85) by the superposition of the field of A with that of many distant absorbers. In terms of such a picture, a remote absorber plays an essential role in governing the emission of radiation; an isolated, accelerated electron radiation with the symmetric field (21–85) would not lose energy.

A plausibility argument can be made for the conclusion that the apparently contradictory observations (1), (2), and (3) are in fact compatible. We shall adopt a simple model* in which the charge A of interest is surrounded by a large absorber containing $n - 1$ individual charges, i.e., the system contains n charges including the charge at A (Fig. 21–3). Let us make the basic assumption that the elementary fields are given by the *time-symmetric solution* Eq. (21–85), i.e.,

$$F_{\text{actual}} = \tfrac{1}{2}(F_{\text{adv}} + F_{\text{ret}}). \tag{21–86}$$

Let us assume, first without proof, that the outer cloud can constitute a complete absorber for the radiation fields, i.e., let

$$\tfrac{1}{2} \sum_{i=0}^{n} (F_{i_{\text{adv}}} + F_{i_{\text{ret}}}) = 0 \quad \text{outside of absorber.} \tag{21–87}$$

---

* For a detailed examination of this model, see J. A. Wheeler and R. P. Feynman, *Revs. Modern Phys.* **17**, 157 (1945).

Since the separate sums over the advanced and retarded fields represent ingoing and outgoing waves, respectively, which cannot interfere with one another destructively everywhere, we conclude that the separate sums vanish outside the absorber and that therefore also

$$\tfrac{1}{2} \sum_{i=0}^{n} (F_{i_{\mathrm{adv}}} - F_{i_{\mathrm{ret}}}) = 0 \quad \text{outside of absorber.} \tag{21-88}$$

We have noted [see observation (2) above] that the combination $F_{\mathrm{adv}} - F_{\mathrm{ret}}$ is nonsingular at any one of the n point charges; according to Eq. (21-88), the sum

$$S = \tfrac{1}{2} \sum_{i=0}^{n} (F_{i_{\mathrm{adv}}} - F_{i_{\mathrm{ret}}}) \tag{21-89}$$

is thus nonsingular and vanishes over a closed boundary; by our general theorems S thus vanishes *everywhere*. We can thus write S as

$$0 = S = \tfrac{1}{2} \sum_{i=0}^{n} (F_{i_{\mathrm{adv}}} - F_{i_{\mathrm{ret}}})$$

$$= \sum_{\substack{i=0 \\ i \neq A}}^{n-1} [\tfrac{1}{2}(F_{i_{\mathrm{adv}}} + F_{i_{\mathrm{ret}}}) - F_{i_{\mathrm{ret}}}] + \tfrac{1}{2}(F^{A}_{i_{\mathrm{adv}}} - F^{A}_{i_{\mathrm{ret}}}), \tag{21-90}$$

where we have separated the field of A from that of the other charges. If we now evaluate the terms of Eq. (21-90) at the point A, using Eqs. (21-81) and (21-86), we obtain

$$\sum_{\substack{i=0 \\ i \neq A}}^{n-1} F_{i_{\mathrm{actual}}} = \sum_{\substack{i=0 \\ i \neq A}}^{n-1} F_{i_{\mathrm{ret}}} + F^{A}_{\mathrm{react}}. \tag{21-91}$$

Physically this equation means that the actual force per unit charge acting on A (using the time-symmetric formalism) equals the sum of the radiation field from the absorber (using the usual retarded field formalism) plus the nonsingular radiation reaction term computed from Eq. (21-81). Hence, all our past calculations of radiation fields (using *retarded* potentials) presumably remain applicable so far as the behavior of a "test charge" in an external field is concerned. The correct radiation reaction in the low-velocity approximation, however, is Eq. (21-82) computed from the nonsingular difference between retarded and advanced fields; Eq. (21-79) presumably gives the correct relativistic expression for the equation of motion. Hence, the contradiction between the observations (1), (2), and (3) is resolved.

The prescription of using (21–81) to compute the radiation reaction, as justified by the discussion of the absorber model, effectively subtracts out the divergent mass terms. Such a procedure is called "renormalization" of the mass; it permits use of the empirical mass without regard for the problem of its nature.

21–13 Direct calculation of the relativistic radiation reaction. According to Section 21–12 we can compute the radiation reaction from Eq. (21–81) without encountering divergence problems, even for a point electron. Specifically, we can write the covariant radiation reaction as

$$F_{\text{react}}^i = \frac{eu_j}{2} (F_{\text{ret}}^{ij} - F_{\text{adv}}^{ij}), \tag{21–92}$$

where the F^{ij} are the fields of the charge computed at the world point of the charge.

The Liénard-Wiechert potentials are, from Eq. (18–19),

$$\varphi^i = \frac{e}{4\pi\epsilon_0} \frac{u^i}{u^j R_j}, \tag{18–19}$$

where

$$R_j R^j = 0; \quad R_j = x_j - x'_j(s). \tag{21–93}$$

The quadratic equation (21–93) will have two roots which will determine whether we are dealing with the advanced or retarded solution. If we differentiate Eq. (21–93) we obtain

$$(dx_j - u_j \, ds)(x^j - x'^j) = 0, \tag{21–94}$$

or

$$ds = \frac{dx_j R^j}{u_k R^k}. \tag{21–95}$$

To obtain the fields we must compute the derivatives of Eq. (18–19). For instance,

$$\frac{\partial \varphi^i}{\partial x^l} = \frac{e}{4\pi\epsilon_0} \frac{R_l}{(u_k R^k)} \frac{d}{ds}\left[\frac{u^i}{u_j R^j}\right] \tag{21–96}$$

$$= \frac{e}{4\pi\epsilon_0} \frac{R_l}{(u_k R^k)^2} \left\{\frac{du^i}{ds} - \frac{u^i[1 + R^j(du_j/ds)]}{(u_k R^k)}\right\}; \tag{21–97}$$

$$\therefore F^{il} = \frac{e}{4\pi\epsilon_0} \frac{1}{(u_k R^k)^3} \left\{\left(R^l \frac{du^i}{ds} - R^i \frac{du^l}{ds}\right)(u_k R^k) \right.$$
$$\left. - (R^l u^i - R^i u^l)\left[1 + R^j \frac{du_j}{ds}\right]\right\}. \tag{21–98}$$

This expression holds for both the retarded and advanced solutions, depending on the choice of R^j; R^j is given by

$$R^j = (\mathbf{r}, \pm r) \quad \text{and} \quad u^j R_j = \left(1 - \frac{u^2}{c^2}\right)^{-1/2}\left(\pm r - \frac{\mathbf{r} \cdot \mathbf{u}}{c}\right),$$

$$(21\text{-}99)$$

where the $+$ refers to the retarded and $-$ to the advanced solution. At $R^j = 0$, Eq. (21–98) becomes singular, as is to be expected if the retarded and advanced solutions are evaluated separately. Subtraction of the two solutions removes the singularity so that the limit $R_j \to 0$ remains finite. The evaluation of this limit can be made by expanding Eq. (21–98) in powers of the components of R_j, and then letting the components go to zero after the retarded and advanced solutions have been subtracted. This expansion is straightforward but involves many steps;* the result is

$$F^{ij}_{\text{react}} = \frac{F^{ij}_{\text{ret}} - F^{ij}_{\text{adv}}}{2} = \frac{e}{6\pi\epsilon_0}\left(u^i\frac{d^2u^j}{ds^2} - u^j\frac{d^2u^i}{ds^2}\right). \quad (21\text{-}100)$$

Equation (21–100) corresponds exactly to the equation of motion (21–79). We thus conclude that this equation of motion, subject to the boundary conditions avoiding the run-away solutions, is a valid description of the classical motion of a point charge.

SUGGESTED REFERENCES

C. Møller, *The Theory of Relativity*. This work contains an excellent detailed discussion of the conservation laws in general closed systems, and the problems involved in their extension to the partial fields of macroscopic electrodynamics.

R. Becker, *Theorie der Elektrizität*. Abraham's symmetrical energy-momentum tensor, designed to replace the unsymmetrical form derived by Minkowski for electromagnetic media, is presented in this work, in addition to the standard formulation of the macroscopic field tensor.

H. A. Lorentz, *The Theory of Electrons*. Gives detailed discussion of the model discussed in Section 21–7.

L. Landau and E. Lifshitz, *The Classical Theory of Fields*. Chapter 9 discusses the radiation reaction briefly.

A. Sommerfeld, *Electrodynamics*. Section 36 discusses the radiation reaction in covariant form.

J. A. Wheeler and R. P. Feynman. I. *Revs. Modern Phys.* **17**, 157 (1945). II. *Revs. Modern Phys.* **21**, 425 (1949).

The first article discusses in considerable detail the action of an absorber surrounding a radiator in controlling the nature of the fields (see Section 21–12).

---

* For the algebraic details, see P. A. M. Dirac, *loc. cit.*, pp. 163 ff.

The second article shows how an "action at a distance" theory can be formulated consistently if the symmetrical combination of advanced and retarded fields is adopted.

F. ROHRLICH, *Lectures on Classical Electrodynamics*. Discusses covariant formulation of the energy-momentum of the field. See also *Am. J. Phys.* **28,** 639 (1960).

G. PLASS, *Revs. Modern Phys.* **33,** 37 (1961). Discusses solutions of the complete relativistic equation of motion with radiation reaction.

Earlier discussions of the electromagnetic mass concept are contained in:

R. BECKER, *Theorie der Elektricität*, Band II.

J. H. JEANS, *The Mathematical Theory of Electricity and Magnetism*.

M. MASON AND W. WEAVER, *The Electromagnetic Field*.

EXERCISES

1. Check the algebraic correctness of the conservation laws as derived from the tensor of Eq. (21–9).

2. Prove that the quantity $F^{ik}F_k^j$ in Eq. (21–9) is symmetrical in i and j.

3. Show that $T_i^i = 0$.

4. Find the radiation reaction force on an electron being accelerated in a betatron (i.e., while moving in a circle of fixed radius). What is the dependence of the reaction force on the particle energy?

5. Given an electron moving initially in a circle of radius a about a proton. If energy were lost by radiation according to the classical law, what would happen? How long would the system last?

6. Two dipole antennas are excited sinusoidally 90° out of phase and separated by a quarter-wavelength in space. Find the net force on the antennas. Take the dipole axes parallel to their relative displacement.

7. Show that in an arbitrary frame the relativistic Dirac equation reduces to

$$m_0 \frac{du}{dt} - \frac{2}{3}\tau_0 m_0 \frac{d^2u}{dt^2} + \frac{2}{3}\frac{\tau_0 m_0 u\,(du/dt)^2}{u^2 + c^2} = \left(1 + \frac{u^2}{c^2}\right)^{1/2} F(\tau),$$

for one-dimensional motion under a force $F(\tau)$ given as a function of the proper time τ in the frame of the particle.

8. Show that an integral of the differential equation of Exercise 7 for the velocity u is given by

$$\frac{u}{c} = \sinh\left(\sinh^{-1}\left(\frac{u_0}{c}\right) + \frac{1}{m_0 c}\right.$$
$$\left. \times \left\{\int_0^\tau F(\tau')\,d\tau' + \int_0^\infty \left[\exp\left(-\frac{3}{2}\frac{\tau'}{\tau_0}\right)\right][F(\tau + \tau') - F(\tau')]\,d\tau'\right\}\right).$$

Compare with the result of Exercise 8, Chapter 17.

(*Hint:* Transform the differential equation to a new variable $w(\tau)$ in place of $u(\tau)$ by the substitution $u(\tau)/c = \sinh\,[w(\tau)/c]$.)

CHAPTER 22

RADIATION, SCATTERING, AND DISPERSION

We have seen in Chapter 21 that in the nonrelativistic case Eq. (21–57), and in the relativistic case Eq. (21–76), constitute the correct differential equation of motion. We arrived at these equations by various routes; one, the calculation in Section 21–7 based on Lorentz's spherically symmetric model, imposed restrictions on the rates at which dynamical quantities may change. The more general discussion of Section 21–12 makes it plausible that the equations of motion have general validity, provided only that the auxiliary constants in the integration are restricted to avoid the "run-away" solutions. In this section we shall examine various consequences of the equations of motion.

22–1 Radiative damping of a charged harmonic oscillator. Consider an electron bound with a harmonic force $\mathbf{F} = -k\mathbf{x}$, corresponding to a natural frequency $\omega_0 = \sqrt{k/m}$. The equation of motion, including the self-force of the electron, is then

$$\dddot{x} + \omega_0^2 x = \frac{e^2 x}{6\pi\epsilon_0 c^3 m} = \frac{2}{3}\tau_0 x. \tag{22-1}$$

We have assumed nonrelativistic motion only; m is the empirically determined mass. We shall assume that the right-hand side of Eq. (22–1) is small in comparison with the binding term, so that we can make the approximation

$$\dddot{x} \simeq -\omega_0^2 \dot{x}. \tag{22-2}$$

If for convenience we set*

$$\gamma = \frac{2}{3}\tau_0\omega_0^2 = \frac{e^2\omega_0^2}{6\pi\epsilon_0 c^3 m}, \tag{22-3}$$

then Eq. (22–1) becomes

$$\ddot{x} + \gamma\dot{x} + \omega_0^2 x = 0. \tag{22-4}$$

The solution of Eq. (22–1), valid for small γ, is

$$x = Ae^{-i\omega_0 t}e^{-\gamma t/2}. \tag{22-5}$$

---

* Not to be confused with $\gamma = (1 - v^2/c^2)^{1/2}$ of earlier chapters.

401

Both the potential and the kinetic energy of the oscillator thus fall off as $e^{-\gamma t}$:

$$U \simeq \frac{mA^2\omega_0^2}{2} e^{-\gamma t} , \qquad (22\text{--}6)$$

and

$$-\frac{dU}{dt} = \gamma U = \frac{e^2\omega_0^4 A^2}{12\pi\epsilon_0 c^3} e^{-\gamma t}. \qquad (22\text{--}7)$$

But Eq. (22–7) is just the rate of radiation, averaged over a cycle, so that energy is conserved in accord with our first derivation of $\mathbf{F}_{rad}$. Equation (22–7) corresponds to a damped wave train emitted by the oscillator after a given amplitude A has been excited by an external impulse. It is clear from Eq. (22–7) that $1/\gamma$ is the mean duration of the radiated pulse when it is averaged over the energy. This is the classical quantity corresponding to the quantum-mechanical "lifetime" of an excited state produced by an external impulse.

The limitation on the frequency imposed by the condition that $\gamma \ll \omega_0$ is entirely unimportant except for high-energy γ-rays. In terms of the quantum energy of the outgoing radiation, the inequality becomes

$$\hbar\omega_0 \ll \frac{4\pi\epsilon_0 \hbar c}{e^2} mc^2 = 137 mc^2 \simeq 70 \text{ Mev}. \qquad (22\text{--}8)$$

Because of the radiative damping, the radiation emitted by the oscillator is not monochromatic. The line width can be obtained by making a Fourier analysis of the fields, which are proportional to Eq. (22–5). If

$$E = \int_{-\infty}^{\infty} E_\omega \, e^{-i\omega t} \, d\omega = E_0 \, e^{-i\omega_0 t} e^{-\gamma t/2}, \qquad (22\text{--}9)$$

then

$$E_\omega = \frac{E_0}{2\pi} \int_0^{\infty} e^{-i\omega_0 t} e^{-\gamma t/2} e^{i\omega t} \, dt$$

$$= \frac{E_0}{2\pi} \frac{1}{i(\omega - \omega_0) - \gamma/2}. \qquad (22\text{--}10)$$

Equation (22–10) corresponds (see Fig. 22–1) to a radiation intensity

$$I_\omega = \frac{I_0\gamma}{2\pi} \frac{1}{(\omega - \omega_0)^2 + \gamma^2/4} \qquad (22\text{--}11)$$

normalized in such a way that

$$\int_{-\infty}^{\infty} I_\omega \, d\omega = I_0. \qquad (22\text{--}12)$$

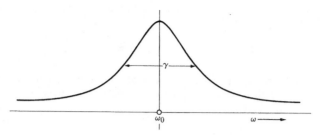

FIG. 22–1. Radiation intensity plotted against frequency as given in Eq. (22–11).

The full frequency width at half intensity is therefore

$$\Delta\omega \simeq \gamma = \tfrac{2}{3}\tau_0\omega_0^2. \qquad (22\text{–}13)$$

The corresponding width expressed in wavelength is

$$\Delta\lambda = \frac{2\pi\,\Delta\omega c}{\omega_0^2} = \frac{4\pi}{3}\,c\tau_0 = \frac{4\pi}{3}\,r_0 \simeq 10^{-12}\ \text{cm}, \qquad (22\text{–}14)$$

which is constant, independent of the frequency of the oscillator. In practice, the radiation damping is, of course, not the only source of width for a spectral line: interruptions of the wave train by atomic collisions and Doppler shifts also contribute to line width.

The relation between line width and mean life given by Eq. (22–13), namely, $\Delta\omega\gamma^{-1} = 1$, is equivalent to the relation

$$\Delta E\,\Delta t \sim \hbar, \qquad (22\text{–}15)$$

where $\Delta E = \hbar\,\Delta\omega$. Equation (22–15) is the quantum-mechanical relation between the lifetime and the energy width of a state.

22–2 Forced vibrations. In the previous section, we have discussed the motion and consequent radiation of a bound electron following a transient disturbing impulse; this is the classical theory of spectral emission. Let us now consider the steady state motion of a harmonically bound electron forced to vibrate by an external electromagnetic wave. The classical theories of the scattering, absorption, and dispersion of light are based on such a consideration.

For the treatment of optical frequencies and even soft x-rays, we may assume that the velocity of the electron is well below the velocity of light, and therefore we may neglect the effect of the magnetic vector in the incoming radiation. The equation of motion of a bound electron in an external (polarized) plane wave field $\mathbf{E} = \mathbf{E}_0 e^{-i\omega t}$ is then

$$\ddot{\mathbf{x}} - \frac{3}{2}\tau_0\dddot{\mathbf{x}} + \omega_0^2\mathbf{x} = \frac{e}{m}\mathbf{E}_0\ e^{-i\omega t}, \qquad (22\text{--}16)$$

with τ_0 as defined in Eq. (22–1). We shall be interested in the steady state solution, which is proportional to $e^{-i\omega t}$, and so we may put

$$\dddot{\mathbf{x}} \simeq -\omega^2\dot{\mathbf{x}}. \qquad (22\text{--}17)$$

We may define

$$\gamma = \tfrac{2}{3}\tau_0\ \omega^2, \qquad (22\text{--}18)$$

although it does not matter very much whether we use Eq. (22–18) or Eq. (22–4); τ_0 is sufficiently small so that the term is important only when $\omega \sim \omega_0$, i.e., only near resonance. The steady state solution of Eq. (22–16) is then given by

$$\mathbf{x} = \mathbf{E}_0\frac{e}{m}\left(\frac{1}{\omega_0^2 - \omega^2 - i\omega\gamma}\right)e^{-i\omega t}. \qquad (22\text{--}19)$$

The acceleration of the charge in terms of the external field,

$$\ddot{\mathbf{x}} = \frac{e}{m}\left(\frac{-\omega^2}{\omega_0^2 - \omega^2 - i\omega\gamma}\right)\mathbf{E}, \qquad (22\text{--}20)$$

may be substituted in Eqs. (19–20) for the dipole radiation fields to give the scattered radiation. We shall examine in detail several applications of Eq. (22–20).

22–3 Scattering by an individual free electron. For an unbound and weakly accelerated electron, $\gamma \simeq 0$, $\omega_0 \simeq 0$, so that

$$\ddot{\mathbf{x}} = \frac{e}{m}\mathbf{E}. \qquad (22\text{--}21)$$

This acceleration gives rise to a radiation field of magnitude

$$E(\mathbf{r}) = \frac{e\,(\sin\alpha)\ddot{x}}{4\pi\epsilon_0 rc^2}, \qquad (22\text{--}22)$$

where α is the angle between $\mathbf{E}$ and $\mathbf{r}$, as shown in Fig. 22–2. The rate at which the re-radiated energy crosses a unit area is

$$|\mathbf{N}| = \sqrt{\frac{\epsilon_0}{\mu_0}}\left[\frac{e\,(\sin\alpha)\ddot{x}}{4\pi\epsilon_0 rc^2}\right]^2, \qquad (22\text{--}23)$$

$$|\mathbf{N}| = r_0^2\sin^2\alpha\left(\frac{I_0}{r^2}\right), \qquad (22\text{--}24)$$

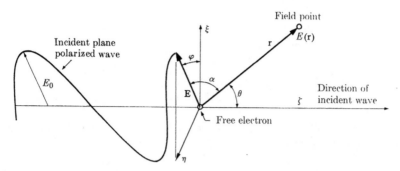

FIG. 22-2. A plane wave to be scattered by an electron, showing the polarization and scattering angles.

where $r_0 = e^2/4\pi\epsilon_0 mc^2$ is the classical electron radius, and

$$I_0 = \sqrt{\frac{\epsilon_0}{\mu_0}}\, E^2 \tag{22-25}$$

is the primary intensity. Then

$$-\frac{dU}{dt} = \int r^2 N\, d\Omega = \frac{8\pi I_0 r_0^2}{3}. \tag{22-26}$$

If we divide Eq. (22-26) by I_0, the intensity of the incoming radiation, we obtain the effective scattering cross section per electron,

$$\sigma_0 = \frac{8\pi}{3}\, r_0^2. \tag{22-27}$$

Equation (22-26) is known as the Thomson scattering formula, and Eq. (22-27) is called the cross section for Thomson scattering. Both are constant, independent of the frequency.

Comparison with experiment is facilitated if we express Eq. (22-24) in terms of the scattering angle θ and the polarization angle φ, shown in Fig. 22-2. The angles α, θ, and φ are related by

$$\cos \alpha = \cos \varphi \sin \theta,$$

i.e., $\sin^2 \alpha = 1 - \cos^2 \varphi (1 - \cos^2 \theta)$.

If the primary wave is unpolarized, i.e., randomly polarized, we must average over φ, and thus we obtain

$$\overline{\sin^2 \alpha} = \tfrac{1}{2}(1 + \cos^2 \theta),$$

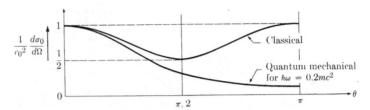

FIG. 22–3. Differential scattering cross section as a function of scattering angle, showing deviations from Thomson scattering for moderately high frequencies.

since the average of $\cos^2 \varphi$ is $\frac{1}{2}$. In terms of θ, the total rate of scattered energy is

$$- \frac{dU}{dt} = I_0 r_0^2 \int \frac{(1 + \cos^2 \theta)}{2} \, d\Omega = \frac{8 \pi I_0 r_0^2}{3}. \qquad (22\text{–}28)$$

The differential scattering cross section per unit solid angle is given by

$$\frac{d\sigma_0}{d\Omega} = \frac{r_0^2 (1 + \cos^2 \theta)}{2}. \qquad (22\text{–}29)$$

This cross section is shown graphically as a function of the scattering angle in Fig. 22–3. The quantum-mechanical formula for the scattered radiation approaches Eq. (22–29) as the frequency goes to zero, but the scattered intensity is not symmetrical. In general, the scattered radiation is more concentrated in the forward direction, as indicated by the curve for radiation whose primary energy is 100 kilo-electron-volts.

There is another feature of Thomson scattering which is modified by quantum considerations. Classically, the scattered radiation has the same frequency as that of the incoming wave. One qualification must be made: momentum, as well as energy, is removed from the incident beam, and from the symmetry of the scattered radiation it is evident that this momentum is given to the electron. This is equivalent to an average force in the direction of the incoming beam, and equal to $(I_0/c)\sigma_0$, which would gradually impart a net velocity to the electron. Actually this is not observed. What happens instead is that each quantum of radiation imparts some of its energy and momentum to the electron, and that the frequency of the scattered quantum depends on the angle of scattering. For a discussion of Compton scattering and the quantum-mechanical Klein-Nishina formula, see, for example, W. Heitler, *The Quantum Theory of Radiation*, third edition, pp. 211–224.

22–4 Scattering by a bound electron. For the general case of a harmonically bound electron, the acceleration given in Eq. (22–20) must be used in Eq. (22–23) to determine the re-radiated energy per unit area in a given direction. With this substitution, the total cross section for scattering becomes

$$\sigma = \sigma_0 \frac{\omega^4}{(\omega_0^2 - \omega^2)^2 + (\gamma\omega)^2}, \qquad (22\text{--}30)$$

where σ_0 is the Thomson cross section given in Eq. (22–27). The angular distribution is the same as in the free electron case.

The maximum value of Eq. (22–30) is obtained if $\omega \simeq \omega_0$, i.e., for resonance scattering. In that case, the scattering cross section can become very large: the cross section

$$\sigma = \sigma_0 \left(\frac{\omega_0}{\gamma}\right)^2 \simeq 6\pi\lambda_0^2 \qquad (22\text{--}31)$$

is greatly in excess of the classical area of the electron.

For strong binding, $\omega \ll \omega_0$, $\gamma \ll \omega_0$, and Eq. (22–30) becomes

$$\sigma = \sigma_0 \left(\frac{\omega}{\omega_0}\right)^4, \qquad (22\text{--}32)$$

giving a cross section that depends on the inverse fourth power of the incident wavelength. In Chapter 13 we obtained scattering from a conducting sphere which behaved in the same way for the case that the dimensions of the sphere were small compared with λ, and it is found that a small dielectric sphere scatters radiation according to the same frequency dependence. The inverse fourth power law was first derived by Rayleigh in his investigation of the blueness of the sky; he assumed that the individual molecules scatter in a wholly random fashion, so that the phases are random and the intensities add. Not only does Eq. (22–32) account for the blue color; with average empirical values of ω_0 it leads quantitatively to generally good agreement with observed atmospheric scattering, both in total amount and in polarization. Justification for the assumption of random phases is discussed in Section 22–8.

22–5 Absorption of radiation by an oscillator. The consequences of Eq. (22–19) have thus far been discussed principally in terms of the effect of a bound electron on the propagation and scattering of electromagnetic radiation. Another question of fundamental interest concerns the amount of energy abstracted from the wave and converted to energy of the bound electron system.

Consider a somewhat more complicated case than that of a monochromatic plane wave. Let E_ω represent the Fourier component of a general time dependent $E(t)$, i.e., $E = \int E_\omega e^{-i\omega t}\, d\omega$, and let x_ω be the Fourier component of the resultant displacement $x(t)$. Clearly, as with a monochromatic wave,

$$x_\omega = \frac{eE_\omega}{m}\, \frac{1}{\omega_0^2 - \omega^2 - i\omega\gamma}. \tag{22-33}$$

Let us now consider the rate of work done by the electric field on the electron, and compute the time integral of that work. To make this computation, we must utilize the theorem of Eq. (14–45). With this theorem, it follows from Eq. (22–33) that the energy ΔU absorbed by the electron due to the passage of the electric field pulse is

$$\Delta U = \int_{-\infty}^{\infty} eE(t)\dot{x}(t)\, dt = \frac{2\pi e^2}{m} \int_0^\infty |E_\omega|^2 \frac{2\omega^2\gamma}{(\omega_0^2 - \omega^2)^2 + \omega^2\gamma^2}\, d\omega. \tag{22-34}$$

This integral can be evaluated under the assumption that the resonance term is sharply peaked relative to the frequency distribution of the external field. In that case,

$$\Delta U = \frac{2\pi e^2}{m} |E_{\omega_0}|^2 2\omega_0^2 \gamma \int_{-\omega_0}^{\infty} \frac{d\xi}{(2\omega_0\xi)^2 + \omega_0^2\gamma^2}, \tag{22-35}$$

where $\xi = \omega - \omega_0$. We may replace the lower limit by $-\infty$ to obtain the simple result

$$\Delta U = \frac{2\pi^2 e^2}{m} |E_{\omega_0}|^2. \tag{22-36}$$

Note that the result is independent of γ; the energy absorbed from an incident field satisfying the assumption above is thus independent of line width, regardless of the physical cause of such a width.

The result stated in Eq. (22–36) can be brought into a different form. Let S be the energy flow per unit area in the incident beam, which is given by

$$S = \int_{-\infty}^{\infty} \epsilon_0 c E^2\, dt = 4\pi\epsilon_0 c \int_0^\infty |E_\omega|^2\, d\omega, \tag{22-37}$$

according to Eq. (14–49′). Hence, if we define S_ω by

$$S = \int_0^\infty S_\omega\, d\omega, \tag{22-38}$$

Eq. (22–36) becomes simply, in terms of the classical electron radius r_0,

$$\Delta U = 2\pi^2 r_0 c S_{\omega_0}. \tag{22-39}$$

Then, since ΔU represents the energy abstracted from a given energy flow S per unit area, we can write Eq. (22–39) as an integral of the absorption cross section over the frequency:

$$\int_0^\infty \sigma(\omega) \, d\omega = 2\pi^2 r_0 c. \tag{22–40}$$

This is the classical "sum rule," giving the total absorption cross section integrated over the frequency. The presence of more than one binding frequency for a single charge will not modify Eq. (22–40).

Some further remarks are appropriate here. Equation (22–40) gives the integrated cross section for "dipole absorption" only, since the "excursion" x of the oscillator is taken to be small compared with the wavelength. Note also that we assumed the displacement $\mathbf{x}$ to be parallel to $\mathbf{E}$. This would be the case if the incident radiation is polarized and directed onto a one-dimensional oscillator so that $\mathbf{E}$ is parallel to $\mathbf{x}$, or if an arbitrary wave group is incident on a three-dimensional oscillator. If a randomly directed and polarized radiation field is incident on a one-dimensional oscillator, then Eqs. (22–36), (22–39), and (22–40) should be divided by 3, because Eq. (22–34) would then contain a factor $\cos^2 \theta$, where θ is the angle between $\mathbf{x}$ and $\mathbf{E}$, and the mean value of $\cos^2 \theta$ averaged over a sphere is $\frac{1}{3}$.

Equation (22–40) can be generalized to any mechanism of electric dipole absorption. We must remember, however, that if this equation is applied to a system where the charged particles are not light compared with the total mass of the system, e must be replaced by $e_{\text{effective}}$, where

$$\frac{e_{\text{eff}}}{e} = \frac{\text{displacement of charge relative to center of mass}}{\text{absolute displacement of charge}} \tag{22–41}$$

and m becomes the "reduced" mass of the absorbing system. If more than one charged particle is present, the sum rule can be generalized to

$$\int_0^\infty \sigma(\omega) \, d\omega = 2\pi^2 c \sum_i \frac{e_{i\text{eff}}^2}{4\pi\epsilon_0 m_i c^2}. \tag{22–42}$$

22–6 Equilibrium between an oscillator and a radiation field. In the previous section, we obtained an expression for the energy absorbed by an elastically bound electron in an external energy flux S of radiation whose spectral distribution is $S_\omega \, d\omega$. Let us now consider a linear oscillator in a randomly directed and polarized radiation field whose spectral distribution of energy density U_v is given by

$$U_v = \int_0^\infty U_{v\omega} \, d\omega. \tag{22–43}$$

Let ΔU now be the energy absorbed in a given time Δt. Clearly,

$$U_{v\omega} = S_\omega/c \, \Delta t,$$

and in order to make use of Eq. (22–39) we must divide the right side by 3 for reasons discussed above. Hence,

$$\frac{\Delta U}{\Delta t} = \frac{2\pi^2}{3} r_0 c^2 U_{v\omega}. \qquad (22\text{–}44)$$

From Eq. (20–19) we know that the energy radiated by such an oscillator in a time Δt is given by

$$\frac{\Delta U_{\text{rad}}}{\Delta t} = \frac{2}{3} r_0 \frac{m\dot{u}^2}{c}. \qquad (22\text{–}45)$$

Averaged over a cycle of the oscillation, Eq. (22–45) becomes

$$\frac{\overline{\Delta U_{\text{rad}}}}{\Delta t} = \frac{2}{3} \frac{r_0 \omega^2}{c} \overline{U}, \qquad (22\text{–}46)$$

since $\overline{m\dot{u}^2} = m\dot{u}_0^2/2 = \omega^2 \overline{U}$, where $\dot{u}_0$ is the crest value of $\dot{u}$ and where $\overline{U}$ is the energy of the oscillator. Now consider the oscillator to be surrounded by a wall (at the same temperature as the oscillator) which absorbs the emitted radiation and re-emits it randomly. Under these conditions, the oscillator will establish equilibrium with its surroundings, and hence Eqs. (22–46) and (22–44) can be equated. Thus

$$U_{v\omega} = \frac{\omega^2}{\pi^2 c^3} \overline{U}. \qquad (22\text{–}47)$$

This relation governs the frequency distribution of the energy density of electromagnetic radiation in a cavity with perfectly "black" walls; the distribution is controlled by the statistical behavior of the oscillator at a given temperature. It is this conclusion which leads to the paradoxical Rayleigh-Jeans spectral distribution of blackbody radiation if classical statistical mechanics is applied to the oscillator, i.e., if we put $\overline{U} = kT$. Obviously, the energy density corresponding to a particular frequency cannot increase indefinitely with increasing frequency at a given temperature.

The distribution law, Eq. (22–47), can be arrived at in an entirely different manner. It will be shown in Chapter 24 that the field equations of a bounded radiation field can be transformed in such a way that they are equivalent to the equations for simple harmonic oscillators. In fact, there is a one-to-one correspondence between equivalent oscillators and the normal modes of the field in the enclosure. The coefficient $\omega^2/\pi^2 c^3$ in Eq.

(22–47) then corresponds to the number of normal modes per unit circular frequency interval (compare with the answer to problem 8 in the exercises at the end of Chapter 13).

22–7 Effect of a volume distribution of scatterers. If we have N electronic scatterers per unit volume, each of them will scatter an incoming plane wave in accord with Eq. (22–30). The scattered radiation will, in general, combine coherently with the external field, and thus modify the effective velocity of the wave. It is possible to obtain the effect of a great number of scatterers, electrons or molecules, by superposition of the scattered wavelets. For the study of refraction, however, it is mathematically simpler to treat the re-radiation as being due to the electric polarization of an entire volume element, rather than to the combination of single electron displacements. The polarization is related to the individual electron displacement by

$$\mathbf{P} = Ne\mathbf{x}. \tag{22–48}$$

The effect of the volume polarization is to add the polarization current $\partial \mathbf{P}/\partial t$ to the vacuum displacement current $\epsilon_0\, \partial \mathbf{E}/\partial t$ as a circulation source of magnetic field. Equation (9–6)(4) becomes

$$\boldsymbol{\nabla} \times \mathbf{B} = \mu_0 \left(\epsilon_0 \frac{\partial \mathbf{E}}{\partial t} + \frac{\partial \mathbf{P}}{\partial t} \right). \tag{22–49}$$

This equation, when combined with $\boldsymbol{\nabla} \times \mathbf{E} = -\partial \mathbf{B}/\partial t$ in the usual way, yields the homogeneous wave equations

$$\boldsymbol{\nabla}^2 \mathbf{E} - \frac{n^2}{c^2} \frac{\partial^2 \mathbf{E}}{\partial t^2} = 0,$$
$$\boldsymbol{\nabla}^2 \mathbf{B} - \frac{n^2}{c^2} \frac{\partial^2 \mathbf{B}}{\partial t^2} = 0, \tag{22–50}$$

where

$$n = \sqrt{1 + \frac{|\mathbf{P}|}{\epsilon_0 |\mathbf{E}|}} = \sqrt{\kappa} \tag{22–51}$$

is the index of refraction.

For dilute systems, in which the scattered energy is very small compared with that in the incident beam, it is a good approximation to take as the local field at the scatterer just the field of the incoming wave. Therefore, from Eq. (22–19),

$$n^2 = \kappa = 1 + \frac{Ne^2}{\epsilon_0 m} \frac{1}{\omega_0^2 - \omega^2 - i\omega\gamma}. \tag{22–52}$$

Equation (22–52) is the dispersion formula for a dilute gas of electrons with a single binding frequency. If $\omega < \omega_0$, we have what is called "normal dispersion," in which the refractive index diminishes with increasing wavelength. At $\omega \simeq \omega_0$ there is an absorption band, at which n has an appreciable imaginary part. On the short wavelength side of a resonance frequency $n < 1$, but we note that c/n is the *phase* velocity: the group velocity, $d\omega/dk$, is less than c on both sides of the resonance frequency. For frequencies in the immediate neighborhood of ω_0 the question of transmission is more complicated, and we shall consider it further.

A monochromatic plane wave solution would have the form

$$\mathbf{E}(t) = \mathbf{E}_0 \exp\left[i\omega t - n(\mathbf{k} \cdot \mathbf{r})\right], \qquad (22\text{--}53)$$

where $\mathbf{k}$ is the wave vector in the direction of propagation. Hence, if n has an imaginary part, $\mathbf{E}$ will contain a factor corresponding to an exponential absorption.

If the medium is not dilute, the local field is not equal to the external field, but may be approximated by Eq. (2–36):

$$\mathbf{E}_{\text{eff}} = \mathbf{E} + (\mathbf{P}/3\epsilon_0).$$

It is this field that must be substituted in Eq. (22–9) to obtain the polarization as defined in Eq. (22–48). When the fields are eliminated, we obtain the analog of the Clausius-Mossotti relation, Eq. (2–39):

$$\frac{\kappa - 1}{\kappa + 2} = \frac{n^2 - 1}{n^2 + 2} = \frac{Ne^2}{3\epsilon_0 m} \frac{1}{\omega_0^2 - \omega^2 - i\omega\gamma}. \qquad (22\text{--}54)$$

The resemblance to Eq. (2–39) is even closer if we express Eq. (22–54) in terms of the classical electron radius r_0, and the wave numbers $k_0 = \omega_0/c$, $k = \omega/c$, $\Gamma = \gamma/c$; and put $N = \rho_m N_0/M$ in terms of the density ρ_m, the molecular weight M, and Avogadro's number N_0:

$$\frac{3M}{\rho_m}\left(\frac{n^2 - 1}{n^2 + 2}\right) = N_0\alpha = \frac{4\pi N_0 r_0}{k_0^2 - k^2 - ik\Gamma}. \qquad (22\text{--}55)$$

For a given frequency, the right side of Eq. (22–55) depends only on the atomic properties of the material and is thus independent of the density. Since it is referred to a gram molecule, the quantity on the left is called the "molar refraction," and we see that it is equal to the polarizability per gram molecule.

Equations (22–52), (22–54), and (22–55) can be generalized to take account of the fact that all electrons may not have the same binding energy. If a fraction of the electrons f_i has a binding frequency ω_{0i} and a damping width γ_i, Eq. (22–54), for example, becomes

$$\frac{n^2 - 1}{n^2 + 2} = \frac{Ne^2}{3\epsilon_0 m} \sum_i \frac{f_i}{\omega_{0_i}^2 - \omega^2 + i\omega\gamma_i}. \tag{22–56}$$

There are several practical examples of the case where the volume distribution consists of free electrons: refraction of electromagnetic waves by the ionosphere, re-radiation by a "plasma" in an electric discharge, and even the refraction of x-rays, since the resonance frequencies of electrons in light atoms are so small in comparison with x-ray frequencies that the electrons can be considered free. For free electron systems that are dilute, Eq. (22–52) becomes

$$n^2 = 1 - \frac{Ne^2}{\epsilon_0 m\omega^2}. \tag{22–57}$$

In terms of the reduced free-space wavelength $\lambda = c/\omega$, and the classical electron radius $r_0 = e^2/4\pi\epsilon_0 mc^2$, Eq. (22–57) can be written

$$n^2 = 1 - 4\pi N r_0 \lambda^2. \tag{22–58}$$

The index of refraction of a volume distribution of free electrons is always less than unity, although the group velocity is less than c. However, for wavelengths longer than a certain value, $\lambda > (4\pi N r_0)^{-1/2}$, n^2 is negative and the refractive index pure imaginary. For such frequencies, free electron media can exhibit total reflection at all angles, behaving somewhat like a waveguide for frequencies lower than the cutoff frequency.

In Section 11–8 we introduced the concept of phase and group velocity; since the index n is a measure of the phase velocity, and since its frequency dependence has been discussed above, we can calculate the group velocity by Eq. (11–69).

If the phase velocity is a slowly varying function of the frequency, a pulse may travel through a dispersive medium with relatively little change, and the group velocity, defined for its average of maximum amplitude frequency, is the rate of propagation of energy. In the region of anomalous dispersion, however, $dn/d\lambda$ is positive, and the group velocity may easily be greater than c. For any particular frequency it is more reasonable to define the mean velocity of energy transport as the ratio of the time average of the Poynting vector to that of the energy density. This velocity, which is called the energy velocity, actually has a minimum at ω_0.

The practical difficulty of describing the transmission of a signal in the region of anomalous dispersion arises from the fact that the group, or packet, is highly distorted, so that the concept of group velocity as defined in Section 11–8 is no longer valid. The problem corresponding to a plane wave that starts through a medium at a particular time has been worked

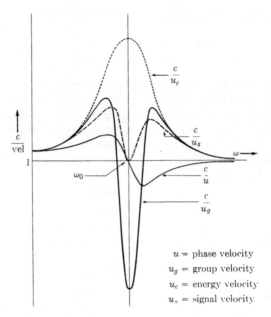

FIG. 22–4. Behavior of various velocities near a resonance frequency (after Brillouin).

out by Sommerfeld and by Brillouin, and details may be found in Volume II of *Congrès International d'Electricité*, Paris, 1932.* The definition of a signal velocity depends on both the nature of the pulse and the nature of the signal detector, but careful analysis shows that it is never greater than c. Figure 22–4 indicates the behavior of the four velocities in the neighborhood of an absorption frequency ω_0.

22–8 Scattering from a volume distribution. Rayleigh scattering. That part of the scattered radiation coherent with the incoming wave and in the same direction has been successfully treated by means of the total polarization of the assemblage. For the scattering at any other angle it is, in general, necessary to superpose the individual wavelets with the proper phases. If the electrons are distributed completely at random it may be assumed that the phase differences are as often positive as negative and thus average out, so that the total scattered intensity is simply

---

* The method is to make a Fourier analysis of the incident pulse, let each component be propagated with the phase velocity $u(\omega)$, and resynthesize the pulse.

the sum of the individual scattered intensities. If the scatterers are regularly placed, as in a crystal, this is no longer true, and the effect is precisely that of a three-dimensional diffraction grating: in any direction for which successive contributions from neighboring scattering centers are in phase there is a maximum of intensity. For x-rays in ordinary crystals there may be many such directions, giving rise to what is known as Bragg diffraction.* If the wavelength is long in comparison with the distance between neighbors, however, the condition for a maximum cannot be fulfilled except in the forward direction, just as a one-dimensional diffraction grating does not produce a pattern if the grating space is much smaller than λ.

In connection with this last point, there arises a question about random scatterers. Consider a cubical volume whose edge is half a wavelength, and which contains many atoms or molecules. Even though the electrons may be distributed randomly within the cube, its radiation will be on the average out of phase with that from a similar neighboring volume in a line of observation at right angles to the beam. The net observed scattering from many such volumes should then be zero from any angle, except, of course, the straightforward direction. Now these are the conditions on atmospheric scattering of visible light—why, then, is Rayleigh's result in accord with observation?

The answer to this question has been given by Einstein and Smolukowski on the basis of a study of density fluctuations in the atmosphere. The number of scatterers per "cell," or per cube of dimension $\lambda/2$, is constant only on the average, and the fluctuations in this number are responsible for the scattering. If N_j is the number of scattering electrons per cell, and δN_j is the difference of the actual number from the mean, the scattering intensity is proportional to

$$\left| \sum_j e^{i\varphi_j} \, \delta N_j \right|^2 = \sum_j (\delta N_j)^2 + \sum_{j \neq k} e^{i(\varphi_j - \varphi_k)} \, \delta N_j \, \delta N_k, \qquad (22\text{--}59)$$

where φ_j is the phase of the jth cell. To make a comparison with observation, we must take the average of Eq. (22–59). In dilute gases the fluctuations of different cells are independent, and since the average of δN_j is zero, by definition, the second term on the right cancels out. If the fluc-

---

* Actually the three-dimensional nature of a crystal "grating" has the consequence that constructive interference is obtained only at a given wavelength *and* a given angle from any set of planes containing regularly spaced scatterers. Thus a single crystal must be rotated during the exposure to monochromatic x-rays if all the regularities are to be evidenced in the diffraction pattern. In a one- or two-dimensional grating there is always an angle at which constructive interference at a given wavelength is obtained if the wavelength is sufficiently short.

tuations are small the "dispersion" (the square of what is called in probability theory the root-mean-square deviation) is given by

$$\overline{(\delta N_j)^2} = N_j. \tag{22-60}$$

Therefore the total scattering is simply proportional to $\Sigma N_j = N$, the total number of scattering centers, in agreement with Rayleigh's assumption. For denser media, the evaluation of the average of Eq. (22-59) is more complicated, and near a chemical phase change these fluctuations give rise to what is called critical opalescence.

22-9 The dispersion relation. A general relation between refraction and absorption (i.e., between the real and imaginary parts of the dielectric constant) may be derived from a consideration of the Fourier analysis of the polarization produced by an external field, in which the polarizability is allowed to depend on the frequency in an unspecified way. This relation was first derived from classical theory by Kramers,[*] after being suggested by analogy with quantum theory considerations.

The proof is most straightforward in terms of the real and imaginary parts of the dielectric constant. Let

$$\kappa = \kappa_1 + i\kappa_2, \tag{22-61}$$

where κ_1 and κ_2 are real. Assume that a particular molecule or other particle is subject to a field

$$\mathbf{E}(t) = \int_{-\infty}^{\infty} \mathbf{E}_\omega e^{-i\omega t} \, d\omega. \tag{22-62}$$

Since $\mathbf{E}(t)$ is real, $\mathbf{E}_{-\omega} = \mathbf{E}_\omega^*$, the electric dipole moment produced in the particle is

$$\mathbf{p}(t) = \frac{\epsilon_0}{N} \int_{-\infty}^{\infty} [\kappa(\omega) - 1]\mathbf{E}_\omega e^{-i\omega t} \, d\omega, \tag{22-63}$$

where N is the number of particles per unit volume. The condition which must be satisfied if $\mathbf{p}(t)$ is to be real is

$$\kappa(-\omega) = [\kappa(\omega)]^*,$$

or

$$\kappa_1(-\omega) = \kappa_1(\omega), \tag{22-64}$$

$$\kappa_2(-\omega) = -\kappa_2(\omega).$$

---

[*] H. A. Kramers, *Atti. Congr. Internat. Fisici, Como* **2**, 545–557 (1927); or *Collected Scientific Papers*, Amsterdam, North Holland Pub. Co., 1956.

An additional condition is that the polarization must not precede in time the field which produces it. It is this condition which leads to the desired relation, as a result of the fact that it amounts to a demand that $\kappa(\omega)$ have no singularities in the upper half of the complex frequency plane. This can be seen by taking the field as a very short pulse at $t = 0$, for which the Fourier inversion

$$\mathbf{E}_\omega = \frac{1}{2\pi} \int_{-\infty}^{\infty} \mathbf{E}(t) e^{i\omega t}\, dt \tag{22–65}$$

is a constant. The demand is then that

$$\int_{-\infty}^{\infty} [\kappa(\omega) - 1] e^{-i\omega t}\, d\omega = 0, \qquad \text{for} \qquad t < 0. \tag{22–66}$$

We have noted in Section 22–5 that for our model of polarization of a system of oscillators, $\kappa \to 1$ as $\omega \to \infty$. We can infer on physical grounds that this relation is general, since it states that for frequencies above any binding frequency the electrons act as if they are effectively free and thus the amplitude of their motion is limited by their inertia only. For negative t the path of integration may be extended to an infinite semicircle encompassing the upper half-plane, since the real negative exponential will cause the integrand to vanish over an infinite circular path in the upper half of the complex plane. Since this is true for all negative t, the sum of the residues of the function $(\kappa_1 + i\kappa_2 - 1)e^{-i\omega t}$ in the upper half-plane must vanish, and therefore $(\kappa_1 + i\kappa_2 - 1)$ is analytic throughout the half plane. A rigorous proof of this conclusion is given by Titchmarsh,* and has been discussed by Jauch and Rohrlich.† A discussion of the asymptotic behavior of the dielectric constant required for validity of this derivation will also be found in these references; the conditions are satisfied by the actual behavior of the dielectric constant as given by Eq. (22–52).

Let us choose a real positive value of ω, and for convenience designate the variable of integration by ω'. If the function $[\kappa_1(\omega') + i\kappa_2(\omega') - 1]$ is analytic, it follows from Cauchy's theorem that

$$\oint \frac{\kappa_1 + i\kappa_2 - 1}{\omega' - \omega}\, d\omega' = 0 \tag{22–67}$$

if the contour does not include the point $\omega' = \omega$. If we choose the contour

---

* E. C. Titchmarsh, *Introduction to the Theory of Fourier Integrals*, Oxford University Press, 1937.

† J. M. Jauch and F. Rohrlich, *The Theory of Photons and Electrons*, Appendix A7, pp. 470–473.

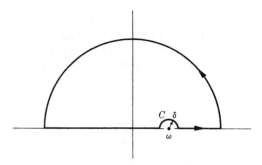

FIG. 22–5. Path of integration in the complex ω' plane for $t < 0$. The contour C is a semicircle of radius δ about the point $\omega' = \omega$ in the upper half-plane.

of Fig. 22–5 it is clear that

$$\int_{-\infty}^{\infty} \frac{\kappa(\omega') - 1}{\omega' - \omega}\, d\omega' = 0, \tag{22–67'}$$

where the integral from $-\infty$ to $+\infty$ is along the real axis except for the small semicircle of radius δ which excludes from the contour the singularity at $\omega' = \omega$. Condition (22–67') then becomes

$$\int_{-\infty}^{0} \frac{\kappa(\omega') - 1}{\omega' - \omega}\, d\omega' + \int_{0}^{\infty} \frac{\kappa(\omega') - 1}{\omega' - \omega}\, d\omega' + \int_{C} \frac{\kappa(\omega') - 1}{\omega' - \omega}\, d\omega' = 0, \tag{22–68}$$

where the Cauchy principal value is understood by the integral from 0 to ∞, i.e.,

$$\int_{0}^{\infty} f(\omega, \omega')\, d\omega' = \lim_{\delta \to 0} \left[\int_{0}^{\omega-\delta} f(\omega, \omega')\, d\omega' + \int_{\omega+\delta}^{\infty} f(\omega, \omega')\, d\omega' \right]; \tag{22–69}$$

and where C is the semicircle of radius δ in the upper half-plane about the point $\omega' = \omega$. By utilizing the relations (22–64) and the smallness of δ, we can write Eq. (22–68) as

$$\int_{0}^{\infty} \frac{\kappa_1(\omega') - i\kappa_2(\omega') - 1}{-\omega' - \omega}\, d\omega' + \int_{0}^{\infty} \frac{\kappa_1(\omega') + i\kappa_2(\omega') - 1}{\omega' - \omega}\, d\omega'$$

$$+ [\kappa_1(\omega) + i\kappa_2(\omega) - 1] \int_{C} \frac{d\omega'}{\omega' - \omega}$$

$$= 2\int_{0}^{\infty} \frac{\omega[\kappa_1(\omega') - 1] + i\omega'\kappa_2(\omega')}{\omega'^2 - \omega^2}\, d\omega' - \pi i[\kappa_1(\omega) + i\kappa_2(\omega) - 1] = 0, \tag{22–70}$$

since the integral around C is $-\pi i$ by Cauchy's theorem. Equating the real and imaginary parts of Eq. (22–70) to zero yields the desired relations:

$$\kappa_1(\omega) - 1 = \frac{2}{\pi} \int_0^\infty \frac{\omega' \kappa_2(\omega')}{\omega'^2 - \omega^2} \, d\omega', \qquad (22\text{--}71)$$

$$\kappa_2(\omega) = -\frac{2\omega}{\pi} \int_0^\infty \frac{\kappa_1(\omega') - 1}{\omega'^2 - \omega^2} \, d\omega'. \qquad (22\text{--}72)$$

Again we note that the integrals are to be interpreted as principal values, in the sense of Eq. (22–69). It is clear that the relations depend only on the reality of the polarization and the condition that the polarization cannot precede in time the field which produces it.

Equations (22–71) and (22–72) are known as the dispersion relations. We shall see in the following section that the imaginary part κ_2 is related to the total cross sections for scattering and absorption of the incident waves; Eq. (22–71) thus expresses the real part of the dielectric constant as an integral over the total cross sections for scattering and absorption over all frequencies.

22–10 A general theorem on scattering and absorption. Many of the results obtained in this chapter may also be derived from the "shadow theorem," a general theorem concerning the scattering and absorption of radiation by an object or a collection of objects. Let us assume a complex representation for the fields, such that the time dependence for a single frequency may be represented by the factor $e^{-i\omega t}$. We may use the macroscopic field equations and allow for the possibility of absorption through the conductivity σ. In general, with the assumption of linear constitutive equations, the Maxwell relations

$$\boldsymbol{\nabla} \times \mathbf{E} = -\mu \frac{\partial \mathbf{H}}{\partial t},$$

$$\boldsymbol{\nabla} \times \mathbf{H} = \epsilon \frac{\partial \mathbf{E}}{\partial t} + \sigma \mathbf{E}, \qquad (22\text{--}73)$$

are also true for $\mathbf{E}^*$ and $\mathbf{H}^*$, the complex conjugates of $\mathbf{E}$ and $\mathbf{H}$. By a procedure entirely analogous to that of Section 10–5, it can be shown that

$$\boldsymbol{\nabla} \cdot \left(\frac{\mathbf{E} \times \mathbf{H}^* + \mathbf{E}^* \times \mathbf{H}}{4} \right) = -\frac{\partial}{\partial t} \left(\frac{\mu |\mathbf{H}|^2 + \epsilon |\mathbf{E}|^2}{4} \right) - \frac{\sigma |\mathbf{E}|^2}{2}. \qquad (22\text{--}74)$$

The numerical factor has been adjusted so that $\frac{1}{4}(\mathbf{E} \times \mathbf{H}^* + \mathbf{E}^* \times \mathbf{H})$ represents the (real) density of energy flow; $\frac{1}{4}(\mu |\mathbf{H}|^2 + \epsilon |\mathbf{H}|^2)$ is the energy density; and $\frac{1}{2}\sigma |\mathbf{E}|^2$ is the rate of energy absorption per unit volume.

Now let us consider a plane wave, traveling in the positive z-direction, polarized so that $\mathbf{E}$ is along x, which encounters scattering and absorbing objects in a finite region of space near the origin, as indicated in Fig. 22–6. The coordinate unit vectors are $\hat{\mathbf{x}}$, $\hat{\mathbf{y}}$, $\hat{\mathbf{z}}$; and $\mathbf{n}$ is a unit vector in the direction specified by the polar angle θ and the azimuthal angle φ. For a scattered wave making an angle θ with the z-axis, $\mathbf{n} \cdot \hat{\mathbf{z}} = \cos \theta$. Asymptotically the total field at large distances from the origin may be written

$$
\begin{aligned}
\mathbf{E} &= E_0\left[\hat{\mathbf{x}}e^{ik\hat{\mathbf{z}}\cdot\mathbf{r}} + \frac{1}{kr}\mathbf{F}(\theta, \varphi)e^{ikr}\right], \\
\mathbf{H} &= \left(\frac{\epsilon_0}{\mu_0}\right)^{1/2}E_0\left\{(\hat{\mathbf{z}} \times \hat{\mathbf{x}})e^{ik\hat{\mathbf{z}}\cdot\mathbf{r}} + \frac{1}{kr}[\mathbf{n} \times \mathbf{F}(\theta, \varphi)]e^{ikr}\right\},
\end{aligned}
\tag{22-75}
$$

where $\mathbf{F}(\theta, \varphi)$ is the vector amplitude of the scattered electric field, in general complex, relative to the magnitude E_0 of the incoming wave. Note that $\mathbf{n} \cdot \mathbf{F}(\theta, \varphi) = 0$.

Using the fields of Eq. (22–75), let us now integrate Eq. (22–74) over the volume of a sphere containing the objects which affect the incoming wave, and sufficiently large that the asymptotic form of the scattered wave is valid. Under conditions of a steady state the time rate of change of energy density of the fields within the sphere is zero, so that

$$
\begin{aligned}
&\int \frac{1}{4}(\mathbf{E} \times \mathbf{H}^* + \mathbf{E}^* \times \mathbf{H}) \cdot d\mathbf{S} = -\int \frac{1}{2}\sigma|\mathbf{E}|^2\,dv \\
&= \frac{1}{4}\left(\frac{\epsilon_0}{\mu_0}\right)^{1/2}E_0^2\int_S\left[2\hat{\mathbf{z}} + \{[\hat{\mathbf{x}} \times (\mathbf{n} \times \mathbf{F}^*)] + [\mathbf{F}^* \times (\hat{\mathbf{z}} \times \hat{\mathbf{x}})]\}\,\frac{e^{ikr(1-\cos\theta)}}{kr}\right. \\
&\quad + \{[\mathbf{F} \times (\hat{\mathbf{z}} \times \hat{\mathbf{x}})] + [\hat{\mathbf{x}} \times (\mathbf{n} \times \mathbf{F})]\}\,\frac{e^{ikr(1-\cos\theta)}}{kr} \\
&\quad \left. + \left(\frac{1}{kr}\right)^2\{[\mathbf{F}^* \times (\mathbf{n} \times \mathbf{F})] + [\mathbf{F} \times (\mathbf{n} \times \mathbf{F}^*)]\}\right] \cdot d\mathbf{S}.
\end{aligned}
\tag{22-76}
$$

The surface integral is readily simplified. The integral of $\hat{\mathbf{z}}$ vanishes over a closed sphere. The terms linear in $\mathbf{F}$ and $\mathbf{F}^*$ do not differ appreciably from zero except in the forward direction ($\theta = 0$) since the exponential factor $\exp[ikr(1 - \cos\theta)]$ oscillates rapidly for large kr as a function of θ except near $\cos\theta = 1$ [cf. Eq. (14–47)]. Therefore,

$$
\begin{aligned}
&\int_0^{2\pi}\int_0^\pi \{[\hat{\mathbf{x}} \times (\mathbf{n} \times \mathbf{F})] + [\mathbf{F} \times (\hat{\mathbf{z}} \times \hat{\mathbf{x}})]\}\,\frac{e^{ikr(1-\cos\theta)}}{kr}\,r^2\sin\theta\,d\theta\,d\varphi \\
&\cong -\frac{2}{ik^2}\int_0^{2\pi}d\varphi(\mathbf{F} \cdot \hat{\mathbf{x}})_{\theta=0} = \frac{4\pi i}{k^2}(\mathbf{F} \cdot \hat{\mathbf{x}})_{\theta=0},
\end{aligned}
\tag{22-77}
$$

since

$$[\hat{\mathbf{x}} \times (\mathbf{n} \times \mathbf{F})] + [\mathbf{F} \times (\hat{\mathbf{z}} \times \hat{\mathbf{x}})] = 2(\mathbf{F} \cdot \hat{\mathbf{x}}), \quad \text{for} \quad \theta = 0. \quad (22\text{–}78)$$

Analogous considerations apply to the terms linear in $\mathbf{F}^*$, and

$$\left(\frac{1}{kr}\right)^2 \int \{[\mathbf{F}^* \times (\mathbf{n} \times \mathbf{F})] + [\mathbf{F} \times (\mathbf{n} \times \mathbf{F}^*)]\} \cdot d\mathbf{S}$$
$$= \frac{2}{k^2} \int_0^{2\pi} d\varphi \int_0^{\pi} |\mathbf{F}|^2 \sin\theta \, d\theta, \quad (22\text{–}79)$$

since

$$\mathbf{F} \times (\mathbf{n} \times \mathbf{F}^*) = \mathbf{n}(\mathbf{F} \cdot \mathbf{F}^*) = \mathbf{F}^* \times (\mathbf{n} \times \mathbf{F}). \quad (22\text{–}80)$$

Thus Eq. (22–76) becomes

$$\left(\frac{\epsilon_0}{\mu_0}\right)^{1/2} E_0^2 \left[\frac{1}{k^2} \int_0^{2\pi} d\varphi \int_0^{\pi} |\mathbf{F}|^2 \sin\theta \, d\theta + \frac{2\pi i}{k^2} \hat{\mathbf{x}} \cdot (\mathbf{F} - \mathbf{F}^*)_{\theta=0}\right]$$
$$= - \int \sigma |\mathbf{E}|^2 \, dv. \quad (22\text{–}81)$$

The scattering cross section per unit solid angle is the energy scattered per unit solid angle per unit incident energy per unit area. Inspection of Eq. (22–75) shows that this quantity is

$$\frac{d\sigma_{\text{scat}}}{d\Omega} = \frac{1}{k^2} |\mathbf{F}(\theta, \varphi)|^2, \quad (22\text{–}82)$$

and thus,

$$\sigma_{\text{scat}} = \frac{1}{k^2} \int_0^{2\pi} d\varphi \int_0^{\pi} |\mathbf{F}|^2 \sin\theta \, d\theta. \quad (22\text{–}83)$$

The cross section for absorption σ_{abs} is the energy dissipated (assumed here by conduction) per unit incident energy per unit area: thus,

$$\sigma_{\text{abs}} = \left(\frac{\mu_0}{\epsilon_0}\right)^{1/2} \frac{1}{E_0^2} \int \sigma |\mathbf{E}|^2 \, dv \quad (22\text{–}84)$$

is the cross section for absorption by these same objects. Therefore,

$$\sigma_{\text{total}} = \sigma_{\text{scat}} + \sigma_{\text{abs}} = \frac{2\pi}{ik^2} \hat{\mathbf{x}} \cdot (\mathbf{F} - \mathbf{F}^*)_{\theta=0}$$
$$= \frac{4\pi}{k^2} \, \text{Im} \, (\mathbf{F})_{\theta=0}, \quad (22\text{–}85)$$

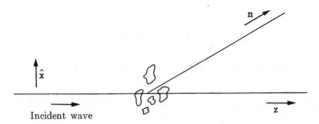

FIG. 22–6. Scattering and absorption of a plane wave by objects near the origin of coordinates.

i.e., the sum of the cross sections for scattering and absorption is proportional to the imaginary part of the scattered amplitude in the forward direction. The corresponding theorem for particle waves is well known in quantum mechanics.* The theorem of Eq. (22–85) is often called the "optical theorem."

The forward scattering amplitudes are related to the refractive index. Consider, for example, the scatterer in Fig. 22–6 to be in the form of a thin rod parallel to the $\hat{\mathbf{x}}$-axis and to have an induced dipole moment $\mathbf{p}$ given by

$$\mathbf{p} = \epsilon_0(\kappa - 1)V\mathbf{E}_{\text{incident}} = \epsilon_0(\kappa - 1)VE_0\hat{\mathbf{x}}, \qquad (22\text{–}86)$$

where V is the volume of the scatterer; it is assumed that the wavelength is large compared to the dimensions of the scatterer. The radiation field from the scatterer in the forward direction can be calculated from the dipole radiation formulas of Section 14–7; comparison of Eqs. (14–79) and (22–75) yields, for the forward direction,

$$E_0 \frac{\mathbf{F}(\theta, \varphi)}{kr} = \frac{\mathbf{p}k^2}{4\pi\epsilon_0 r} ; \qquad (22\text{–}87)$$

or, from Eq. (22–86),

$$\mathbf{F} = \left(\frac{\kappa - 1}{4\pi}\right) k^3 V\hat{\mathbf{x}}. \qquad (22\text{–}88)$$

Hence, if we write $\kappa = \kappa_1 + i\kappa_2$ as in Section 22–6, we can write (22–85) simply as

$$\sigma_{\text{total}} = Vk\kappa_2. \qquad (22\text{–}89)$$

---

* In fact, the theorem was first derived for particle waves: E. Feenberg, *Phys. Rev.* **40**, 40 (1932).

The dispersion relation (22–71) is thus:

$$\kappa_1(\omega) = 1 + \frac{2c}{\pi V} \int_0^\infty \frac{\sigma_{\text{total}}(\omega')}{\omega'^2 - \omega^2}\, d\omega', \qquad (22\text{–}90)$$

where the integral is to be interpreted as its principal value, and where σ_{total} is the total cross section of a scatterer of volume V and of shape as described above, but made of a material whose "bulk" dielectric constant has a real part κ_1.

Suggested References

R. Becker, *Theorie der Elektrizität*, Band II. Sections B, on the elastically bound electron, and C III, on dispersion and absorption, are highly recommended.

L. Landau and E. Lifshitz, *The Classical Theory of Fields*. An excellent account of radiation and scattering will be found in Chapter 9.

L. Landau and E. Lifshitz, *Electrodynamics of Continuous Media*. The dispersion relation is derived in Chapter IX, and Chapter XIV is devoted to scattering.

H. A. Lorentz, *Problems of Modern Physics*. Many of the subjects covered in this series of lectures given in 1922 are still of great interest, but the work is cited here because it includes the application of Rayleigh's formula to the atmosphere.

L. Rosenfeld, *Theory of Electrons*. Probably the best account from a modern point of view of the classical theory of dispersion, including critical opalescence.

J. A. Stratton, *Electromagnetic Theory*. Includes a treatment of ionosphere transmission.

H. W. Bode, *Network Analysis and Feedback Amplifier Design*. This book discusses in a practical way how real and imaginary parts of any transfer function are related by equations similar to the dispersion relations discussed here.

Exercises

1. Show that the scattering cross section of a free electron for an elliptically polarized wave (electric field $\mathbf{E} = \mathbf{A} \cos \omega t + \mathbf{B} \sin \omega t$, where $\mathbf{A}$ and $\mathbf{B}$ are mutually perpendicular vector amplitudes) is given by

$$\frac{d\sigma}{d\Omega} = r_0^2 \left[\frac{(\mathbf{A} \times \mathbf{n})^2 + (\mathbf{B} \times \mathbf{n})^2}{A^2 + B^2} \right],$$

where $\mathbf{n}$ is a unit vector in the direction of the radiation.

2. What is the minimum frequency that can be propagated in free space without attenuation, assuming one electron per cubic centimeter? Find the group velocity for higher frequencies.

3. Show that in the presence of a magnetic induction field $\mathbf{B}$ in the direction of wave propagation, an electron gas exhibits two velocities with indexes of refraction given by

$$n_\pm^2 = 1 - \frac{(Ne^2/m\epsilon_0\omega^2)}{1 \pm (e|\mathbf{B}|/m\omega)} = 1 - \frac{4r_0N/k^2}{1 \pm (\omega_L/\omega)},$$

where ω_L is the Larmor frequency.

4. Make a sufficiently accurate plot of the wave and group velocities and the absorption coefficient near a resonance frequency to indicate the essential features, showing that the maximum and minimum of the index of refraction occur at frequencies where the absorption coefficient has half its maximum value.

5. Compute the rate of energy transport, i.e., the time average of the Poynting vector over that of the energy density, at $\omega = \omega_0$.

6. How much polarization is to be expected in sky light scattered at angle θ? Assume unpolarized light from the sun and take account only of single scattering.

7. Consider a nucleus of N neutrons and Z protons, and let $A = N + Z$. Show that under the assumption of pure electric dipole absorption, the "integrated cross section" is given by

$$\int \sigma \, dE = 2\pi^2 \frac{NZ}{A} \alpha \left(\frac{\hbar}{Mc}\right)^2 Mc^2,$$

where α is the fine-structure constant, $\simeq 1/137$, and M is the proton mass, approximately equal to that of the neutron. Evaluate this expression numerically.

8. Derive Eq. (22–77).

9. From Eqs. (22–19) and (22–22), calculate the scattering amplitude $\mathbf{F}$ of a bound electron. Compute the total cross section from the "optical theorem," Eq. (22–85), and compare with Eq. (22–30).

CHAPTER 23

THE MOTION OF CHARGED PARTICLES
IN ELECTROMAGNETIC FIELDS

The fundamental problem of electromagnetic theory deals with the motion of charged particles interacting via electromagnetic forces. The field concept is not necessary in formulating equations for such motion: formulation in terms of action at a distance is possible in a relativistically covariant way,[*] provided that the time-symmetric formalism involving the mean between advanced and retarded fields given by Eq. (21–86) is adopted. Separation of the problem into "radiation" (production of a field by a charge system) and "motion of charged particles in a field" is in a certain sense artificial and introduces some difficulties. In particular, this separation implies that the radiation can be computed from the motion of the system as driven by external forces. As we have seen, there is also the "radiation reaction" which must be taken into account in order that the conservation laws be satisfied. The reactions can be computed from the radiation and applied as corrections to the external forces for a more accurate description, and this process of successive approximations converges unless the variability of the motion is too high. In case the reactions are negligible, however, the separation is possible and generally advantageous, since the sources of the field need not be known in detail for a description of the motion of a charge. We shall deal here with the motion of charged particles only under the conditions that reactions may be neglected.

23–1 World-line description. In the covariant language of Chapters 17 and 18, the equation of motion for a particle of charge e, rest mass m_0, and four-velocity u^i, moving in an external field F^{ij}, is given by

$$m_0 c^2 \frac{du^i}{ds} = e u_j F^{ij}, \qquad (23\text{–}1)$$

corresponding to the three-dimensional form

$$\frac{d\mathbf{p}}{dt} = e(\mathbf{E} + \mathbf{u} \times \mathbf{B}). \qquad (23\text{–}2)$$

---

[*] See, e.g., J. A. Wheeler and R. P. Feynman, *Revs. Modern Phys.* **21,** 425 (1949).

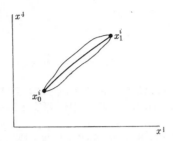

FIG. 23–1. World-line variation of the path of a particle.

Here, as before,

$$\mathbf{p} = m_0\mathbf{u}/\sqrt{1 - u^2/c^2}. \tag{23-3}$$

To find the equations of motion in canonical form, it is customary to proceed by choosing a Lagrangian which yields the correct equations. Here we shall do this in such a way as to emphasize the required invariance properties.

Let the world line of the particle be given by $x^i(s)$, where s is the "proper time" for the particle. Let us look for a Lagrangian L_4 in four-space such that the motion is given by a variational principle,

$$\delta \int_{x_0^i}^{x_1^i} L_4(x^i, u^i) \, ds = 0. \tag{23-4}$$

The variation is taken between fixed world points x_0^i and x_1^i, as indicated in Fig. 23–1. Invariance requires that L_4 be a scalar, and in order to yield a linear equation of motion, the invariants making up L_4 must be not higher than second order in u^i. For a free particle, we are led to choose

$$L_4 = \frac{m_0 c^2}{2} u^i u_i, \tag{23-5}$$

explicit dependence on x^i being excluded, since all world points are equivalent for a free particle. (The coefficient of $u^i u_i$ is arbitrary in this case.) We may introduce a scalar interaction with the field described by a four-potential $\phi^i(x^i)$ by putting for the total Lagrangian

$$L_4 = \frac{m_0 c^2}{2} u^i u_i + e u_i \phi^i. \tag{23-6}$$

[The constants in Eq. (23–6) have been adjusted so that the equations of motion will agree with the usual convention for rest mass and charge.]

Let us now apply the variational principle, Eq. (23–4), to obtain the Eulerian equations:

$$\delta \int_{x_0^i}^{x_1^i} L_4(x^i, u^i)\, ds = \int_{x_0^i}^{x_1^i} \left[m_0 c^2 u_i\, \delta u^i + e \left(\phi_i\, \delta u^i + \frac{\partial \phi_i}{\partial x^j}\, \delta x^j u^i \right) \right] ds.$$

(23–7)

Since

$$\delta u^i = \frac{d(\delta x^i)}{ds} \quad \text{and} \quad \frac{d\phi_i}{ds} = \frac{\partial \phi_i}{\partial x^j}\, u^j,$$

we can integrate by parts to obtain

$$\delta \int_{x_0^i}^{x_1^i} L_4(x^i, u^i)\, ds = \int_{x_0^i}^{x_1^i} \left[-\frac{d}{ds}\, (m_0 c^2 u_j) + e \left(\frac{\partial \phi_i}{\partial x^j} - \frac{\partial \phi_j}{\partial x^i} \right) u^i \right] \delta x^j\, ds.$$

(23–8)

The integrated part does not appear, since the variation of the function vanishes at x_0^i and x_1^i. Since $\delta x^j(s)$ is arbitrary along the path, the bracketed expression must be equal to zero, and hence

$$m_0 c^2 \frac{du_j}{ds} = e F_{ji} u^i, \tag{23–9}$$

which is the covariant equivalent of the contravariant Eq. (23–1).

23–2 Hamiltonian formulation and the transition to three-dimensional formalism. A Hamiltonian in the ordinary sense cannot be readily introduced in covariant formulation, since in the ordinary Hamiltonian equations,

$$\frac{\partial H}{\partial p_\alpha} = \frac{dq_\alpha}{dt}, \qquad \frac{\partial H}{\partial q_\alpha} = -\frac{dp_\alpha}{dt}, \tag{23–10}$$

the time enters asymmetrically. On the other hand, if we introduce a four-Hamiltonian, H_4, which is to obey the relations

$$\frac{\partial H_4}{\partial p^i} = \frac{dq_i}{ds}, \qquad \frac{\partial H_4}{\partial q^i} = -\frac{dp_i}{ds}, \tag{23–11}$$

then H_4 (which is different from H) will be a scalar. To establish a connection between Eq. (23–11) and Eq. (23–10), we must recognize that (23–11) represents eight equations, while (23–10) represents six. We shall show that if we find a scalar $H_4(p^i, q^i)$ such that (23–11) yields the correct equation of the world point, then the magnitude of this scalar is a universal constant for a given particle. If we then solve this equation for $p^4(x_\alpha, p_\alpha, x^4)$, we can show that p^4 obeys Eqs. (23–10) for H. Since p^4 is thus generated by a covariant process, the resultant equations, although not "manifestly" covariant, are nevertheless relativistically correct. We can, of course, add the final test that the correct relativistic equations of motion shall result.

The four-Hamiltonian obeying Eq. (23–11) is defined by

$$H_4 = p^i u_i - L_4, \qquad p^i = \frac{\partial L_4}{\partial u_i}, \qquad (23\text{--}12)$$

or, from Eq. (23–6),

$$p^i = m_0 c^2 u^i + e\phi^i. \qquad (23\text{--}13)$$

Hence,

$$H_4 = \frac{1}{2} m_0 c^2 u^i u_i = \frac{1}{2m_0 c^2} \{[p^i - e\phi^i(x^i)][p_i - e\phi_i(x^i)]\} \qquad (23\text{--}14)$$

gives the explicit functional dependence of H_4 on p^i and x^i. Numerically, however, $u^i u_i = 1$, as can be easily verified by means of the components of u^i, namely, $u^i = (\mathbf{u}/c, 1)\,(1 - u^2/c^2)^{-1/2}$.

The four-Hamiltonian, Eq. (23–14), together with Eq. (23–11), gives the correct equation of motion. By direct differentiation,

$$\frac{\partial H_4}{\partial q^i} = -\frac{e}{m_0 c^2}\,(p^j - e\phi^j)\frac{\partial \phi_j}{\partial x^i}$$

$$= -\frac{dp_i}{ds} = -m_0 c^2 \frac{du_i}{ds} - e\frac{\partial \phi_i}{\partial x^j}\,u^j. \qquad (23\text{--}15)$$

Hence,

$$eu^j\left(\frac{\partial \phi_j}{\partial x^i} - \frac{\partial \phi_i}{\partial x^j}\right) = m_0 c^2 \frac{du_i}{ds}, \qquad (23\text{--}16)$$

in agreement with Eq. (23–1).

Let us now turn to the ordinary three-dimensional representation of the Hamiltonian. We postulate

$$H(p_\alpha, q_\alpha, t) = p^4(p_\alpha, q_\alpha, x^4), \qquad (23\text{--}17)$$

where p^4 is the solution of Eq. (23–14) for $H_4 = \frac{1}{2}m_0 c^2$, i.e.,

$$H(p_\alpha, q_\alpha, t) = e\phi + c\sqrt{(\mathbf{p} - e\mathbf{A})^2 + (m_0 c)^2}. \qquad (23\text{--}18)$$

If we put

$$L = -H + \mathbf{p} \cdot \mathbf{u},$$ (23–19)

$$\mathbf{p} - e\mathbf{A} = \frac{m_0\mathbf{u}}{\sqrt{1 - u^2/c^2}} = m\mathbf{u},$$

we find that

$$L(u, q_\alpha, t) = -e\phi + e(\mathbf{A} \cdot \mathbf{u}) - m_0c^2\sqrt{1 - u^2/c^2}$$ (23–20)

is the ordinary three-dimensional Lagrangian. Note that this Lagrangian fills the necessary relativistic requirement that the action integral,

$$\int L\,dt = -\int (u^i\phi_i + m_0c^2)\,ds,$$ (23–21)

shall be a scalar invariant. It can be easily shown that the three-dimensional Hamiltonian and Lagrangian, Eqs. (23–18) and (23–20) respectively, lead to the equation of motion, (23–2).

It remains to be shown that if $p^4 = H$, then H satisfies Eq. (23–10). According to Eq. (23–11), we can establish the relation between x^4 and s, i.e., between "time" and "proper time":

$$\frac{dx^4}{ds} = +\frac{\partial H_4}{\partial p^4}.$$ (23–22)

Hence, in general,

$$\frac{dx_\alpha}{dx^4} = -\frac{\partial H_4/\partial p_\alpha}{\partial H_4/\partial p_4}, \qquad \frac{dp_\alpha}{dx^4} = +\frac{\partial H_4/\partial x_\alpha}{\partial H_4/\partial p_4}.$$ (23–23)

[Note the change in sign due to the presence of covariant components in Eq. (23–11).] If we set the total derivative of the invariant in Eq. (23–14) equal to zero, we obtain

$$\frac{\partial H_4}{\partial x_\alpha} + \frac{\partial H_4}{\partial p_4}\frac{\partial p_4}{\partial x_\alpha} = 0,$$ (23–24)

$$\frac{\partial H_4}{\partial p_\alpha} + \frac{\partial H_4}{\partial p_4}\frac{\partial p_4}{\partial p_\alpha} = 0.$$ (23–25)

Hence, with Eq. (23–23),

$$\frac{dx_\alpha}{dt} = \frac{\partial p_4}{\partial p_\alpha},$$ (23–26)

$$\frac{dp_\alpha}{dt} = -\frac{\partial p_4}{\partial x_\alpha}.$$ (23–27)

Thus $p_4(x_\alpha, p_\alpha)$ satisfies the ordinary three-dimensional Hamiltonian equation, and the postulated Eq. (23–17) leads to a relativistically correct Hamiltonian and to a correspondingly correct Lagrangian.

23–3 Equations for the trajectories. Thus far we have assembled the various relativistically correct expressions for integrating the equations of motion with respect to the time. Frequently, however, there are cases where it is desirable to obtain equations for the particle orbits, without regard to the time involved in the course of the path. These can be found by integrating the equations of motion in time and then eliminating the time but, in general, it is less tedious to obtain directly differential equations for the orbits in space. This can be accomplished most easily by means of the "principle of least action."

It should be recalled that in classical mechanics there are two variational principles in general use: (1) Hamilton's principle,

$$\delta \int_{t_1}^{t_2} L \, dt = 0,$$

and (2) the principle of least action,

$$\delta \int_{x_{\alpha_1}}^{x_{\alpha_2}} \mathbf{p} \cdot d\mathbf{l} = 0.$$

These principles are quite different in their physical content. We have already made use of Hamilton's principle, in which the variation is taken between two unvaried points in space and time; the varied paths will not obey the equations of motion or the conservation laws. In the principle of least action, the end points are fixed points in space but *not* in time; however, the varied path does obey the law of conservation of energy.

Together with Eq. (23–19) the principle of least action gives

$$\delta \int_{x_{\alpha_1}}^{x_{\alpha_2}} \mathbf{p} \cdot d\mathbf{l} = \delta \int_{x_{\alpha_1}}^{x_{\alpha_2}} (m\mathbf{u} + e\mathbf{A}) \cdot d\mathbf{l}$$

$$= \delta \int_{x_{\alpha_1}}^{x_{\alpha_2}} \left(\frac{m_0 \mathbf{u}}{\sqrt{1 - u^2/c^2}} + e\mathbf{A} \right) \cdot d\mathbf{l} = 0. \quad (23\text{–}28)$$

Let us now make one of the three coordinates, say x, the independent variable, and let y', z' be the slopes dy/dx, dz/dx, respectively. Equation (23–28) then becomes

$$\delta \int_{x_1}^{x_2} [mu \sqrt{1 + y'^2 + z'^2} + e(A_x + A_y y' + A_z z')] \, dx = 0. \quad (23\text{–}29)$$

The Eulerian equations of this system will yield the differential equations for the orbit if the magnitude of the velocity u is known as a function of x. Since in the principle of least action the varied paths obey the law of conservation of energy, we can express u in terms of the kinetic energy T and the rest energy E_0 of the particle, i.e.,

$$\frac{mcu}{E_0} = \frac{u}{c} \frac{1}{\sqrt{1 - u^2/c^2}} = \sqrt{T^2 + 2TE_0}/E_0.$$

In terms of $\tau = T/E_0$ the variational equation is then given by

$$\delta \int_{x_1}^{x_2} \left[\sqrt{2\tau(1 + \tau/2)(1 + y'^2 + z'^2)} + \frac{ec}{E_0} (A_x + A_y y' + A_z z') \right] dx = 0. \tag{23-30}$$

The principal application of these considerations is for the calculation of orbits in electric and magnetic lens structures. This is a fairly specialized subject, and only solutions of general interest will be discussed here. For more detailed information the reader is referred to the electron optics literature, some of which is listed at the end of the chapter.

In electron and ion optics one usually deals with orbits in the proximity of an axis. Let us first consider the case of radial symmetry about such an axis, i.e., the only component of $\mathbf{A}$ is A_θ, which is independent of θ. If, in cylindrical coordinates, we make z the independent variable, with $r' = dr/dz$ and $\theta' = d\theta/dz$, Eq. (23-30) becomes

$$\delta \int_{z_1}^{z_2} \left[\sqrt{2\tau(1 + \tau/2)}(r'^2 + r^2\theta'^2 + 1)^{1/2} + \frac{ec}{E_0} A_\theta r\theta' \right] dz = 0.$$

Eulerian equations for $r(z)$ and $\theta(z)$ can then be generated in the usual way. A first integral of the equation in $\theta(z)$ can be found immediately and leads to the law of "conservation of canonical angular momentum" in the form

$$\frac{r^2\theta'\sqrt{2\tau(1 + \tau/2)}}{(r'^2 + r^2\theta'^2 + 1)^{1/2}} + \frac{ec}{E_0} A_\theta r = C. \tag{23-31}$$

It can be shown that Eq. (23-31) is equivalent to

$$mr^2\dot{\theta} + eA_\theta r = C', \tag{23-32}$$

where m is the relativistic mass. Hence a particle emitted in a magnetic field will attain mechanical angular momentum when leaving the field; a particle passing through a magnetic field will experience a change in mechanical angular momentum, but this will be restored to its initial value on leaving the field.

Equation (23–31) can be substituted into the Eulerian differential equation for $r(z)$ in order to eliminate dependence on θ'. The resulting equation is algebraically complicated, but it can be stated simply by means of a change in parameters. Let

$$\xi = 2\tau(1 + \tau/2),$$

$$\eta = \frac{1}{\xi^{1/2}}\left(\frac{C}{r} - \frac{ec}{E_0}A_\theta\right), \qquad (23\text{–}33)$$

$$\lambda = \xi(1 - \eta^2),$$

where C is the constant of Eq. (23–31). Then, without any approximation, the Eulerian equation becomes

$$r'' = \frac{1 + r'^2}{2\lambda}\left[(1 + r'^2)\frac{\partial\lambda}{\partial r} - r'\frac{\partial\lambda}{\partial z}\right]. \qquad (23\text{–}34)$$

This equation is more general than is usually necessary. In most cases, λ has a simple form; e.g., for a purely electrostatic system and low velocity motion λ is just the kinetic energy and hence the electrostatic potential.

Equation (23–34) reduces to a much simpler form if one considers "paraxial" motion, i.e., motion in which the inclination of rays to the axis never becomes large. In this case, two approximations are possible: we can ignore r'^2, and we can relate $\partial/\partial r$ to $\partial/\partial z$ by the field equations. Specifically, we have, near the axis,

$$\frac{\partial\phi}{\partial r} = -\frac{r}{2}\frac{\partial^2\phi}{\partial z^2}, \qquad (23\text{–}35)$$

and

$$A_\theta = \frac{Br}{2}, \qquad (23\text{–}36)$$

correct to order r^2, and we need consider the values of ϕ and B only on the axis. With these approximations, Eq. (23–34) becomes

$$r'' + r'\frac{\tau'}{2\tau}\left(\frac{1 + \tau}{1 + \tau/2}\right) + \frac{r}{4}\frac{\tau''}{\tau}\left(\frac{1 + \tau}{1 + \tau/2}\right) + \frac{r}{8}\frac{(eB/m_0c)^2}{\tau(1 + \tau/2)}$$

$$-\frac{C^2}{2\tau(1 + \tau/2)r^3} = 0. \quad (23\text{–}37)$$

If we put $\beta = u/c$ and $\gamma = 1/\sqrt{1 - u^2/c^2}$, Eq. (23–37) takes the simpler form

$$r'' + \frac{r'\gamma'}{\beta^2 r} + \frac{r\gamma''}{2\beta^2\gamma} + \frac{r}{4}\left(\frac{Be}{m_0c\beta\gamma}\right)^2 - \frac{C^2}{\beta^2\gamma^2 r^3} = 0. \qquad (23\text{–}38)$$

The physical content of Eq. (23–38) is clear. The first three terms correspond to an electric lens system with orbits of zero angular momentum. The last term adds a "centrifugal potential" or an effective "repulsive core" to the system in case the canonical angular momentum is different from zero. The fourth term represents the lens action of a magnetic field. We shall consider the effects separately.

23–4 Applications.

1. *Electrostatic lens, no "eccentric" rays, $B = 0 = C$.* In this case, Eq. (23–28) becomes

$$r'' + \frac{r'\gamma'}{\beta^2\gamma} + \frac{r\gamma''}{2\beta^2\gamma} = 0, \tag{23–39}$$

or

$$(\beta\gamma r')' + \frac{r\gamma''}{2\beta} = 0. \tag{23–40}$$

Qualitatively, the action of an electrostatic lens depends on the change in velocity of a particle traversing the system. In Fig. 23–2 it is assumed that a positive charge travels near the axis from left to right. In an accelerating lens the particle is faster and hence spends less time in the diverging part of the field; conversely, if the lens produces deceleration the particle is slower in the now converging second part of the field. In either case, therefore, the lens is converging. This can be seen quantitatively if we apply the transformation

$$R = rT^{1/4} \tag{23–41}$$

to Eq. (23–39) in the nonrelativistic approximation. We obtain

$$R'' + \frac{3}{16}\left(\frac{T'}{T}\right)^2 R = 0, \tag{23–42}$$

which demonstrates the convergence property of the lens. By integrating Eq. (23–42) through a region of variable potential, we can obtain formulas for the focal properties of electrostatic lenses.

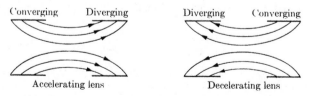

FIG. 23–2. Accelerating and decelerating electrostatic lenses, both of net convergence.

The velocity of the particle, and hence the (nonrelativistic) square root of the electrostatic potential, is analogous to the refractive index, n, in optics. This can be understood immediately if we note the correspondence between Fermat's principle in optics,

$$\delta \int_{x_1}^{x_2} n \, dl = 0,$$

and the nonrelativistic principle of least action,

$$\delta \int_{x_1}^{x_2} \sqrt{T} \, dl = 0.$$

Note also that electrostatic lenses are "second order" devices in the sense that the focusing action depends quadratically on T' and hence quadratically on the axial field strength; physically, this is due to the fact that the lens action depends on *both* the change in velocity caused by the field *and* the radial component of the field.

For simplicity, we have considered only the nonrelativistic case. The same arguments apply to the general case, except that the transformation leading to Eq. (23–42) is somewhat more complicated than Eq. (23–41).

2. *Magnetostatic lens, no "eccentric" rays, $C = 0$.* If there are no electrostatic fields to produce changes in energy, Eq. (23–28) becomes simply

$$r'' + \frac{r}{4}\left(\frac{B}{B\rho}\right)^2 = 0, \qquad (23\text{–}43)$$

where $B\rho = cm_0\beta\gamma/e$ is the "magnetic rigidity" of the particle, i.e., the product of the magnetic field and the radius of curvature. No approximation involving small velocities needs to be made in this case. It is obvious that a magnetic lens, like an electrostatic lens, is always a converging device. Equation (23–43) can be readily integrated to give the focal length.

The lens action is also of second order in this case, i.e., it is quadratic in the field. Let us try to understand this physically. Consider a magnetic lens, say in the form of a solenoid, and let a particle enter with zero canonical angular momentum (Fig. 23–3). While entering the fringing field, the particle will acquire a mechanical angular momentum

$$-p_\theta = eA_\theta \simeq \frac{eBr}{2}, \qquad (23\text{–}44)$$

according to Eq. (23–32); this is also qualitatively evident by inspection of the Lorentz force on the charge in this part of the field. This angular velocity then interacts with the longitudinal component of **B** to produce a radially inward force. Thus **B** has acted twice, and the net effect is quad-

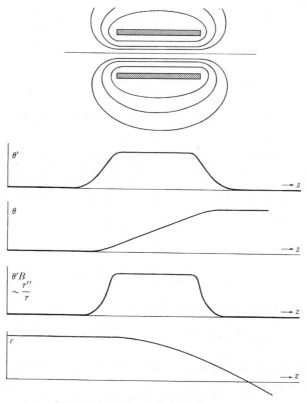

Fig. 23-3. Effect of a magnetic lens.

ratic. The reader can easily convince himself that the direction of the net radial force is always opposite to the direction of the radius vector.

3. *General case.* Equation (23–38) can be written as

$$\frac{d}{dz}(\beta\gamma r') + r\left\{\frac{\gamma''}{2\beta} + \frac{1}{4}\left(\frac{eB}{m_0c}\right)^2\frac{1}{\beta\gamma} - \frac{C^2}{\beta\gamma r^4}\right\} = 0, \qquad (23\text{–}45)$$

which, except for the last term, is the equation of motion corresponding to a harmonic oscillator of variable mass and variable force constant. The last term corresponds to an r^{-2} repulsive core in such a potential; finite canonical angular momentum makes it impossible for the particle to cross the axis.

In general, the solution of Eq. (23–45) requires numerical or graphical integration. Note that β and γ are assumed to be given functions of z

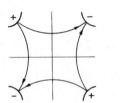 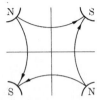

FIG. 23-4. Electric and magnetic quadrupole lenses.

through the value of the electrostatic potential on the axis. If $C = 0$ and the parameters vary slowly with respect to the period of radial oscillations, the equation can be integrated for extended systems by means of the so-called "adiabatic theorem." In that case the equation has the form

$$(Fr')' + Gr = 0. \tag{23-46}$$

If F and G are functions of z fulfilling the conditions that

$$F'/F \ll \sqrt{G/F}, \qquad G'/G \ll \sqrt{G/F},$$

solutions of Eq. (23-46) are represented by

$$r = (FG)^{-1/4} \exp\left[\pm i \int^z (G/F)^{1/2} dz\right], \tag{23-47}$$

and the exponent may be integrated analytically or numerically, depending on the nature of the fields.

4. *First-order lenses.* It can be easily shown that it is impossible to build a first-order lens, either electric or magnetic, which has cylindrical symmetry. In an electrostatic field of cylindrical symmetry the mean value of E_r will always vanish in charge-free regions because of the divergence condition. Similarly, the mean value of the azimuthal component B_θ of the magnetic field will vanish in a current-free region, since in this case **B** is irrotational. Hence, no net radial impulse of first order in **E** or **B** can be given to a particle in a field of cylindrical symmetry.

First-order focusing effects can be produced by asymmetrical lenses. The prime example of such a lens is the electric or magnetic quadrupole, which is essentially a "fully astigmatic" lens. It can be seen from Fig. 23-4 that such a structure, if acting as a thin lens, will have equal but opposite focal strengths in two mutually perpendicular planes; hence the theorem that the mean radial momentum imparted to a particle (averaged over azimuthal angles) shall vanish is still not violated. The combination of a diverging lens and a converging lens of equal strength separated in space is converging; hence lens combinations made up of alternating elements of quadrupoles are converging. This is the basis of the "strong-focusing" principle in electron and ion optics.

5. *Solutions simplified by change of Lorentz frame.* Frequently, integration of the equations of motion may be simplified by choosing a Lorentz frame other than the frame of physical interest. Most notable are the cases where the combined effect of electric and magnetic fields can be reduced to the effect of a purely electric or a purely magnetic field. The possibilities for such transformations can be ascertained by examination of the two invariants, scalar and pseudoscalar,

$$E^2 - c^2 B^2 = I_1, \tag{23-48}$$

$$\mathbf{E} \cdot \mathbf{B} = I_2. \tag{23-49}$$

We have seen in Chapter 18 that if $I_1 = 0 = I_2$, we have the case of a plane wave which is invariant in all its properties. If $I_2 = 0$ but $I_1 \neq 0$, it is possible to transform away either $\mathbf{B}$ or $\mathbf{E}$. If $E < cB$, a frame can be found for which $\mathbf{E}$ vanishes. If $\mathbf{B}$ is uniform, the orbits are then circles (or helixes) and thus become cycloids in the laboratory frame. If $cB < E$, a transformation may be effected to a frame in which $\mathbf{B}$ vanishes, and the orbits are catenaries if $\mathbf{E}$ is uniform. A component of $\mathbf{E}$ parallel to $\mathbf{B}$ ($I_2 \neq 0$) cannot be transformed away, but if $\mathbf{E}$ is parallel to $\mathbf{B}$, motion in the plane defined by the fields and the initial velocity depends only on $\mathbf{E}$ and can be integrated first; a solution for the motion at right angles to this plane then follows.

23-5 The motion of a particle with magnetic moment in an electromagnetic field. In Chapter 7 we derived the nonrelativistic precessional motion of a particle with angular momentum $\mathbf{s}$ and gyromagnetic ratio

$$\Gamma = ge/2m_0 \tag{23-50}$$

in a magnetic field. The resultant motion was governed by Eqs. (7-68) and (7-69). Let us now examine the relativistic generalization of this motion for a particle of charge e, rest mass m_0, and arbitrary g-factor.

In three dimensions the magnetic moment $\mathbf{m}$ and the angular momentum $\mathbf{s}$ are axial vectors. For an arbitrary classical object there is no basic reason why $\mathbf{s}$ and $\mathbf{m}$ should be parallel. Since the theory is applied principally to elementary particles, however, we shall restrict the discussion to the case in which only one preferred axis of a particle can be used; this feature is a quantum-mechanical result, and we utilize it here only as an apparently restrictive assumption. Hence we shall assume that

$$\mathbf{m}_0 = \Gamma \mathbf{s}_0 = (ge/2m_0)\mathbf{s} \tag{23-51}$$

in the rest frame of the object, without demanding that the relation between $\mathbf{m}_0$ and $\mathbf{s}_0$ be a general linear relation among the components.

If we allow a particle to carry both an electric and a magnetic moment, then (see Chapter 18) it is no longer possible to consider the magnetic moment $\mathbf{m}$ as the 1, 2, 3 component of a four-component quantity; a covariant description would involve a second-rank antisymmetric matrix,

$$m^{ij} = \begin{pmatrix} 0 & -m_z/c & m_y/c & p_x \\ m_z/c & 0 & -m_x/c & p_y \\ -m_y/c & m_x/c & 0 & p_z \\ -p_x & -p_y & -p_z & 0 \end{pmatrix}, \qquad (23\text{-}52)$$

which combines the magnetic and electric moments of the particle, and the angular momentum would correspond to a similar antisymmetric matrix s^{ij}. Therefore we write

$$m^{ij} = \Gamma s^{ij}. \qquad (23\text{-}53)$$

as a covariant relation. The magnetization-polarization density M^{ij} [see Eq. (18–62)] transforms like a tensor. Hence, if m^{ij} is to be a tensor, we define m^{ij} to be M^{ij} multiplied by the *proper* volume (i.e., the volume in the rest frame) of the particle. This must be taken into account when interpreting the components of m^{ij} in a given frame. The angular momentum tensor s^{ij} of a structure of finite proper three-dimensional volume v^0 is here defined as

$$s^{ij} = \int_{v^0} (p^i x^j - p^j x^i)\, dv^0 \qquad (23\text{-}54)$$

$$= \left(1 - \frac{u^2}{c^2}\right)^{1/2} \int_v (p^i x^j - p^j x^i)\, dv. \qquad (23\text{-}55)$$

where p^i is the usual four-momentum and u is the velocity. Note that both m^{ij} and s^{ij} are defined as tensors, but that the integral $\int_v M^{ij}\, dv$ and the integral which occurs on the right-hand side of Eq. (23–55) are not.

Let us now make the simplifying assumption that $\mathbf{p}^0 = 0$, i.e., that the particle in its rest frame has no electric moment. This assumption is justified for all elementary particles to the accuracy with which experiments have thus far been carried out. If desired, this restriction can be later removed by interchanging the roles of all electric and magnetic quantities [i.e., by replacing $\mathbf{m}$ by $\mathbf{p}$ and the electromagnetic field term F^{ij} by its dual G^{ij} defined in Eq. (17–37)] and then superposing the resultant equations of motion; since the equations of motion are linear this procedure is justified.

We are seeking a covariant equation which has the property of reducing to Eq. (7–68) in a proper frame ($\mathbf{u} = 0$), i.e.,

$$d\mathbf{s}^0/dt = \Gamma(\mathbf{s}^0 \times \mathbf{B}^0), \tag{23–56}$$

in a frame in which $u^i = (\mathbf{0}, 1)$ and in which

$$(s^{ij})^0 = \begin{pmatrix} 0 & -s_z & s_y & 0 \\ s_z & 0 & -s_x & 0 \\ -s_y & s_x & 0 & 0 \\ 0 & 0 & 0 & 0 \end{pmatrix}. \tag{23–57}$$

The assumption that the time components of the tensors m^{ij} and s^{ij} vanish in a proper frame permits us to reduce the complexity of the equations of motion by replacing the 2nd-rank tensors s^{ij} and m^{ij} with the pseudo-four-vectors s^i and m^i, such that in a proper frame

$$s^{i0} = (\mathbf{s}^0, 0), \tag{23–58}$$

$$m^{i0} = (\mathbf{m}^0, 0). \tag{23–59}$$

In a general frame we can therefore write

$$s^i = (\mathbf{s}, \boldsymbol{\beta} \cdot \mathbf{s}), \tag{23–60}$$

$$m^i = (\mathbf{m}, \boldsymbol{\beta} \cdot \mathbf{m}), \tag{23–61}$$

where the vector $\mathbf{s}$ is defined to have the components $(\gamma s_x^0, s_y^0, s_z^0)$ and the vector $\mathbf{m}$ the components $(\gamma m_x^0, m_y^0, m_z^0)$. Note that since

$$u^i s_i = 0 \tag{23–62}$$

in the rest frame, it is zero in all frames. The equation of motion of the particle due to its charge is, as usual,

$$m_0 c^2 (du^i/ds) = eF^{ij} u_j \tag{23–63}$$

in a *uniform* electromagnetic field F^{ij}. If the field is nonuniform, terms coupling magnetic or electric moments to the field gradients should be introduced. The complete equation of motion, including the force on the magnetic moment in the magnetic field, would have an additional term corresponding to the three-dimensional force

$$(\mathbf{m} \cdot \boldsymbol{\nabla})\mathbf{B}. \tag{23–64}$$

Since $|\mathbf{m}|$ is of order (cea), where a is a length whose order of magnitude is the size of the structure, we see that this term is negligible unless the fields vary appreciably across a distance a. If the fields do change so abruptly, classical methods are not applicable.

A covariant expression whose space component reduces to Eq. (23–56) in a rest frame is

$$\frac{ds^i}{ds} = \frac{\Gamma}{c} F^{ij} s_j = \frac{ge}{2m_0 c^2} F^{ij} s_j, \qquad (23\text{–}65)$$

which is similar in form to Eq. (23–63). But Eq. (23–65) cannot be the complete covariant generalization of Eq. (23–56); this can be shown by considering the derivative of the invariant expression (23–62):

$$s_i(du^i/ds) + u^i(ds_i/ds) = 0, \qquad (23\text{–}66)$$

and hence, from Eq. (23–63),

$$u^i(ds_i/ds) = -(e/m_0 c^2) F^{ij} u_j s_i = +(e/m_0 c^2) F^{ij} u_i s_j. \qquad (23\text{–}67)$$

According to Eq. (23–65), however,

$$u^i(ds_i/ds) = (ge/2m_0 c^2) F^{kl} u_k s_l, \qquad (23\text{–}68)$$

which agrees with Eq. (23–67) only if

$$g = 2. \qquad (23\text{–}69)$$

Hence the simple equation (23–65), in which the form of the interaction of the magnetic moment with the electromagnetic field is the same as that of a four-current, is complete only if $g = 2$. In a pure magnetic field the equation of motion obeyed by **u** and that obeyed by **s** are exactly the same if $g = 2$; hence the angle between **u** and **s** will remain fixed for arbitrary motion of the particle in a magnetic field. If an electric field is also present this is of course no longer true, since the fourth component of s^i, namely, $s^4 = \boldsymbol{\beta} \cdot \mathbf{s}$, itself involves the velocity. (We may note that $g = 2$ for a particle with spin equal to $\frac{1}{2}$ which obeys the Dirac equation.)

If $g \neq 2$, Eq. (23–56) must still be valid in the rest frame, but some supplementary term must be added to Eq. (23–65) so that the consequences of Eqs. (23–62) and (23–63) are compatible. In order to bring Eq. (23–68) into agreement with the generally valid equation (23–67), we must modify Eq. (23–65) in such a way that Eq. (23–68) becomes

$$u^i(ds_i/ds) = (e/m_0 c^2) F^{kl} u_k s_l$$

$$= (ge/2m_0 c^2) F^{kl} u_k s_l - (g-2)(e/2m_0 c^2) F^{kl} u_k s_l, \qquad (23\text{–}70)$$

without, however, modifying the space components of Eq. (23–56) in a rest frame. Hence Eq. (23–65) should be changed to become

$$(ds^i/ds) = (e/2m_0c^2)[gF^{ij}s_j - (g - 2)(F^{kl}u_ks_l)u^i].$$ (23–71)

An interesting consequence of this equation is its prediction for a particle of zero magnetic moment, $g = 0$; for this case

$$\frac{ds^i}{ds} = \frac{e}{m_0c^2} (F^{kj}s_ju_k)u^i = - \left(\frac{du^j}{ds} s_j\right)u^i.$$ (23–72)

Equation (23–72) corresponds to the so-called "Thomas precession" of a particle carrying a defined axis moving (in a given frame) with an acceleration not parallel to its velocity. This phenomenon is a purely kinematic effect and can be derived directly without reference to the specific cause of the acceleration. To see this, let us consider three sets of reference frames:

(1) A frame Σ^0 in which an object is at rest such that a fixed direction of a vector s_0 in this object is parallel to the x_0-axis.

(2) A frame Σ' in which Σ^0 undergoes an infinitesimal velocity change which is *not* along x_0.

(3) A frame Σ (the laboratory frame) in which Σ' (and Σ^0) move with a velocity $\mathbf{u}$ along the x-axis, and in which the infinitesimal change of Σ^0 in Σ' is observed to be $d\mathbf{u}$.

It can be shown[*] by successive Lorentz transformations that the angle between the s_0-axis and the s-axis in Σ is given by

$$d\boldsymbol{\theta} = - \frac{(\gamma - 1)}{u^2} (d\mathbf{u} \times \mathbf{u}),$$ (23–73)

where the magnitude and the axis of rotation are indicated by the axial vector $d\boldsymbol{\theta}$. Hence if the particle has an acceleration $d\mathbf{u}/dt$, the direction of the spin appears in the laboratory frame to rotate at a rate

$$\boldsymbol{\Omega}_T = - \left(\frac{\gamma - 1}{u^2}\right)\left(\frac{d\mathbf{u}}{dt} \times \mathbf{u}\right).$$ (23–74)

Equation (23–72) is equivalent to Eq. (23–74); we shall demonstrate the equivalence only to terms in the second order in u^2/c^2. Equation (23–74) is then

$$\boldsymbol{\Omega}_T = - \frac{1}{2c^2} \frac{d\mathbf{u}}{dt} \times \mathbf{u}.$$ (23–75)

---

[*] See, e.g., C. Moller, *The Theory of Relativity*, Oxford, Clarendon Press, 1952, pp. 53 ff.

The space components of (23–72) are

$$\frac{d\mathbf{s}}{dt} = \left(\frac{du^j}{dt}\, s_j\right)\gamma\boldsymbol{\beta} = \left[\mathbf{s}\cdot\frac{d}{dt}\,(\gamma\boldsymbol{\beta}) - (\mathbf{s}\cdot\boldsymbol{\beta})\,\frac{d\gamma}{dt}\right]\gamma\boldsymbol{\beta}$$

$$= \frac{\gamma^2}{c^2}\left(\mathbf{s}\cdot\frac{d\mathbf{u}}{dt}\right)\mathbf{u}. \tag{23–76}$$

In tensor notation,

$$\frac{ds_\alpha}{dt} = \frac{\gamma^2}{c^2}\, s_\beta\left(u_\alpha\,\frac{du_\beta}{dt}\right) = \frac{\gamma^2}{2c^2}\, s_\beta\left[\left(u_\alpha\,\frac{du_\beta}{dt} - u_\beta\,\frac{du_\alpha}{dt}\right) + \frac{d}{dt}\,(u_\alpha u_\beta)\right],$$

$$\tag{23–77}$$

where the term in brackets gives separately the antisymmetric and symmetric parts of the transformation matrix from ds_α/dt to s_β. The first term on the right-hand side is a pure rotation, corresponding to the Thomas precession; this term can be written as

$$\frac{d\mathbf{s}}{dt} = \frac{\gamma^2}{2c^2}\left[\mathbf{s}\times\left(\frac{d\mathbf{u}}{dt}\times\mathbf{u}\right)\right] = \boldsymbol{\Omega}_T\times\mathbf{s}, \tag{23–78}$$

in agreement with (23–75). This expression gives the rotation rate of the vector $\mathbf{s}$ rather than $\mathbf{s}_0$; the difference is of higher order in β.*

Let us analyze the physical meaning of Eq. (23–71) in more detail. As pointed out previously, the equation of motion for s^i is identical to the equation of motion (23–63) for u^i if $g = 2$. Hence it is useful to write an equation of motion which describes the s^i motion *relative* to the u^i motion. To do this, we introduce a "tetrad"† of unit orthogonal four-vectors $_a e^i$ which are *instantaneously co-moving* with the rest frame of the particle, and the orientation of the e^i stays fixed relative to the orbit. The subscript preceding the symbol refers to the sequence of the four unit vectors and has no relation to specific transformation properties. We thus put

$$s^i = {_a\xi}\,{_a e^i}, \tag{23–79}$$

where the $_a e^i$ obey Eq. (23–63), but where the orientation of the $_a e^i$ must

---

* The Thomas precession has also been derived by iteration of an infinitesimal Lorentz transformation; see W. H. Furry, *Am. J. Phys.* **23**, 517 (1955).

† This method is due to L. Michel; see V. Bargmann, L. Michel, and V. L. Telegdi, *Phys. Rev. Letters* **2**, 435 (1959).

be chosen to fit a specific problem. The time variation of the $_a\xi$ thus describes the relative motion. Let us substitute (23–79) into (23–71); for simplicity a dot over a symbol is used to indicate d/ds. We obtain

$$_a\dot\xi \, _ae^i + \, _a\xi \, _a\dot e^i = \frac{e}{2m_0c^2} \{gF^{ij} \, _be_j \, _b\xi - (g - 2)[F^{kl}u_k \, _be_l \, _b\xi]u^i\}. \quad (23\text{–}80)$$

Without losing generality we can orient the $_ae^i$ axes such that $_4e^i = u^i$, since u^i satisfies Eq. (23–63). Using the equation of motion (23–63), we can simplify Eq. (23–80) to read

$$_a\dot\xi \, _ae^i = \frac{e}{2m_0c^2} (g - 2)\{F^{ij} \, _be_j \, _b\xi - F^{kl}u_k \, _be_l \, _b\xi u^i\}. \quad (23\text{–}81)$$

If we multiply both sides of (23–81) by $_ae_i$ for $a = 1, 2, 3$, but not 4, the second term on the right-hand side will vanish, since $u^i = \, _4e^i$, and thus $_4e^i \, _ae_i = 0$. Hence

$$_a\dot\xi = \frac{e}{2m_0c^2} (g - 2)_b\xi\{F^{ij} \, _be_j \, _ae_i\}. \quad (23\text{–}82)$$

This is the general equation of motion for the $_a\xi$.

Let us interpret Eqs. (23–81) and (23–82) in the rest frame Σ^0 of the particles where $u^i = (0, 1)$. If $_a\xi$ the components of a vector $\boldsymbol{\xi}$ referred to the $_ae$, the space components of Eq. (23–81) become

$$\dot{\boldsymbol{\xi}} = \frac{e}{2m_0c^2} (g - 2)(\mathbf{B}^0 \times \boldsymbol{\xi})$$

$$= \boldsymbol{\Omega} \times \boldsymbol{\xi}, \quad (23\text{–}83)$$

where

$$\boldsymbol{\Omega} = (e/2m_0c^2)(g - 2)\mathbf{B}^0. \quad (23\text{–}84)$$

Expressed in terms of the laboratory fields the fields in the rest frame are, from Eqs. (18–41) and (18–43),

$$B^0_{||} = B_{||},$$

$$\mathbf{B}^0_\perp = \gamma\left(\mathbf{B}_\perp - \frac{\boldsymbol{\beta} \times \mathbf{E}}{c}\right); \quad (23\text{–}85)$$

$$\mathbf{B}^0 = \frac{\mathbf{u}(\mathbf{u} \cdot \mathbf{B})}{u^2} + \gamma\mathbf{B}_\perp - \gamma(\boldsymbol{\beta} \times \mathbf{E})$$

$$= \gamma\left[\mathbf{B} - \left(\frac{\gamma}{\gamma + 1}\right)\frac{(\mathbf{u} \cdot \mathbf{B})\mathbf{u}}{c^2} - \frac{(\mathbf{u} \times \mathbf{E})}{c^2}\right]. \quad (23\text{–}86)$$

Equation (23–86), in combination with Eq. (23–83), expresses the precession rate in the *rest frame* in terms of the *laboratory* fields **E** and **B**. Using *laboratory* time t, but orientation ξ in the rest frame of the particle, we have

$$\frac{d\xi}{dt} = \frac{e(g - 2)}{2m_0c^2}\left[\mathbf{B} - \left(\frac{\gamma}{\gamma + 1}\right)\frac{(\mathbf{u} \cdot \mathbf{B})\mathbf{u}}{c^2} - \frac{\mathbf{u} \times \mathbf{E}}{c^2}\right] \times \xi. \quad (23\text{–}87)$$

By expressing the result in terms of the coordinates of ξ in a co-moving rest frame, we avoid the problem of the meaning of the direction of **s** in the laboratory, which is complicated by the Thomas precession, as discussed previously. Integration of the spin motion in the laboratory can be accomplished by first choosing an appropriate set of the basic vectors e^i which obey the equation of motion (23–63).

Suggested References

H. Goldstein, *Classical Mechanics.* There are many adequate treatments of Lagrangian and Hamiltonian particle mechanics, but the reader may find this modern work most useful.

H. C. Corben and P. M. Stehle, *Classical Mechanics.* Another modern treatment of classical mechanics. There is very little relativistic mechanics in this work, but the distinction between covariant and contravariant components of vectors and tensors is maintained so that the formulation is immediately useful in relativistic theory.

L. Landau and E. Lifshitz, *The Classical Theory of Fields.* Chapter 3 is devoted to charges in fields, and covers much of the material we have included here.

O. Klemperer, *Electron Optics,* 2nd edition. In addition to a discussion of the principles and many applications, this expanded version of an early work also contains an excellent bibliography.

V. K. Zworykin, G. A. Morton, E. G. Ramberg, J. Hillier, and A. W. Vance, *Electron Optics and the Electron Microscope.* A fairly complete treatment of general interest as of 1945.

Two bibliographies of electron microscopy appeared in 1950, one edited by V. E. Coslett, the other by Claire Marton *et al.*

Exercises

1. Verify that the Lagrangian

$$L = -m_0c^2\sqrt{1 - u^2/c^2} - e\phi + e\mathbf{A} \cdot \mathbf{u}$$

yields the correct equation of motion.

2. Show that the orbit of a particle of charge e, rest mass m_0, in a uniform electrostatic field $\mathbf{E}$ along the x-axis, is

$$x = \frac{U_0}{eE} \cosh \frac{eEy}{p_0 c},$$

where $U_0 = \sqrt{(m_0 c^2)^2 + c^2 p_0^2}$ and p_0 is the initial momentum in the y-direction. The momentum in the x-direction at $y = 0$ has been taken equal to zero.

3. Consider an accelerating column for charged particles such as might be installed in an electrostatic generator. As a result, the energy of the particle is given approximately by

$$e\phi = k\left(z + \frac{g\alpha}{2\pi} \sin \frac{2\pi z}{g}\right),$$

$$eE = e\phi' = k\left(1 + \alpha \cos \frac{2\pi z}{g}\right),$$

where z is the axial distance and k is the mean accelerating field. Since

$$1 + (\alpha^2/2) = \overline{E^2}/\overline{E}^2,$$

α is a measure of the fluctuation of the axial field. Assume that g, the wavelength of these variations, is small compared with the focal length of each lens of thickness g. Hence, ϕ'^2 may be represented by its mean square and the fluctuation term in ϕ may be neglected. What is the minimum value of α such that the orbits will be oscillatory?

4. An electrostatic "strong-focusing" lens produces a field corresponding to the complex potential $W(z) = W(x + iy)$:

$$W = \frac{E_0 z^2}{2a},$$

where E_0 is the electric field at radius a. How could such a field be realized? If the length of the lens is L, calculate the focal length in the x- and y-planes. Do *not* assume the lens to be thin.

5. Consider two coaxial conducting circular cylinders of equal radii charged to potentials $-V_0/2$, $+V_0/2$, respectively. Let the separation between the cylinders be small compared with the radius a of the cylinders. Calculate the potential within the space bounded by the cylinders.

6. In the preceding problem, let $V_0 = A \cos \omega t$, where $\omega \ll c/a$, so that the static solution is valid. Let a particle of charge e cross the field along a trajectory parallel to the axis at a distance b from it. What is the energy gain?

7. Consider a particle of energy $U_0 \gg eV_0$ passing through the system of problem 5. What is the focal length?

8. Prove Eq. (23–47) as a solution of Eq. (23–46).

CHAPTER 24

HAMILTONIAN FORMULATION OF MAXWELL'S EQUATIONS

In Chapter 23 we found the covariant equations of motion for a point particle in an external field, using the Lagrangian and Hamiltonian formulations. Let us now formulate the equations of motion of the field, i.e., Maxwell's equations, in terms of Lagrangian and Hamiltonian functions. No new physical information concerning classical electrodynamics is arrived at in this way, but the transition to quantum electrodynamics can be accomplished through canonical formulation of the field. Moreover, the methods facilitate the investigation of various types of fields other than the electromagnetic.

In extending the Hamiltonian formulation to a field, we are faced with a new consideration, namely, the fact that the number of degrees of freedom, in the mechanical sense, is infinite. We shall begin, therefore, by studying the transition of a mechanical system of N degrees of freedom to a system in which N becomes infinite.

24–1 Transition to a one-dimensional continuous system. Let us consider a set of N point particles of equal mass m, connected by springs of equal length a and force constant k. (See Fig. 24–1.) Let η_i be the displacement from equilibrium of the ith mass. The solution of this problem rests on finding a suitable Lagrangian L such that the equation

$$\delta \int L(\eta_i, \dot{\eta}_i, t)\, dt = 0 \qquad (24\text{–}1)$$

represents the correct equation of motion. In classical mechanics, we know that $L = T - V$, where T and V are the kinetic and potential energies respectively. By the geometry of the problem at hand, we thus have (neglecting end contributions):

$$L = \tfrac{1}{2} \sum_{i=1}^{N} [m\dot{\eta}_i^2 - k(\eta_{i+1} - \eta_i)^2]. \qquad (24\text{–}2)$$

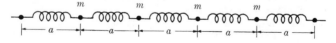

FIG. 24–1. A system of mass points connected by springs.

Equation (24–2) can be written in the form

$$L = \sum_{i=1}^{N} a\mathcal{L}_i, \qquad (24\text{–}3)$$

where

$$\mathcal{L}_i = \frac{1}{2}\left[\frac{m}{a}\,\dot{\eta}_i^2 - ka\left(\frac{\eta_{i+1} - \eta_i}{a}\right)^2\right] \qquad (24\text{–}4)$$

is a quantity which we might call the "linear Lagrangian density." Equation (24–1), with the Lagrangian of Eq. (24–2), gives the usual equation of motion for a set of coupled oscillators; orthogonalization of Eq. (24–2) yields the set of normal modes.

Our principal interest here is to allow the number of degrees of freedom to tend to infinity; this we shall do by letting

$$a \rightarrow dx,$$

$$m/a \rightarrow \mu,$$

$$ka \rightarrow Y, \qquad (24\text{–}5)$$

$$\frac{\eta_{i+1} - \eta_i}{a} \rightarrow \frac{\partial \eta}{\partial x},$$

where μ is the linear mass density and Y is Young's modulus. By means of this transition, we have replaced the discrete index i by the continuous variable x. The Lagrangian then becomes

$$L = \frac{1}{2}\int\left[\mu\dot{\eta}^2 - Y\left(\frac{\partial \eta}{\partial x}\right)^2\right]dx = \int \mathcal{L}\,dx, \qquad (24\text{–}6)$$

with

$$\mathcal{L}\left(\eta, \dot{\eta}, \frac{\partial \eta}{\partial x}, t\right) = \frac{1}{2}\left[\mu\dot{\eta}^2 - Y\left(\frac{\partial \eta}{\partial x}\right)^2\right]. \qquad (24\text{–}7)$$

The Lagrangian density thus acquires an explicit dependence on the spatial derivatives of the "field coordinate" η. We will not consider the case where $\mathcal{L}$ depends explicitly on t.

Let us now derive the Eulerian equation corresponding to the variational principle of Eq. (24–1), which has become

$$\delta\int L\,dt = \delta\iint \mathcal{L}\,dx\,dt = 0. \qquad (24\text{–}8)$$

On substituting Eq. (24–7) and integrating by parts in the usual way, we obtain (ignoring explicit time dependence)

$$\delta \int L \, dt = \iint \left\{ \frac{\partial \mathcal{L}}{\partial \eta} \, \delta \eta + \frac{\partial \mathcal{L}}{\partial (\partial \eta / \partial x)} \, \delta \left(\frac{\partial \eta}{\partial x} \right) + \frac{\partial \mathcal{L}}{\partial (\partial \eta / \partial t)} \, \delta \left(\frac{\partial \eta}{\partial t} \right) \right\} dx \, dt$$

$$= \iint \left\{ \frac{\partial \mathcal{L}}{\partial \eta} - \frac{\partial}{\partial x} \left[\frac{\partial \mathcal{L}}{\partial (\partial \eta / \partial x)} \right] - \frac{\partial}{\partial t} \left[\frac{\partial \mathcal{L}}{\partial (\partial \eta / \partial t)} \right] \right\} \delta \eta \, dx \, dt = 0. \quad (24\text{-}9)$$

The integrated part has been made to vanish by the condition that $\delta \eta = 0$ at the end points of integration over t. Since $\delta \eta$ is an arbitrary function of x, we obtain the partial differential equation

$$\frac{\partial \mathcal{L}}{\partial \eta} - \frac{\partial}{\partial x} \left[\frac{\partial \mathcal{L}}{\partial (\partial \eta / \partial x)} \right] - \frac{\partial}{\partial t} \left[\frac{\partial \mathcal{L}}{\partial (\partial \eta / \partial t)} \right] = 0. \quad (24\text{-}10)$$

Equation (24–10) is often written in the form

$$\frac{\delta \mathcal{L}}{\delta \eta} - \frac{\partial}{\delta t} \left[\frac{\partial \mathcal{L}}{\partial (\partial \eta / \partial t)} \right] = 0, \quad (24\text{-}11)$$

where

$$\frac{\delta \mathcal{L}}{\delta \eta} = \frac{\partial \mathcal{L}}{\partial \eta} - \frac{\partial}{\partial x} \left[\frac{\partial \mathcal{L}}{\partial (\partial \eta / \partial x)} \right] \quad (24\text{-}12)$$

is called the "variational" or "functional" derivative. Note that in going to the limit of a continuous variable we have replaced a system of N ordinary differential Lagrangian equations by a partial differential equation.

With the substitution of the Lagrangian density, Eq. (24–7), Eq. (24–11) yields immediately the wave equation,

$$\mu \frac{\partial^2 \eta}{\partial t^2} - Y \frac{\partial^2 \eta}{\partial x^2} = 0, \quad (24\text{-}13)$$

corresponding to compressional waves traveling with velocity $\sqrt{Y/\mu}$.

24–2 Generalization to a three-dimensional continuum. These considerations may be generalized to a three-dimensional field η, where η may be any covariant parameter. The variational principle corresponding to Eq. (24–1) is given by

$$\delta \iint \mathcal{L} \, dv \, dt = 0, \quad (24\text{-}14)$$

or

$$\delta \iint \mathcal{L} \left(\eta, \frac{\partial \eta}{\partial x^i} \right) d^4 x = 0. \quad (24\text{-}15)$$

This formulation is evidently covariant if the Lagrangian density $\mathcal{L}$ is a scalar. Let us again vary the functional dependence of $\mathcal{L}$ on η and $\partial\eta/\partial x^i$, but consider the x^i as fixed independent coordinates. Partial integration gives

$$\int \left\{ \frac{\partial\mathcal{L}}{\partial\eta} - \frac{\partial}{\partial x^i}\left[\frac{\partial\mathcal{L}}{\partial(\partial\eta/\partial x^i)} \right] \right\} d^4x \, \delta\eta = 0, \qquad (24\text{--}16)$$

leading to the covariant Lagrangian equations

$$\frac{\partial\mathcal{L}}{\partial\eta} = \frac{\partial}{\partial x^i}\left[\frac{\partial\mathcal{L}}{\partial(\partial\eta/\partial x^i)} \right]. \qquad (24\text{--}17)$$

To exhibit the time dependence explicitly, we may write these equations in the form

$$\frac{\delta\mathcal{L}}{\delta\eta} = \frac{\partial}{\partial t}\left(\frac{\partial\mathcal{L}}{\partial\dot\eta} \right), \qquad (24\text{--}18)$$

where

$$\frac{\delta\mathcal{L}}{\delta\eta} = \frac{\partial\mathcal{L}}{\partial\eta} - \frac{\partial}{\partial x^\alpha}\left[\frac{\partial\mathcal{L}}{\partial(\partial\eta/\partial x^\alpha)} \right], \qquad (24\text{--}19)$$

with $\dot\eta = \partial\eta/\partial t$ and with x^α representing the three space variables. The operation $\delta/\delta\eta$ is now the variational derivative in three dimensions. Our program is to find an $\mathcal{L}$ such that Eq. (24–17) will lead, in the electromagnetic case, to Maxwell's equations. Clearly, the preceding discussion will apply to any field theory.

It has been possible to state the action principle and write a Lagrangian partial differential equation in obviously covariant form. To introduce a Hamiltonian, we must single out the time among the x^i as has been done in Eqs. (24–18) and (24–19). We can then define a "momentum density" conjugate to η,

$$\pi(x^\alpha, t) = \frac{\partial\mathcal{L}}{\partial\dot\eta}, \qquad (24\text{--}20)$$

and a Hamiltonian density,

$$\mathcal{H}\left(\eta, \frac{\partial\eta}{\partial x_\alpha}, \pi; t \right) = \pi\dot\eta - \mathcal{L}. \qquad (24\text{--}21)$$

Let us treat only the case in which the Hamiltonian does not depend explicitly on the time. The Hamiltonian equations follow in the usual way. Consider an increment δH of the total Hamiltonian $H = \int \mathcal{H} \, dv$:

$$\delta H = \int \left[\dot\eta \, \delta\pi + \pi \, \delta\dot\eta - \frac{\partial\mathcal{L}}{\partial(\partial\eta/\partial x^i)} \, \delta\left(\frac{\partial\eta}{\partial x^i} \right) - \frac{\partial\mathcal{L}}{\partial\eta} \, \delta\eta \right] dv.$$

With Eq. (24–17) and the definition of π, this becomes, on partial integration,

$$\delta H = \int (\dot{\eta}\, \delta\pi - \dot{\pi}\, \delta\eta)\, dv. \tag{24–22}$$

Since

$$H = \int \mathcal{3C}\left(\pi,\, \eta,\, \frac{\partial \eta}{\partial x^\alpha}\right) dv, \tag{24–23}$$

we may also write

$$\delta H = \int \left[\frac{\partial \mathcal{3C}}{\partial \pi}\, \delta\pi + \frac{\partial \mathcal{3C}}{\partial \eta}\, \delta\eta + \frac{\partial \mathcal{3C}}{\partial(\partial \eta/\partial x^\alpha)}\, \delta\left(\frac{\partial \eta}{\partial x^\alpha}\right)\right] dv. \tag{24–24}$$

If we integrate Eq. (24–24) by parts and identify the result with Eq. (24–22), we obtain, in the notation of Eq. (24–19),

$$\dot{\eta} = \frac{\partial \mathcal{3C}}{\partial \pi}, \qquad \dot{\pi} = -\frac{\delta \mathcal{3C}}{\delta \eta} \tag{24–25}$$

as the new form of Hamilton's equations.

Hamilton's equations lead to the usual meaning of the time rate of change in terms of Poisson brackets. If Λ is the density of a physical variable L, i.e., if $L = \int \Lambda\, dv$, then by means of the process used for obtaining Eq. (24–25), we find

$$\frac{dL}{dt} = \int \left(\frac{\delta \Lambda}{\delta \eta}\, \dot{\eta} + \frac{\delta \Lambda}{\delta \pi}\, \dot{\pi}\right) dv = \int \left(\frac{\delta \Lambda}{\delta \eta}\, \frac{\partial \mathcal{3C}}{\partial \pi} - \frac{\delta \Lambda}{\delta \pi}\, \frac{\delta \mathcal{3C}}{\delta \eta}\right) dv$$

$$= \int [\Lambda,\, \mathcal{3C}]\, dv, \tag{24–26}$$

which is analogous to the usual Poisson bracket. These expressions lead to convenient starting points for quantization.

The discussion above leads to definite field equations if a Lagrangian density $\mathcal{L}$ is given. In order that the field equations be linear, the Lagrangian must not contain higher powers than the second of either η or $\partial \eta/\partial x^i$. As the simplest example, we might consider

$$\mathcal{L} = \frac{1}{2}\left(\frac{\partial \eta}{\partial x^i}\, \frac{\partial \eta}{\partial x_i} - \mu^2 \eta^2\right), \tag{24–27}$$

which leads, by Eq. (24–11), to the field equations

$$(\Box - \mu^2)\eta = 0. \tag{24–28}$$

The corresponding momentum density is given by

$$\pi = \frac{1}{c^2}\, \frac{\partial \eta}{\partial t}, \tag{24–29}$$

and we are led to a positive definite Hamiltonian,

$$\mathcal{H} = \tfrac{1}{2}[c^2\pi^2 + (\nabla\eta)^2 + \mu^2\eta^2]. \tag{24–30}$$

This is the scalar meson field of Yukawa, of which the point solution is

$$\eta = e^{i(\mathbf{k}\cdot\mathbf{r}-\omega t)}\frac{e^{-\mu r}}{r}. \tag{24–31}$$

24–3 The electromagnetic field. The electromagnetic field demands a more complicated formulation. Presumably, we are now dealing with a vector field, i.e.,

$$\eta \rightarrow \phi^i. \tag{24–32}$$

The total Lagrangian will have three parts: (1) a mass term for the motion of material particles; (2) an interaction term relating particle to field; and (3) a field term corresponding to the field equation.

Terms (1) and (2) can be written down from our former considerations. If we take

$$\mathcal{L}_1 = \tfrac{1}{2}\rho_m^0 c^2 u_i u^i, \tag{24–33}$$

where ρ_m^0 is the proper mass density, and

$$\mathcal{L}_2 = j^i\phi_i, \tag{24–34}$$

we know that the correct motions of point particles in an external field will result. In writing down L_3 we may be guided by a classical analogy: let us choose L_3 such that $\mathcal{L}$ is an invariant, and also is analogous to the classical difference between potential and kinetic energy. In an electromagnetic oscillation, energy oscillates between electric and magnetic energy, just as in a mechanical oscillation energy oscillates between kinetic and potential energy. The only scalar which is quadratic in the first derivatives of the four-potential is proportional to this difference, namely, $F^{ij}F_{ij} = 2(c^2B^2 - E^2)$. (The pseudoscalar quantity $F^{ij}G_{ij}$, where G_{ij} is the dual of F_{ij}, is excluded because it would transform differently from the remaining terms of the Lagrangian, as we have seen in Section 18–2.) It is, of course, not excluded that a Lagrangian density might also include terms quadratic in the four-potentials themselves, such as that found in Eq. (24–27), but it is just this term that introduces an exponential dependence on r into the point solution, Eq. (24–31). Thus, while a Lagrangian density analogous to Eq. (24–27) yields a possible field theory from the point of view of the transformation character of the terms, it does not correspond to the experimental properties of the electromagnetic field. When translated into quantum mechanics, a theory in which the Lagrangian density contains terms quadratic in the potentials as well as their derivatives predicts photons of finite rest mass, instead of the zero rest mass photons of electromagnetic theory.

Thus, for the electromagnetic field, we are led to take

$$\mathcal{L}_3 = \frac{\epsilon_0(E^2 - c^2B^2)}{2} = -\frac{\epsilon_0}{4} F^{ij}F_{ij}. \tag{24-35}$$

Note that the field equations,

$$\mathbf{\nabla} \cdot \mathbf{B} = 0, \qquad \mathbf{\nabla} \times \mathbf{E} = -\frac{\partial \mathbf{B}}{\partial t},$$

are already implied in the connecting equations

$$F^{ij} = \frac{\partial \phi^j}{\partial x_i} - \frac{\partial \phi^i}{\partial x_j} \tag{17-30}$$

corresponding to

$$\mathbf{E} = -\mathbf{\nabla}\phi - \frac{\partial \mathbf{A}}{\partial t}, \qquad \mathbf{B} = \mathbf{\nabla} \times \mathbf{A}.$$

With Eq. (24–35), the total Lagrangian density is given by

$$\mathcal{L} = -\frac{\epsilon_0}{4} F^{ij}F_{ij} + eu^i\phi_i + \frac{1}{2}\rho_m^o c^2 u^i u_i. \tag{24-36}$$

A further reason for this choice of the Lagrangian is the connection between Lagrangian density and the energy-momentum tensor, Eq. (21–9). If we consider the Hamiltonian as retaining its meaning as an energy density, we can put

$$\frac{\partial \mathcal{H}}{\partial t} + \mathbf{\nabla} \cdot \mathbf{N} = 0. \tag{24-37}$$

Now

$$\frac{\partial \mathcal{H}}{\partial t} = \frac{\partial}{\partial t}\left(\dot{\eta}\frac{\partial \mathcal{L}}{\partial \dot{\eta}} - \mathcal{L}\right) = \dot{\eta}\frac{\partial}{\partial t}\left(\frac{\partial \mathcal{L}}{\partial \dot{\eta}}\right) - \dot{\eta}\frac{\partial \mathcal{L}}{\partial \eta} - \frac{\partial \mathcal{L}}{\partial(\partial \eta/\partial x^\alpha)}\frac{\partial^2 \eta}{\partial x^\alpha \partial t}$$

$$= -\frac{\partial}{\partial x^\alpha}\left[\dot{\eta}\frac{\partial \mathcal{L}}{\partial(\partial \eta/\partial x^\alpha)}\right]. \tag{24-38}$$

Hence, from Eq. (24–37),

$$N^\alpha = cT^{4\alpha} = \dot{\eta}\frac{\partial \mathcal{L}}{\partial(\partial \eta/\partial x^\alpha)}. \tag{24-39}$$

The covariant generalization of Eqs. (24–39) and (24–21) is

$$T^i_j = \frac{\partial \eta}{\partial x^j} \frac{\partial \mathcal{L}}{\partial (\partial \eta / \partial x^i)} - \mathcal{L} \, \delta^i_j. \qquad (24\text{–}40)$$

Thus the choice of Lagrangian fulfills both Eq. (24–39) and $T^4_4 = \mathcal{K}$, as required.

It can be easily shown that the Lagrangian $\mathcal{L}_2 + \mathcal{L}_3$ leads to Maxwell's equations. Note that this choice of $\mathcal{L}$ is not unique. Since the equations of motion will depend on the fields F^{ij}, which in turn do not depend on the quantity $\partial \phi^i / \partial x^i$, any function of $\partial \phi^i / \partial x^i$ may be added to $\mathcal{L}$. It is customary but not mandatory to use the Lorentz condition $\partial \phi^i / \partial x^i = 0$. From Eq. (24–17) we have, using $\mathcal{L} = \mathcal{L}_2 + \mathcal{L}_3$ and $\partial \mathcal{L} / \partial \phi_j = j^j$,

$$\frac{\partial}{\partial x^i} \frac{\partial \mathcal{L}}{\partial (\partial \phi_j / \partial x^i)} = \frac{\epsilon_0}{4} \frac{\partial}{\partial x^i} \left\{ \frac{\partial}{\partial (\partial \phi_j / \partial x^i)} \left[\left(\frac{\partial \phi^l}{\partial x_k} - \frac{\partial \phi^k}{\partial x_l} \right) \left(\frac{\partial \phi_l}{\partial x^k} - \frac{\partial \phi_k}{\partial x^l} \right) \right] \right\}$$

$$= \epsilon_0 \frac{\partial F^{ij}}{\partial x^i}, \qquad (24\text{–}41)$$

and thus

$$\frac{\partial F^{ij}}{\partial x^i} = \frac{j^j}{\epsilon_0}, \qquad (17\text{–}34)$$

which are Maxwell's source equations.

This calculation can, of course, also be carried out using the three-dimensional form,

$$\mathcal{L} = -\tfrac{1}{2}\epsilon_0 (c^2 B^2 - E^2) - \mathbf{A} \cdot \mathbf{j} + \rho \phi, \qquad (24\text{–}42)$$

and Eq. (24–19). In the language of three dimensions, we may take

$$\eta^\alpha = A^\alpha, \qquad (24\text{–}43)$$

$$\pi^\alpha = -\epsilon_0 E^\alpha, \qquad (24\text{–}44)$$

i.e., $\mathbf{A}$ and $\mathbf{E}$ are canonically conjugate. In terms of these variables, the Hamiltonian density of the field in the absence of sources is given by

$$\mathcal{K} = \epsilon_0 \left(\frac{\partial \mathbf{A}}{\partial t} \right)^2 + \epsilon_0 \left(\boldsymbol{\nabla} \phi \cdot \frac{\partial \mathbf{A}}{\partial t} \right) + \frac{(\boldsymbol{\nabla} \times \mathbf{A})^2}{2\mu_0} - \frac{\epsilon_0}{2} \left(\frac{\partial \mathbf{A}}{\partial t} \right)^2$$

$$- \epsilon_0 \left(\boldsymbol{\nabla} \phi \cdot \frac{\partial \mathbf{A}}{\partial t} \right) - \frac{\epsilon_0}{2} (\boldsymbol{\nabla} \phi)^2$$

$$= \frac{\pi^2}{2\epsilon_0} + \frac{(\boldsymbol{\nabla} \times \boldsymbol{\eta})^2}{2\mu_0} - (\boldsymbol{\pi} \cdot \boldsymbol{\nabla} \phi). \qquad (24\text{–}45)$$

The last term can be made zero by a particular choice of gauge; in any case, since $\nabla \cdot \mathbf{E} = 0$, the volume integral of the last term vanishes, and thus does not contribute to the energy.

24–4 Periodic solutions in a box. Plane wave representation.

An alternative way of treating Maxwell's equations in Hamiltonian form is to make the number of degrees of freedom finite by confining the field to a box of dimensions L. We may take the boundary conditions such that the field functions shall be periodic with period L in the three dimensions. The method is of great historical significance in the derivation of the Rayleigh-Jeans law (see Section 22–6), and leads to a representation of the field which is readily quantized.

Let us take as the Hamiltonian

$$H = \int \left(\frac{\epsilon_0 E^2 + \mu_0 H^2}{2} \right) dv = \int \left[\frac{\pi^2}{2\epsilon_0} + \frac{(\nabla \times \mathbf{A})^2}{2\mu_0} \right] dv. \quad (24\text{–}46)$$

Since the solutions are periodic, they can be expanded in space by a Fourier series. That is to say, if we put

$$\mathbf{u}_{k\lambda} = \frac{1}{L^{3/2}} \, \boldsymbol{\epsilon}_{k\lambda} e^{i\mathbf{k}\cdot\mathbf{r}}, \quad (24\text{–}47)$$

then in terms of these functions any vector function can be expanded as a sum over $\mathbf{k}$ and over the unit vectors, $\boldsymbol{\epsilon}_{k\lambda}$. The boundary conditions restrict $\mathbf{k}$ to the values $\mathbf{k} = (2\pi/L)(l\hat{\mathbf{x}} + m\hat{\mathbf{y}} + n\hat{\mathbf{z}})$, with l, m, n representing all positive and negative integers, and the $\boldsymbol{\epsilon}_{k\lambda}(\lambda = 1, 2, 3)$ permit an arbitrary choice of polarization. For example, we can write

$$\mathbf{A}(x^\alpha, t) = \sqrt{\mu_0} \sum_{k,\lambda} \mathbf{u}_{k\lambda}(r) q_{k\lambda}(t), \quad (24\text{–}48)$$

$$\boldsymbol{\pi}(x^\alpha, t) = \sqrt{\epsilon_0} \sum_{k,\lambda} \mathbf{u}_{k\lambda}(r) p_{k\lambda}(t). \quad (24\text{–}49)$$

The functions of Eq. (24–47) satisfy the auxiliary relations,

$$\begin{aligned} \nabla \cdot \mathbf{u}_{k\lambda} &= i\mathbf{k} \cdot \mathbf{u}_{k\lambda}, \\ \nabla \times \mathbf{u}_{k\lambda} &= i\mathbf{k} \times \mathbf{u}_{k\lambda}. \end{aligned} \quad (24\text{–}50)$$

Let us apply these relations to a pure, i.e., transverse, radiation field. We can choose a gauge in which $\phi = 0$, and thus take $\nabla \cdot \mathbf{A} = 0$, $\nabla \cdot \mathbf{E} = 0$. In that case $\mathbf{k} \cdot \boldsymbol{\epsilon}_{k\lambda} = 0$, and hence the summation over possible polarizations will include only two components, $\boldsymbol{\epsilon}_{k1}$ and $\boldsymbol{\epsilon}_{k2}$, for each $\mathbf{k}$, where $\boldsymbol{\epsilon}_{k\lambda}$

is perpendicular to $\mathbf{k}$. Thus

$$\nabla \times \mathbf{u}_{k1} = ik\mathbf{u}_{k2}. \tag{24-51}$$

For the first term in the Hamiltonian, we must evaluate

$$\int \pi^2 \, dv = \frac{\epsilon_0}{L^3} \int \sum_k \sum_{k'} e^{i(\mathbf{k}+\mathbf{k}')\cdot\mathbf{r}} p_k p_{k'} \, dv$$

$$= \epsilon_0 \sum_k \sum_{k'} \delta(k + k') p_k p_{k'} = \sum_k \epsilon_0 p_k p_{-k}. \tag{24-52}$$

Since π is real, Eq. (24-49) is equal to its complex conjugate, and since $\mathbf{u}_{k\lambda} = \mathbf{u}^*_{-k\lambda}$ we must have $p_k = p^*_{-k}$. Hence,

$$\frac{1}{\epsilon_0} \int \pi^2 \, dv = \sum_k \sum_{\lambda=1}^2 |p_{k\lambda}|^2. \tag{24-53}$$

Also, from Eqs. (24-51) and (24-48),

$$\frac{1}{\mu_0} \int (\nabla \times \mathbf{A})^2 \, dv = \sum_{k,\lambda} |ikq_{k\lambda}|^2 = \sum_{k,\lambda} k^2 |q_{k\lambda}|^2. \tag{24-54}$$

Hence,

$$H = \tfrac{1}{2} \sum_{k,\lambda} (|p_{k\lambda}|^2 + k^2 |q_{k\lambda}|^2). \tag{24-55}$$

Each term in Eq. (24-55) is the same as the Hamiltonian for a harmonic oscillator. Thus the equations of the electromagnetic field are equivalent to the equations of motion for a set of harmonic oscillators. The number of oscillators, $(\omega^2/\pi^2 c^3) \, d\omega$, corresponding to an angular frequency interval $d\omega$, results from taking the length of the box large in comparison with the wavelength of the radiation considered. By the application of statistical mechanics the mean energy of each oscillator which it is in equilibrium with walls at temperature T may be found. We have already seen that the resulting distribution law, Eq. (22-47), is in profound disagreement with experiment; it was Planck's effort to correct this treatment which originally led to the quantum hypothesis.

SUGGESTED REFERENCES

H. GOLDSTEIN, *Classical Mechanics*. Chapter 11 of this work forms an excellent introduction to the Hamiltonian formulation for fields. Our treatment of the transition from discrete to continuous systems parallels that in Goldstein's Section 11-1.

G. WENTZEL, *Introduction to the Quantum Theory of Fields*. The *raison d'être* for the Lagrangian and Hamiltonian formulation of classical electrodynamics is to facilitate quantization, and discussions are rare except as a preliminary to quantum-mechanical treatments. Wentzel's Chapter 1 is largely devoted to classical fields.

L. I. SCHIFF, *Quantum Mechanics*. The alternative treatments we have outlined are to be found in Chapters XIII and XIV of Schiff's very useful book on quantum-mechanics.

EXERCISE

1. Show that the transformation of the Lagrangian density

$$\mathcal{L}' = \mathcal{L} + \frac{\partial}{\partial x^i} [C^i(\eta)],$$

where C^i is an arbitrary vector function of η, does not affect the Eulerian equation. Hence the choice of Lagrangian density to represent a given field is not unique.

APPENDIXES

APPENDIX I

UNITS AND DIMENSIONS IN ELECTROMAGNETIC THEORY

Classical mechanics is characterized by the fact that its mathematical formulation does not contain any fundamental constants inherent in the theory. Hence all physical laws in classical mechanics "scale" perfectly for any change in parameters over any arbitrary range of magnitudes. It is customary, although not at all mandatory, to formulate classical mechanics in terms of three-dimensional entities: mass (M), length (L), and time (T). The number could be increased, for instance, by choosing the constant K in the equation Volume $= K$ (Length)^3 to be different from unity and to have dimensions. The dimension L^3 is customarily identified with volume by choice, not by necessity. Similarly, the number of fundamental units can be decreased by arbitrarily defining certain constants to be unity and dimensionless. The convention $c = \hbar = 1$ frequently used in quantum-mechanical calculations is such an example. In these units $L = T = M^{-1}$ arbitrarily.

We mention these examples only to indicate that the number of independent dimensions is arbitrary even in classical mechanics, although convenience suggests a specific choice. In general, the greater the number of dimensional entities chosen, the more independent units can be chosen to suit the orders of magnitude convenient for a particular purpose. It should be remembered, however, that changing units or even numbers of dimensions does not affect the physical content of any equation if it is correctly interpreted.

In classical mechanics the MLT system is used conventionally, and hence the issues discussed above are usually not of interest. For electromagnetic theory, the conventions are of more recent origin and appear more controversial.

Electromagnetic theory differs from classical mechanics by the fact that one constant, c, the velocity of light *in vacuo*, appears as a fundamental constant of the theory. Physical laws thus "scale" correctly over arbitrary magnitudes only if ratios of length and time are held constant. In this property electromagnetic theory exhibits a feature which special relativity extends to all laws of physics.

If the MLT system is used in the mechanical quantities in electromagnetic theory, the constant c having dimensions LT^{-1} will appear explicitly. Whether any additional dimensional units are introduced is entirely a matter of convention. As an example, if in Coulomb's law in the form $F = Kq_1q_2/r^2$ the constant K is chosen arbitrarily to be dimensionless, then the charge q automatically acquires the dimensions $L^{3/2}M^{1/2}T^{-1}$ and no basic units beyond those of M, L, and T need be specified. The justification for this procedure is analogous to that for setting K in the equation $V = KL^3$ equal to a dimensionless constant and thus giving a volume the dimension L^3.

If we do not choose a dimensionless constant in one of the equations relating mechanical and electromagnetic quantities, then we retain the freedom of choosing one of the electrical units arbitrarily and assigning to it a dimension. This has been done in what is called the mks system.

The particular set of units employed in this text is the mks system now in fairly general use. The principal convenience of this system is that it incorporates the common technical units—volt, ampere, coulomb, etc.—and is thus particularly suitable for treating applications that involve both "lumped" circuit parameters and fields. Since these technical units imply that the unit of time is the second and also define the power simply in watts, the natural choice of mechanical units are the meter, kilogram, and second. With this choice of electrical and mechanical units the constants ϵ_0 and μ_0 in the equations

$$\mathbf{F} = \frac{q_1 q_2}{4\pi\epsilon_0} \frac{\mathbf{r}_{12}}{r_{12}^3}, \tag{1}$$

$$\mathbf{F} = \frac{\mu_0}{4\pi} \iint \frac{\mathbf{j}_1 \times (\mathbf{j}_2 \times \mathbf{r}_{12})}{r_{12}^3} \, dv_1 \, dv_2 \tag{2}$$

can be determined once we have selected a basic electrical unit. We shall postpone their numerical determination until we have seen how the electrical unit was chosen, but we note that two constants, ϵ_0 and μ_0, are carried in the equations when they are written in mks units, although only one constant, c, is fundamental to the theory.

Historically, a set of units (esu) was defined by using unity in place of $4\pi\epsilon_0$ in Coulomb's law, Eq. (1), and cgs units for mechanical quantities. This defines the electrostatic unit of charge with mechanical dimensions indicated above, and from this the units of potential, electric field, etc., are defined. On the other hand, if we set $\mu_0 = 4\pi$ in Eq. (2), and if cgs mechanical units are used, the equation defines a unit of current (with dimensions $M^{1/2}L^{1/2}T^{-1}$) called the electromagnetic unit (emu) or abampere. Units for other electrical quantities can be derived from the abampere and the cgs relations. The charge densities and current densities thus defined obey the relation

$$\mathbf{j}_{\text{emu}} = \rho_{\text{esu}} \frac{\mathbf{u}}{c}, \tag{3}$$

where c appears here as the measured ratio of the units. This ratio was first determined by Weber and Kohlrausch by measuring the discharge of a capacitor whose electrostatic capacity was known. A consistent set of units (Gaussian units) is obtained by using the electrical quantities derived from ρ_{esu} and the magnetic quantities derived from j_{emu} and carrying c in the equations as the only constant.

As indicated in Eqs. (1) and (2), the mks system is commonly used in its rationalized form; the cgs units have been quoted in their unrationalized form. A rational system of units contains the factor 4π in Coulomb's and Ampère's laws of force so as to eliminate it in the Maxwell field equations which involve sources. General vector relations do not contain such factors. The appearance of the geometrical factor is indicated mathematically by the form of the Green's function. If a source (such as a point source) defines a problem in spherical symmetry, then in rational units one obtains 4π explicitly in the resulting solution; if the source (such as a line source) defines a field structure of circular symmetry, then 2π appears. A system of units analogous to the Gaussian system but in rationalized form is known as the Heaviside-Lorentz system.

Historically, the ampere was taken to be exactly $\frac{1}{10}$ of the abampere, or emu, of current. This fact enables us to determine the magnitude of μ_0 in Eq. (2), since we have seen that the abampere is defined by setting $\mu_0/4\pi = 1$. It is customary to use the coulomb, not the ampere, as the basic electrical quantity, and we need only transform the defining equation of the electromagnetic system with all quantities of unit size into mks units. Explicitly, in the electromagnetic cgs system,

$$1 \text{ dyne} = 1 \text{ (abampere)}^2$$

[since all lengths cancel on the right side of Eq. (2)], which in mks units for which 1 abampere $= 10$ coulomb/second, becomes

$$10^{-5} \text{ newton} \equiv 10^{-5} \frac{\text{kilogram-meter}}{\text{second}^2} = \frac{\mu_0}{4\pi} \frac{10^2 \text{ (coulomb)}^2}{\text{(second)}^2}$$

or

$$\mu_0 = 4\pi \times 10^{-7} \frac{\text{kilogram-meter}}{\text{(coulomb)}^2} = 4\pi \times 10^{-7} \frac{\text{henry}}{\text{meter}}.$$

The constant ϵ_0 is now obtained from the relation $\epsilon_0 \mu_0 = 1/c^2$,

$$\epsilon_0 = \left(\frac{10^7}{4\pi c^2} \simeq \frac{1}{36\pi \times 10^9} \right) \left(\frac{\text{(coulomb)}^2 \text{ (second)}^2}{\text{kilogram (meter)}^3} = \frac{\text{farad}}{\text{meter}} \right).$$

What has been done here is to define μ_0 in terms of the arbitrarily chosen size and dimension of an electrical unit, whereupon ϵ_0 is automatically fixed if the system is to be consistent with the mechanical units and the *experimental* value of the fundamental constant c.

Electrical units which are referred to mechanical standards via defining relations containing fixed numerical constants are called *absolute* electrical units. Actually, the accuracy with which the emu and esu could be realized in terms of their defining equations was until recently insufficient for practical purposes: during the period when the accuracy of verification of absolute units was inferior to the reproducibility of standards, the practical units were based on such standards* as the *international* ampere and the *international* ohm. Improvements in techniques have resulted in greatly improved absolute electrical measurements, so that the former international standards have been relegated to the role of secondary standards. Hence the value $\mu_0 = 4\pi \times 10^{-7}$ as an exact numerical constant refers to the practical units as absolute rather than international units. Note that ϵ_0 in the mks system depends on the experimental relation between the velocity of electromagnetic radiation to the length and time standards (although μ_0 does not); this corresponds to the explicit presence of c in the Gaussian and Heaviside-Lorentz systems. In the "natural" system the velocity of light itself constitutes a standard. It is clear that an experimental measurement of the velocity of light can only provide a measure of the ratio of the velocity of propagation of electromagnetic radiation to the ratio of length and time standards. Hence the resultant number can never have any fundamental significance in an absolute sense, but is of great practical utility in providing independent accurate means of relating the length and time standards.

Clearly, the physical content of the fundamental relations is the same in all systems of units. It is easy to translate the laws of electrodynamics from one system to another: for vacuum conditions, the relations in this book are written so that the transformation

$$cB_{\text{mks}} \rightarrow B_{\text{Gaussian}},$$

$$\epsilon_0 \rightarrow (4\pi)^{-1},$$

$$\mu_0 \epsilon_0 \rightarrow 1/c^2$$

will effect a reduction to their Gaussian equivalents. Table I–2 contains a brief summary of fundamental electromagnetic relations valid *in vacuo* as they appear in the various systems of units. For convenience in the numerical conversion of units, a list of the most important conversion factors is given in Table I–1. Table I–4 contains other numerical constants and functional relations useful in applications involving atomic particles.

---

* For example, the international ampere was defined as "the value of the unvarying current which on passing through a solution of silver nitrate in water in accordance with standard specifications deposits silver at the rate of 0.001118 gm per second."

The situation regarding the equations in material media is somewhat more complex than that for vacuum conditions and is frequently misunderstood. To gain some insight into the problem, let us consider the fundamental process by which Maxwell's equations in media are generated from the vacuum relations.

The vacuum source equations have the general form

$$D(F) = S, \tag{4}$$

where D is a linear differential operator acting on the field F and S is the source. In media S is broken up in terms of an "accessible" (macroscopic) source S and an "inaccessible" source S_p, i.e.,

$$D(F) = S + S_p. \tag{5}$$

S_p is then derived from an auxiliary quantity F_p by the same differential operator D such that

$$S_p = D(F_p) \tag{6}$$

and hence

$$D(F - F_p) = S. \tag{7}$$

A partial field $F_H = F - F_p$ can thus be defined such that

$$D(F_H) = S, \tag{8}$$

i.e., such that this field is derived from the "accessible" (often called "true") sources only. In Table I–3 this general statement is illustrated in terms of the actual electrodynamic quantities.

Ambiguity arises because in this general formulation quantities of the type F_p (i.e., **P** and **M**) appear in a dual role. On the one hand, in the relations

$$\rho_P = -\mathbf{\nabla} \cdot \mathbf{P}, \tag{9}$$

$$\mathbf{j}_M = \mathbf{\nabla} \times \mathbf{M}, \tag{10}$$

P and **M** represent purely *source* quantities—they describe certain charge and current distributions. In both relations, on the other hand, **P** and **M** can be viewed as *fields*, namely, those electric or magnetic fields whose sources are ρ_P or $\mathbf{j}_M$ respectively. In relations of the type

$$\epsilon_0 \mathbf{E} = \mathbf{D} - \mathbf{P}, \tag{11}$$

$$\mathbf{B}/\mu_0 = \mathbf{H} + \mathbf{M}, \tag{12}$$

as written in the now conventional mks system, **P** and **E** or **B** and **M** are given different units: P is measured in coulombs/meter^2, while E is in volts/meter; M is in amperes/meter, while B is in webers/meter^2. This convention emphasizes the roles of M and P as current and charge descriptions. The "partial field" aspect is, however, equally valid, and is somewhat obscured by the constants in Eqs. (11) and (12). In cgs units, Eqs. (11) and (12) become

$$E = D - 4\pi P, \qquad B = H + 4\pi M \text{ (Gaussian)},$$

$$E = D - P, \qquad B = H + M \text{ (Heaviside-Lorentz)}.$$

Here the units of all quantities are the same, and hence the "partial field" aspect is emphasized.

To solve field problems, it is in general necessary to specify the "constitutive equation" giving $P(E)$ and $M(B)$. If these are of linear form, such as

$$P = \epsilon_0(\kappa - 1)E, \tag{13}$$

it is often said that E represents an "intensive variable" and P an "extensive variable," i.e., E is cause and P effect. This point of view is emphasized by the fact that $E \cdot \delta P$ represents the differential of work done in this case. From this aspect the use of different units for E and P appears justified. Actually the cause-effect situation is very much less clear when permanent polarization (electrets) or permanent magnets are considered. It should be noted that the basic relations (11) and (12) are *additive*; on the other hand, relations of type (13) are not at all general.

We may summarize by saying that the question of whether E, P, and D (or B, M, and H) should have the same units is fairly irrelevant; in fact, an understanding of the *dual* physical function of P and M is the principal requirement for clarity in the classical theory of electric and magnetic media.

TABLE I-1

CONVERSION FACTORS

Multiply the number of mks units below	by	to obtain the number of Gaussian (cgs) units of
ampere	10^{-1}	current in abamperes
ampere/meter^2	10^{-5}	current density in abampere/cm^2
coulomb	$3 \times 10^{9*}$	charge in esu
coulomb/meter^3	$3 \times 10^{3*}$	charge density in esu/cm^3
farad $=$ coulomb/volt	$9 \times 10^{-11*}$	capacitance in cm
henry $=$ volt sec/ampere	10^9	inductance in emu
joule	10^7	energy in ergs
newton	10^5	force in dynes
ohm $=$ volt/ampere	$\frac{1}{30}*$	resistance in esu of potential per abampere
	$\frac{1}{9} \times 10^{11*}$	resistance in esu
volt	$\frac{1}{300}*$	potential in esu
volt/meter	$\frac{1}{3} \times 10^{-4*}$	electric field intensity **E** in esu
coulomb/meter^2	$12\pi \times 10^{5*}$	electric displacement **D** in esu
weber $=$ volt second	10^8	magnetic flux in maxwells
weber/meter^2	10^4	flux density **B** in gauss
ampere-turns/meter	$4\pi/10^3$	field intensity **H** in oersteds
mho/meter	$\frac{3}{10}*$	conductivity in abamperes/cm^2/esu of field intensity
ampere turns	$4\pi/10$	magnetomotive force in gilberts

* In all conversion factors marked by an asterisk 3 is used for $c/10^{10}$, where in this definition c is measured in cgs units. If higher accuracy is desired, a more precise value of c must be substituted. DuMond and Cohen give $c = 299792.5$ km/sec. (J. W. M. DuMond and E. R. Cohen, 1961 adjustment of natural constants, to be published in *Annals of Physics* and *Nuovo Cimento*.)

Table I-2

FUNDAMENTAL ELECTROMAGNETIC RELATIONS VALID "IN VACUO" AS THEY APPEAR IN THE VARIOUS SYSTEMS OF UNITS

mks (rationalized)	Gaussian* (cgs)	Heaviside-Lorentz (cgs)	"Natural" units $c = \hbar = 1$ (rationalized)
$\nabla \cdot \mathbf{E} = \rho/\epsilon_0$	$\nabla \cdot \mathbf{E} = 4\pi\rho$	$\nabla \cdot \mathbf{E} = \rho$	$\nabla \cdot \mathbf{E} = \rho$
$\mathbf{E} = \dfrac{1}{4\pi\epsilon_0} \displaystyle\int \dfrac{\rho \mathbf{r}}{r^3}\, dv$	$\mathbf{E} = \displaystyle\int \dfrac{\rho \mathbf{r}}{r^3}\, dv$	$\mathbf{E} = \dfrac{1}{4\pi} \displaystyle\int \dfrac{\rho \mathbf{r}}{r^3}\, dv$	$\mathbf{E} = \dfrac{1}{4\pi} \displaystyle\int \dfrac{\rho \mathbf{r}}{r^3}\, dv$
$\nabla \times \mathbf{B} = \mu_0 \left(\mathbf{j} + \epsilon_0 \dfrac{\partial \mathbf{E}}{\partial t}\right)$ $\mathbf{B} = \dfrac{\mu_0}{4\pi} \displaystyle\int \dfrac{\left(\mathbf{j} + \epsilon_0 \dfrac{\partial \mathbf{E}}{\partial t}\right) \times \mathbf{r}}{r^3}\, dv$	$\nabla \times \mathbf{B} = 4\pi\mathbf{j} + \dfrac{1}{c}\dfrac{\partial \mathbf{E}}{\partial t}$ $\mathbf{B} = \displaystyle\int \dfrac{\left(\mathbf{j} + \dfrac{1}{4\pi c}\dfrac{\partial \mathbf{E}}{\partial t}\right) \times \mathbf{r}}{r^3}\, dv$	$\nabla \times \mathbf{B} = \mathbf{j} + \dfrac{1}{c}\dfrac{\partial \mathbf{E}}{\partial t}$ $\mathbf{B} = \dfrac{1}{4\pi} \displaystyle\int \dfrac{\left(\mathbf{j} + \dfrac{1}{c}\dfrac{\partial \mathbf{E}}{\partial t}\right) \times \mathbf{r}}{r^3}\, dv$	$\nabla \times \mathbf{B} = \mathbf{j} + \dfrac{\partial \mathbf{E}}{\partial t}$ $\mathbf{B} = \dfrac{1}{4\pi} \displaystyle\int \dfrac{\left(\mathbf{j} + \dfrac{\partial \mathbf{E}}{\partial t}\right) \times \mathbf{r}}{r^3}\, dv$
$\nabla \cdot \mathbf{B} = 0$	$\nabla \cdot \mathbf{B} = 0$	$\nabla \cdot \mathbf{B} = 0$	$\nabla \cdot \mathbf{B} = 0$
$\nabla \times \mathbf{E} = -\dfrac{\partial \mathbf{B}}{\partial t}$	$c\nabla \times \mathbf{E} = -\dfrac{\partial \mathbf{B}}{\partial t}$	$c\nabla \times \mathbf{E} = -\dfrac{\partial \mathbf{B}}{\partial t}$	$\nabla \times \mathbf{E} = -\dfrac{\partial \mathbf{B}}{\partial t}$
$\mathbf{F} = e(\mathbf{E} + \mathbf{u} \times \mathbf{B})$	$\mathbf{F} = e\left(\mathbf{E} + \dfrac{\mathbf{u}}{c} \times \mathbf{B}\right)$	$\mathbf{F} = e\left(\mathbf{E} + \dfrac{\mathbf{u}}{c} \times \mathbf{B}\right)$	$\mathbf{F} = e(\mathbf{E} + \mathbf{u} \times \mathbf{B})$

$\mathbf{B} = \nabla \times \mathbf{A}$	$\mathbf{B} = \nabla \times \mathbf{A}$	$\mathbf{B} = \nabla \times \mathbf{A}$	$\mathbf{B} = \nabla \times \mathbf{A}$
$\mathbf{E} = -\nabla\phi - \dfrac{1}{c^2}\dfrac{\partial \mathbf{A}}{\partial t}$	$\mathbf{E} = -\nabla\phi - \dfrac{1}{c}\dfrac{\partial \mathbf{A}}{\partial t}$	$\mathbf{E} = -\nabla\phi - \dfrac{1}{c}\dfrac{\partial \mathbf{A}}{\partial t}$	$\mathbf{E} = -\nabla\phi - \dfrac{\partial \mathbf{A}}{\partial t}$
$\nabla \cdot \mathbf{j} + \dfrac{\partial \rho}{\partial t} = 0$	$\nabla \cdot \mathbf{j} + \dfrac{1}{c}\dfrac{\partial \rho}{\partial t} = 0$	$\nabla \cdot \mathbf{j} + \dfrac{1}{c}\dfrac{\partial \rho}{\partial t} = 0$	$\nabla \cdot \mathbf{j} + \dfrac{\partial \rho}{\partial t} = 0$
$\nabla \cdot \mathbf{A} + \dfrac{1}{c^2}\dfrac{\partial \phi}{\partial t} = 0$	$\nabla \cdot \mathbf{A} + \dfrac{1}{c}\dfrac{\partial \phi}{\partial t} = 0$	$\nabla \cdot \mathbf{A} + \dfrac{1}{c}\dfrac{\partial \phi}{\partial t} = 0$	$\nabla \cdot \mathbf{A} + \dfrac{\partial \phi}{\partial t} = 0$
$F^{ij} = \begin{pmatrix} 0 & -cB_z & cB_y & +E_x \\ cB_z & 0 & -cB_x & +E_y \\ -cB_y & cB_x & 0 & +E_z \\ -E_x & -E_y & -E_z & 0 \end{pmatrix}$	$F^{ij} = \begin{pmatrix} 0 & -B_z & B_y & +E_x \\ B_z & 0 & -B_x & +E_y \\ -B_y & B_x & 0 & +E_z \\ -E_x & -E_y & -E_z & 0 \end{pmatrix}$	F^{ij} has same form as in Gaussian units	F^{ij} has same form as in Gaussian units
$\dfrac{\partial F^{ij}}{\partial x^i} = \dfrac{j^j}{\epsilon_0}$	$\dfrac{\partial F^{ij}}{\partial x^i} = 4\pi j^j$	$\dfrac{\partial F^{ij}}{\partial x^i} = j^j$	$\dfrac{\partial F^{ij}}{\partial x^i} = j^j$

* In some textbooks employing Gaussian units **j** is measured in esu. In that case the equations are those given here except that **j** appears with a factor $1/c$.

TABLE I-3

DEFINITION OF FIELDS FROM SOURCES (MKS SYSTEM)

	Electric	Magnetic	Equivalent covariant description
Vacuum (all sources "accessible")	$\nabla \cdot \mathbf{E} = \rho/\epsilon_0$	$\nabla \times \mathbf{B} = \mu_0\left(\mathbf{j} + \epsilon_0 \frac{\partial \mathbf{E}}{\partial t}\right)$	$\frac{\partial F^{ij}}{\partial x^i} = \frac{j^j}{\epsilon_0}$
Material media, sources separated	$\nabla \cdot \mathbf{E} = \frac{\rho + \rho_P}{\epsilon_0}$	$\nabla \times \mathbf{B} = \mu_0\left(\mathbf{j} + \mathbf{j}_M + \mathbf{j}_P + \epsilon_0 \frac{\partial \mathbf{E}}{\partial t}\right)$	$\frac{\partial F^{ij}}{\partial x^i} = \frac{1}{\epsilon_0}\left(j^j + j^j_M\right)$
Inaccessible sources defined from auxiliary function	$\nabla \cdot \mathbf{E} = \frac{\rho}{\epsilon_0} + \frac{(-\nabla \cdot \mathbf{P})}{\epsilon_0}$	$\nabla \times \mathbf{B} = \mu_0\left(\mathbf{j} + \nabla \times \mathbf{M} + \frac{\partial \mathbf{P}}{\partial t} + \epsilon_0 \frac{\partial \mathbf{E}}{\partial t}\right)$	$\frac{\partial F^{ij}}{\partial x^i} = \frac{1}{\epsilon_0}\left(j^j + \frac{\partial M^{ij}}{\partial x^i}\right)$
Definition of partial field	$\mathbf{D} = \epsilon_0\mathbf{E} - (-\mathbf{P})$	$\mathbf{H} = \frac{\mathbf{B}}{\mu_0} - \mathbf{M}$	$H^{ij} = \epsilon_0 F^{ij} - M^{ij}$
Field equations in media	$\nabla \cdot \mathbf{D} = \rho$	$\nabla \times \mathbf{H} = \mathbf{j} + \frac{\partial \mathbf{D}}{\partial t}$	$\frac{\partial H^{ij}}{\partial x^i} = j^j$

TABLE I–4.   USEFUL NUMERICAL RELATIONS

A. *Some atomic constants.**

cgs Gaussian	cgs Heaviside-Lorentz	mks	Name
$\dfrac{e^2}{\hbar c} = \dfrac{1}{137.039}$	$\dfrac{e^2}{4\pi\hbar c} = \dfrac{1}{137.039}$	$\dfrac{e^2}{4\pi\hbar c\epsilon_0} = \dfrac{1}{137.039}$	Fine structure constant
$\dfrac{e^2}{mc^2} = 2.81776$ $\times 10^{-13}$ cm	$\dfrac{e^2}{4\pi mc^2} = 2.81776$ $\times 10^{-13}$ cm	$\dfrac{e^2}{4\pi\epsilon_0 mc^2} = 2.81776$ $\times 10^{-15}$ m	Classical electron radius
$\dfrac{\hbar^2}{me^2} = 5.29166$ $\times 10^{-9}$ cm	$\dfrac{4\pi\hbar^2}{me^2} = 5.29166$ $\times 10^{-9}$ cm	$\dfrac{4\pi\epsilon_0\hbar^2}{me^2} = 5.29166$ $\times 10^{-11}$ m	Bohr radius

B. *Relations useful if energy of a particle is measured in electron volts.**

$$E_0 = m_0c^2 = 0.51097 \text{ Mev} \quad \text{for electron}$$
$$E_0 = m_0c^2 = 938.21 \text{ Mev} \quad \text{for proton}$$

1. "Magnetic rigidity" $B\rho$ of particle of kinetic energy T carrying charge e.

cgs Gaussian	mks
$B\rho = \dfrac{10^8}{c}\sqrt{T^2 + 2TE_0} \sim \dfrac{\sqrt{T^2 + 2TE_0}}{300}$	$B\rho = \dfrac{\sqrt{T^2 + 2TE_0}}{c} \sim \dfrac{\sqrt{T^2 + 2TE_0}}{3 \times 10^8}$

2. Tension τ of wire carrying current I having same orbit in magnetic field as particle of kinetic energy T carrying electronic charge e.

cgs Gaussian	mks
$\tau = I\sqrt{T^2 + 2TE_0} \times \dfrac{10^8}{c} \sim \dfrac{I\sqrt{T^2 + 2TE_0}}{300}$	$\tau c = I\sqrt{T^2 + 2TE_0}$

3. Velocity u, momentum p, kinetic energy T, total energy E.

$$\frac{E}{E_0} = \cosh\theta = \frac{T}{E_0} + 1 \qquad = \sqrt{\left(\frac{cp}{E_0}\right)^2 + 1} \quad = \frac{1}{\sqrt{1 - (u^2/c^2)}}$$

$$\frac{cp}{E_0} = \sinh\theta = \sqrt{\left(\frac{T}{E_0}\right)^2 + 2\frac{T}{E_0}} \qquad = \sqrt{\left(\frac{E}{E_0}\right)^2 - 1} \quad = \frac{(u/c)}{\sqrt{1 - (u^2/c^2)}}$$

$$\frac{u}{c} = \tanh\theta = \frac{\sqrt{2TE_0 + T^2}}{T + E_0} \qquad = \frac{\sqrt{E^2 - E_0^2}}{E} \quad = \frac{cp}{\sqrt{(cp)^2 + E_0^2}}$$

$$\frac{T}{E_0} = \cosh\theta - 1 = \frac{1}{\sqrt{1 - (u^2/c^2)}} - 1 \quad = \frac{E}{E_0} - 1 \quad = \sqrt{\left(\frac{cp}{E_0}\right)^2 + 1} - 1$$

* The numerical values here are consistent with those given by J. W. M. DuMond and E. R. Cohen, *loc. cit.*

APPENDIX II

USEFUL VECTOR RELATIONS

A summary of the more useful formulas from vector analysis for application to electromagnetic theory is given for convenience in Table II–1. Note that $\mathbf{A}$, $\mathbf{B}$, $\mathbf{C}$, $\mathbf{D}$ are arbitrary vector fields, ϕ and ψ arbitrary scalar fields. It is clear that the relations involving the vector operator ∇ may be derived formally from the vector identities (1) through (5) if one remembers that ∇ is a differential operator as well as a vector and thus does not commute with functions of the coordinates. Thus (7) and (10) may be derived by writing, respectively,

$$\nabla \cdot (\phi\mathbf{A}) = \nabla \cdot (\phi_c\mathbf{A}) + \nabla \cdot (\phi\mathbf{A}_c),$$

and

$$\nabla \times (\mathbf{A} \times \mathbf{B}) = (\nabla \cdot \mathbf{B})\mathbf{A} - (\nabla \cdot \mathbf{A})\mathbf{B}$$

$$= (\nabla \cdot \mathbf{B}_c)\mathbf{A} + (\nabla \cdot \mathbf{B})\mathbf{A}_c - (\nabla \cdot \mathbf{A}_c)\mathbf{B} - (\nabla \cdot \mathbf{A})\mathbf{B}_c,$$

where the subscript c indicates that the function is held constant and may be permuted with the vector operator, with due regard to sign changes if such changes are indicated by the ordinary vector relations.

In this book we use interchangeably the Gibbs notation involving the ∇ operator and tensor notation in which we write the αth component in place of the vector. In the latter notation the relations (7) and (10) would arise as follows:

$$\nabla \cdot (\phi\mathbf{A}) = \frac{\partial}{\partial x_\alpha}(\phi A_\alpha) = \phi\frac{\partial A_\alpha}{\partial x_\alpha} + A_\alpha\frac{\partial \phi}{\partial x_\alpha},$$

$$\nabla \times (\mathbf{A} \times \mathbf{B})|_\alpha = (\nabla \cdot \mathbf{B})A_\alpha - (\nabla \cdot \mathbf{A})B_\alpha = \frac{\partial(B_\beta A_\alpha)}{\partial x_\beta} - \frac{\partial(A_\beta B_\alpha)}{\partial x_\beta}.$$

Note that a repeated index implies summation over all components.

TABLE II–1

VECTOR FORMULAS

General relations:

(1) $\mathbf{A} \cdot \mathbf{B} \times \mathbf{C} = \mathbf{A} \times \mathbf{B} \cdot \mathbf{C} \equiv (\mathbf{A}, \mathbf{B}, \mathbf{C}) = (\mathbf{B}, \mathbf{C}, \mathbf{A}) = (\mathbf{C}, \mathbf{A}, \mathbf{B})$
$= -(\mathbf{A}, \mathbf{C}, \mathbf{B}) = -(\mathbf{C}, \mathbf{B}, \mathbf{A}) = -(\mathbf{B}, \mathbf{A}, \mathbf{C})$

(2) $\mathbf{A} \times (\mathbf{B} \times \mathbf{C}) = (\mathbf{A} \cdot \mathbf{C})\mathbf{B} - (\mathbf{A} \cdot \mathbf{B})\mathbf{C}$

(3) $\mathbf{A} \times (\mathbf{B} \times \mathbf{C}) + \mathbf{B} \times (\mathbf{C} \times \mathbf{A}) + \mathbf{C} \times (\mathbf{A} \times \mathbf{B}) = 0$

(4) $(\mathbf{A} \times \mathbf{B}) \cdot (\mathbf{C} \times \mathbf{D}) = \mathbf{A} \cdot [\mathbf{B} \times (\mathbf{C} \times \mathbf{D})]$
$= (\mathbf{A} \cdot \mathbf{C})(\mathbf{B} \cdot \mathbf{D}) - (\mathbf{A} \cdot \mathbf{D})(\mathbf{B} \cdot \mathbf{C})$

(5) $(\mathbf{A} \times \mathbf{B}) \times (\mathbf{C} \times \mathbf{D}) = (\mathbf{A} \times \mathbf{B} \cdot \mathbf{D})\mathbf{C} - (\mathbf{A} \times \mathbf{B} \cdot \mathbf{C})\mathbf{D}$

(6) $\text{grad } (\phi\psi) \equiv \nabla(\phi\psi) = \phi\nabla\psi + \psi\nabla\phi$

(7) $\text{div } (\phi\mathbf{A}) \equiv \nabla \cdot (\phi\mathbf{A}) = \mathbf{A} \cdot \nabla\phi + \phi\nabla \cdot \mathbf{A}$

(8) $\text{curl } (\phi\mathbf{A}) \equiv \nabla \times (\phi\mathbf{A}) = \phi\nabla \times \mathbf{A} - \mathbf{A} \times \nabla\phi$

(9) $\text{div } (\mathbf{A} \times \mathbf{B}) \equiv \nabla \cdot (\mathbf{A} \times \mathbf{B}) = \mathbf{B} \cdot (\nabla \times \mathbf{A}) - \mathbf{A} \cdot (\nabla \times \mathbf{B})$

(10) $\text{curl } (\mathbf{A} \times \mathbf{B}) \equiv \nabla \times (\mathbf{A} \times \mathbf{B})$
$= \mathbf{A}(\nabla \cdot \mathbf{B}) - \mathbf{B}(\nabla \cdot \mathbf{A}) + (\mathbf{B} \cdot \nabla)\mathbf{A} - (\mathbf{A} \cdot \nabla)\mathbf{B}$

(11) $\text{grad } (\mathbf{A} \cdot \mathbf{B}) \equiv \nabla(\mathbf{A} \cdot \mathbf{B})$
$= \mathbf{A} \times (\nabla \times \mathbf{B}) + \mathbf{B} \times (\nabla \times \mathbf{A}) + (\mathbf{B} \cdot \nabla)\mathbf{A} + (\mathbf{A} \cdot \nabla)\mathbf{B}$

(12) $\nabla^2\phi = \nabla \cdot \nabla\phi$

(13) $\nabla^2\mathbf{A} = \nabla(\nabla \cdot \mathbf{A}) - \nabla \times (\nabla \times \mathbf{A})$

(14) $\nabla \times \nabla\phi = 0$

(15) $\nabla \cdot (\nabla \times \mathbf{A}) = 0$

Special relations:

Let $\mathbf{r}$ be the radius vector, of magnitude r, from the origin to the point x, y, z, and let $\mathbf{J}$ be any constant vector. Then

(16) $\nabla \cdot \mathbf{r} = 3$

(17) $\nabla \times \mathbf{r} = 0$

(18) $\nabla r = \mathbf{r}/r$

(19) $\nabla(1/r) = -\mathbf{r}/r^3$

(20) $\nabla \cdot (\mathbf{r}/r^3) = -\nabla^2(1/r) = 0$ if $r \neq 0$

(21) $\nabla \cdot (\mathbf{J}/r) = \mathbf{J} \cdot [\nabla(1/r)] = -(\mathbf{J} \cdot \mathbf{r})/r^3$

(22) $\nabla \times [\mathbf{J} \times (\mathbf{r}/r^3)] = -\nabla[(\mathbf{J} \cdot \mathbf{r})/r^3]$ if $r \neq 0$

(23) $\nabla^2(\mathbf{J}/r) = \mathbf{J}\nabla^2(1/r) = 0$ if $r \neq 0$

(24) $\nabla \times (\mathbf{J} \times \mathbf{B}) = \mathbf{J}(\nabla \cdot \mathbf{B}) + \mathbf{J} \times (\nabla \times \mathbf{B}) - \nabla(\mathbf{J} \cdot \mathbf{B})$

Integral relations:

If V is a volume bounded by a closed surface S, with $d\mathbf{S}$ positive outward from the enclosed volume,

(25) $\oint_S \mathbf{A} \cdot d\mathbf{S} = \int_V (\nabla \cdot \mathbf{A}) \, dv$

(26) $\displaystyle\oint_S \phi \, d\mathbf{S} = \int_V (\boldsymbol{\nabla}\phi) \, dv$

(27) $\displaystyle\oint_S d\mathbf{S} \times \mathbf{A} = \int_V (\boldsymbol{\nabla} \times \mathbf{A}) \, dv$

If S is an open surface bounded by the contour C, of which the line element is $d\mathbf{s}$,

(28) $\displaystyle\oint_C \phi \, d\mathbf{s} = \int_S d\mathbf{S} \times \boldsymbol{\nabla}\phi$

(29) $\displaystyle\oint_C \mathbf{A} \cdot d\mathbf{s} = \int_S (\boldsymbol{\nabla} \times \mathbf{A}) \cdot d\mathbf{S}$

APPENDIX III

VECTOR RELATIONS IN CURVILINEAR COORDINATES

A list of vector operators expressed in the common orthogonal curvilinear coordinates is often useful in the solution of physical problems. For the derivation of these relations, it is possible to procede quite formally from the definition of the operator ∇ in Cartesian coordinates and the transformation equations to other coordinate systems, but for physical applications it is advantageous to work from the geometrical definitions of gradient, divergence, and curl. One may first specify the coordinate system and derive the required expressions, or make a general derivation valid for any curvilinear coordinates and only then specify the coordinates. We shall follow the latter plan, first outlining a derivation valid for any right-handed system of orthogonal coordinates for which the line element is known, and then writing the particular forms for Cartesian, cylindrical, and spherical polar coordinates.

A line element in three dimensions is an infinitesimal displacement in space. If only one of three orthogonal coordinates q_1, q_2, q_3 is varied, the corresponding line element may be written

$$ds_1 = h_1 \, dq_1, \tag{1}$$

together with similar expressions in q_2 and q_3. For any infinitesimal displacement,

$$ds^2 = h_1^2 \, dq_1^2 + h_2^2 \, dq_2^2 + h_3^2 \, dq_3^2. \tag{2}$$

Now the gradient of a scalar function ψ is defined by the requirement that

$$\nabla \psi \cdot d\mathbf{s} = d\psi, \tag{3}$$

giving the change in ψ corresponding to the space displacement $d\mathbf{s}$. Then

$$(\nabla \psi)_1 = \lim_{ds_1 \to 0} \frac{\psi(q_1 + dq_1) - \psi(q_1)}{ds_1} = \frac{1}{h_1} \frac{\partial \psi}{\partial q_1} \tag{4}$$

is the general form of a gradient component.

To find the divergence, we shall consider an infinitesimal volume $dv = ds_1 \, ds_2 \, ds_3$ bounded by the surfaces $q_1 = $ constant, $q_1 + dq_1 = $ constant, etc., as indicated in Fig. III–1. Let us apply Gauss's divergence theorem, $\int \nabla \cdot \mathbf{A} \, dv = \int \mathbf{A} \cdot d\mathbf{S}$, to a vector $\mathbf{A}(q_1, q_2, q_3)$ with components A_1, A_2, A_3, integrating over this infinitesimal volume. The integral of the outward normal component of $\mathbf{A}$ over the two surfaces

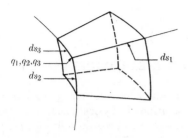

FIG. III–1. Element of volume for computing the divergence.

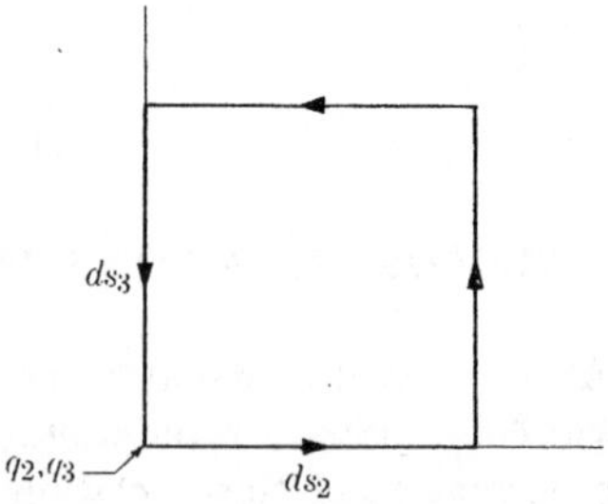

FIG. III–2. Element of area in the q_2, q_3 plane for finding the q_1 component of the curl. Arrows show the direction of the path of integration.

perpendicular to the direction of increasing q_1 is

$$(A_1 \, ds_2 \, ds_3)_{q_1+dq_1} - (A_1 \, ds_2 \, ds_3)_{q_1} = \frac{\partial}{\partial q_1} (A_1 \, ds_2 \, ds_3) \, dq_1$$

$$= \frac{\partial(h_2 h_3 A_1)}{\partial q_1} \, dq_1 \, dq_2 \, dq_3$$

$$= \frac{1}{h_1 h_2 h_3} \frac{\partial(h_2 h_3 A_1)}{\partial q_1} \, dv,$$

and analogous expressions hold for the other two sets of surfaces. Since the sum of these three terms is, by Gauss's theorem, equal to $\nabla \cdot \mathbf{A} \, dv$, the divergence is given explicitly by

$$\nabla \cdot \mathbf{A} = \frac{1}{h_1 h_2 h_3} \left(\frac{\partial(h_2 h_3 A_1)}{\partial q_1} + \frac{\partial(h_3 h_1 A_2)}{\partial q_2} + \frac{\partial(h_1 h_2 A_3)}{\partial q_3} \right). \tag{5}$$

The Laplacian of a scalar function can be written down immediately, since it is just the divergence of the gradient:

$$\nabla^2 \psi = \frac{1}{h_1 h_2 h_3} \left[\frac{\partial}{\partial q_1} \left(\frac{h_2 h_3}{h_1} \frac{\partial \psi}{\partial q_1} \right) + \frac{\partial}{\partial q_2} \left(\frac{h_3 h_1}{h_2} \frac{\partial \psi}{\partial q_2} \right) + \frac{\partial}{\partial q_3} \left(\frac{h_1 h_2}{h_3} \frac{\partial \psi}{\partial q_3} \right) \right]. \tag{6}$$

To obtain a particular component of the curl of a vector, we may apply Stokes' theorem to an infinitesimal area at right angles to the direction of the desired component. Consider the area defined by ds_2 and ds_3, as in Fig. III–2. By Stokes' theorem, $\oint \mathbf{A} \cdot d\mathbf{s} = \int (\nabla \times \mathbf{A}) \cdot d\mathbf{S}$, which, in this application, becomes

$$(A_2 \, ds_2)_{q_3} + (A_3 \, ds_3)_{q_2+dq_2} - (A_2 \, ds_2)_{q_3+dq_3} - (A_3 \, ds_3)_{q_2}$$
$$= (\nabla \times \mathbf{A})_1 \, ds_2 \, ds_3.$$

TABLE III-1 COORDINATE SYSTEMS

Cartesian coordinates	Cylindrical coordinates	Spherical polar coordinates
	Orthogonal line elements	
dx, dy, dz	$dr, r\,d\varphi, dz$	$dr, r\,d\theta, r\sin\theta\,d\varphi$
	Components of gradient	
$(\nabla\psi)_x = \dfrac{\partial\psi}{\partial x}$	$(\nabla\psi)_r = \dfrac{\partial\psi}{\partial r}$	$(\nabla\psi)_r = \dfrac{\partial\psi}{\partial r}$
$(\nabla\psi)_y = \dfrac{\partial\psi}{\partial y}$	$(\nabla\psi)_\varphi = \dfrac{1}{r}\dfrac{\partial\psi}{\partial\varphi}$	$(\nabla\psi)_\theta = \dfrac{1}{r}\dfrac{\partial\psi}{\partial\theta}$
$(\nabla\psi)_z = \dfrac{\partial\psi}{\partial z}$	$(\nabla\psi)_z = \dfrac{\partial\psi}{\partial z}$	$(\nabla\psi)_\varphi = \dfrac{1}{r\sin\theta}\dfrac{\partial\psi}{\partial\varphi}$
	The divergence of $\mathbf{A} = \nabla\cdot\mathbf{A}$	
$\dfrac{\partial A_x}{\partial x} + \dfrac{\partial A_y}{\partial y} + \dfrac{\partial A_z}{\partial z}$	$\dfrac{1}{r}\dfrac{\partial(rA_r)}{\partial r} + \dfrac{1}{r}\dfrac{\partial A_\varphi}{\partial\varphi} + \dfrac{\partial A_z}{\partial z}$	$\dfrac{1}{r^2}\dfrac{\partial(r^2 A_r)}{\partial r} + \dfrac{1}{r\sin\theta}\dfrac{\partial(\sin\theta\,A_\theta)}{\partial\theta} + \dfrac{1}{r\sin\theta}\dfrac{\partial A_\varphi}{\partial\varphi}$
	Components of curl $\mathbf{A}$	
$(\nabla\times\mathbf{A})_x = \left(\dfrac{\partial A_z}{\partial y} - \dfrac{\partial A_y}{\partial z}\right)$	$(\nabla\times\mathbf{A})_r = \left(\dfrac{1}{r}\dfrac{\partial A_z}{\partial\varphi} - \dfrac{\partial A_\varphi}{\partial z}\right)$	$(\nabla\times\mathbf{A})_r = \dfrac{1}{r\sin\theta}\left(\dfrac{\partial(\sin\theta\,A_\varphi)}{\partial\theta} - \dfrac{\partial A_\theta}{\partial\varphi}\right)$
$(\nabla\times\mathbf{A})_y = \left(\dfrac{\partial A_x}{\partial z} - \dfrac{\partial A_z}{\partial x}\right)$	$(\nabla\times\mathbf{A})_\varphi = \left(\dfrac{\partial A_r}{\partial z} - \dfrac{\partial A_z}{\partial r}\right)$	$(\nabla\times\mathbf{A})_\theta = \dfrac{1}{r\sin\theta}\dfrac{\partial A_r}{\partial\varphi} - \dfrac{1}{r}\dfrac{\partial(rA_\varphi)}{\partial r}$
$(\nabla\times\mathbf{A})_z = \left(\dfrac{\partial A_y}{\partial x} - \dfrac{\partial A_x}{\partial y}\right)$	$(\nabla\times\mathbf{A})_z = \dfrac{1}{r}\left(\dfrac{\partial(rA_\varphi)}{\partial r} - \dfrac{\partial A_r}{\partial\varphi}\right)$	$(\nabla\times\mathbf{A})_\varphi = \dfrac{1}{r}\left(\dfrac{\partial(rA_\theta)}{\partial r} - \dfrac{\partial A_r}{\partial\theta}\right)$
	Laplacian of $\psi = \operatorname{div}\operatorname{grad}\psi = \nabla^2\psi$	
$\dfrac{\partial^2\psi}{\partial x^2} + \dfrac{\partial^2\psi}{\partial y^2} + \dfrac{\partial^2\psi}{\partial z^2}$	$\dfrac{1}{r}\dfrac{\partial}{\partial r}\left(r\dfrac{\partial\psi}{\partial r}\right) + \dfrac{1}{r^2}\dfrac{\partial^2\psi}{\partial\varphi^2} + \dfrac{\partial^2\psi}{\partial z^2}$	$\dfrac{1}{r^2}\dfrac{\partial}{\partial r}\left(r^2\dfrac{\partial\psi}{\partial r}\right) + \dfrac{1}{r^2\sin\theta}\dfrac{\partial}{\partial\theta}\left(\sin\theta\dfrac{\partial\psi}{\partial\theta}\right) + \dfrac{1}{r^2\sin^2\theta}\dfrac{\partial^2\psi}{\partial\varphi^2}$

Therefore

$$(\nabla \times \mathbf{A})_1 = \frac{1}{h_2 h_3} \left[\frac{\partial (h_3 A_3)}{\partial q_2} - \frac{\partial (h_2 A_2)}{\partial q_3} \right], \tag{7}$$

and the other two components are obtained by cyclic interchange of the coordinate indices.

In Table III–1 are listed the explicit forms for gradient, divergence, curl, and Laplacian in the three most common coordinate systems. For the definition of other othogonal coordinates see, e.g., Margenau and Murphy, *The Mathematics of Physics and Chemistry*.

BIBLIOGRAPHY

BIBLIOGRAPHY

WORKS DEALING PRIMARILY WITH CLASSICAL ELECTRODYNAMICS
OR SOME OF ITS ASPECTS

ABRAHAM, M., AND R. BECKER, *The Classical Theory of Electricity and Magnetism*, from the 8th German ed. (1930), Blackie and Son, 1932; from the 14th German ed. (1949), Hafner Publishing Co., 1950.

ALFVÈN, H., *Cosmical Electrodynamics*, Oxford University Press, 1950.

BECKER, R., *Theorie der Elektrizität*, Band II, Teubner, 1933 (reprinted by J. W. Edwards, Ann Arbor, 1946).

BLEANEY, B. AND B. I. BLEANEY, *Electricity and Magnetism*, Oxford, Clarendon Press, 1957.

BODE, H. W., *Network Analysis and Feedback Amplifier Design*, Van Nostrand, 1945.

CHANDRASEKHAR, S., *Plasma Physics*, Chicago University Press, 1960.

CONDON, E. U., Principles of Microwave Radio, *Revs. Modern Phys.* **14,** 341 (1942).

COSLETT, V. E., *Bibliography of Electron Microscopy*, Arnold, 1950.

COWLING, T. G., *Magnetohydrodynamics*, Interscience Publishers, 1957.

ELSASSER, W. M., Hydrodynamic Dynamo Theory, *Revs. Modern Phys.* **28,** 135 (1956).

HARNWELL, G. P., *Principles of Electricity and Electromagnetism*, 2nd ed., McGraw-Hill, 1949.

HERTZ, H., *Electric Waves*, Macmillan (London), 1893.

JEANS, J. H., *The Mathematical Theory of Electricity and Magnetism*, 5th ed., Cambridge University Press, 1925.

KLEMPERER, O., *Electron Optics*, 2nd ed., Cambridge University Press, 1953.

KRAUS, T. D., *Electromagnetics*, McGraw-Hill, 1953.

LAMONT, H. R. L., *Wave Guides*, 3rd ed., Methuen, 1950.

LANDAU, L. D. AND E. M. LIFSHITZ, *The Classical Theory of Fields* (trans. from the Russian), Addison-Wesley, 1951.

LANDAU, L. D. AND E. M. LIFSHITZ, *Electrodynamics of Continuous Media* (trans. from the Russian), Addison-Wesley, 1960.

LORENTZ, H. A., *The Theory of Electrons*, 2nd ed., Teubner, 1908 (reprinted by Dover, 1952).

MARTON, C., AND OTHERS, *Bibliography of Electron Microscopy*, U. S. Government Printing Office, 1950.

MASON, M., AND W. WEAVER, *The Electromagnetic Field*, University of Chicago Press, 1929.

MAXWELL, J. C., *A Treatise on Electricity and Magnetism*, 3rd ed., Oxford University Press, 1904.

PLANCK, M. K. E. L., *Theory of Electricity and Magnetism*, Macmillan, 1932.

ROHRLICH, F., *The Classical Electron*, in *Lectures on Theoretical Physics*, Interscience Publishers, 1960.

ROSENFELD, L., *Theory of Electrons*, North-Holland Publishing Co. (Amsterdam), 1951.

SCHELKUNOFF, S. A., *Electromagnetic Waves*, Van Nostrand, 1943.

SCHOTT, G. A., *Electromagnetic Radiation*, Cambridge University Press, 1912.

SLATER, J. C., *Microwave Electronics*, Van Nostrand, 1950.

SLATER, J. C., Microwave Electronics, *Revs. Modern Phys.* **18**, 441 (1946).

SLATER, J. C., AND N. H. FRANK, *Electromagnetism*, McGraw-Hill, 1947.

SMYTHE, W. R., *Static and Dynamic Electricity*, 2nd ed., McGraw-Hill, 1950.

SOMMERFELD, A., *Electrodynamics; Lectures on Theoretical Physics*, Vol. III, Academic Press, 1952.

SPITZER, L., *Physics of Fully Ionized Gases*, Interscience Publishers, 1956.

STRATTON, J. A., *Electromagnetic Theory*, McGraw-Hill, 1941.

VAN VLECK, J. H., *The Theory of Electric and Magnetic Susceptibilities*, Clarendon Press, Oxford, 1932.

WEBER, E., *Electromagnetic Fields, Theory and Applications*, Vol. 1, *Mapping of Fields*, Wiley, 1950.

ZWORYKIN, V. K., G. A. MORTON, E. G. RAMBERG, J. HILLIER, AND A. W. VANCE, *Electron Optics and the Electron Microscope*, Wiley, 1945.

MATHEMATICAL WORKS

BYERLY, W. E., *Fourier's Series and Spherical Harmonics*, Ginn, 1893.

CHURCHILL, R. V., *Fourier Series and Boundary Value Problems*, McGraw-Hill, 1941.

COURANT, R., AND D. HILBERT, *Methoden der Mathematischen Physik*, I, Springer, 1931; *Methods of Mathematical Physics*, Vol. I, Interscience Publishers, 1953.

INCE, E. L., *Ordinary Differential Equations*, Longmans, 1927 (reprinted by Dover, 1944).

JAHNKE, E., AND F. EMDE, *Funktionentafeln mit Formeln und Kurven* (Tables of Functions), Teubner, 1938 (reprinted from 4th rev. ed. in English and German by Dover, 1945).

JEFFREYS, H. AND B. S., *Methods of Mathematical Physics*, 3rd. ed., Cambridge University Press, 1956.

KELLOGG, O. D., *Foundations of Potential Theory*, Springer, 1929 (reprinted by Frederick Ungar Publishing Co., n.d.).

MACROBERT, T. M., *Spherical Harmonics*, 2nd ed., Dover, 1948.

MARGENAU, H., AND G. M. MURPHY, *The Mathematics of Physics and Chemistry*, Van Nostrand, 1943.

MORSE, P., AND H. FESHBACH, *Methods of Mathematical Physics*, McGraw-Hill, 1953.

PIPES, L. A., *Applied Mathematics for Engineers and Physicists*, McGraw-Hill, 1946.

SYNGE, J. L., AND A. SCHILD, *Tensor Calculus*, University of Toronto Press, 1949.

WATSON, G. N., *Theory of Bassel Functions*, 2nd ed., Macmillan, 1944.

WEATHERBURN, C. E., *An Introduction to Riemannian Geometry and the Tensor Calculus*, Cambridge University Press, 1938.

WHITTAKER, E. T., AND G. N. WATSON, *A Course of Modern Analysis*, 4th ed., Cambridge University Press, 1927.

WORKS ON RELATIVITY THEORY

BERGMANN, P. G., *An Introduction to the Theory of Relativity*, Prentice-Hall, 1942.

EINSTEIN, A., *The Meaning of Relativity*, 4th ed., Princeton University Press, 1953.

FOCK, V., *The Theory of Space Time and Gravitation*, Pergamon Press, 1959.

LORENTZ, H. A., A. EINSTEIN, H. MINKOWSKI, AND H. WEYL, *The Principle of Relativity*, Methuen, 1923 (reprinted by Dover).

MØLLER, C., *The Theory of Relativity*, Oxford University Press, 1952.

PAULI, W., Relativitätstheorie, *Encyclopädie der Mathematischen Wissenschaften* V, Pt. 2, 539, Teubner, 1921.

TOLMAN, R. C., *Relativity, Thermodynamics and Cosmology*, Oxford University Press, 1934.

TOLMAN, R. C., *The Theory of the Relativity of Motion*, University of California Press, 1917.

WORKS CONTAINING MATERIAL ON OR RELATED TO CLASSICAL ELECTRODYNAMICS

BERGMANN, P. G., *Basic Theories of Physics: Mechanics and Electrodynamics*, Prentice-Hall, 1949.

BLATT, J. M., AND V. F. WEISSKOPF, *Theoretical Nuclear Physics*, Wiley, 1952.

BORN, M., *Optik*, Springer, 1933.

CORBEN, H. C., AND P. M. STEHLE, *Classical Mechanics*, Wiley, 1950.

FOWLER, R. H., *Statistical Mechanics*, 2nd ed., Cambridge University Press, 1936.

GOLDSTEIN, H., *Classical Mechanics*, Addison-Wesley, 1950.

HEITLER, W., *The Quantum Theory of Radiation*, 3rd ed., Oxford University Press, 1954.

JAUCH, J. M., AND F. ROHRLICH, *The Theory of Photons and Electrons*, Addison-Wesley Publishing Co., 1955.

JOOS, G., *Theoretical Physics*, 2nd rev. ed. (translated from the German ed. of 1934), Hafner Publishing Co., 1951.

LAMB, H., *Hydrodynamics*, 6th ed. rev., Macmillan, 1932 (reprinted by Dover, 1945).

LINDSAY, R. B., AND H. MARGENAU, *Foundations of Physics*, Wiley, 1936.

LORENTZ, H. A., *Problems of Modern Physics*, Ginn, 1927.

MAYER, J. E., AND M. G. MAYER, *Statistical Mechanics*, Wiley, 1940.

MORSE, P. M., *Vibration and Sound*, 2nd ed., McGraw-Hill, 1948.

PAGE, L., *Introduction to Theoretical Physics*, 3rd ed., Van Nostrand, 1952.

SCHIFF, L. I., *Quantum Mechanics*, McGraw-Hill, 1949.

SLATER, J. C., AND N. H. FRANK, *Introduction to Theoretical Physics*, McGraw-Hill, 1933.

SOMMERFELD, A., *Optics; Lectures on Theoretical Physics*, Vol. IV, Academic Press, 1954.

WENTZEL, G., *Quantentheorie der Wellenfelder* (Quantum Theory of Fields), F. Deuticke (Vienna), 1943 (reprinted by J. W. Edwards, 1946, and in English by Interscience Publishers, 1949).

INDEX

Aberration of light, 279, 302, 380
Absolute electrical units, 462
Absolute permeability, 145
Absolute temperature, 100
Absorption, 403, 416
Absorption band, 412
Absorption cross section, 421
Absorption of radiation, 407
Accelerated charge, 354
Accelerated frame, 337
Accessible charge, 31
Accessible source, 463
Action at a distance, 396
Action integral, 429
Adiabatic theorem, 436
Advanced potential, 244, 394
Ampère's law, 123, 127
Analytic function, 61
Angular distributions, 369
Angular momentum, 130, 133, 269, 270, 431, 437
  orbital, 270
Angular width of radiation, 367
Anisotropic media, 30, 99
Anomalous dispersion, 413
Antisymmetric tensor, 100
Associated Legendre function, 82
Atomic constants, 469
Atomic field, 38
Attentuation, wave guide, 225
Axial vector, 327

Bessel function, 88, 230, 256
  of imaginary argument, 90, 154, 350
  spherical, 230, 256
Bessel's equation, 88
Betatron, 169, 400
Biot and Savart law, 125
Blackbody, 191, 410
Blue sky, 407
Bound electron, 407
Boundary conditions, electrostatic, 33
  for plane waves, 195
  magnetostatic, 145

  metallic, 212
Boundary value, 11, 46, 148
Bouwkamp, 265
Bragg diffraction, 415
Bremsstrahlung, 361
  cross section, 371
Brewster's angle, 198
Brillouin, 414

Canonical angular momentum, 431
Canonical equations of motion, 426
Canonical field coordinates and momenta, 453
Canonically conjugate, 453
Capacity, 63, 121
Cartesian coordinates, 149, 473, 475
Casimir, 265
Cauchy principal value, 418
Cauchy-Riemann equations, 62
Cauchy's theorem, 417
Causality, 390, 416
Cavity, 217
  circular cylindrical, 222
  cylindrical, 219
  rectangular, 218
  spherical, 233
Cavity definition of field, 34, 35
Center of mass frame, 318
Centrifugal potential, 433
Čerenkov radiation, 373
Characteristic impedance, 187
Charge, accelerated, 354
  accessible, 31
  conservation of, 118
  electronic, 325, 344
  free, 28
  inaccessible, 31
  line, 13, 64
  point, 13, 398
  polarization, 28, 129
  surface, 13
  total, 29
  true, 28, 129, 335
Charge layer, 21

Charged ring, 87
Circular cylinder, 64
  scattering by, 226
Circular cylindrical cavity, 222
Circular orbits, 363
Circularly polarized, 270
Circulation density, 2, 5
Classical mechanics, 272
Clausius-Mossotti electrical equation
  of state, 39, 109, 412
Clock, 289
Coaxial cable, 184
Coaxial cylinder, 223
Coefficient, reflection, 195, 198, 201
  transmission, 198, 200, 204
Coherent radiation, 370
Collision, 318
  elastic, 318, 321
  inelastic, 320
Complete set, 58
Complex conjugate, 65
Complex potential, 65, 152
Complex transformation, 66
Complex variable, 61
Conducting disk, 94
Conducting fluids, 205
Conductivity, electrical, 119
Conformal transformation, 61, 66
Conjugate function, 61
Conservation, of charge, 118
  of electromagnetic energy, 178, 377
  of momentum, 184
Conservation laws, 310
Conservative field, 10
Constitutive equations, 31, 160, 164,
  336
Contraction of a tensor, 308
Contravariant four-vector, 306
Convection potential, 348
Convective current, 129, 163
Conversion factors, 465
Coordinates, Cartesian, 149, 473,
  475
  curvilinear, 473
  cylindrical, 88, 153, 473, 475
  non-Cartesian, 149
  plane polar, 73

  rectangular, 53
  spherical polar, 57, 81, 473, 475
Coulomb's law, 8, 114
Covariance, 305
Covariant conservation laws, 377
Covariant electrodynamics, 324
Covariant energy-momentum vector,
  391
Covariant four-vector, 306
Covariant Lagrangian equations,
  449
Critical opalescence, 416
Cross section, absorption, 412
  bremsstrahlung, 371
  differential scattering, 406
  for Rayleigh scattering, 407, 414
  resonance scattering, 407
  scattering, 228, 236, 405, 407, 421
    by cylinder, 229
    by sphere, 237
  for Thomson scattering, 405
Cubic crystal, 38
Current, convective, 129, 163
  covariant, 324, 333
  displacement, 135, 140, 163, 170, 195
  magnetization, 129, 134
  polarization, 129, 140, 163, 195
  true, 129, 140, 163
  vacuum displacement, 135, 140
Current density, 119
Current element, 123
Current loop, 131, 154
Curvilinear coordinates, 473
Cutoff frequency, 224, 413
Cylindrical cavity, 219
Cylindrical coordinates, 88, 153, 473,
  475

D'Alembertian, 241, 324
Damping, radiative, 401
Debye potentials, 231, 265
Degenerate solution, 217
Degrees of freedom, 447
Delta-function, Dirac, 3, 92, 243, 249
Density fluctuations, 415
Depolarization factor, 85
Derivative, substantial (total), 109

Dielectric, 36
  nonisotropic, 30, 99
Dielectric constant, 30
Dielectric constant tensor, 30, 99
Dielectric cylinder, 74
Dielectric equation of state, 109
Dielectric half-space, 50
Dielectric liquid, 111, 114
Dielectric solid, 114
Dielectric sphere, 83, 84
Differential scattering cross section, 406
Diffusion equation, 186, 206
Dimensions, 459
Dipole, 18
  electric, 19
  magnetic, 133
  oscillating, 257
Dipole absorption, 409
Dipole interaction, 19
Dipole layer, 12, 126
Dipole moment, 14, 85
Dipole radiation, electric, 257
  magnetic, 261
Dipole volume distribution, 23
Dirac delta function, 3, 92, 243, 249
Dirac radiation reaction, 393, 399
Dispersion, 203, 401, 403, 412
  anomalous, 413
Dispersion relation, 416, 419
Displacement current, 135, 140, 163, 170, 195
Displacement vector, 28
Distant absorbers, 396
Doppler shift, 380
Double layer, 20
Dual of a tensor, 330, 438

Earnshaw's theorem, 26
Eclipsing binary stars, 281
Edge effect, 80
Eigenfunction, 214
Eigenvalue, 214
Einstein, 283, 305
Elastic collision, 318, 321
Electret, 30, 41
Electric dipole, 19

Electric dipole radiation, 257
Electric flux, 8
Electric moment, 334, 337
Electric neutrality, 333
Electric quadrupole, 262, 436
Electric quadrupole radiation, 262
Electric susceptibility, 30
Electric volume force, 104, 107
Electrical conductivity, 119
Electrical images, 49
Electromagnetic field tensor, 327
Electromagnetic mass, 381, 387
Electromagnetic unit, 460
Electromotive force, 119, 145, 158, 173, 179
Electron optics, 431
Electron radius, 341
  classical, 342
Electronic charge, 325, 344
Electrostatic field, 1
  in a medium, 33
Electrostatic lens, 431, 433
Electrostatic potential, 10
Electrostatic unit, 460
Electrostriction, 111
Elliptic cylinders, 72
Elliptical polarization, 200, 423
Emission theory, 280
Energy, of a dipole, 19
  in electrostatic field, 95
  in magnetostatic field, 171
  velocity of propagation, 413
Energy balance, 180
Energy density, 98, 102, 377
Energy flow, 180
Energy integral, 185
Energy-momentum tensor, 377, 452
Energy-momentum vector, covariant, 391
Entropy, 100, 286
Equation, of continuity, 109, 118, 135, 254, 325
  of motion, 332, 389, 401, 425
  of state, 111
  of telegraphy, 189
Equilibrium, in electrostatic field, 26

between oscillator and field, 409
invariance of, 317, 379
Equipotential surfaces, 63
Equivalent photons, 325
Ether, 191, 273
Ether drag, 279
Eulerian equation, 427, 432, 447
Event, defined, 298
Extensive variable, 100

Faraday disk, 338
Faraday's law of induction, 158, 162
Feenberg, 422
Fermat's principle, 434
Fermi, 350
Ferromagnetic substances, 140, 171
Feynman, 396
Field, cavity definition of, 34, 35
  concept, 1
  conservative, 10
  electrostatic, 1
  harmonic, 7
  induction, 246, 258
  macroscopic, 28
  magnetic induction, 125
  molecular, 35
  multipole, 233
  pseudoscalar, 2, 8
  pseudovector, 2, 8
  radiation, 245, 246, 258
  scalar, 1, 2, 8
  scalar meson, 451
  tensor, 1
  transformation properties of, 1
  transition, 258
  transverse, 189
  vector, 1, 2, 8
Field energy, 95, 96
Field intensity, magnetic, 139
Field tensor, electromagnetic, 327
Fine structure constant, 325
Fizeau, 195
Flip coil, 168
Fluctuations, density, 415
Flux, 8, 63, 160
Flux density, 147
Flux linkages, 176

Force, on current systems, 123, 175
  electric volume, 104, 107
  electromagnetic volume, 110, 182
  Lorentz, 183, 191, 331, 378
  magnetic volume, 177, 182
  Minkowski, 316, 392
Forced vibrations, 403
Forward scattering amplitude, 422
Four-dimensional space, 297
Four-momentum, 312
Four-potential, 324
Four-vector, 305, 320
  contravariant, 306
  covariant, 306
Four-velocity, 312
Fourier analysis, 242
Fourier transform, 26, 56, 202, 242, 249
Free charge, 28
Free electron, 413
Free energy, 100
Free energy density, 101
Frequency spectrum, 361, 366
Fresnel, 193, 198, 203
Fresnel-Fizeau convection coefficient, 280
Frozen lines, 205
Functional derivative, 448
Furry, 442

g-factor, 131, 437
Galilean relativity, 272, 273
Galilean transformation, 273
Gauge transformation, 241, 325
Gaussian units, 25, 461
Gauss's theorem, 4, 8, 309
  in four dimensions, 309
General relativity, 283, 339
Gibb's notation, 470
Gradient, 10
Green's function, concept, 44, 53, 92
  for conducting plane, 52
  for conducting sphere, 51
  for cylinder, 91
  for radiation field, 243
  for wedge, 78
Green's reciprocity theorem, 43, 46

Green's theorem, 21, 45
  two-dimensional, 79
Group of Lorentz transformation, 296
Group velocity, 202, 209, 224, 413
Guide impedance, 226
Guide propagation constant, 224
Gyromagnetic ratio, 131, 133, 437

Half-wave antenna, 254
Hall effect, 118
Hamiltonian, 427
Hamiltonian density, 449
Hamiltonian formulation, 446
Hankel function, 90, 230
Harmonic function, 7, 62
Harmonic oscillator, 401, 455
Heaviside-Lorentz units, 461
Helical orbit, 376
Hertz vector, 254
Hughes, 144
Hydrostatic pressure, 112, 207
Hyperbolic cylinders, 71
Hysteresis, 172

Images, method of, 49
Imaginary part, 190
Impedance, wave guide, 226
Improper transformation, 1, 330
Inaccessible charge, 31
Inaccessible source, 463
Incompressible fluid, 112
Index of refraction, 189, 194, 202, 411
Inductance, 174
Induction field, 246, 258
  magnetic, 125
Inelastic collision, 320
Inertial frame, 273, 283
Inhomogeneous wave equation, 240
Intensive variable, 100
Interaction energy, 172
Interference, 276
Interferometer, 277
Internal energy, 100
International units, 462
Invariant, 296, 307
Inversion, method of, 47, 79
Isothermal process, 100

Isotropic, 38

Joule heat, 171, 173, 179

Kramers, 416

Lagrangian, 426, 444, 446
Lagrangian density, 447
Landé g-factor, 131
Langevin formula, 40
Laplace's equation, 11, 53
  in dielectrics, 43
Laplacian operator, 3, 474
Larmor frequency, 424
Least action, principle of, 430, 434
Legendre function, 15
  associated, 82
Legendre polynomial, 15, 83, 86
Liénard-Wiechert potentials, 327, 341, 355
Lifetime of state, 402
Light cone, 301
Line charge, 13, 64
Line current, 123, 152
Line element, 473
Line width, 402
Linear antenna, 251
Linear circuit, 173
Linearly polarized wave, 190
Lines of force, 27, 257
Logarithmic potential, 65, 75
Longitudinal velocity addition, 302
Lorentz condition, 240, 254, 325
Lorentz contraction, 291
Lorentz-Fitzgerald contraction, 278
Lorentz force, 182, 183, 191, 331, 378
Lorentz transformation, 286, 293, 305
Losses, Joule heat, 171, 173, 179
  at a metallic surface, 219
Loss-less transmission line, 252

Macroscopic field description, 8, 28, 33, 332
Magnetic boundary problems, 139
Magnetic circuit, 145
Magnetic dipole, 133
Magnetic dipole radiation, 261

Magnetic field energy, 171, 173
Magnetic field intensity, 139
Magnetic flux, 146, 158
Magnetic induction field, 125
Magnetic interaction, 123
Magnetic lens, 431, 434
Magnetic materials, 139
Magnetic moment, 130, 261, 334, 437
Magnetic pole, 141, 330
Magnetic quadrupole, 436
Magnetic pressure, 207
Magnetic Reynolds number, 206
Magnetic rigidity, 434, 469
Magnetic scalar potential, 125, 146
Magnetic susceptibility, 144
Magnetic vector potential, 127
Magnetic viscosity, 207
Magnetic volume force, 177, 182
Magnetization, 134, 139, 140
Magnetization current, 129, 134
Magnetization potential, 264
Magnetized media, 144
Magnetohydrodynamics, 205
Magnetohydrodynamic waves, 207
Mapping in complex plane, 48
Mass, electromagnetic, 381, 387
  variation with velocity, 313
Mass density, 379
Mass-energy equivalence, 314
Material media, 332, 380
Matrix, 105
Maxwell stress tensor, 103, 178, 181, 191, 377
Maxwell's equations, 158, 160, 185, 329
Mechanical momentum, 183, 311
Mechanical work, 173
Meson field, scalar, 451
Metallic boundary conditions, 212
Metallic reflection, 200
Metric tensor, 309
Michel, 442
Michelson-Morley experiment, 275
Microscopic field description, 8, 28
Minkowski diagram, 299
Minkowski force, 316, 392
mks system, 7, 460
Modes of vibration, 217

Molar refraction, 412
Molecular field, 35
Moment, electric, 334, 337
  magnetic, 130, 261, 334, 437
  multipole, 15
Moment tensor, 335
Momentum, electromagnetic, 183, 378, 390
  mechanical, 183, 311
Momentum balance, 181
Momentum density, 184, 316, 377, 449
Monochromatic sources, 245
Motion of charged particles, 425
Motion of conductor in magnetic field, 165, 336
Motional field, 163
Moving dielectric, 165
Moving magnet, 167
Moving media, 160, 163, 165, 193, 194, 280, 302
Multipole, 15
Multipole expansion, 17, 131, 255
Multipole moment, 15
Multipole potential, 14, 131, 233, 260
Multipole radiation, 233, 260, 267
Mutual inductance, 175

Natural units, 462
Net point method, 59
Neumann's formula, 175
Newton's third law applied to magnetic interaction, 124
Nisbet, 230, 264
Nonconservative potential, 23
Nonisotropic dielectric, 30, 99
Normal derivative, 12, 32, 43
Normal modes, 410
Normalizing factor, 86

Octupole, 16
Ohm's law, 118
Open circuit voltage, 120
Optical theorem, 422
Orbital angular momentum, 270
Orthogonal coordinates, 53
Orthogonality, 54, 58, 86

Orthogonalization, 217
Oscillating dipole, 257

Parallel plate capacitor, 36, 122
Paraxial motion, 432
Periodic solutions, 454
Permanent magnet, 140, 142, 337
Permeability, absolute, 145
  relative, 145
Permeable media, 144
Phase, invariance of, 380
Phase velocity, 197, 202, 224, 412, 413
Photon, 270
Planck, 363
Plane of incidence, 196
Plane polar coordinates, 73
Plane polarized wave, 190
Plane wave, 185
Plane wave expansion in spherical wave
  functions, 234
Plane wave in moving medium, 193
Poincaré, 385, 392
Point charge, 13, 398
Poisson bracket, 450
Poisson's equation, 11, 149
Polar vector, 327
Polarizability, 38, 412
Polarization, 24, 28, 38, 411
Polarization charge, 28, 114, 129
Polarization current, 129, 140, 163,
  195
Polarization potential, 255
Polarized medium in motion, 336
Polarizing angle, 199
Pole, magnetic, 141, 330
Polygonal boundaries, 67
Postulates of special relativity, 283
Potential, advanced, 244, 394
  centrifugal, 433
  convection, 348
  Debye, 231, 265
  electrostatic, 10
  four-vector, 325
  Liénard-Wiechert, 327, 341, 355
  magnetization, 264
  polarization, 255
  retarded, 244

scalar, 5
  magnetic, 125, 146
vector, 5, 128, 152
  magnetic, 127
Yukawa, 60, 451
Potential problems, 42
Poynting vector, 180, 249, 268, 377
Precession, 133, 439
  Thomas, 441
Pressure, in dielectric liquid, 111
  magnetic, 207
  radiation, 191
Principal axes, 18, 105, 263
Principal coordinates, 30, 105
Principal value of singular integral
  (Cauchy), 418
Propagation vector, 189, 196
Proper frame, 288
Proper length, 291
Proper quantity, 286
Proper time, 290
Proper transformation, 1
Pseudoscalar field, 2, 8
Pseudotensor, 330
Pseudovector field, 2, 8

Q of oscillating system, 219
Quadrupole, 16, 18
  electric, 262, 436
  magnetic, 436
Quadrupole moment, 18, 264
Quadrupole tensor, 18
Quantum of radiation, 363, 380

Radiated energy, 248, 250
Radiated energy, rate of, 251, 359, 365
Radiation, absorption of, 407
  from accelerated charge, 356
  Čerenkov, 373
  from circular orbit, 363
  from current systems, 251
  deceleration, 361
  from electric dipole, 257
  from electric quadrupole, 262
  energy and angular momentum of,
    267

from magnetic dipole, 261
scattering, by bound electron, 407
  by circular cylinder, 226
  by free electron, 404
  by sphere, 233
Radiation field, 245, 246
Radiation loss, 180
Radiation pressure, 191
Radiation pulse, 249
Radiation reaction, 377, 386, 387, 394, 425
Radiation resistance, 253
Radiative damping, 401
Random scatterers, 415
Rank of a tensor, 105
Rasetti, 143
Rate of radiation, 251, 359, 365
Rationalized units, 461
Rayleigh-Jeans distribution, 410
Rayleigh scattering, 407, 414
Reaction force, 383
Real part, 190
Reciprocity theorem, Green's, 43, 46
Rectangular cavity, 218
Rectangular coordinates, 53
Reflection coefficient, 195, 198, 201
Refraction, 195, 411, 416
Refractive index, 189, 411, 416
Relative permeability, 145
Relativistic equations of motion, 392
Relativistic kinematics, 286
Relativistic mechanics, 305
Relativity, Galilean, 272, 273
  general, 283, 339
  postulates of, 283
  special theory of, 183, 272, 283, 288, 293
Relaxation method, 59
Relaxation time, 112, 123, 186
Reluctance, 146
Residue, 417
Resistance, 121, 145
Resonance scattering cross section, 407
Rest-energy, 314
Retarded potential, 244, 342
Reynolds number, magnetic, 206
Rigid body, 287, 349

Ritz emission theory, 280
Rohrlich, 390
Rotating frames, 337
"Run-away" solution, 389, 394

Saturation, magnetic, 147
Scalar, 307
Scalar field, 1, 2, 8, 470
Scalar meson field, 451
Scalar potential, 5
  magnetic, 125, 146
Scalar superpotential, 264
Scattering, multiple Coulomb, 376
  of radiation, 401, 403, 404
    by bound electron (Rayleigh), 414
    by circular cylinder, 226
    by free electron (Thomson), 405
    by sphere, 233
Scattering cross section, 228, 236, 405, 407, 421
  differential, 406
Schiff, 339
Schwarz transformation, 67, 152
Secular determinant, 105
Self-energy, 97, 98, 173, 175
Self-inductance, 175, 177
Sense of time, 286, 395
Separation constant, 54, 81, 215
Separation of variables, 53
Shadow theorem, 419
Simultaneity, 287
Single-angle transformation, 70
Skin-depth, 118, 201, 214
Snell's law, 197
Solid angle, 4, 20, 53, 126
Source, accessible, 463
  inaccessible, 463
Source density, 2, 5
Source equations, 329, 335, 463
Space-like interval, 296
Space-time average of field quantity, 34
Space-time interval, 296
Special relativity, postulates of, 283, 293
Special theory of relativity, 183, 272, 288
Specific inductive capacity, 30

Spherical Bessel function, 230, 256
Spherical cavity, 233
Spherical harmonic, 81
Spherical polar coordinates, 57, 81, 473, 475
Spherical shell, 116
Spherical waves, 229
Spin motion, 440, 444
Stationary current, 120
Stationary flow, 118
Stationary media, 159
Stokes' theorem, 10, 127, 159
Stream function, 63, 152
Streamline, 62
Stress tensor, Maxwell, 103, 178, 181, 191, 377
Strong-focusing, 437
Substantial derivative, 109
Sum rule, 409
Summation convention, 17, 103, 470
Superposition principle, 9, 46
Superpotential, 212, 255
Surface charge, 13
Surface dipole layer, 12, 20, 126
Surface harmonics, 82
Surface resistance, 219
Susceptibility, electric, 30
Susceptibility tensor, 30, 99
Symmetric tensor, 100, 105
Symmetry properties, 311
Synchronization, 292
Systems of units, 466

Taylor expansion, 15
Temperature dependence of specific inductive capcity, 39, 40, 100
Tensor, antisymmetric, 100
  contraction of, 308
  contravariant, 307
  covariant, 308
  dielectric constant, 30, 99
  dual of a, 330, 438
  electromagnetic field, 327
  energy-momentum, 377, 452
  Maxwell stress, 103, 178, 181, 191, 377
  metric, 309
  moment, 335
  quadrupole, 18
  rank of a, 105, 307
  susceptibility, 30, 99
  symmetric, 100, 105
Tensor analysis, 307
Tensor divergence, 104, 182
Tensor field, 1
Test charge, 7
Thermodynamics, 100
Thomas precession, 441
Thomson scattering, 405
Thomson's theorem, 101
Three-dimensional vector field, 1
Time constant, 123
Time dilation, 290
Time-like interval, 296
Time symmetric field solution, 396
Torque, on electric dipole, 19
  on magnetic dipole, 133
  on moving electric dipole (Trouton-Noble), 274, 349
  on relativistic lever, 317
Total charge, 29
Total derivative, 109
Total energy, 100
Total reflection, 199
Trajectory equations, 430
Transformation, 47, 61
  complex, 66
  conformal, 61
  Galilean, 273
  gauge, 325
  improper, 1, 330
  Lorentz, 286
  multiple-angle, 71
  proper 1,
  Schwarz, 67
  single-angle, 70
Transformation properties, of charge, 324
  of current, 324
  of distance, 294
  of electromagnetic field, 330
  of energy, 315
  of field, 1
  of force, 317

of four-vector, 305
of general tensor, 308
of magnetization, 336
of momentum, 315
of polarization, 336
of potential, 324
of time, 295
of velocity, 301
Transition field, 258
Transmission coefficient, 198, 200, 204
Transverse electric-magnetic modes (TEM), 223
Transverse electric modes (TE), 220
Transverse field, 189
Transverse magnetic modes (TM), 220
Trouton-Noble experiment, 274, 349
True charge, 28, 129, 335
True current, 129, 140, 163
Two-dimensional equipotential, 59

Uniform field, 64, 84
Uniformly moving electron, 341
Uniqueness theorem, 6, 42, 147
Unipolar magnetic induction, 337
Units, 459
  absolute electric, 462
  electromagnetic, 460
  electrostatic, 460
  Gaussian, 461
  Heaviside-Lorentz, 461
  international, 462
  natural, 462
  rationalized, 461
  systems of, 466
  unrationalized, 461
Universal time scale, 273, 287
Unpolarized light, 199
Unrationalized units, 461

Vacuum displacement current, 135, 140
Variable, extensive, 100
  intensive, 100
Variation of mass with velocity, 313
Variational derivative, 448
Variational principle, 426
Vector, axial, 327
  covariant and contravariant, 306
  Hertz, 254
  polar, 327
  Poynting, 180, 249, 268, 377
Vector field, 1, 2, 8, 470
  three-dimensional, 1
Vector potential, 5, 128, 152
  magnetic, 127
Vector product, 327
Vector relations, 470
Velocity, energy, 413
  group, 202, 209, 224, 413
  phase, 197, 202, 224, 412, 413
Velocity of light, 8, 272, 282, 459
  numerical value, 465
Virtual displacement, 108
Virtual photon, 350
Virtual process, 99, 101, 107, 172
Volume force, electric, 104, 107
  electromagnetic, 110, 182
  magnetic, 177, 182

Wannier, 144
Wave equation, 185, 194, 214
Wave guides, 223
Wave packet, 202
Weight function, 216
Wheeler, 396
World line, 299, 425

X-ray, 361

Yukawa potential, 60, 451

A CATALOG OF SELECTED
DOVER BOOKS
IN SCIENCE AND MATHEMATICS

Astronomy

BURNHAM'S CELESTIAL HANDBOOK, Robert Burnham, Jr. Thorough guide to the stars beyond our solar system. Exhaustive treatment. Alphabetical by constellation: Andromeda to Cetus in Vol. 1; Chamaeleon to Orion in Vol. 2; and Pavo to Vulpecula in Vol. 3. Hundreds of illustrations. Index in Vol. 3. 2,000pp. 6⅛ x 9¼.
Vol. I: 23567-X
Vol. II: 23568-8
Vol. III: 23673-0

EXPLORING THE MOON THROUGH BINOCULARS AND SMALL TELESCOPES, Ernest H. Cherrington, Jr. Informative, profusely illustrated guide to locating and identifying craters, rills, seas, mountains, other lunar features. Newly revised and updated with special section of new photos. Over 100 photos and diagrams. 240pp. 8¼ x 11. 24491-1

THE EXTRATERRESTRIAL LIFE DEBATE, 1750–1900, Michael J. Crowe. First detailed, scholarly study in English of the many ideas that developed from 1750 to 1900 regarding the existence of intelligent extraterrestrial life. Examines ideas of Kant, Herschel, Voltaire, Percival Lowell, many other scientists and thinkers. 16 illustrations. 704pp. 5⅜ x 8½. 40675-X

THEORIES OF THE WORLD FROM ANTIQUITY TO THE COPERNICAN REVOLUTION, Michael J. Crowe. Newly revised edition of an accessible, enlightening book recreates the change from an earth-centered to a sun-centered conception of the solar system. 242pp. 5⅜ x 8½. 41444-2

A HISTORY OF ASTRONOMY, A. Pannekoek. Well-balanced, carefully reasoned study covers such topics as Ptolemaic theory, work of Copernicus, Kepler, Newton, Eddington's work on stars, much more. Illustrated. References. 521pp. 5⅜ x 8½.
65994-1

A COMPLETE MANUAL OF AMATEUR ASTRONOMY: Tools and Techniques for Astronomical Observations, P. Clay Sherrod with Thomas L. Koed. Concise, highly readable book discusses: selecting, setting up and maintaining a telescope; amateur studies of the sun; lunar topography and occultations; observations of Mars, Jupiter, Saturn, the minor planets and the stars; an introduction to photoelectric photometry; more. 1981 ed. 124 figures. 26 halftones. 37 tables. 335pp. 6½ x 9¼.
42820-6

AMATEUR ASTRONOMER'S HANDBOOK, J. B. Sidgwick. Timeless, comprehensive coverage of telescopes, mirrors, lenses, mountings, telescope drives, micrometers, spectroscopes, more. 189 illustrations. 576pp. 5⅝ x 8¼. (Available in U.S. only.)
24034-7

STARS AND RELATIVITY, Ya. B. Zel'dovich and I. D. Novikov. Vol. 1 of *Relativistic Astrophysics* by famed Russian scientists. General relativity, properties of matter under astrophysical conditions, stars, and stellar systems. Deep physical insights, clear presentation. 1971 edition. References. 544pp. 5⅝ x 8¼. 69424-0

Chemistry

THE SCEPTICAL CHYMIST: The Classic 1661 Text, Robert Boyle. Boyle defines the term "element," asserting that all natural phenomena can be explained by the motion and organization of primary particles. 1911 ed. viii+232pp. 5⅜ x 8½.
42825-7

RADIOACTIVE SUBSTANCES, Marie Curie. Here is the celebrated scientist's doctoral thesis, the prelude to her receipt of the 1903 Nobel Prize. Curie discusses establishing atomic character of radioactivity found in compounds of uranium and thorium; extraction from pitchblende of polonium and radium; isolation of pure radium chloride; determination of atomic weight of radium; plus electric, photographic, luminous, heat, color effects of radioactivity. ii+94pp. 5⅜ x 8½. 42550-9

CHEMICAL MAGIC, Leonard A. Ford. Second Edition, Revised by E. Winston Grundmeier. Over 100 unusual stunts demonstrating cold fire, dust explosions, much more. Text explains scientific principles and stresses safety precautions. 128pp. 5⅜ x 8½. 67628-5

THE DEVELOPMENT OF MODERN CHEMISTRY, Aaron J. Ihde. Authoritative history of chemistry from ancient Greek theory to 20th-century innovation. Covers major chemists and their discoveries. 209 illustrations. 14 tables. Bibliographies. Indices. Appendices. 851pp. 5⅜ x 8½. 64235-6

CATALYSIS IN CHEMISTRY AND ENZYMOLOGY, William P. Jencks. Exceptionally clear coverage of mechanisms for catalysis, forces in aqueous solution, carbonyl- and acyl-group reactions, practical kinetics, more. 864pp. 5⅜ x 8½.
65460-5

ELEMENTS OF CHEMISTRY, Antoine Lavoisier. Monumental classic by founder of modern chemistry in remarkable reprint of rare 1790 Kerr translation. A must for every student of chemistry or the history of science. 539pp. 5⅜ x 8½. 64624-6

THE HISTORICAL BACKGROUND OF CHEMISTRY, Henry M. Leicester. Evolution of ideas, not individual biography. Concentrates on formulation of a coherent set of chemical laws. 260pp. 5⅜ x 8½. 61053-5

A SHORT HISTORY OF CHEMISTRY, J. R. Partington. Classic exposition explores origins of chemistry, alchemy, early medical chemistry, nature of atmosphere, theory of valency, laws and structure of atomic theory, much more. 428pp. 5⅜ x 8½. (Available in U.S. only.) 65977-1

GENERAL CHEMISTRY, Linus Pauling. Revised 3rd edition of classic first-year text by Nobel laureate. Atomic and molecular structure, quantum mechanics, statistical mechanics, thermodynamics correlated with descriptive chemistry. Problems. 992pp. 5⅜ x 8½. 65622-5

FROM ALCHEMY TO CHEMISTRY, John Read. Broad, humanistic treatment focuses on great figures of chemistry and ideas that revolutionized the science. 50 illustrations. 240pp. 5⅜ x 8½. 28690-8

Engineering

DE RE METALLICA, Georgius Agricola. The famous Hoover translation of greatest treatise on technological chemistry, engineering, geology, mining of early modern times (1556). All 289 original woodcuts. 638pp. 6¾ x 11. 60006-8

FUNDAMENTALS OF ASTRODYNAMICS, Roger Bate et al. Modern approach developed by U.S. Air Force Academy. Designed as a first course. Problems, exercises. Numerous illustrations. 455pp. 5⅜ x 8½. 60061-0

DYNAMICS OF FLUIDS IN POROUS MEDIA, Jacob Bear. For advanced students of ground water hydrology, soil mechanics and physics, drainage and irrigation engineering, and more. 335 illustrations. Exercises, with answers. 784pp. 6⅛ x 9¼. 65675-6

THEORY OF VISCOELASTICITY (Second Edition), Richard M. Christensen. Complete, consistent description of the linear theory of the viscoelastic behavior of materials. Problem-solving techniques discussed. 1982 edition. 29 figures. xiv+364pp. 6⅛ x 9¼. 42880-X

MECHANICS, J. P. Den Hartog. A classic introductory text or refresher. Hundreds of applications and design problems illuminate fundamentals of trusses, loaded beams and cables, etc. 334 answered problems. 462pp. 5⅜ x 8½. 60754-2

MECHANICAL VIBRATIONS, J. P. Den Hartog. Classic textbook offers lucid explanations and illustrative models, applying theories of vibrations to a variety of practical industrial engineering problems. Numerous figures. 233 problems, solutions. Appendix. Index. Preface. 436pp. 5⅜ x 8½. 64785-4

STRENGTH OF MATERIALS, J. P. Den Hartog. Full, clear treatment of basic material (tension, torsion, bending, etc.) plus advanced material on engineering methods, applications. 350 answered problems. 323pp. 5⅜ x 8½. 60755-0

A HISTORY OF MECHANICS, René Dugas. Monumental study of mechanical principles from antiquity to quantum mechanics. Contributions of ancient Greeks, Galileo, Leonardo, Kepler, Lagrange, many others. 671pp. 5⅜ x 8½. 65632-2

STABILITY THEORY AND ITS APPLICATIONS TO STRUCTURAL MECHANICS, Clive L. Dym. Self-contained text focuses on Koiter postbuckling analyses, with mathematical notions of stability of motion. Basing minimum energy principles for static stability upon dynamic concepts of stability of motion, it develops asymptotic buckling and postbuckling analyses from potential energy considerations, with applications to columns, plates, and arches. 1974 ed. 208pp. 5⅜ x 8½. 42541-X

METAL FATIGUE, N. E. Frost, K. J. Marsh, and L. P. Pook. Definitive, clearly written, and well-illustrated volume addresses all aspects of the subject, from the historical development of understanding metal fatigue to vital concepts of the cyclic stress that causes a crack to grow. Includes 7 appendixes. 544pp. 5⅜ x 8½. 40927-9

ROCKETS, Robert Goddard. Two of the most significant publications in the history of rocketry and jet propulsion: "A Method of Reaching Extreme Altitudes" (1919) and "Liquid Propellant Rocket Development" (1936). 128pp. 5⅜ x 8½.   42537-1

STATISTICAL MECHANICS: Principles and Applications, Terrell L. Hill. Standard text covers fundamentals of statistical mechanics, applications to fluctuation theory, imperfect gases, distribution functions, more. 448pp. 5⅜ x 8½.   65390-0

ENGINEERING AND TECHNOLOGY 1650–1750: Illustrations and Texts from Original Sources, Martin Jensen. Highly readable text with more than 200 contemporary drawings and detailed engravings of engineering projects dealing with surveying, leveling, materials, hand tools, lifting equipment, transport and erection, piling, bailing, water supply, hydraulic engineering, and more. Among the specific projects outlined–transporting a 50-ton stone to the Louvre, erecting an obelisk, building timber locks, and dredging canals. 207pp. 8⅜ x 11¼.   42232-1

THE VARIATIONAL PRINCIPLES OF MECHANICS, Cornelius Lanczos. Graduate level coverage of calculus of variations, equations of motion, relativistic mechanics, more. First inexpensive paperbound edition of classic treatise. Index. Bibliography. 418pp. 5⅜ x 8½.   65067-7

PROTECTION OF ELECTRONIC CIRCUITS FROM OVERVOLTAGES, Ronald B. Standler. Five-part treatment presents practical rules and strategies for circuits designed to protect electronic systems from damage by transient overvoltages. 1989 ed. xxiv+434pp. 6⅛ x 9¼.   42552-5

ROTARY WING AERODYNAMICS, W. Z. Stepniewski. Clear, concise text covers aerodynamic phenomena of the rotor and offers guidelines for helicopter performance evaluation. Originally prepared for NASA. 537 figures. 640pp. 6⅛ x 9¼.
64647-5

INTRODUCTION TO SPACE DYNAMICS, William Tyrrell Thomson. Comprehensive, classic introduction to space-flight engineering for advanced undergraduate and graduate students. Includes vector algebra, kinematics, transformation of coordinates. Bibliography. Index. 352pp. 5⅜ x 8½.   65113-4

HISTORY OF STRENGTH OF MATERIALS, Stephen P. Timoshenko. Excellent historical survey of the strength of materials with many references to the theories of elasticity and structure. 245 figures. 452pp. 5⅜ x 8½.   61187-6

ANALYTICAL FRACTURE MECHANICS, David J. Unger. Self-contained text supplements standard fracture mechanics texts by focusing on analytical methods for determining crack-tip stress and strain fields. 336pp. 6⅛ x 9¼.   41737-9

STATISTICAL MECHANICS OF ELASTICITY, J. H. Weiner. Advanced, self-contained treatment illustrates general principles and elastic behavior of solids. Part 1, based on classical mechanics, studies thermoelastic behavior of crystalline and polymeric solids. Part 2, based on quantum mechanics, focuses on interatomic force laws, behavior of solids, and thermally activated processes. For students of physics and chemistry and for polymer physicists. 1983 ed. 96 figures. 496pp. 5⅜ x 8½.   42260-7

Mathematics

FUNCTIONAL ANALYSIS (Second Corrected Edition), George Bachman and Lawrence Narici. Excellent treatment of subject geared toward students with background in linear algebra, advanced calculus, physics, and engineering. Text covers introduction to inner-product spaces, normed, metric spaces, and topological spaces; complete orthonormal sets, the Hahn-Banach Theorem and its consequences, and many other related subjects. 1966 ed. 544pp. 6⅛ x 9¼. 40251-7

ASYMPTOTIC EXPANSIONS OF INTEGRALS, Norman Bleistein & Richard A. Handelsman. Best introduction to important field with applications in a variety of scientific disciplines. New preface. Problems. Diagrams. Tables. Bibliography. Index. 448pp. 5⅜ x 8½. 65082-0

VECTOR AND TENSOR ANALYSIS WITH APPLICATIONS, A. I. Borisenko and I. E. Tarapov. Concise introduction. Worked-out problems, solutions, exercises. 257pp. 5⅝ x 8¼. 63833-2

THE ABSOLUTE DIFFERENTIAL CALCULUS (CALCULUS OF TENSORS), Tullio Levi-Civita. Great 20th-century mathematician's classic work on material necessary for mathematical grasp of theory of relativity. 452pp. 5⅝ x 8¼. 63401-9

AN INTRODUCTION TO ORDINARY DIFFERENTIAL EQUATIONS, Earl A. Coddington. A thorough and systematic first course in elementary differential equations for undergraduates in mathematics and science, with many exercises and problems (with answers). Index. 304pp. 5⅜ x 8½. 65942-9

FOURIER SERIES AND ORTHOGONAL FUNCTIONS, Harry F. Davis. An incisive text combining theory and practical example to introduce Fourier series, orthogonal functions and applications of the Fourier method to boundary-value problems. 570 exercises. Answers and notes. 416pp. 5⅜ x 8½. 65973-9

COMPUTABILITY AND UNSOLVABILITY, Martin Davis. Classic graduate-level introduction to theory of computability, usually referred to as theory of recurrent functions. New preface and appendix. 288pp. 5⅜ x 8½. 61471-9

ASYMPTOTIC METHODS IN ANALYSIS, N. G. de Bruijn. An inexpensive, comprehensive guide to asymptotic methods–the pioneering work that teaches by explaining worked examples in detail. Index. 224pp. 5⅜ x 8½ 64221-6

APPLIED COMPLEX VARIABLES, John W. Dettman. Step-by-step coverage of fundamentals of analytic function theory–plus lucid exposition of five important applications: Potential Theory; Ordinary Differential Equations; Fourier Transforms; Laplace Transforms; Asymptotic Expansions. 66 figures. Exercises at chapter ends. 512pp. 5⅜ x 8½. 64670-X

INTRODUCTION TO LINEAR ALGEBRA AND DIFFERENTIAL EQUATIONS, John W. Dettman. Excellent text covers complex numbers, determinants, orthonormal bases, Laplace transforms, much more. Exercises with solutions. Undergraduate level. 416pp. 5⅜ x 8½. 65191-6

CALCULUS OF VARIATIONS WITH APPLICATIONS, George M. Ewing. Applications-oriented introduction to variational theory develops insight and promotes understanding of specialized books, research papers. Suitable for advanced undergraduate/graduate students as primary, supplementary text. 352pp. 5⅜ x 8½.
64856-7

COMPLEX VARIABLES, Francis J. Flanigan. Unusual approach, delaying complex algebra till harmonic functions have been analyzed from real variable viewpoint. Includes problems with answers. 364pp. 5⅜ x 8½.
61388-7

AN INTRODUCTION TO THE CALCULUS OF VARIATIONS, Charles Fox. Graduate-level text covers variations of an integral, isoperimetrical problems, least action, special relativity, approximations, more. References. 279pp. 5⅜ x 8½.
65499-0

COUNTEREXAMPLES IN ANALYSIS, Bernard R. Gelbaum and John M. H. Olmsted. These counterexamples deal mostly with the part of analysis known as "real variables." The first half covers the real number system, and the second half encompasses higher dimensions. 1962 edition. xxiv+198pp. 5⅜ x 8½.
42875-3

CATASTROPHE THEORY FOR SCIENTISTS AND ENGINEERS, Robert Gilmore. Advanced-level treatment describes mathematics of theory grounded in the work of Poincaré, R. Thom, other mathematicians. Also important applications to problems in mathematics, physics, chemistry, and engineering. 1981 edition. References. 28 tables. 397 black-and-white illustrations. xvii+666pp. 6⅛ x 9¼.
67539-4

INTRODUCTION TO DIFFERENCE EQUATIONS, Samuel Goldberg. Exceptionally clear exposition of important discipline with applications to sociology, psychology, economics. Many illustrative examples; over 250 problems. 260pp. 5⅜ x 8½.
65084-7

NUMERICAL METHODS FOR SCIENTISTS AND ENGINEERS, Richard Hamming. Classic text stresses frequency approach in coverage of algorithms, polynomial approximation, Fourier approximation, exponential approximation, other topics. Revised and enlarged 2nd edition. 721pp. 5⅜ x 8½.
65241-6

INTRODUCTION TO NUMERICAL ANALYSIS (2nd Edition), F. B. Hildebrand. Classic, fundamental treatment covers computation, approximation, interpolation, numerical differentiation and integration, other topics. 150 new problems. 669pp. 5⅜ x 8½.
65363-3

THREE PEARLS OF NUMBER THEORY, A. Y. Khinchin. Three compelling puzzles require proof of a basic law governing the world of numbers. Challenges concern van der Waerden's theorem, the Landau-Schnirelmann hypothesis and Mann's theorem, and a solution to Waring's problem. Solutions included. 64pp. 5⅜ x 8½.
40026-3

THE PHILOSOPHY OF MATHEMATICS: An Introductory Essay, Stephan Körner. Surveys the views of Plato, Aristotle, Leibniz & Kant concerning propositions and theories of applied and pure mathematics. Introduction. Two appendices. Index. 198pp. 5⅜ x 8½.
25048-2

INTRODUCTORY REAL ANALYSIS, A.N. Kolmogorov, S. V. Fomin. Translated by Richard A. Silverman. Self-contained, evenly paced introduction to real and functional analysis. Some 350 problems. 403pp. 5⅜ x 8½. 61226-0

APPLIED ANALYSIS, Cornelius Lanczos. Classic work on analysis and design of finite processes for approximating solution of analytical problems. Algebraic equations, matrices, harmonic analysis, quadrature methods, more. 559pp. 5⅜ x 8½. 65656-X

AN INTRODUCTION TO ALGEBRAIC STRUCTURES, Joseph Landin. Superb self-contained text covers "abstract algebra": sets and numbers, theory of groups, theory of rings, much more. Numerous well-chosen examples, exercises. 247pp. 5⅜ x 8½. 65940-2

QUALITATIVE THEORY OF DIFFERENTIAL EQUATIONS, V. V. Nemytskii and V.V. Stepanov. Classic graduate-level text by two prominent Soviet mathematicians covers classical differential equations as well as topological dynamics and ergodic theory. Bibliographies. 523pp. 5⅜ x 8½. 65954-2

THEORY OF MATRICES, Sam Perlis. Outstanding text covering rank, nonsingularity and inverses in connection with the development of canonical matrices under the relation of equivalence, and without the intervention of determinants. Includes exercises. 237pp. 5⅜ x 8½. 66810-X

INTRODUCTION TO ANALYSIS, Maxwell Rosenlicht. Unusually clear, accessible coverage of set theory, real number system, metric spaces, continuous functions, Riemann integration, multiple integrals, more. Wide range of problems. Undergraduate level. Bibliography. 254pp. 5⅜ x 8½. 65038-3

MODERN NONLINEAR EQUATIONS, Thomas L. Saaty. Emphasizes practical solution of problems; covers seven types of equations. ". . . a welcome contribution to the existing literature. . . . "–*Math Reviews.* 490pp. 5⅜ x 8½. 64232-1

MATRICES AND LINEAR ALGEBRA, Hans Schneider and George Phillip Barker. Basic textbook covers theory of matrices and its applications to systems of linear equations and related topics such as determinants, eigenvalues, and differential equations. Numerous exercises. 432pp. 5⅜ x 8½. 66014-1

MATHEMATICS APPLIED TO CONTINUUM MECHANICS, Lee A. Segel. Analyzes models of fluid flow and solid deformation. For upper-level math, science, and engineering students. 608pp. 5⅜ x 8½. 65369-2

ELEMENTS OF REAL ANALYSIS, David A. Sprecher. Classic text covers fundamental concepts, real number system, point sets, functions of a real variable, Fourier series, much more. Over 500 exercises. 352pp. 5⅜ x 8½. 65385-4

SET THEORY AND LOGIC, Robert R. Stoll. Lucid introduction to unified theory of mathematical concepts. Set theory and logic seen as tools for conceptual understanding of real number system. 496pp. 5⅜ x 8¼. 63829-4

TENSOR CALCULUS, J.L. Synge and A. Schild. Widely used introductory text covers spaces and tensors, basic operations in Riemannian space, non-Riemannian spaces, etc. 324pp. 5⅝ x 8¼. 63612-7

ORDINARY DIFFERENTIAL EQUATIONS, Morris Tenenbaum and Harry Pollard. Exhaustive survey of ordinary differential equations for undergraduates in mathematics, engineering, science. Thorough analysis of theorems. Diagrams. Bibliography. Index. 818pp. 5⅜ x 8½. 64940-7

INTEGRAL EQUATIONS, F. G. Tricomi. Authoritative, well-written treatment of extremely useful mathematical tool with wide applications. Volterra Equations, Fredholm Equations, much more. Advanced undergraduate to graduate level. Exercises. Bibliography. 238pp. 5⅜ x 8½. 64828-1

FOURIER SERIES, Georgi P. Tolstov. Translated by Richard A. Silverman. A valuable addition to the literature on the subject, moving clearly from subject to subject and theorem to theorem. 107 problems, answers. 336pp. 5⅜ x 8½. 63317-9

INTRODUCTION TO MATHEMATICAL THINKING, Friedrich Waismann. Examinations of arithmetic, geometry, and theory of integers; rational and natural numbers; complete induction; limit and point of accumulation; remarkable curves; complex and hypercomplex numbers, more. 1959 ed. 27 figures. xii+260pp. 5⅜ x 8½. 42804-4

POPULAR LECTURES ON MATHEMATICAL LOGIC, Hao Wang. Noted logician's lucid treatment of historical developments, set theory, model theory, recursion theory and constructivism, proof theory, more. 3 appendixes. Bibliography. 1981 ed. ix+283pp. 5⅜ x 8½. 67632-3

CALCULUS OF VARIATIONS, Robert Weinstock. Basic introduction covering isoperimetric problems, theory of elasticity, quantum mechanics, electrostatics, etc. Exercises throughout. 326pp. 5⅜ x 8½. 63069-2

THE CONTINUUM: A Critical Examination of the Foundation of Analysis, Hermann Weyl. Classic of 20th-century foundational research deals with the conceptual problem posed by the continuum. 156pp. 5⅜ x 8½. 67982-9

CHALLENGING MATHEMATICAL PROBLEMS WITH ELEMENTARY SOLUTIONS, A. M. Yaglom and I. M. Yaglom. Over 170 challenging problems on probability theory, combinatorial analysis, points and lines, topology, convex polygons, many other topics. Solutions. Total of 445pp. 5⅜ x 8½. Two-vol. set.
Vol. I: 65536-9 Vol. II: 65537-7

INTRODUCTION TO PARTIAL DIFFERENTIAL EQUATIONS WITH APPLICATIONS, E. C. Zachmanoglou and Dale W. Thoe. Essentials of partial differential equations applied to common problems in engineering and the physical sciences. Problems and answers. 416pp. 5⅜ x 8½. 65251-3

THE THEORY OF GROUPS, Hans J. Zassenhaus. Well-written graduate-level text acquaints reader with group-theoretic methods and demonstrates their usefulness in mathematics. Axioms, the calculus of complexes, homomorphic mapping, *p*-group theory, more. 276pp. 5⅜ x 8½. 40922-8

Math–Decision Theory, Statistics, Probability

ELEMENTARY DECISION THEORY, Herman Chernoff and Lincoln E. Moses. Clear introduction to statistics and statistical theory covers data processing, probability and random variables, testing hypotheses, much more. Exercises. 364pp. 5⅜ x 8½. 65218-1

STATISTICS MANUAL, Edwin L. Crow et al. Comprehensive, practical collection of classical and modern methods prepared by U.S. Naval Ordnance Test Station. Stress on use. Basics of statistics assumed. 288pp. 5⅜ x 8½. 60599-X

SOME THEORY OF SAMPLING, William Edwards Deming. Analysis of the problems, theory, and design of sampling techniques for social scientists, industrial managers, and others who find statistics important at work. 61 tables. 90 figures. xvii +602pp. 5⅜ x 8½. 64684-X

LINEAR PROGRAMMING AND ECONOMIC ANALYSIS, Robert Dorfman, Paul A. Samuelson and Robert M. Solow. First comprehensive treatment of linear programming in standard economic analysis. Game theory, modern welfare economics, Leontief input-output, more. 525pp. 5⅜ x 8½. 65491-5

PROBABILITY: An Introduction, Samuel Goldberg. Excellent basic text covers set theory, probability theory for finite sample spaces, binomial theorem, much more. 360 problems. Bibliographies. 322pp. 5⅜ x 8½. 65252-1

GAMES AND DECISIONS: Introduction and Critical Survey, R. Duncan Luce and Howard Raiffa. Superb nontechnical introduction to game theory, primarily applied to social sciences. Utility theory, zero-sum games, n-person games, decision-making, much more. Bibliography. 509pp. 5⅜ x 8½. 65943-7

INTRODUCTION TO THE THEORY OF GAMES, J. C. C. McKinsey. This comprehensive overview of the mathematical theory of games illustrates applications to situations involving conflicts of interest, including economic, social, political, and military contexts. Appropriate for advanced undergraduate and graduate courses; advanced calculus a prerequisite. 1952 ed. x+372pp. 5⅜ x 8½. 42811-7

FIFTY CHALLENGING PROBLEMS IN PROBABILITY WITH SOLUTIONS, Frederick Mosteller. Remarkable puzzlers, graded in difficulty, illustrate elementary and advanced aspects of probability. Detailed solutions. 88pp. 5⅜ x 8½. 65355-2

PROBABILITY THEORY: A Concise Course, Y. A. Rozanov. Highly readable, self-contained introduction covers combination of events, dependent events, Bernoulli trials, etc. 148pp. 5⅜ x 8¼. 63544-9

STATISTICAL METHOD FROM THE VIEWPOINT OF QUALITY CONTROL, Walter A. Shewhart. Important text explains regulation of variables, uses of statistical control to achieve quality control in industry, agriculture, other areas. 192pp. 5⅜ x 8½. 65232-7

Math–Geometry and Topology

ELEMENTARY CONCEPTS OF TOPOLOGY, Paul Alexandroff. Elegant, intuitive approach to topology from set-theoretic topology to Betti groups; how concepts of topology are useful in math and physics. 25 figures. 57pp. 5⅜ x 8½. 60747-X

COMBINATORIAL TOPOLOGY, P. S. Alexandrov. Clearly written, well-organized, three-part text begins by dealing with certain classic problems without using the formal techniques of homology theory and advances to the central concept, the Betti groups. Numerous detailed examples. 654pp. 5⅜ x 8½. 40179-0

EXPERIMENTS IN TOPOLOGY, Stephen Barr. Classic, lively explanation of one of the byways of mathematics. Klein bottles, Moebius strips, projective planes, map coloring, problem of the Koenigsberg bridges, much more, described with clarity and wit. 43 figures. 210pp. 5⅜ x 8½. 25933-1

CONFORMAL MAPPING ON RIEMANN SURFACES, Harvey Cohn. Lucid, insightful book presents ideal coverage of subject. 334 exercises make book perfect for self-study. 55 figures. 352pp. 5⅝ x 8¼. 64025-6

THE GEOMETRY OF RENÉ DESCARTES, René Descartes. The great work founded analytical geometry. Original French text, Descartes's own diagrams, together with definitive Smith-Latham translation. 244pp. 5⅜ x 8½. 60068-8

PRACTICAL CONIC SECTIONS: The Geometric Properties of Ellipses, Parabolas and Hyperbolas, J. W. Downs. This text shows how to create ellipses, parabolas, and hyperbolas. It also presents historical background on their ancient origins and describes the reflective properties and roles of curves in design applications. 1993 ed. 98 figures. xii+100pp. 6½ x 9¼. 42876-1

THE THIRTEEN BOOKS OF EUCLID'S ELEMENTS, translated with introduction and commentary by Thomas L. Heath. Definitive edition. Textual and linguistic notes, mathematical analysis. 2,500 years of critical commentary. Unabridged. 1,414pp. 5⅜ x 8½. Three-vol. set. Vol. I: 60088-2 Vol. II: 60089-0 Vol. III: 60090-4

GEOMETRY OF COMPLEX NUMBERS, Hans Schwerdtfeger. Illuminating, widely praised book on analytic geometry of circles, the Moebius transformation, and two-dimensional non-Euclidean geometries. 200pp. 5⅝ x 8¼. 63830-8

DIFFERENTIAL GEOMETRY, Heinrich W. Guggenheimer. Local differential geometry as an application of advanced calculus and linear algebra. Curvature, transformation groups, surfaces, more. Exercises. 62 figures. 378pp. 5⅜ x 8½. 63433-7

CURVATURE AND HOMOLOGY: Enlarged Edition, Samuel I. Goldberg. Revised edition examines topology of differentiable manifolds; curvature, homology of Riemannian manifolds; compact Lie groups; complex manifolds; curvature, homology of Kaehler manifolds. New Preface. Four new appendixes. 416pp. 5⅜ x 8½. 40207-X

History of Math

THE WORKS OF ARCHIMEDES, Archimedes (T. L. Heath, ed.). Topics include the famous problems of the ratio of the areas of a cylinder and an inscribed sphere; the measurement of a circle; the properties of conoids, spheroids, and spirals; and the quadrature of the parabola. Informative introduction. clxxxvi+326pp; supplement, 52pp. 5⅜ x 8½. 42084-1

A SHORT ACCOUNT OF THE HISTORY OF MATHEMATICS, W. W. Rouse Ball. One of clearest, most authoritative surveys from the Egyptians and Phoenicians through 19th-century figures such as Grassman, Galois, Riemann. Fourth edition. 522pp. 5⅜ x 8½. 20630-0

THE HISTORY OF THE CALCULUS AND ITS CONCEPTUAL DEVELOP-MENT, Carl B. Boyer. Origins in antiquity, medieval contributions, work of Newton, Leibniz, rigorous formulation. Treatment is verbal. 346pp. 5⅜ x 8½. 60509-4

THE HISTORICAL ROOTS OF ELEMENTARY MATHEMATICS, Lucas N. H. Bunt, Phillip S. Jones, and Jack D. Bedient. Fundamental underpinnings of modern arithmetic, algebra, geometry, and number systems derived from ancient civilizations. 320pp. 5⅜ x 8½. 25563-8

A HISTORY OF MATHEMATICAL NOTATIONS, Florian Cajori. This classic study notes the first appearance of a mathematical symbol and its origin, the competition it encountered, its spread among writers in different countries, its rise to popularity, its eventual decline or ultimate survival. Original 1929 two-volume edition presented here in one volume. xxviii+820pp. 5⅜ x 8½. 67766-4

GAMES, GODS & GAMBLING: A History of Probability and Statistical Ideas, F. N. David. Episodes from the lives of Galileo, Fermat, Pascal, and others illustrate this fascinating account of the roots of mathematics. Features thought-provoking references to classics, archaeology, biography, poetry. 1962 edition. 304pp. 5⅜ x 8½. (Available in U.S. only.) 40023-9

OF MEN AND NUMBERS: The Story of the Great Mathematicians, Jane Muir. Fascinating accounts of the lives and accomplishments of history's greatest mathematical minds–Pythagoras, Descartes, Euler, Pascal, Cantor, many more. Anecdotal, illuminating. 30 diagrams. Bibliography. 256pp. 5⅜ x 8½. 28973-7

HISTORY OF MATHEMATICS, David E. Smith. Nontechnical survey from ancient Greece and Orient to late 19th century; evolution of arithmetic, geometry, trigonometry, calculating devices, algebra, the calculus. 362 illustrations. 1,355pp. 5⅜ x 8½. Two-vol. set. Vol. I: 20429-4 Vol. II: 20430-8

A CONCISE HISTORY OF MATHEMATICS, Dirk J. Struik. The best brief history of mathematics. Stresses origins and covers every major figure from ancient Near East to 19th century. 41 illustrations. 195pp. 5⅜ x 8½. 60255-9

Physics

OPTICAL RESONANCE AND TWO-LEVEL ATOMS, L. Allen and J. H. Eberly. Clear, comprehensive introduction to basic principles behind all quantum optical resonance phenomena. 53 illustrations. Preface. Index. 256pp. 5⅜ x 8½. 65533-4

QUANTUM THEORY, David Bohm. This advanced undergraduate-level text presents the quantum theory in terms of qualitative and imaginative concepts, followed by specific applications worked out in mathematical detail. Preface. Index. 655pp. 5⅜ x 8½. 65969-0

ATOMIC PHYSICS: 8th edition, Max Born. Nobel laureate's lucid treatment of kinetic theory of gases, elementary particles, nuclear atom, wave-corpuscles, atomic structure and spectral lines, much more. Over 40 appendices, bibliography. 495pp. 5⅜ x 8½. 65984-4

A SOPHISTICATE'S PRIMER OF RELATIVITY, P. W. Bridgman. Geared toward readers already acquainted with special relativity, this book transcends the view of theory as a working tool to answer natural questions: What is a frame of reference? What is a "law of nature"? What is the role of the "observer"? Extensive treatment, written in terms accessible to those without a scientific background. 1983 ed. xlviii+172pp. 5⅜ x 8½. 42549-5

AN INTRODUCTION TO HAMILTONIAN OPTICS, H. A. Buchdahl. Detailed account of the Hamiltonian treatment of aberration theory in geometrical optics. Many classes of optical systems defined in terms of the symmetries they possess. Problems with detailed solutions. 1970 edition. xv+360pp. 5⅜ x 8½. 67597-1

PRIMER OF QUANTUM MECHANICS, Marvin Chester. Introductory text examines the classical quantum bead on a track: its state and representations; operator eigenvalues; harmonic oscillator and bound bead in a symmetric force field; and bead in a spherical shell. Other topics include spin, matrices, and the structure of quantum mechanics; the simplest atom; indistinguishable particles; and stationary-state perturbation theory. 1992 ed. xiv+314pp. 6⅛ x 9¼. 42878-8

LECTURES ON QUANTUM MECHANICS, Paul A. M. Dirac. Four concise, brilliant lectures on mathematical methods in quantum mechanics from Nobel Prize–winning quantum pioneer build on idea of visualizing quantum theory through the use of classical mechanics. 96pp. 5⅜ x 8½. 41713-1

THIRTY YEARS THAT SHOOK PHYSICS: The Story of Quantum Theory, George Gamow. Lucid, accessible introduction to influential theory of energy and matter. Careful explanations of Dirac's anti-particles, Bohr's model of the atom, much more. 12 plates. Numerous drawings. 240pp. 5⅜ x 8½. 24895-X

ELECTRONIC STRUCTURE AND THE PROPERTIES OF SOLIDS: The Physics of the Chemical Bond, Walter A. Harrison. Innovative text offers basic understanding of the electronic structure of covalent and ionic solids, simple metals, transition metals and their compounds. Problems. 1980 edition. 582pp. 6⅛ x 9¼. 66021-4

HYDRODYNAMIC AND HYDROMAGNETIC STABILITY, S. Chandrasekhar. Lucid examination of the Rayleigh-Benard problem; clear coverage of the theory of instabilities causing convection. 704pp. 5⅜ x 8¼. 64071-X

INVESTIGATIONS ON THE THEORY OF THE BROWNIAN MOVEMENT, Albert Einstein. Five papers (1905–8) investigating dynamics of Brownian motion and evolving elementary theory. Notes by R. Fürth. 122pp. 5⅜ x 8½. 60304-0

THE PHYSICS OF WAVES, William C. Elmore and Mark A. Heald. Unique overview of classical wave theory. Acoustics, optics, electromagnetic radiation, more. Ideal as classroom text or for self-study. Problems. 477pp. 5⅜ x 8½. 64926-1

PHYSICAL PRINCIPLES OF THE QUANTUM THEORY, Werner Heisenberg. Nobel Laureate discusses quantum theory, uncertainty, wave mechanics, work of Dirac, Schroedinger, Compton, Wilson, Einstein, etc. 184pp. 5⅜ x 8½. 60113-7

ATOMIC SPECTRA AND ATOMIC STRUCTURE, Gerhard Herzberg. One of best introductions; especially for specialist in other fields. Treatment is physical rather than mathematical. 80 illustrations. 257pp. 5⅜ x 8½. 60115-3

AN INTRODUCTION TO STATISTICAL THERMODYNAMICS, Terrell L. Hill. Excellent basic text offers wide-ranging coverage of quantum statistical mechanics, systems of interacting molecules, quantum statistics, more. 523pp. 5⅜ x 8½. 65242-4

THEORETICAL PHYSICS, Georg Joos, with Ira M. Freeman. Classic overview covers essential math, mechanics, electromagnetic theory, thermodynamics, quantum mechanics, nuclear physics, other topics. xxiii+885pp. 5⅜ x 8½. 65227-0

PROBLEMS AND SOLUTIONS IN QUANTUM CHEMISTRY AND PHYSICS, Charles S. Johnson, Jr. and Lee G. Pedersen. Unusually varied problems, detailed solutions in coverage of quantum mechanics, wave mechanics, angular momentum, molecular spectroscopy, more. 280 problems, 139 supplementary exercises. 430pp. 6½ x 9¼. 65236-X

THEORETICAL SOLID STATE PHYSICS, Vol. I: Perfect Lattices in Equilibrium; Vol. II: Non-Equilibrium and Disorder, William Jones and Norman H. March. Monumental reference work covers fundamental theory of equilibrium properties of perfect crystalline solids, non-equilibrium properties, defects and disordered systems. Total of 1,301pp. 5⅜ x 8½. Vol. I: 65015-4 Vol. II: 65016-2

WHAT IS RELATIVITY? L. D. Landau and G. B. Rumer. Written by a Nobel Prize physicist and his distinguished colleague, this compelling book explains the special theory of relativity to readers with no scientific background, using such familiar objects as trains, rulers, and clocks. 1960 ed. vi+72pp. 23 b/w illustrations. 5⅜ x 8½. 42806-0 $6.95

A TREATISE ON ELECTRICITY AND MAGNETISM, James Clerk Maxwell. Important foundation work of modern physics. Brings to final form Maxwell's theory of electromagnetism and rigorously derives his general equations of field theory. 1,084pp. 5⅜ x 8½. Two-vol. set. Vol. I: 60636-8 Vol. II: 60637-6

QUANTUM MECHANICS: Principles and Formalism, Roy McWeeny. Graduate student–oriented volume develops subject as fundamental discipline, opening with review of origins of Schrödinger's equations and vector spaces. Focusing on main principles of quantum mechanics and their immediate consequences, it concludes with final generalizations covering alternative "languages" or representations. 1972 ed. 15 figures. xi+155pp. 5⅜ x 8½. 42829-X

INTRODUCTION TO QUANTUM MECHANICS WITH APPLICATIONS TO CHEMISTRY, Linus Pauling & E. Bright Wilson, Jr. Classic undergraduate text by Nobel Prize winner applies quantum mechanics to chemical and physical problems. Numerous tables and figures enhance the text. Chapter bibliographies. Appendices. Index. 468pp. 5⅜ x 8½. 64871-0

METHODS OF THERMODYNAMICS, Howard Reiss. Outstanding text focuses on physical technique of thermodynamics, typical problem areas of understanding, and significance and use of thermodynamic potential. 1965 edition. 238pp. 5⅜ x 8½. 69445-3

TENSOR ANALYSIS FOR PHYSICISTS, J. A. Schouten. Concise exposition of the mathematical basis of tensor analysis, integrated with well-chosen physical examples of the theory. Exercises. Index. Bibliography. 289pp. 5⅜ x 8½. 65582-2

THE ELECTROMAGNETIC FIELD, Albert Shadowitz. Comprehensive undergraduate text covers basics of electric and magnetic fields, builds up to electromagnetic theory. Also related topics, including relativity. Over 900 problems. 768pp. 5⅝ x 8¼. 65660-8

GREAT EXPERIMENTS IN PHYSICS: Firsthand Accounts from Galileo to Einstein, Morris H. Shamos (ed.). 25 crucial discoveries: Newton's laws of motion, Chadwick's study of the neutron, Hertz on electromagnetic waves, more. Original accounts clearly annotated. 370pp. 5⅜ x 8½. 25346-5

RELATIVITY, THERMODYNAMICS AND COSMOLOGY, Richard C. Tolman. Landmark study extends thermodynamics to special, general relativity; also applications of relativistic mechanics, thermodynamics to cosmological models. 501pp. 5⅜ x 8½. 65383-8

STATISTICAL PHYSICS, Gregory H. Wannier. Classic text combines thermodynamics, statistical mechanics, and kinetic theory in one unified presentation of thermal physics. Problems with solutions. Bibliography. 532pp. 5⅜ x 8½. 65401-X

Paperbound unless otherwise indicated. Available at your book dealer, online at **www.doverpublications.com**, or by writing to Dept. GI, Dover Publications, Inc., 31 East 2nd Street, Mineola, NY 11501. For current price information or for free catalogs (please indicate field of interest), write to Dover Publications or log on to **www.doverpublications.com** and see every Dover book in print. Dover publishes more than 500 books each year on science, elementary and advanced mathematics, biology, music, art, literary history, social sciences, and other areas.